D1361816

continued on back

Selecting and Ordering Populations:
A New Statistical Methodology

A WILEY PUBLICATION IN APPLIED STATISTICS

Selecting and Ordering Populations: A New Statistical Methodology

JEAN DICKINSON GIBBONS
The University of Alabama

INGRAM OLKIN
Stanford University

MILTON SOBEL
University of California at Santa Barbara

John Wiley & Sons, New York · London · Sydney · Toronto

Library of Congress Cataloging in Publication Data:

Gibbons, Jean Dickinson, 1938-
  Selecting and ordering populations.

  (Wiley series in probability and mathematical statistics)
  Bibliography: p.
  Includes index.
  1. Order statistics. I. Olkin, Ingram, 1924-joint author.
II. Sobel, Milton, 1919- joint author. III. Title.

QA278.7.G5   519.5′4   77-3700
ISBN 0-471-02670-0

To   Jack
     Vivian
     Marc

Preface

Users of statistical methods as well as teachers and students of statistics need to become acquainted with ranking and selection procedures, since these techniques answer a question that is raised in many investigations but seldom answered by the more traditional methods of analysis. In layman's terms, the question might be one of the following: Which one (or ones) of several well-defined groups is best (in some well-defined sense of best)? Which of several alternative courses of action is best? How do the choices rank? The statistical answer to such problems must be given in such a way that the probability of a correct selection is controlled. In the language of statistics, we have a well-defined family of populations that differ with respect to the value of a characteristic as represented by a single parameter. We define a best population and ask for methods to identify the best one of several given members of such a family, or we define a method of ordering the values of the parameter and ask for methods of ranking several given members of such a family.

Statisticians have been developing the theory of ordering (or ranking) and selection procedures for over 25 years. In the last 10 to 15 years the rate of development has increased considerably. Many publications in the statistical literature have been devoted to various theoretical aspects of ordering and selection problems. However, there seems to have been little effort to encourage applied statisticians to use this body of theoretical knowledge while designing experiments and/or analyzing data. This book makes these methods accessible to a large number of users of statistical methods. Investigators and researchers who, although not statisticians, use statistical techniques—psychologists, engineers, biologists, management scientists, and the like—should find no obstacle to applying these techniques. Until recently the literature describing these procedures was available only in dispersed statistical journals and in advanced books. Recently some elementary mathematical statistics books are treating the subject (cf., Dudewicz (1976)). Moreover, the very essential tables required for carrying out the procedures have not been put into convenient collections of statistical tables. Indeed, it has taken considerable effort to prepare the tables in a manner that is usable to the practitioner. Because of this it is not surprising that these procedures have been used seldom in the application of statistical methods.

This book attempts to provide both applied statisticians and other users of statistical methods with a compendium of the applied aspects of those ordering and selection procedures that have been well developed theoretically as of this date. The authors have collected the statistical literature on the subject and present it here at a level that can be understood even by those with little or no formal training in statistical theory. The applications of these procedures are described in a nonmathematical way, and all the tables needed for application are included here. The authors have expended a great deal of time and effort in organizing, collecting, and developing these tables, and show here how they may be applied to many examples and problems from a wide variety of specific fields of application.

The authors feel that selection and ordering procedures cannot be discussed properly without including an outline of the history of their development as well as the names of those persons who made contributions to the field. This material is given at the end of each chapter in a separate section called Notes, References, and Remarks, so that there is no interruption in the flow of the text. A complete bibliography is given at the end of the book.

We expect this book to be valuable as a handbook or reference book for the researcher or investigator who uses the basic methods of statistical analysis, as well as for the applied statistician. The book can also be used in the classroom by graduate or undergraduate students in statistics, business, education, engineering, and all fields of the behavioral, social, and physical sciences where quantitative techniques of analysis are considered relevant to the curriculum. The book does not presuppose the knowledge of calculus, and the only statistical prerequisite is a standard first-year statistics course.

The essential ingredients of the theory are explained in Chapters 1 and 2. These together with parts of Chapters 3, 4, and 5 would be ample for a short quarter course, or as supplementary reading for a standard course. A full quarter or semester course should also include parts or all of Chapters 6, 7, 9, 11, 12 and 13. Chapters 10, 14, and 15 are a bit more specialized, and could be used as reference material, or as part of a full year's course. A few sections are more difficult than the general level of the book; these are indicated by an asterisk and can be omitted without interrupting the flow of the material.

We want to express our appreciation to John Petkau for comments, suggestions, and help; to David Pasta and Ronald Thisted, for assistance

PREFACE

in preparation of some of the tables; to Peter O. Anderson, E. J. Dudewicz, Thomas P. McWilliams, M. Haseeb Rizvi, and Rand Wilcox, for their comments; to Cynthia Copeland, Judi Davis, Kitty Elledge, for typing the manuscript; to the Board of Visitors of the Graduate School of Business and the Research Grants Committee of the University of Alabama, for partial support.

JEAN DICKINSON GIBBONS
INGRAM OLKIN
MILTON SOBEL

University, Alcbama
Stanford, California
Santa Barbara, California
February 1977

Contents

CHAPTER 1

The Philosophy of Selecting and Ordering Populations

1.1 INTRODUCTION

Everyone is constantly faced with the problem of choosing among several alternatives. This choice is a decision about which alternative is best, in some well-defined sense of *best*. Some decisions are of relatively little import in the long run, such as when an individual is selecting a pair of shoes to purchase. Other decisions are of greater significance to the individual, such as the choice of which college to attend or of whom (or whether) to marry. Still other decisions are of considerable consequence to society, such as whether cyclamates should be banned or which drugs are safe for use during pregnancy.

In any such decision-making situation we must select one or more groups from some number k of groups under consideration, or one or more courses of action from some number k of possible courses of action. Sometimes the selection is made in a completely arbitrary or subjective way. However, in many cases the selection is based on quantitative evaluations of the relative "merits" of the groups or actions. The term *merits* is used here in the broadest sense; it includes whatever factors are considered important by the person making the selection. If these quantitative evaluations are based on the results of a statistical experiment, we have a *statistical selection procedure*. In the language of statistics the k groups are called *populations* and the problem is to design an experiment that is not biased toward any subset of the populations and then take enough observations to select the one (or sometimes several) population that is best; here the term *best* is assumed to be well defined before the data are taken. The experiment consists of taking random samples from each of these k populations (usually the same number from each), and the selection is then based on the data obtained. The problem of determining how many observations are "enough" is quite important and we devote a good deal of attention to it.

1

In other cases we may want to order or rank the groups or alternatives according to their bestness. If this ranking is based on a statistical experiment as described earlier, we have a *statistical ranking procedure*.

In summary, then, ranking and selection procedures are statistical techniques for comparing some number k of populations. We assume at the outset that the populations are not all the same and can be ordered in some meaningful way, from worst to best, or from smallest to largest. These selection procedures are designed specifically to identify the best single population, or the best subset of populations, or some subset of populations that includes the best population, or the like.

The practical situations in which such procedures are appropriate are many and varied. The first application of a ranking and selection procedure to a substantive field apparently was reported in the area of poultry science, specifically in relation to the problem of providing impartial information to breeders, hatcherymen and potential buyers of chicks about which poultry stock is best. Becker (1961) considered the problem of selecting the one chicken stock that has the largest hen-house egg production. These methods, if well known, can have wide applicability in other fields. In drug studies the researcher may want to identify the one drug that produces the "best" response. In choosing a medium for advertising some product, the decision maker may want to select the one that reaches the highest proportion of potential buyers of that product. (Advertising applications are considered in Dalal and Srinivasan (1975).) In choosing a common stock for investment, the purchaser may seek the company with the lowest average ratio of price to earnings. In choosing a supplier of kits for use in laboratory analysis, the purchaser may want to identify the supplier whose kits have the greatest precision. Examples of practical situations where ranking and selection procedures are needed for proper statistical analysis are given throughout this book.

As with any statistical methods designed to compare some k populations, ranking and selection techniques are based on the data obtained when random samples are drawn from each of the k populations. Since the classical procedures for comparing k populations are familiar to most persons acquainted with elementary statistical methods and these procedures are based on the same kind of data, a brief discussion of these familiar procedures is included here so that we can point out the similarities and differences between the two types of procedures.

In the framework of testing hypotheses, the classical procedure attempts to determine whether k parameters all have a common value. Each parameter represents the same type of description, attribute, or response for all populations, but the populations may differ. For example, if we have k different drugs for treatment of a certain condition, the concern is whether these drugs differ in their therapeutic value. If therapeutic value is measured by some parameter θ_j for the jth drug, the classical procedure permits us to decide about the following null hypothesis, sometimes called the *homogeneity hypothesis*:

$$H_0: \quad \theta_1 = \theta_2 = \cdots = \theta_k.$$

Tests of this null hypothesis are frequently called *tests of homogeneity*. The alternative hypothesis, which may be implicit or explicit, is usually that these drugs do not all have equal θ values. This means that the θ values are unequal for at least one pair of drugs.

The hypothesis of homogeneity is at the core of a vast amount of classical statistical theory and methodology. For example, the analysis of variance test (Fisher's F test) is used to test homogeneity between means under the assumption of normal distributions with a common variance. References to other tests of homogeneity are given in Section 1.7. All these classical tests have been well developed and investigated; they are not discussed further in this book.

In the general situation in which the family of probability distributions is specified except for the respective values of certain parameters, a test can be developed for the homogeneity hypothesis using any one of a number of techniques. (For example, a general procedure that is used frequently is known as the likelihood ratio test.) Also, when the distribution family is not fully specified or the assumed family cannot be confirmed empirically, nonparametric tests are frequently applied.

If a test of homogeneity is the primary and final goal of an investigation or experiment, alternative methods of statistical analysis are not needed. However, there are many practical situations in which other kinds of information or other goals are of interest. For example, suppose that the null hypothesis of homogeneity is rejected. The investigator is seldom satisfied with terminating the analysis with this decision; it only whets the appetite for trying to make additional assertions. In particular, he may want (*a*) to determine which populations differ from which others, and in what direction, or (*b*) to see which populations can be considered best in some well-defined sense of the term *best*. In case (*a*), techniques of multiple comparisons or simultaneous inference are frequently appropriate. The methods of multiple comparisons may also provide information that is relevant for case (*b*), but ranking and selection procedures are more appropriate for this purpose.

The discussion in the previous paragraph may imply to the reader that ranking and selection procedures are simply one more solution to the problem of what to do next when the null hypothesis of homogeneity is rejected. This is not the case at all. When the goal is, say, to select the one best population out of k populations, a test of homogeneity of all k populations is really inadequate for, or at least not pertinent to, the primary problem at hand. In the case of k drugs, for example, suppose that we wish to (or are forced to) choose one (or possibly two) of the k drugs and to assert that it is the best of the k in therapeutic value. The test of homogeneity can only tell us whether or not the drugs are equivalent; this test is not set up to resolve the problem of choosing the single best drug (or possibly the best two drugs in order of effectiveness or without order). Although some modifications and extensions of the test of homogeneity have been formulated to provide further information, no modification can be appropriate if we assume at the outset that for any two different drugs, differences in therapeutic value must surely exist. Moreover, if we *must* make a choice among the k drugs, the conclusion corresponding to the null hypothesis H_0, namely, that all k drugs have equivalent therapeutic value, is not realistic, not useful, and not even available to the investigator. The ranking and selection procedures have been designed specifically to resolve such problems.

1.2 POSSIBLE GOALS FOR RANKING AND SELECTION PROCEDURES

The methods known generally as *ranking and selection procedures* include techniques appropriate for many different goals, although each different goal requires a careful formulation of the corresponding problem. For any given set of k populations some of the goals that can be accomplished by these methods are listed below:

1. Selecting the *one* best population.
2. Selecting a *random* number of populations such that all populations better than a control population or a standard are included in the selected subset.
3. Selecting the t best populations for $t \geq 2$, (a) in an ordered manner or (b) in an unordered manner.
4. Selecting a *random* number of populations, say r, that includes the t best populations.
5. Selecting a *fixed* number of populations, say r, that includes the t best populations.
6. Ordering *all* the k populations from best to worst (or vice versa).
7. Ordering a *fixed-size* subset of the populations from best to worst (or vice versa).

Procedures appropriate for the first goal are the primary subject of this book. (Chapters 10, 11, 12, and 13 discuss some methods appropriate for most of the other goals.) As a result, the remainder of this introductory chapter is concerned with certain general aspects of the ranking and selection procedures appropriate for the goal of identifying the one population that is best among a given set of k populations.

1.3 SPECIFICATION OF A DISTANCE MEASURE AND THE INDIFFERENCE ZONE

To repeat the problem, we assume that among some k specified populations there is a best population. If the best one differs by at least some minimal threshold value from all the others, then we have a strong preference to identify it. Thus the goal is to identify this best population, and the selection technique is to be based on the results of random samples taken from each of the k populations. As with any decision based on sample data, we must be concerned about making an error. The error that could occur in this problem is making an erroneous choice, that is, asserting that a population *is* the best one when it actually is *not* the best one. The statistical methodology is designed to control the probability of making this kind of error.

Before we can discuss this error in any detail, we must orient ourselves to the assumption that some differences do exist between the populations. From the discussion in Section 1.1 it is clear that this assumption necessitates some new thinking and some new approaches to statistical methodology. In particular, we must define an appropriate *distance measure* or a *distance function* to serve as a measure of the differences between the population we want to identify and the remaining populations. We must also separate the class of all possible parameter states, called the *parameter space*, into two types: (1) those with large distances, called the *preference zone*, and (2) those with small distances, called the *indifference zone*.

These new ideas are developed in this section, primarily in the framework of a specific experimental situation that we now describe in detail. Many drugs purport to have effects that are similar in beneficial value. Suppose we are given a method of making a quantitative evaluation of the effectiveness of any tranquilizing drug according to its ability to induce a relaxed state, the time it takes to accomplish this, the absence of serious side effects, or the like. That is, on observation, a single numerical score can be given to any drug, and that score is regarded as the measure of its overall effectiveness. We assume that the proper interpretation of these scores is that larger values indicate more effective drugs. Then we are interested in finding that one drug among k specific tranquilizing drugs that has the largest score.

Suppose it is known that if any one of these k drugs is administered to a large number of individuals in some class, the scores X have a probability distribution $F(x;\theta)$ indexed by a parameter θ. This parameter might represent the expected score when that drug is applied to a randomly selected individual in this class. For example, $F(x;\theta)$ might represent the normal distribution with mean θ and a known variance, or the binomial distribution with parameter θ, or the Poisson distribution with parameter θ, and so on. Then the differences between the k scores reflect the differences between the respective k parameters, which we denote by $\theta_1, \theta_2, \ldots, \theta_k$. Formally we assume we are given k drugs and by administering any one drug to different individuals we obtain sampling results from one of our k populations, say the jth one corresponding to the jth drug; the latter depends on the parameter θ_j. Observations from each of these k populations are taken independently and are distributed as follows:

Population	1	2	$\cdots$	k
Distribution	$F(x;\theta_1)$	$F(x;\theta_2)$	$\cdots$	$F(x;\theta_k)$

Our goal is to select the one drug (or, equivalently, the one population) with the largest value of θ. Recall that the θ values are completely unknown and that every observation made contributes some information about these θ values and hence about their relative magnitudes, or, equivalently, about their ranking. We need a notation to distinguish between θ_1, the parameter that represents the first drug, and the smallest θ value, and similarly for every other drug. A standard notation is to write the ordered values of θ, that is, the values rearranged in increasing order of magnitude, as

$$\theta_{[1]} \leqslant \theta_{[2]} \leqslant \cdots \leqslant \theta_{[k]}. \qquad (1.3.1)$$

For example, the fifth drug has parameter value θ_5, but if this drug is the best one among these k drugs and $k=8$, then θ_5 is equal to $\theta_{[8]}$. For an arbitrary k our goal can be stated under this notation as being "To select that drug whose θ value is $\theta_{[k]}$." Note that the ordering in (1.3.1) permits equalities to hold between two or more values of θ. If an equality holds for $\theta_{[k]}$, that is, if some two or more drugs have the same largest score, we will be content with selecting any one drug whose θ value is equal to $\theta_{[k]}$.

In this experiment suppose that a sample of N different individuals is to serve as subjects, and each subject is to be given one and only one drug, but the same drug may be given to several subjects. Let x_{ij} denote the score

Table 1.3.1   Schematic presentation of the model

Population (drug)	1	2	$\cdots$	k
True unknown parameter value	θ_1	θ_2	$\cdots$	θ_k
Number of observations	n_1	n_2	$\cdots$	n_k
Scores	x_{11} x_{21} . . . $x_{n_1 1}$	x_{12} x_{22} . . . $x_{n_2 2}$	$\cdots$ $\cdots$ $\cdots$ $\cdots$ $\cdots$ $\cdots$	x_{1k} x_{2k} . . . $x_{n_k k}$

resulting for the ith individual among those that are given the jth drug, and suppose we have n_j individuals taking the jth drug. Then the $N = n_1 + n_2 + \cdots + n_k$ observations collected can be recorded in the schematic format of Table 1.3.1.

If we have control over the number of individuals and the allocation of drugs, then the simplest procedure is to make N a multiple of k and choose $n_1 = n_2 = \cdots = n_k = N/k$. (We may also want to equalize other factors, such as the number of females and males taking each drug, the number of individuals under and over 50 years of age taking each drug, the number of individuals with a particular genetic background, or the like, but such considerations are not taken into account in the present discussion.) In some applications we cannot (or do not) have any control over either N or the allocation; it may also happen that equal sample sizes are not the most efficient allocation. Therefore, when the methodology is explained specifically, we must include procedures that can be used for arbitrary n_j, even though it is simplest, and often most efficient, to use equal sample sizes. However, at present, we are still formulating the problem.

Given the sample scores on the drugs (as in Table 1.3.1), the goal is to select the drug with the largest θ value. Intuitively it seems quite reasonable and appropriate to compute an estimator $\hat{\theta}_j$ of each θ_j from the corresponding set of sample data and to designate the best population as that one for which the value of the sample estimate is largest. For example, if θ is a population mean, we might calculate the separate sample means and select as the best population that one which gives rise to the largest sample mean. Although this procedure is quite reasonable (and perhaps is even optimal), the possibility of an error is always present because the population with the largest θ value, $\theta_{[k]}$, does not always produce the largest value among the sample estimates. For example, it is not always the case that the *best* student in a class scores the highest grade on an examination. Consequently, although we know which population produced the sample with the largest value of $\hat{\theta}$, we do not know that this same population has parameter value $\theta_{[k]}$.

To give a specific example in which this type of thinking was used explicitly, we refer to the study by Becker (1961). Among 10 strains (or stocks) of chickens $A, B, C, \ldots, J$, strain A is known to be better than any of the other nine. The characteristic used to score and thereby compare the strains is hen-house egg production, and a sample is taken from each of the 10 strains. Becker showed that, regardless of the sizes of the 10 samples, it is possible for a poor strain to win the test. Indeed, as one may suspect, the chance of this error occurring increases as one or more of the true (or population) differences between the best and the inferior strains become smaller. (In the process of making these differences smaller, we have to be sure that no difference between the best and an inferior strain is allowed to get larger.)

In order to gain the proper perspective on the risks involved in a ranking and selection procedure, we first review the risks in a classical test of any null hypothesis H_0. Recall that the risks are assessed in terms of the probability of each of two types of error. These errors are rejecting H_0 when it is true, called a *Type I error*, and accepting H_0 when it is false, called a *Type II error* (see Table 1.3.2).

Table 1.3.2   Errors in making a statistical decision

		True situation	
		H_0 true	H_0 false
Statistician's	Accept H_0	Correct	Error (Type II)
decision	Reject H_0	Error (Type I)	Correct

The probability of the Type I error, usually denoted by α, is also called the *level of significance* of the test. The probability of the Type II error is usually denoted by β, and $1 - \beta$ is called the *power* of the test since it is the probability of a correct decision when H_0 is false. For any given test, level and power together provide an indication of the performance of the test for the particular model assumed and the sample sizes used.

Consider a classical test of the null hypothesis of homogeneity for the k parameters. Then a Type I error can be committed only at a point in the parameter space where all parameters have the same value, that is, when $\theta_1 = \theta_2 = \cdots = \theta_k$, and a Type II error can occur only at a point in the parameter space where equality among all parameters does not hold.

In a ranking and selection problem, such as the drug example under discussion here, we define a selection to be correct if the θ value of the population selected is the largest θ value, that is, $\theta_{[k]}$. Hence whenever $\theta_{[k]} > \theta_{[k-1]}$, there is only one selection that is correct. Note that our goal is not to estimate the value of $\theta_{[k]}$, nor to make a decision about the value of $\theta_{[k]}$, but only to select the population whose θ value is equal to $\theta_{[k]}$. Hence an error can occur *only* if the selection is incorrect. Further, in most practical applications the investigator is willing to assume at the outset that the parameters are not all equal and indeed that no two of them are exactly equal. (However, we do not need this assumption and it is never used, so that it need not be any cause for concern.) Thus we have no analogue for the Type I error and have little interest in it. However, we do have a counterpart for the Type II error because we can select a population and assert that it has the largest θ value when in fact the θ value for this population is smaller than $\theta_{[k]}$. Therefore the counterpart of β is the probability of making this kind of incorrect selection. Correspondingly, the counterpart of $1 - \beta$ (or the power) is the probability of making a correct selection when $\theta_{[k]} > \theta_{[k-1]}$, and this probability provides an indication of the performance of the ranking and selection procedure for the particular model assumed and the sample sizes used.

Suppose we have two different procedures (or rules) R_1 and R_2 for selecting the best population (that is, the one with the largest θ value) and we want to evaluate and/or compare the rules. One basis for comparison is the probability of making a correct selection using each of these rules. Suppose these probabilities are denoted by

$$P_1 = P \{ \text{correct selection when using } R_1 \},$$

$$P_2 = P \{ \text{correct selection when using } R_2 \}.$$

Each of these probabilities depends on the true vector $\theta = (\theta_1, \theta_2, \ldots, \theta_k)$ of θ values. (Note that we are using θ to denote both a vector and a scalar; the meaning is clear from the context.) Of course, if $P_1 \geqslant P_2$ for all vectors θ then we prefer rule R_1, and if $P_2 \geqslant P_1$ for all vectors θ then we prefer rule R_2. If we have in mind one specific alternative configuration of parameter values, say the vector $\theta_0 = (\theta_{10}, \theta_{20}, \ldots, \theta_{k0})$ where each θ_{j0} is a specified value, then the rules need to be compared only at θ_0. However, this is usually not the case, since a whole set (possibly infinite) of vector values of θ_0 is normally of concern. In general, rule R_1 is better for some configurations of the θ values and rule R_2 is better for others.

To set up criteria for a good rule, we have to analyze the entire set of θ values, that is, the parameter space, and separate out those regions where we have a strong preference for making a correct selection from those regions where we are indifferent between two or more different selections.

The former region is referred to as the *preference zone* (PZ), and the latter region is the *indifference zone* (IZ).

For $k=2$ populations these two regions of the parameter space can be illustrated in two dimensions where the axes are the (unordered) θ_1 and θ_2. Suppose that θ_1 and θ_2 can be any real numbers. Then these zones might appear as in Figure 1.3.1. The indifference zone pictured (the shaded area) is bounded by straight lines with intercepts equal in absolute value. This zone is written in symbols as $-\delta^* < \theta_2 - \theta_1 < \delta^*$, or equivalently, $|\theta_2 - \theta_1| < \delta^*$. This says that we are indifferent about the choice between θ_1 and θ_2 when $|\theta_2 - \theta_1|$ is close to zero, that is, when θ_2 is close to θ_1 in either direction. Note that in Figure 1.3.1 the line $\theta_1 = \theta_2$ is interior to the indifference zone, as it should always be for $k=2$. In general, the indifference zone can have different shapes; the shape is of course determined by the function of θ_1 and θ_2 used to define the indifference zone.

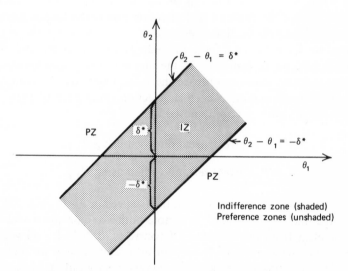

Figure 1.3.1 Breakdown of the parameter space.

If we were to replace the unordered parameters θ_1 and θ_2 by the ordered parameters $\theta_{[1]} \leqslant \theta_{[2]}$, the entire parameter space is cut in half because the line $\theta_{[2]} = \theta_{[1]}$ is the lower boundary for points in the space; then the indifference zone in Figure 1.3.1 would be only half as wide. In particular, the indifference zone is $0 < \theta_{[2]} - \theta_{[1]} < \delta^*$; this representation is given in Figure 1.3.2.

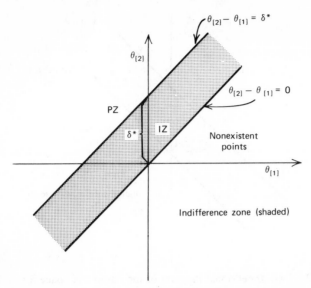

Figure 1.3.2 Breakdown of the (ordered) parameter space.

We now consider the indifference and preference zones for an arbitrary k, where the goal is still to select the best one of k populations. (Obviously, k is at least 2 for the problem to be meaningful.) Then the entire parameter space is a region in k dimensions. However, in selecting the best one, we need not be overly concerned with the sizes of $\theta_{[1]}, \theta_{[2]}, \ldots, \theta_{[k-2]}$, since we know that they are no larger than the two largest θ values. In fact, our primary concern is with the relation between the largest and next-to-largest values of θ, that is, between $\theta_{[k]}$ and $\theta_{[k-1]}$. Suppose as before that we are indifferent about the selection when $\theta_{[k]}$ is not too much larger than $\theta_{[k-1]}$. Then the definition of the indifference zone (and its complement, the preference zone) need not depend on any θ values except these two largest, and the zones can again be depicted using a two-dimensional diagram, with axes called $\theta_{[k]}$ and $\theta_{[k-1]}$. If the indifference zone is defined as $\theta_{[k]} - \theta_{[k-1]} < \delta^*$, where δ^* is a specified number, then the diagram for selecting the best one of k populations (with general $k \geqslant 2$) is given in Figure 1.3.3 for the ordered parameters $\theta_{[k-1]}$ and $\theta_{[k]}$. As in the case of two parameters, the indifference zone could be defined in many other ways, and the shaded area in the corresponding diagram would then take on different shapes.

In general, the quantity δ, which is used (with a threshold value δ^*) to separate the indifference and preference zones, is called the *measure of*

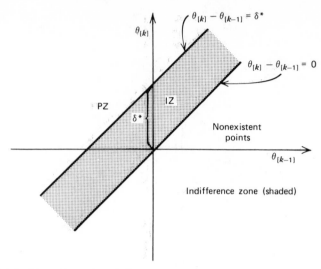

Figure 1.3.3   Two-dimensional diagram of the parameter space for arbitrary k.

distance in the formulation of ordering and selection procedures. The distance measure δ is some given function of the parameters $\theta_{[k]}$ and $\theta_{[k-1]}$. We use a Greek letter, specifically δ, as a reminder that the distance measure usually depends on unknown parameters. Suppose that, as in the preceding paragraph, the investigator has a strong preference for selecting the population with parameter $\theta_{[k]}$ over some other population if δ is large, and is indifferent if δ is small because the error then committed is not serious and can be neglected. Then the indifference zone is defined as the region where $\delta < \delta^*$ and the preference zone by $\delta \geqslant \delta^*$, where δ^* is a specified positive number.

In Figures 1.3.1, 1.3.2, and 1.3.3 the parameter θ can take on any real value; that is, it has the infinite range $-\infty < \theta < \infty$. This is the case, for example, when θ represents the mean of a normal distribution. However, in many situations the range of θ is restricted. For the binomial distribution the relevant parameter is the probability of success and its values are limited to numbers between 0 and 1. As a result, certain points of the parameter space are nonexistent in the binomial model, and the figure representing the indifference and preference zones is somewhat changed, even if the measure of distance is the same. Figure 1.3.4 shows the indifference and preference zones when the distance measure is $\delta = \theta_{[k]} - \theta_{[k-1]}$, as before, but now θ refers to the binomial parameter.

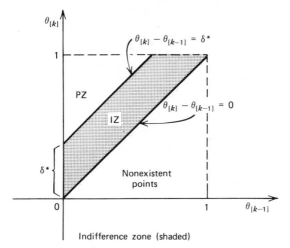

Figure 1.3.4 Two-dimensional diagram of the parameter space for the binomial problem.

In the binomial model we could alternatively define the measure of distance between $\theta_{[k]}$ and $\theta_{[k-1]}$ by defining the (so-called) odds ratio,

$$\delta_{OR} = \frac{\theta_{[k]}(1 - \theta_{[k-1]})}{\theta_{[k-1]}(1 - \theta_{[k]})}. \qquad (1.3.2)$$

Note that this measure can be interpreted as the ratio of $\theta_{[k]}/(1 - \theta_{[k]})$ to $\theta_{[k-1]}/(1 - \theta_{[k-1]})$, that is, the quotient of the "odds for success" in the best and next-to-best populations. Since we are still indifferent when δ_{OR} is small, the indifference zone is again defined by $\delta_{OR} < \delta_{OR}^{*}$ and the preference zone by $\delta_{OR} \geqslant \delta_{OR}^{*}$. The indifference and preference zones using this odds-ratio measure of distance for the problem of selecting the one best binomial population out of k are represented as in Figure 1.3.5. Note that here the lower boundary of the preference zone is not a straight line, but is a curve that passes through $(0,0)$ and $(1,1)$.

It is clear from the preceding discussion that the partition of the parameter space can be depicted by using either ordered vectors $\theta_{[\cdot]} = (\theta_{[1]}, \theta_{[2]}, \ldots, \theta_{[k]})$ or unordered vectors, $\theta = (\theta_1, \theta_2, \ldots, \theta_k)$, whichever is more convenient. Either of these vectors can be referred to as a point, or as a configuration of points with some specified property.

Because the term *correct selection* appears repeatedly, we frequently abbreviate it to CS. Then the probability of a correct selection under an

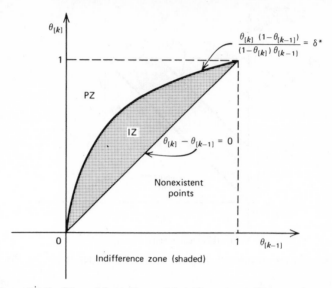

Figure 1.3.5 Binomial problem with odds ratio as distance measure.

arbitrary vector or configuration θ can be denoted by $P\{CS|\theta\}$. When θ is in the preference zone, this probability should be large; otherwise we are indifferent. Hence we can restrict our attention to the preference zone only. Since the number of points in the preference zone is usually infinite, at first this does not seem to simplify the problem of finding the probability of a correct selection. However, there may be some special configuration in the preference zone for which the probability of a correct selection is a minimum over all θ in the preference zone. This configuration is called a *least favorable configuration*, abbreviated as LF configuration, and denoted by $\theta_{LF} = (\theta_{1,LF}, \theta_{2,LF}, \ldots, \theta_{k,LF})$. When such a configuration exists, we may confine ourselves only to this configuration (rather than to the entire preference zone). It yields a lower bound to the probability of a correct selection over all possible configurations in the preference zone; then we achieve a considerable simplification of the problem.

The probability of a correct selection for the least favorable configuration θ_{LF}, represented by $P\{CS|\theta_{LF}\}$, may be called the *least favorable probability of a correct selection*. Then we have

$$P\{CS|\theta\} \geqslant P\{CS|\theta_{LF}\} \tag{1.3.3}$$

for all vectors θ in the preference zone.

The significance of the inequality in (1.3.3) is that if we make sure that the probability of a correct selection for the least favorable configuration, $P\{CS|\theta_{LF}\}$, is at least some specified value, say P^*, then we know that the probability of a correct selection, $P\{CS|\theta\}$, is also at least P^* for any θ in the preference zone. Thus P^* gives a conservative lower bound of $P\{CS|\theta\}$ for all those configurations for which we strongly prefer a correct selection. We should point out that the least favorable configuration need not be a specific vector, as, for example, $\theta_{LF} = (\theta_{1,LF}, \theta_{2,LF}, \ldots, \theta_{k,LF})$. It could also be a region or a set of vectors, which explains why we use the term *configuration*. In general, the probability $P\{CS|\theta\}$ should be the same for all vectors θ in the least favorable configuration.

REMARK.   By confining ourselves to a least favorable configuration and obtaining the corresponding least favorable probability of a correct selection, we are taking a *conservative* approach to the problem; that is, in the preference zone the probability attained for a correct selection is at least P^*. If we have any prior information concerning the configuration of the parameters of interest, then we can usually achieve a larger probability of a correct selection, provided we know how to use this information. This approach is highly analogous to the considerations of power in classical statistics, but the Type I error concept is no longer used here.   □

In most of the cases that we consider, the least favorable configuration can be defined in terms of the parameters through the distance measure. In the drug example if the component values of the vector θ represent means and the probability model is the normal distribution, the appropriate measure of distance is $\delta = \theta_{[k]} - \theta_{[k-1]}$ and the least favorable configuration results when all components of θ are equal except that the largest component exceeds the next-to-largest component by a specified amount δ^*. Here the range of each component of θ is $(-\infty, \infty)$ and the preference zone is defined by $\delta = \theta_{[k]} - \theta_{[k-1]} \geq \delta^*$. Thus the LF configuration is the set of vectors θ such that $P\{CS|\theta\}$ is minimized over all vectors θ for which $\delta \geq \delta^*$. In this example the minimum occurs at the configuration

$$\theta_{[1]} = \theta_{[2]} = \cdots = \theta_{[k-1]} = \theta' \quad \text{and} \quad \delta = \theta_{[k]} - \theta' = \delta^*. \quad (1.3.4)$$

This configuration contains an infinite number of vector points since θ' is an arbitrary scalar. Moreover this result holds for all sample sizes and does not depend on P^*. The set of vectors that satisfy a relation such as the one in (1.3.4) is what we refer to as a configuration.

The models considered in this book either are of the type described in the previous paragraph or can be reduced to this type by a suitable transformation. The following examples illustrate some common transformations. Consider a problem where the range of each component of θ is $(0, \infty)$. Assume that the ratio of the largest and next-to-largest parameter values, that is, $\delta_R = \theta_{[k]}/\theta_{[k-1]}$, is an appropriate measure of distance. Then the indifference zone is defined by the inequality $1 \leq \delta_R < \delta_R^*$, and the preference zone by $\delta_R \geq \delta_R^*$. In such a case we could define a new parameter $\tilde{\theta} = \log \theta$ and correspondingly $\tilde{\delta}^* = \log \delta_R^*$. (The base of a logarithm is understood to be the constant e, that is, the natural or Napierian logarithm, when no other base is written.) Since $\log \delta_R = \log \theta_{[k]} - \log \theta_{[k-1]} = \tilde{\theta}_{[k]} - \tilde{\theta}_{[k-1]}$, the indifference zone then reduces to a parallel band as in the situation described before (see Figures 1.3.1, 1.3.2, and 1.3.3). The same result holds for the binomial problem when the measure of distance is the odds ratio in (1.3.2) if we define a new parameter $\tilde{\theta} = \log[\theta/(1-\theta)]$. However, this device of reducing the distance function to a simple difference of (transformed) parameters by an appropriate transformation is usually invoked only for large samples when the asymptotic distribution is the normal probability law. For small samples it is better to retain the original parameters and do exact calculations for the probability of a correct selection. We return to this point in Chapter 5 when considering the problem of selecting the population with the smallest variance.

The probability of a correct selection usually depends on all of the parameters and not just on the specified distance measure. As a result, the search for the least favorable configuration is generally not trivial. For example, in the binomial distribution when we use the distance measure $\delta = \theta_{[k]} - \theta_{[k-1]}$, the least favorable configuration depends on all the component values of θ. However, when $k = 2$ and n is arbitrary, and also when k is arbitrary and n is large, the least favorable configuration is the point

$$\theta_{[1]} = \theta_{[2]} = \cdots = \theta_{[k-1]} = .5 - \frac{\delta^*}{2} \quad \text{and} \quad \theta_{[k]} = .5 + \frac{\delta^*}{2}.$$

Further discussion of this problem is deferred until Chapter 4, where ranking and selection procedures are discussed specifically for the binomial distribution.

It should be pointed out that if the true configuration $\theta_0 = (\theta_{10}, \ldots, \theta_{k0})$ is "far removed" from the least favorable configuration $\theta_{LF} = (\theta_{1,LF}, \ldots, \theta_{k,LF})$ in the sense of our distance measure δ, then the true probability of a correct selection, that is, the probability for the configuration θ_0, may be much larger than the least favorable probability of a correct selection. For example, suppose we compute $P\{CS|\theta_{LF}\} = .75$; we may have $P\{CS|\theta_0\} = .95$ for the true situation. Then we will think we are doing only fairly well, whereas actually the performance is very good. Of course, this should be regarded as a bonus, even though usually some price must be (or has been) paid for this bonus. For example, we may have taken more observations than are really needed to identify the best population with some specified level of confidence. This type of inefficiency is tied up with our a priori ignorance about the true value of θ, and should not be confused with inefficiencies of the procedure, that is, using a selection rule R when a better one R' is available. In order to compare different procedures, the function $P\{CS|\theta\}$ must be computed for each selection rule under comparable conditions for many different points θ; that is, the comparisons must be made at the same θ and for the same numbers of observations from each population.

1.4 SELECTION PROCEDURES

We now describe the form of a typical ranking and selection procedure that can be used for a large class of problems involving k populations.

The first step is to determine a function T of the observations that is "appropriate" for the problem at hand. The choice of the statistic T depends on the underlying distribution, on the main parameter of interest as regards selection, on other parameters, and on the goal we have in mind. For example, T might be the sample mean, or the median, or the variance, and so on, and usually involves only one of these. Having chosen the statistic T, we compute a value of T for each of the k samples. Let T_j denote the value of T for the sample from the jth population, for $j = 1, 2, \ldots, k$. For example, in the problem of ordering variances, T_1 might be the variance of the data from the first sample, T_2 the variance for the second sample, and so on. Let $T_{[i]}$ denote the ith from the smallest among $T_1, T_2, \ldots, T_k$. Then the ordered values of these statistics are

$$T_{[1]} \leqslant T_{[2]} \leqslant \cdots \leqslant T_{[k]}.$$

Suppose the goal is to select the one best population. If *best* refers to the one with the largest component value of θ, $\theta_{[k]}$, then the selection rule R is simply to observe which population gives rise to $T_{[k]}$ and assert that this population is the best one. On the other hand, if the situation is such that the *best* population is interpreted as meaning the one with the smallest θ value $\theta_{[1]}$, then the rule is to select the population that gives rise to $T_{[1]}$. (This latter variation is especially useful for the variance problem.) Except when indicated otherwise, *best* here means the population for which $\theta = \theta_{[k]}$, since the formulation when $\theta = \theta_{[1]}$ remains the same with obvious changes.

It may happen that two or more values of T are equal to each other (called a *tie*) and are also larger than any of the other values of T. Therefore the rule R must include a method for making a selection among the populations when ties occur for the largest observed value of our statistic T.

Suppose that there are exactly r ties for the largest sample statistic; that is,

$$T_{[k]} = T_{[k-1]} = \cdots = T_{[k-r+1]} > T_{[k-r]}, \qquad (1.4.1)$$

for some r where r is one of the values $2, 3, \ldots, k-1$. Under the assumption that the sample sizes, once determined, are to remain fixed, we do not have the option of taking more observations in order to break the ties. In this case we recommend breaking the ties by choosing at random one of the samples that produced the r tied values of T in (1.4.1) and asserting that the population from which that sample was drawn has the largest θ value. That is, we choose (or generate) an integer between 1 and r inclusive in such a way that each integer has probability $1/r$ of being chosen, and therefore each of the r tied competitors for first place has common probability $1/r$ of being selected. This could be done by using a table of random numbers (or by rolling a balanced die or cylinder with r sides). For convenience a table of random numbers is provided in Appendix T.

1.5 ANALYTICAL ASPECTS OF THE PROBLEM OF SELECTING THE BEST POPULATION

In this section we discuss three different but related analytical aspects of the problem of selecting the best population. These problems are called the *determination of sample size*, the *calculation of the operating characteristic curve*, and the *estimation of the true probability of a correct selection*. Throughout this section we assume that the best population is to be selected using the selection rule described in the previous section, the distance measure is $\delta = \theta_{[k]} - \theta_{[k-1]} > 0$, and the preference zone is $\delta \geqslant \delta^*$ for some $\delta^* > 0$.

1.5.1 Determination of Sample Size

For the problem of determining the sample size, we assume that the experiment is in the process of being designed because the aim is to determine the sample size required per population in order that we may have a specified confidence that the selection procedure described in Section 1.4 leads to a correct selection. Here we generally assume that each sample is to have the same number n of observations so that the statistics

T_j are all based on the same number of sample values. Then the problem is to compute the smallest integer value of n such that our selection procedure satisfies the requirement that the probability of a correct selection is at least some specified number, say P^*, for any and all parameter points θ that lie in the preference zone. Thus in order to use this approach the investigator must specify in advance (before taking any observations) the number P^* and the definition of the preference zone. When the preference zone is defined by $\delta \geqslant \delta^*$, this means that δ^* must be a prespecified (positive) number. (The value of P^* must always be between $1/k$ and 1, since P^* values less than or equal to $1/k$ can be attained without taking any data.)

In symbols the problem is to find the smallest integer n such that the inequality

$$P\{\text{CS}|\theta\} \geqslant P^* \qquad \text{for all } \theta \text{ in the preference zone,} \qquad (1.5.1)$$

or equivalently,

$$P\{\text{CS}|\theta\} \geqslant P^* \qquad \text{for all } \delta \geqslant \delta^*, \qquad (1.5.2)$$

is satisfied for the prespecified pair of values (δ^*, P^*). As already mentioned, when a least favorable configuration exists, we know that

$$P\{\text{CS}|\theta\} \geqslant P\{\text{CS}|\theta_{\text{LF}}\} \qquad \text{for all } \theta \text{ in the preference zone.}$$

Hence, in practice, we can simply find the number n of observations per sample such that $P\{\text{CS}|\theta_{\text{LF}}\} \geqslant P^*$. Generally we determine the common value of n by using the equality $P\{\text{CS}|\theta_{\text{LF}}\} = P^*$.

Tables given in the appendix of this book can be used to find the common value of n for a range of values of k, δ^*, and P^* for the various ranking and selection problems covered here (for example, selecting the normal population with the smallest variance). The entries in these tables were calculated in the manner described in the previous paragraph. Since in our present problem the value of n is determined before any observations are taken, this approach could also be termed a predata formulation.

An obvious question is how to choose the pair of numbers (δ^*, P^*). The probability P^* is often taken as either .900, .950, .975, or .990, but specification of δ^* may not be so simple. Hence we give here an interpretation of δ^* and P^* that may assist the investigator in making that specification. For the measure of distance $\delta = \theta_{[k]} - \theta_{[k-1]}$, this interpretation is a confidence statement about the difference between the parameter of the population that is selected as best and the parameter of the true best population. Even though such a statement ordinarily is made after the data are obtained, it can be made as soon as the common sample size n is determined by the specified values of δ^* and P^*; as a result, the confidence statement gives further insight into the meaning of δ^*. Suppose that θ_s is the true θ value of the population selected according to the rule R of Section 1.4. Then we know that $\theta_s \leqslant \theta_{[k]}$, the true largest θ value, but how large is the difference between them likely to be? In other words, how serious an error would it be to assert that the sample that gives rise to the statistic $T_{[k]}$ comes from the best population when some other population is really the best one? The confidence statement may be made in either of the following equivalent ways:

1. With confidence level P^* the difference between the chosen and the largest θ values satisfies

$$0 \leqslant \theta_{[k]} - \theta_s \leqslant \delta^*,$$

or equivalently, θ_s satisfies

$$\theta_{[k]} - \delta^* \leqslant \theta_s \leqslant \theta_{[k]}.$$

2. With confidence level P^* the interval $[\theta_s, \theta_s + \delta^*]$ covers the true best θ value, or equivalently, $\theta_s \leqslant \theta_{[k]} \leqslant \theta_s + \delta^*$.

Hence for any P^*, the quantity δ^* can be very roughly and intuitively interpreted as the maximum error likely to be committed where P^* represents the confidence level for this maximum error, or δ^* represents the width of a confidence interval on the true largest θ value where P^* is the level of confidence.

Once n is determined, samples of this size are usually taken from each of the k populations and the selection rule of Section 1.4 is used to identify which population is to be asserted as best. However, the practical aspects of the problem may force the investigator to consider other factors before settling on this value of n as the one to be used. For example, if the n determined by the method above is too large for the resources available or the type of experiment that is required, n has to be reduced. Then it is no longer true that the probability of a correct selection satisfies the inequality in (1.5.1) or (1.5.2) for the previously specified pair of values (δ^*, P^*). In order to determine a smaller value of n, the investigator must either specify a smaller value of P^*, or a larger value of δ^*, or both. In essence, since more precision has been demanded than can be obtained for practical reasons, the investigator must be satisfied with a lower level of confidence that the experiment will produce a correct selection of the best population, or with a larger indifference zone, or with both.

Even if the *n* determined is not too large, the investigator may want to look at the confidence level as a function of δ^* for this value of *n*, or equivalently, at $P\{\text{CS}|\theta_{\text{LF}}\}$ as a function of δ^* for this value of *n*. If the investigator feels that the general performance of the selection procedure based on this value of *n* is not satisfactory, he may want to go back and change *n*.

The important point here is that the investigator may look at the table that gives *n* as a function of (δ^*, P^*) even before deciding on the values to specify for δ^* and P^*. The final choice of a sample size determines the terminal confidence statement that can be made about the selection procedure.

1.5.2 Calculation of the Operating Characteristic Curve

In this subsection we are interested in evaluating the overall performance of the selection procedure for given *k* when the common *n* is fixed. The value of *n* may have been determined by the method of the previous subsection after specifying P^* and δ^*, or it may have been fixed at the outset by other considerations. For example, only a fixed number of subjects may be available for observation, or only a fixed amount of resources may be available to devote to this investigation. The kind of analysis to be described here can be done either before or after the sample data are taken, since the only information used is the value of *k* and the common sample size *n*.

In order to obtain a general evaluation of the selection procedure for a given *n*, we need to know the $P\{\text{CS}|\theta\}$ for values of θ other than θ_{LF}. The values of primary consideration are generally in the preference zone, but not exclusively. One set of θ values that is of special interest is the so-called *generalized least favorable configuration*, abbreviated GLF configuration and denoted by θ_{GLF}. This set is defined by

$$\theta_{[1]} = \theta_{[2]} = \cdots = \theta_{[k-1]} = \theta' \quad \text{and} \quad \delta = \theta_{[k]} - \theta'. \quad (1.5.1)$$

Since from (1.3.4) θ_{LF} is specified by $\theta_{[1]} = \cdots = \theta_{[k-1]} = \theta'$ and $\delta = \delta^*$, the GLF configuration includes the LF configuration as a special case. It is called the *generalized* least favorable configuration because the difference δ between $\theta_{[k]}$ and the common value θ' is not specified, as it was for the LF configuration. The GLF configuration cuts through both the indifference zone and the preference zone to an extent that depends on the values considered for δ.

As an example consider the case of two populations. Then the line $\theta_{[2]} - \theta_{[1]} = \delta^*$ is the LF configuration, whereas the region $\theta_{[1]} \leqslant \theta_{[2]}$ is the GLF configuration. In Figure 1.5.1 the GLF is the set of all existent points and the indifference zone is the shaded region.

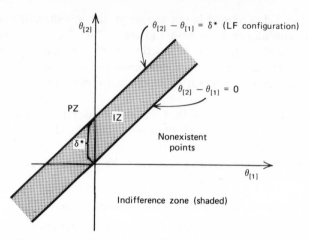

Figure 1.5.1   The generalized least favorable configuration and the least favorable configuration for two parameters ($k = 2$). (For $k = 2$ the GLF configuration is the set of all existent points.)

Suppose we wish to investigate the probability of a correct selection under a GLF configuration; that is, $P\{CS|\theta_{GLF}\}$. In general, this probability will depend on n, $\theta_{[k]}$ and θ', although sometimes it will depend on only n and δ. In any case the probability has a minimum value that approaches $1/k$ when δ approaches 0, and is generally an increasing function of δ for fixed n and θ'. Although this property of monotonicity holds for all the problems that we consider, the reader is cautioned not to assume that it is true in all cases. We regard δ as a "natural" measure of distance for a problem if the probability of a correct selection in the GLF configuration depends on δ, but not on θ'.

The function $P\{CS|\theta_{GLF}\}$ can be calculated under the GLF configuration for any fixed n, as a function of δ and θ' (sometimes only of δ). Here this function is called the *operating characteristic (OC) function* of the selection procedure, and its graph is called the *operating characteristic (OC) curve*. It gives the exact probability of a correct selection under the generalized least favorable configuration, and hence also the minimum probability of a correct selection under all other configurations. For any n the true probability of a correct selection depends on the true values of $\theta_{[k]}$ and θ', and also on the particular true configuration for the θ values. In general, there are no simple formulas for computing these quantities.

1.5.3   Estimation of the True Probability of a Correct Selection

In the "determination of sample size" approach to the problem of selecting the best population, we know that for $\delta \geqslant \delta^*$, the probability of a correct selection under the least favorable configuration is equal to P^*, and the probability under all other configurations in the preference zone is at least P^*. By calculating the operating characteristic function $P\{CS|\theta_{GLF}\}$, we obtain a general evaluation of the procedure with a given common sample size for an arbitrary generalized least favorable configuration, but the true value of δ is, of course, not known to us. Moreover, the true configuration may not be among these generalized least favorable configurations.

Nevertheless, in many problems the data can be used to estimate the true configuration and thereby obtain an estimate of the true probability of a correct selection. Specifically, if the probability is known to depend on the configuration of the θ values through some specific function of the θ values, as, for example, $\delta_1 = \theta_{[k]} - \theta_{[1]}, \delta_2 = \theta_{[k]} - \theta_{[2]}, \ldots, \delta_{k-1} = \theta_{[k]} - \theta_{[k-1]},$ we can use the corresponding function of the values of the sample statistics $T_{[1]}, T_{[2]}, \ldots, T_{[k]},$ to obtain an estimate of the true probability.

Any analysis of the type described in this subsection could be termed a postdata formulation, because it cannot be performed until after the data are obtained.

1.6   SUMMARY

In this chapter we present the general philosophy of ranking and selection procedures. These procedures can be applied to a number of different goals. Since the specification of a distance measure and indifference zone is basic to all these procedures, these concepts are discussed in detail.

After the appropriate statistic is decided on, the usual selection rule for the goal of selecting the one best population is simply to assert that the population giving rise to the largest sample statistic is the best one. However, there are three analytical aspects of the selection problem.

In considering the determination of sample size, the preliminary aim is to determine the minimum number of observations per sample needed in order that the probability of a correct selection is at least some specified value for all configurations in the prescribed preference zone. Whether n is determined by this method or by other considerations, once n is fixed, the performance of the selection procedure can be examined by calculating the $P\{CS|\theta\}$ for various configurations, especially those in the GLF. The operating characteristic curve gives $P\{CS|\theta_{GLF}\}$ as a function of $\theta_{[k]}$ and θ' for any n. This probability is exact for the GLF configuration, and is a lower bound over any other configurations in the preference zone with the same $\theta_{[k]}$ and θ'. Note that this evaluation procedure uses the value of n, but not the observations themselves. The data are used only to identify the population that is asserted as best and not to estimate the true configuration. If we want to estimate the true probability of a correct selection, one way of proceeding is to first estimate the true configuration. Here the data do enter into the analysis as well as the selection.

In the following chapters the general procedures of this chapter are applied to a number of different goals and a variety of specific distribution models.

1.7 NOTES, REFERENCES, AND REMARKS

It seems somewhat surprising that sequential techniques for ranking and selection were developed before nonsequential techniques. The development of sequential analysis by Wald in the 1940s and the subsequent modification of sequential analysis for ranking two populations by Girshick in his 1946 paper constituted the important impetus that started a widespread and long-lasting interest in both sequential and nonsequential ranking and selection problems. In this book only nonsequential aspects of ranking and selection are treated.

The preceding description is quite incomplete, and even for the material treated, there could be differences of opinion. It would be difficult to give credit to all the early writers and forerunners in this area. The exact origin of an idea is rarely completely clear but a list of early authors is useful. In the late 1940s and early 1950s several papers dealt with the so-called slippage model (that is, the movement or shift of one parameter value to the right or left whereas all the other parameter values remain equal), namely, those by Mosteller (1948), Mosteller and Tukey (1950), Bahadur (1950), Bahadur and Robbins (1950), Paulson (1952a, b), and Truax (1953). These were forerunners to the present formulation.

Starting in 1954, a series of fundamental papers developed the present orientation and set the stage for further research during the next 20 years. The first of these papers were those by Bechhofer (1954), Bechhofer, Dunnett, and Sobel (1954), and Bechhofer and Sobel (1954). These three papers contain the main ingredients and ideas that were incorporated into the later work. In addition, papers by Gupta (1956) and Gupta and Sobel (1957) should be mentioned as contributing to the early development of the field. Since this period, numerous papers developing the field have appeared. We comment on these as they arise in later chapters.

Papers concerning the application of ranking and selection to choosing the best chicken strains are by Becker (1961, 1962, 1964) and Soller and Putter (1964, 1965). Other applications that recently came to our attention are comparing effectiveness of advertising by Dalal and Srinavasan (1976) and analyzing simulations of accounting systems in Lin (1976).

A normal-theory test on means often requires the assumption of a common variance, and so a preliminary test for the homogeneity of variances is performed. To test homogeneity of variances (sometimes called homoscedasticity) of k normal populations, the most common procedures are those known as Bartlett's test, Cochran's test, and Hartley's test. Fisher's F test can be used for $k = 2$ populations. To test homogeneity between the probabilities of success in k Bernoulli (or binomial) populations, the usual method is a chi-square test based on a function of the sum of squared deviations between observed and expected frequencies. These tests, as well as the analysis of variance test for homogeneity of means, are discussed in many elementary textbooks and we do not discuss them in this book.

It has always been recognized that the statistician's work does not end with a test of homogeneity. With enough data such a hypothesis is usually rejected if we have even the slightest deviation from exact equality of statistical parameters. This asymptotic inconsistency leads to the need for reassessment of whether the widespread usage of tests of significance is justified and what to do after a significant difference is found. The ranking of populations is a logical alternative and it then becomes desirable to control the probability of a correct ranking or a correct selection (*not* the test of significance). In some cases control can be exercised jointly over the correct selection of the best population and the assertion that a given confidence interval will contain the best population. Recently this type of problem was studied by Alam and Thompson (1973) and by Rizvi and Saxena (1974).

Multiple comparisons comprise a broad area of techniques that is usually concerned more with error rates per comparison than with the overall error associated with any particular string of inequalities that produces a complete or partial ranking of parameters. In a sense all ranking and selection problems deal with multiple comparisons, but the approach is different and multiple comparisons are not treated in this book. Chapter 10 deals with the comparison of k populations with a control and, in particular, Sections 10.6.1 and 10.6.2 show the important distinction that the ranking and selection approach is concerned with joint confidence statements rather than with error rates per comparison. Miller (1966) gives a useful compendium of techniques based on multiple comparisons; a more recent book dealing with this subject is Kleijnen (1975). None of the techniques associated with tests of homogeneity or multiple comparisons can be regarded as forerunners that led to the ideas of ordering and selection.

It is not our intention to make a complete survey of ranking and selection in these notes but to point out some major areas that are not covered in the book and refer the reader to the original papers. For example, selection and ranking procedures based on Bayesian methods and prior distributions are not covered in this book. In this connection we should mention the work of Bross (1949), Grundy, Rees, and Healy (1954), Dunnett (1960), Raiffa and Schlaiffer (1961), Guttman and Tiao (1964), Bland and Bratcher (1968), and Govindarajulu and Gore (1974). Ranking problems with uniform distributions as, for example, in Barr and Rizvi (1966) and Blumenthal, Christie, and Sobel (1971), have also been omitted from this book. Other omissions include applications of ranking and selection to the entropy function (see Gupta and Huang (1976)) and applications of Schur convexity (and concavity) as in Gupta and Wong (1975). However, it should be noted that this last concept does occur in Olkin, Sobel, and Tong (1976), and the latter report forms the theoretical background of new methods introduced for the first time in this book on the estimation of the true probability of a correct selection. The application of empirical Bayes ideas to ranking and selection is due to Robbins (1964), Deely (1965), Deely and Gupta (1968), Bratcher and Bhalla (1974), and Goel and Rubin (1975).

An important reference that is not included in the discussion in the text on applications of these procedures is Kleijnen (1975). This book on statistical techniques in simulation is devoted largely to multiple comparisons and multiple ranking procedures. Routines are developed for assessing the efficiency and the robustness of many ranking procedures. Additional references on multiple ranking procedures are Kleijnen and Naylor (1969) and Kleijnen, Naylor, and Seaks (1972). Some more recent theoretical papers on ranking and selection include Carroll (1976), which deals with the uniformity of sequential ranking procedures, and Chernoff and Yahav (1976), which uses a new criterion for a problem in subset selection. a problem in subset selection.

A fairly complete bibliography up to 1968 is given in Bechhofer, Kiefer, and Sobel (1968); it is of course out of date now. A more recent survey is due to Dudewicz (1968). Finally, we should mention a forthcoming survey of ranking and selection by Gupta (1977). A principal source of table values for the authors is Milton (1963). Some of these values (with less accuracy) appear in Gupta (1963) and another recent table is Gupta, Nagel, and Panchapakesan (1973).

CHAPTER 2

Selecting the One Best Population for Normal Distributions with Common Known Variance

2.1 INTRODUCTION

In this chapter and the next we develop specific methods appropriate for selecting the one population with the largest mean from a set of k independent normal populations; simultaneously we consider the corresponding problem for the smallest mean, since a simple transformation reduces this problem to the former one. The parameter θ used in the general discussion in Chapter 1 is replaced here by μ (the Greek letter mu, which is the symbol commonly used to designate the mean of a normal distribution). The normal family is indexed by two parameters, the mean μ and the standard deviation σ, or, equivalently, the variance σ^2, and is denoted by $N(\mu,\sigma^2)$. Thus, for example, $N(80,40)$ denotes a population with a normal distribution, mean $\mu = 80$ and variance $\sigma^2 = 40$. Here we are dealing with k populations, and we let μ_j and σ_j denote the mean and standard deviation of the jth population, for $j = 1, 2, \ldots, k$. In symbols the general distribution model assumed in this chapter and the next is

$$N\left(\mu_1,\sigma_1^2\right), N\left(\mu_2,\sigma_2^2\right), \ldots, N\left(\mu_k,\sigma_k^2\right). \tag{2.1.1}$$

The ordered μ values are denoted by

$$\mu_{[1]} \leqslant \mu_{[2]} \leqslant \cdots \leqslant \mu_{[k]}. \tag{2.1.2}$$

Then the ultimate goal is to select, on the basis of sample data, the one population that has the largest μ value, $\mu_{[k]}$, or, correspondingly, the one population that has the smallest μ value, $\mu_{[1]}$.

Since we are interested in ranking the populations only according to their μ values, the σ values are regarded as nuisance parameters, especially when they are unknown. Thus, as in the classical theory of hypothesis testing for this model, we need to distinguish the following four different cases:

1. The values of $\sigma_1, \sigma_2, \ldots, \sigma_k$ are assumed equal to a common known value σ, so that the distribution model in (2.1.1) becomes

$$N\left(\mu_1, \sigma^2\right), N\left(\mu_2, \sigma^2\right), \ldots, N\left(\mu_k, \sigma^2\right). \tag{2.1.3}$$

2. The values of $\sigma_1, \sigma_2, \ldots, \sigma_k$ are not known, but are assumed equal to a common unknown value σ, so that the distribution model is (2.1.3) again.

3. The values of $\sigma_1, \sigma_2, \ldots, \sigma_k$ are all known and are not necessarily all equal.

4. The values of $\sigma_1, \sigma_2, \ldots, \sigma_k$ are not known and equality (or any other particular information) is not assumed at the outset.

We treat in this chapter only the situation where the populations have a common known σ, namely case 1. Cases 2 and 3 are covered in the next chapter. Case 4, where the σ values are unknown and unequal, is also covered in the next chapter, but it requires a two-stage procedure.

REMARK.   Case 4 bears a strong resemblance to the so-called Behrens-Fisher problem in classical hypothesis testing, where the model is *two* normal populations with unknown, and not necessarily equal, means and variances and the null hypothesis is that the two means are equal. Roughly speaking, the reason for the difficulty is that under this model the differences in sample means may be due to differences in population variances as well as to differences in population means, and it is not easy to distinguish between the contributions of these two kinds of population differences.   □

Throughout this chapter and the next we assume either that the k populations are independent of each other in the statistical sense or that the k sets of sample observations are drawn in such a way that they are mutually independent; that is, independence holds between samples from different populations as well as within samples from the same population. Thus we can regard the k populations as the components of a k-variate normal population with independent components. For normal populations this condition is satisfied whenever the correlation between any two populations is zero.

Before explaining the statistical procedures for this model and this goal, we give a few concrete illustrations of applications to indicate that this model is a prototype for many practical problems in diverse areas of experimentation and/or general investigation.

Example 2.1.1 (Dairy Farming). An experiment is conducted on a dairy farm to compare the effect of three different types of feed on the yield of milk. Assume that the farmer's goal is really to identify one of these three types of feed as the best one. This goal is quite logical and appropriate if the farmer wants to use only one type of feed, since then he would naturally prefer to use the one that produces the highest yield of milk (assuming other things, like cost, to be equal). It may also be reasonable to assume that milk yield is normally distributed with the same variance for each feed, and that this common variance is known from past records of milk yield. However, such assumptions should be based on "positive" experience with earlier data, and not on guesswork or on a model that is merely convenient.

Example 2.1.2 (Silk Production). Many factors are used to judge the quality of fabrics such as silk, rayon, or nylon. One particular measurable criterion is *denier*, which is defined as the weight of a fixed length of fabric (with standard width). The higher the density of the material, the greater the weight or denier and hence the better the quality. (This measure is similar to *thread count*, which is the number of threads per square inch in woven fabric. Thread count is used to measure the quality of bed linen, for example, and, in particular, to distinguish percale from muslin.) Many factors affect the denier of silk; one of these is the temperature of the bath in which the cocoons are heated. If different baths can be used at several different temperatures (the number of baths is assumed fixed or determined by certain limitations, e.g., the experimental resources available), the denier of silk produced can be measured for each bath. A reasonable goal is to determine which one of these temperatures produces the highest denier on the average. We assume that all the measurements on denier are normally distributed with a common known variance.

Example 2.1.3 (Fastest Route). Taxi drivers frequently travel over a few different routes in going (say) between a downtown area and an airport. Suppose a taxi driver can take any one of six different routes to travel from downtown to the airport. He decides at random which route to take by tossing a balanced die. Suppose he records the traveling time required by each of the six routes for a certain period of days. The obvious goal of primary interest to the driver is to determine which route requires the smallest amount of traveling time on the average. Although it is highly likely that the number of observations collected is not the same for each of the six routes, the differences tend to balance out with a sufficiently large total sample, because each route has the same probability of being the one taken. For this situation we assume a normal distribution and a common variance; in fact, for the procedures of this chapter we go one step further and assume also that the common variance is known. (Chapter 3 deals with methods for normal distributions with a common unknown variance.) If the taxi driver also keeps a record of the time of day, traffic, road conditions, and the like, then these effects on the traveling time can be eliminated. (The method of analysis appropriate for this situation is discussed in Chapter 9.) Treatment of these more complicated situations is postponed until after the simplest cases are thoroughly explained.

In any of these and other examples, goals other than selecting the *one* population with the best μ value may also be of interest. In some situations we may want to find the *two* populations with the best μ values, with or without specifying which one is better than the other; we may want to rank *all* the populations with respect to their μ values; we may want to identify simultaneously the population with the smallest μ value and the one with the largest μ value. All such goals are allowed within the general framework for ranking and selection problems, but any change in the nature of, or the possible number of, decisions permitted requires a new formulation, a new analysis, and (usually) the calculation of a new table. Similarly, if one population represents a control group and we want only to compare each of the other populations with the control group, the problem must be reformulated and appropriate tables developed.

For the time being we restrict consideration to the problem formulated earlier, where the goal is simply to select the one best population among a set of k normal populations with common known variance. To avoid confusion we define *best* as meaning the population with the largest mean in Sections 2.2, 2.3, and 2.4. The problem of selecting the population with the smallest mean is covered in Section 2.5. Throughout, the reader is advised to remember that we are *not* attempting to *estimate* the "best" mean, but only to *identify* the population that has the "best" mean.

In formulating the problem for this model we must first define a distance measure for the difference between the populations with μ values $\mu_{[k]}$ and $\mu_{[k-1]}$. A natural measure, which we call δ, is

$$\delta = \mu_{[k]} - \mu_{[k-1]}. \tag{2.1.4}$$

In view of the ordering in (2.1.2) it is always true that $\delta \geqslant 0$, and $\delta = 0$ if the two largest means are equal. Thus the indifference zone must be defined as $\delta < \delta^*$ and the preference zone as $\delta \geqslant \delta^*$.

The generalized least favorable configuration in terms of the true value of the distance measure δ is

$$\mu_{[1]} = \mu_{[2]} = \cdots = \mu_{[k-1]} \quad \text{and} \quad \delta = \mu_{[k]} - \mu_{[k-1]}, \quad (2.1.5)$$

and the least favorable configuration in terms of the specified value δ^* that separates the indifference and preference zones is the same as (2.1.5) with $\delta = \delta^*$.

2.2   SELECTION RULE

The procedure used to select a population (which we then assert to be the one with the largest mean) is quite simple and straightforward once the k sets of sample data are obtained. Since the parameters of interest are the population means, the analogous sample statistics are the sample means. For any particular sample the mean is found by summing the values of the observations in that sample and dividing by the number of observations in that sample. For the jth sample suppose there are n_j observations. We denote these by $x_{1j}, x_{2j}, \ldots, x_{n_j,j}$. Then the calculation of the mean for the jth sample is described in symbols as

$$\bar{x}_j = \frac{x_{1j} + x_{2j} + \cdots + x_{n_j,j}}{n_j}, \quad (2.2.1)$$

or, in the "shorthand" sum notation, we can write (2.2.1) equivalently as

$$\bar{x}_j = \frac{\sum_{i=1}^{n_j} x_{ij}}{n_j}.$$

These k sample means are then ordered from smallest to largest. Using our standard ordering notation, $\bar{x}_{[j]}$ is the jth smallest sample mean, and the ordered sample means are written as

$$\bar{x}_{[1]} \leqslant \bar{x}_{[2]} \leqslant \cdots \leqslant \bar{x}_{[k]}.$$

The selection rule R is simply to identify the largest sample mean $\bar{x}_{[k]}$ and assert that the population which produced that sample is the population with the largest value of μ, that is, $\mu_{[k]}$. Thus if $\bar{x}_{[k]} = \bar{x}_j$, then the population labeled j is asserted to be best in the sense that μ_j is larger than any other μ_i for $i \neq j$. In the preceding Example 2.1.1 we select the feed that gives the largest average milk yield among those observed; in Example

2.1.2 we select the temperature that produces the largest average denier. In each case we assert that the population with the largest true mean is the one for which the sample mean is largest. Thus in Example 2.1.1 if the first feed produces a larger milk yield than either of the other two feeds observed, then $\bar{x}_{[3]} = \bar{x}_1$ and we assert that μ_1 is larger than either μ_2 or μ_3.

Since the normal distribution is continuous, the probability that two or more observations have the same value (called a tie) is equal to zero. As a result, the existence of ties is not considered in the derivation of the function that gives the probability of a correct selection under the model of normal distributions. Nevertheless, ties can occur in practice, not only among the observations, but also among the sample means, especially if the data are rounded or measurements are made to a small number of significant figures. Accordingly, the rule R must include a method for dealing with the situation that arises when two or more sample means are equal and are also larger than any of the others. When there are ties for first place, one sample should be chosen at random from those samples with tied means. Specifically, when there are r ties for first place among the sample means, one of these samples is selected by using some independent random mechanism that gives probability $1/r$ to each of them. Then the population that produced this mean is asserted to be the best one. To reduce the possibility of ties the data should be obtained with sufficient precision and the meaningful digits should be retained in the calculations whenever possible. The existence of many ties may be an indication that the measurements are not being made with a sufficient degree of precision, or possibly that the assumption of a continuous distribution model is not appropriate.

We illustrate the selection procedure by Example 2.2.1 for equal sample sizes and Example 2.2.2 for unequal sample sizes. The experimental setting and procedure provide a realistic application, although the data are artificial so that the arithmetic calculations can be followed easily.

Example 2.2.1 (Anxiety and IQ). Carrier, Orton, and Malpass (1962) report a study of anxiety levels of four groups of children aged 10–14 years in five southern Illinois communities. The children in this population were classified into groups according to their scores on the Wechsler Intelligence Scale for Children (WISC). The groups were differentiated as Bright (B), Normal (N), and Educable Mentally Handicapped (EMH). The ranges of WISC scores separating these groups are above 115 for B, 85–115 for N, and below 85 for EMH. The EMH category was further subdivided into Noninstitutionalized Educable Mentally Handicapped (NEMH) and Institutionalized Educable Mentally Handicapped (IEMH); thus we have four groups of interest. A random sample of children was then taken from each of these groups and each child was scored on the Children's Manifest Anxiety (CMA) scale. The CMA scale measures anxiety according to the responses to 53 yes or no questions; large scores represent greater anxiety. In this experiment the questions were asked individually and orally. In order to avoid differences related to sex, the data were analyzed separately for boys and girls. The data given in Table 2.2.1 represent the CMA scores for random samples of nine girls from each of the groups. These data were generated (using random normal deviates) to simulate the summary data reported in Carrier, Orton, and Malpass. The goal is to select the group that has the greatest anxiety measured by the CMA scale.

Table 2.2.1   *Simulated scores on the children's manifest anxiety scale for four groups of girls*

	B	N	NEMH	IEMH
	19.0	20.2	24.2	31.7
	8.0	18.2	21.7	27.6
	17.2	22.4	20.2	31.7
	2.0	37.8	21.9	29.9
	9.8	22.7	24.0	24.7
	10.0	19.4	16.7	27.5
	15.2	12.3	25.5	31.4
	14.8	13.6	27.5	31.0
	20.5	24.3	20.0	35.4
Total	116.5	190.9	201.7	270.9
Sample size	9	9	9	9
Sample mean	12.9	21.2	22.4	30.1

The sample mean scores, arranged in increasing order of magnitude, are 12.9, 21.2, 22.4, 30.1 and these correspond respectively to the WISC groups labeled B, N, NEMH, IEMH. Since the IEMH group has the largest sample mean score, we assert that the Institutionalized Educable Mentally Handicapped group has the highest expected degree of anxiety as measured by the CMA scale.

Example 2.2.2   In the research reported by Carrier, Orton, and Malpass (1962) and described in Example 2.2.1, the sample sizes actually used in the experiment were unequal. The number of girls in the respective groups are 11 Bright, 11 Normal, 9 NEMH, and 8 IEMH children. The article gives only summary data on the CMA scores for these samples. The data

Table 2.2.2   Simulated scores on the children's manifest
anxiety scale for four groups of girls

	B	N	NEMH	IEMH
	6.8	17.7	18.1	33.4
	15.4	17.3	21.5	30.6
	15.7	24.8	19.4	31.1
	7.4	18.7	27.7	34.0
	16.7	29.7	26.0	29.6
	15.2	24.7	21.4	20.3
	22.0	28.5	19.9	33.9
	16.0	12.6	25.6	26.0
	13.8	16.4	18.3	
	7.5	27.0		
	16.6	20.7		
Total	153.1	238.1	197.9	238.9
Sample size	11	11	9	8
Sample mean	13.9	21.6	22.0	29.9

given in Table 2.2.2 were generated (using random normal deviates) to simulate the summary data reported (that is, the scores given are simulated but conform to the sample means reported). For these data, which group of children should be asserted as having the highest expected degree of anxiety?

Since the sample mean for these data is largest for the IEMH group of girls, we assert that this group has the highest expected degree of anxiety.

2.3   SAMPLE SIZES EQUAL

In this section we consider three analytical aspects of the problem of selecting the best normal population when there is a common variance, but only for the case where the sample sizes are all equal. From a practical as well as a theoretical point of view, it is preferable for several reasons under this model that all the k sample means be based on the same number n of observations; that is, $n_1 = n_2 = \cdots = n_k = n$. When the normal distributions have a common variance, whether it is known or unknown, selection rules such as R are more efficient when the sample sizes are all equal. (This means that if the total number of observations is fixed at a multiple of k, say N, then the probability of a correct selection is higher if $n_j = N/k$ for each j than if the observations are allocated among the samples in any way other than equally.) Intuitively, the reasoning is as follows. If we knew which two populations corresponded to the largest and next largest means, then we would want to take most of the observations from these. However, we do not have this information and we must use a fixed sample size. If we take unequal sample sizes, then we may be taking more observations from the population with the smallest mean, say, thereby losing information.

Because of this higher efficiency the tables available for ranking and selection procedures generally include only equal sample sizes. Nevertheless, in the general theory the n need not be common, and in practice it is sometimes necessary to deal with unequal sample sizes. Some approximate methods for handling the case of unequal values of n_j are discussed in Section 2.4.

2.3.1   Determination of Sample Size

The problem of determining the sample size is tied in with the design of our experiment for selecting the best population. Having decided to make the sample sizes equal, our present aim is to determine the smallest common sample size n that provides a prescribed level of confidence that the selection rule R will correctly identify the best population once the data are obtained.

The first step then is to choose a pair of constants (δ^*, P^*), where $1/k < P^* < 1$ and $\delta^* > 0$. This choice is interpreted as meaning that the investigator specifies that the probability of a correct selection is to be at least P^* whenever $\delta = \mu_{[k]} - \mu_{[k-1]} \geqslant \delta^*$, that is, whenever the true configuration lies in the preference zone. P^* is then interpreted as the desired confidence that the selection rule will lead to a correct identification when the true difference δ is at least as large as δ^*. As a result, the two values δ^* and P^* are linked together and do not have a unique interpretation separately in relation to the goal.

REMARK.   In this situation a common terminology is to state that δ^* is the smallest difference that we are interested in detecting between the true mean of the best population and the true mean of the second best population, among the k populations under consideration. We prefer to avoid such terminology in this book so that we can stress the idea of *specifying the pair* (δ^*, P^*). The reasons behind this are as follows:

1. The goal is not to detect the value of this true difference.
2. There are many different pairs (δ^*, P^*) that lead to the same common value of n.

Later on, we in fact determine the graph of the curve representing these pairs. This curve can be regarded as the locus of all pairs (δ^*, P^*) that are satisfied by the same common value of n. In summary, although δ^* by itself has a connotation of distance, its value does not fully describe the problem here unless its companion value P^* is also given. □

The values selected for the pair of constants (δ^*, P^*) are arbitrary, but an appropriate specification requires some experience and good judgment on the part of the investigator. It is important to remember that the larger the value of P^* and/or the smaller the value of δ^*, the larger the common value of n that is required.

For a fixed number k of populations and a specified pair (δ^*, P^*), how do we determine the common n value? In the normal distribution model under consideration here and for the goal of selecting the *one* best population from k populations, it so happens (fortunately) that the specified P^* determines a unique quantity, which we call τ_t, irrespective of the value of δ^*. (The subscript t in τ_t is used as a reminder that this value of τ comes from a table, and also to distinguish it from a computed value of τ, denoted by τ_c, to be defined later, and the true value of τ, which has no subscript.) Table A.1 in Appendix A gives the values of τ_t for $P^* = .750$, .900, .950, .975, .990, and .999 when $k = 2, 3, \ldots, 10, 15, 20$, and 25. If the specified value of P^* is not listed in Table A.1, the corresponding value of τ_t can be found by using the method of interpolation explained in Appendix A. Once τ_t is found from Table A.1, we equate τ_t with $\delta^* \sqrt{n} / \sigma$ and solve for n. The explicit expression that gives n is

$$n = \left(\frac{\tau_t \sigma}{\delta^*} \right)^2. \qquad (2.3.1)$$

In order to be sure that the common sample size is large enough to satisfy the (δ^*, P^*) requirement in this type of problem, the value of n computed from (2.3.1) is rounded upward if it is not an integer.

Example 2.3.1. In the silk production experiment described in Example 2.1.2, suppose that different baths are to be used at $k = 6$ different temperatures in order to determine which one produces the highest average denier. It is known from past experience that the denier measurements are normally distributed with common variance $\sigma^2 = 2$. The initial problem is to determine how many measurements should be made at each temperature. The investigator considers all relevant factors and decides to specify the pair $\delta^* = 0.75$ and $P^* = .95$. This means that the probability of a correct selection is to be at least .95 whenever the true difference between the two best temperatures is at least 0.75. Table A.1 with $k = 6$ and $P^* = .95$ indicates that $\tau_t = 3.1591$ corresponds to the pair $(0.75, .95)$. The number of observations needed at each of the six temperatures is calculated from

(2.3.1) as

$$n = \left(\frac{3.1591\sqrt{2}}{0.75} \right)^2 = 35.48.$$

The common n value required is a sample of 36 independent observations at each temperature.

It is of considerable interest that once this method is used to determine n for the specified pair (δ^*, P^*), we can make a confidence statement about the difference between the true largest mean $\mu_{[k]}$ and the true mean of the population selected by the procedure R defined in Section 2.2. Suppose that μ_s is the mean of the population selected as best in accordance with the sample data. Then we know that $\mu_s \leqslant \mu_{[k]}$, but we would also like to know how large the difference $\mu_{[k]} - \mu_s$ is likely to be. The confidence statement that can be made with confidence level P^* is that the largest mean and the next largest mean differ by at most δ^* with confidence level P^*. In symbols, we write

$$0 \leqslant \mu_{[k]} - \mu_s \leqslant \delta^*, \tag{2.3.2}$$

with confidence level P^*. The statement in (2.3.2) is equivalent to the statement that with confidence level P^*, $\mu_{[k]}$ lies in the (closed) interval

$$S = [\mu_s, \mu_s + \delta^*], \tag{2.3.3}$$

and hence the assertion that $\mu_{[k]}$ belongs to S also has confidence level P^*. It should be noted that neither of the confidence statements associated with (2.3.2) or (2.3.3) is a statement about the numerical value of any single true mean. Both are statements about the correctness of the selection made.

It is also important that the reader not confuse the confidence statement S in (2.3.3) with the statement that $\mu_{[k]}$ lies in the interval $S' = [\bar{x}_{[k]}, \bar{x}_{[k]} + \delta^*]$. The statements S and S' are not equivalent, since μ_s is unknown and generally will not be exactly equal to $\bar{x}_{[k]}$. The confidence associated with S' is not equal to P^*; it would have to be calculated (although we do not do so here).

The fact that δ^* and P^* can ultimately be used to make such confidence statements about the selection may be useful as a basis for choosing the pair of constants (δ^*, P^*) which in turn are used to determine the common sample size n. In Example 2.3.1, if we take 36 observations at each temperature, the confidence level is .95 that the mean μ_s of the population selected differs from the true largest mean by no more than 0.75. The important point here is that the investigator can consult the tables before deciding on the (δ^*, P^*) pair to specify. Of course, any terminal statement about the true difference and the associated confidence level is determined by the common sample size actually used.

Note that in Table A.1, for any fixed k, the value of τ_t increases as P^* increases. Using this fact along with the relationship between τ_t, δ^*, and n in (2.3.1), the reader may verify the following observations. For fixed k and δ^*, n increases as P^* increases (since τ_t increases). On the other hand, for fixed k and P^* (so that τ_t remains constant), n increases as δ^* decreases. As δ^* approaches zero, n gets larger and larger. For $\delta^* = 0$ there is no finite value of n that will suffice if $P^* > 1/k$, and if $P^* \leqslant 1/k$ then the simple random selection of one population (for example, by rolling a balanced cylinder with k sides) gives the specified level of confidence without taking any observations at all.

To illustrate these relationships between n and the pair (δ^*, P^*), suppose we have $k = 6$ normal populations with $\sigma^2 = 2$ as in Example 2.3.1. We already know that $n = 36$ observations are needed per population if $\delta^* = 0.75$ and $P^* = .950$. The reader may verify that if we keep P^* fixed at .950, then (2.3.1) shows that we need $n = 20$ for $\delta^* = 1.00$ and we need $n = 80$ for $\delta^* = 0.50$. On the other hand, if we keep δ^* fixed at $\delta^* = 0.75$, then the solution is $n = 45$ for $P^* = .975$, $n = 36$ for $P^* = .950$, and $n = 27$ for $P^* = .900$.

2.3.2 Modification of the Sample Size

In some instances the sample size n determined from (2.3.1) for the specified pair (δ^*, P^*) is not satisfactory. This situation arises frequently when the experimental resources are limited and the determined value of n requires more resources than are available. In this case the investigator can regard any more practical value of n as fixed (temporarily at least), and either

  1. see what happens to the level of confidence if this n value is used for the same specified value of δ^*, or
  2. see what happens to the width of the confidence interval if this n value is used for the same specified value of P^*.

REMARK.   In each of these two cases we are, in effect, specifying the n value, and therefore the notation n^* would be more consistent. However, we do not use n^* since it is clear in this subsection that n is a specified value. □

In case 1, for a fixed k and the particular n value of interest, we have a specified δ^* for the assertion in (2.3.2) or its equivalent in (2.3.3). The confidence level depends on k and a computed value of τ, denoted by τ_c, which is defined as

$$\tau_c = \frac{\delta^* \sqrt{n}}{\sigma}. \tag{2.3.4}$$

Thus τ_c can be interpreted as the "standardized" value of δ^*, or δ^* expressed in the scale of $\sigma/\sqrt{n}$, rather than in the original scale.

For the selection rule R and τ_c computed from (2.3.4), we use $CL(\tau_c|R)$ to denote the confidence level for the specified δ^*, the known σ, and the particular n value of interest. Table A.2 of Appendix A gives the value of $CL(\tau_c|R)$ for $k = 2, 3, \ldots, 10, 15, 20, 25$, and 50. If the computed value of τ_c is not listed in Table A.2, the corresponding value of $CL(\tau_c|R)$ can be found by the method of interpolation explained in Appendix A.

In Example 2.3.1, with $k = 6$, $\sigma^2 = 2$, $\delta^* = 0.75$, suppose that the investigator cannot afford to take 36 observations at each temperature (a total of 216 observations); for the resources available assume that the maximum number he can take is $n = 25$. In order to find the confidence level for the statement $\mu_{[k]} - \mu_s \leqslant \delta^* = 0.75$, τ_c is calculated from (2.3.4) as

$$\tau_c = \frac{0.75\sqrt{25}}{\sqrt{2}} = 2.65.$$

In Table A.2 with $k = 6$ the entries closest to 2.65 are $CL(2.60|R) = .883$ and $CL(2.80|R) = .912$. By interpolation the confidence level here is $CL(2.65|R) = .891$.

Before discussing case 2 we note that this quantity .891 can also be interpreted as the probability of a correct selection, since the entry in Table A.2 for any k (and with τ_c computed by (2.3.4) as a function of δ^*) is the probability of a correct selection under the least favorable configuration with $\delta = \mu_{[k]} - \mu_{[k-1]}$ equal to δ^*. With this interpretation the result is denoted by $P_{LF}\{CS|\tau_c\}$. (The subscript LF is included to emphasize the fact that the probability is calculated under the least favorable configuration.) Thus in the example of the last paragraph where $k = 6$, $n = 25$, and $\sigma^2 = 2$, $P_{LF}\{CS|2.65\} = .891$ is the least favorable probability of a correct selection, that is, the smallest probability of a correct selection that is attainable uniformly over all parameter configurations in which $\mu_{[6]} - \mu_{[5]} \geqslant 0.75$. Thus in this example .891 can be regarded not only as a lower bound for the probability of a correct selection over all configurations in the preference zone, but also as a lower bound for the confidence level where the confidence statement is $\mu_{[6]} - \mu_s \leqslant 0.75$.

In general, for any specified δ^*, n, and σ, we compute τ_c from (2.3.4) and use that value with the appropriate value of k to find the corresponding entry in Table A.2. The result can be interpreted as either $CL(\tau_c|R)$, the confidence level for the statement $\mu_{[k]} - \mu_s \leqslant \delta^*$, or $P_{LF}\{CS|\tau_c\}$, the probability of a correct selection under the least favorable configuration. The numerical operations involved are identical, but the interpretations of the result differ. The first interpretation leads to a lower bound for the confidence level of the assertion $\mu_{[k]} - \mu_s \leqslant \delta^*$, while the second interpretation leads to a lower bound for the probability of a correct selection over all configurations in the preference zone, defined by δ^*.

Unless considerable precision is needed in giving $CL(\tau_c|R)$ or $P_{LF}\{CS|\tau_c\}$ for any computed τ_c, interpolation in Table A.2 is not necessary. Rather, for any τ_c and k these values can be read from the appropriate figure in Appendix B. These figures cover the cases $k = 2, 3, \ldots, 10, 15, 20, 25, 50$. The figures in Appendix B can be interpreted in the two ways explained in the preceding paragraph. Further, for any particular k the curve is the locus of all pairs (τ_c, P^*), and, accordingly (if a δ^* axis is included), all corresponding pairs (δ^*, P^*) for which the probability requirement $P_{LF}\{CS|\tau_c\} \geqslant P^*$ is satisfied when $\delta = \delta^*$.

A different problem arises in case 2 described earlier. The sample size again is fixed (at least temporarily), but the investigator now specifies a value for P^* and drops the specified value δ^*. Then what can we say about the width of the confidence interval in (2.3.2) or (2.3.3)? For any fixed k and specified P^*, Table A.1 gives the corresponding value of τ_t. For a fixed n and σ, τ_t can be regarded as a standardized δ value. As a result, the confidence interval in (2.3.3) can be written in terms of τ_t as

$$S = \left[\mu_s, \mu_s + \frac{\tau_t \sigma}{\sqrt{n}} \right]. \tag{2.3.5}$$

Thus with confidence level P^* we can assert that $\mu_{[k]}$ is included in the interval S specified by (2.3.5). As in (2.3.3) it is important not to equate μ_s in (2.3.5) with $\bar{x}_{[k]}$. It should be noted that what we have in (2.3.5) is a confidence interval on the difference $\mu_{[k]} - \mu_s$, and it is not a confidence interval on $\mu_{[k]}$.

As a numerical example, suppose that $k = 5$, $P^* = .95$. Then from Table A.1 the corresponding value of τ_t is $\tau_t = 3.0552$. We assume a common n and common known σ^2, but leave them arbitrary. Then at level $P^* = .95$, we can assert that $\mu_{[5]}$ lies in the interval $[\mu_s, \mu_s + 3.0552\sigma/\sqrt{n}]$, or equivalently, that $\mu_{[5]} - \mu_s \leqslant 3.0552\sigma\sqrt{n}$.

Note that with the common n fixed, the width of the interval S in (2.3.5) for given k increases as P^* increases because τ_t increases. Thus as P^* approaches 1, the confidence level increases but the confidence interval may become too wide to be useful. These two factors must be balanced in some satisfactory way.

In summary, if the common sample size n determined by the specified pair (δ^*, P^*) is impractical, the investigator can explore the implications of using any other n value with a new P^* and the original δ^*, or with a new δ^* and the original P^*, or with new values for both. The tables in Appendix A can still be used either with the confidence interval approach or with the more classical probability of a correct selection (or indifference zone) approach.

REMARK. Since the value of τ_t and the probability of a correct selection have a one-to-one relationship for fixed k and n, it is clear that specification of a value for τ is equivalent to specification of the corresponding value of P as P^*. Therefore this approach leads to no new formulations or analyses (although then we may have to interpolate in Table A.2 to find P rather than interpolate in Table A.1 to find τ_t).

In the case where P^* is specified but not δ^*, we prefer not to give any interpretation of τ_t in terms of the probability of a correct selection, because the preference zone for a correct selection using the rule R then becomes a function of the fixed value of n. In the development of decision theory and test construction, it is traditional that the preference zone be identified in advance, before anything else is spelled out, and this implies a specification of δ^* that does not depend on n. However, this objection does not apply when the analysis is adapted to the confidence interval, as we did earlier. $\square$

2.3.3 Calculation of the Operating Characteristic Curve

In this subsection we assume that the common sample size n is fixed permanently (either as determined by the method of Section 2.3.1 or by other considerations), and we take k samples of size n, one from each of k normal populations with known variance σ^2. For fixed n and k the selection rule R (of Section 2.2) for identifying the population with mean $\mu_{[k]}$ is determined. The question of interest now is, How much confidence can we have in the selection and how can we evaluate the probability of a correct selection, when we did not specify either δ^* or P^*?

In the last subsection for a specified δ^* and a fixed n and k, we described a method for determining the corresponding confidence coefficient, $CL(\tau_c | R)$, and also for determining the corresponding probability of a correct selection under the least favorable configuration, $P_{LF}\{CS | \tau_c\}$. Here we have no particular δ^* specified. Nevertheless we can still compute the probability of a correct selection by procedure R under the *generalized* least favorable configuration, by considering the true value

τ as a running variable. The true value of τ is related to the true value of $\delta = \mu_{[k]} - \mu_{[k-1]}$ by

$$\tau = \frac{\delta \sqrt{n}}{\sigma}, \qquad (2.3.6)$$

that is, τ is δ expressed in the scale $\sigma/\sqrt{n}$.

For selected values of k Table A.2 gives the probability of a correct selection as a function of the true value τ under the generalized least favorable configuration, that is, $P_{GLF}\{CS|\tau\}$, and the figures in Appendix B present graphs of the pairs $(\tau, P_{GLF}\{CS|\tau\})$ for the true value τ. The probability $P_{GLF}\{CS|\tau\}$ is a function of k and τ, and when τ is used it no longer depends explicitly on n. This function is called the *operating characteristic* (OC) curve for the selection rule R. The curve always approaches the point $P = 1/k$ as τ approaches zero, and the line $P = 1$ is a horizontal asymptote for τ large. For any given k the OC curve indicates how well the selection rule is operating for different values of τ.

As a numerical illustration, suppose that $k = 4$. For $\tau = 1.00$ say, Table A.2 gives $P_{GLF}\{CS|\tau\} = .552$. Table 2.3.1 summarizes $P_{GLF}\{CS|\tau\}$ for running values of τ.

Table 2.3.1　　$P_{GLF}\{CS|\tau\}$ for $k = 4$ *and selected values of* τ

τ	1.00	1.20	1.40	1.60	1.80	2.00	2.40	2.80	3.00	3.40	3.80
$P_{GLF}\{CS|\tau\}$	.552	.614	.674	.729	.779	.823	.893	.940	.956	.978	.990

These values comprise selected points on the operating characteristic curve for the normal means selection problem with $k = 4$ and $\tau \geqslant 1.00$.

The investigator is usually more interested in the properties of the selection rule R in terms of the δ values rather than the τ values, since any δ value is in the original units of the problem and therefore is easier to interpret. In this subsection we are considering only the situation where n is fixed and σ is known, and therefore we can convert any τ value to a δ value by using the relationship given in (2.3.6). We use the symbol $\delta(\tau; \sigma/\sqrt{n})$ to emphasize the fact that this is the value of δ that corresponds to τ for the fixed n and σ, actually for the fixed ratio $\sigma/\sqrt{n}$. The explicit expression from (2.3.6) is

$$\delta(\tau; \sigma/\sqrt{n}) = \frac{\tau\sigma}{\sqrt{n}}, \qquad (2.3.7)$$

Using (2.3.7) for fixed n and σ, the values of $P_{GLF}(CS|\tau)$, or equivalently, the operating characteristic curve, can also be given in terms of running values of $\delta(\tau; \sigma/\sqrt{n})$.

As a numerical illustration, suppose that $k=4$ as in Table 2.3.1, but the specific problem of interest has $n=9$ and $\sigma^2=2$. Then for any τ, from (2.3.7) we have $\delta(\tau; \sqrt{2}/3) = 0.471\tau$. For the running τ values in Table 2.3.1, the corresponding δ values are given in Table 2.3.2. The results in Table 2.3.2, together with Table 2.3.1, can be interpreted as follows. For $k=4$, $n=9$, $\sigma^2=2$, if the true difference δ between $\mu_{[4]}$ and $\mu_{[3]}$ is at least 0.943 say (so that $\tau \geqslant 2.00$), then the probability is at least .823 that data from the population with the largest true mean will produce the largest sample mean.

Table 2.3.2   *Values of* $\delta(\tau; \sigma/\sqrt{n})$ *for* $n=9, \sigma^2=2$

τ	1.00	1.20	1.40	1.60	1.80	2.00	2.40	2.80	3.00	3.40	3.80
$\delta(\tau; \sigma/\sqrt{n})$	0.471	0.566	0.660	0.754	0.848	0.943	1.131	1.280	1.414	1.603	1.791

REMARK.   In making these computations we need to take note of rounding error. For example, if we compute $(\sqrt{2}/3)(1.40)$ in one operation and then round off, we obtain 0.660. However, if we compute $\sqrt{2}/3$ and round off to 0.471, and then multiply by 1.40, we obtain 0.659. □

The results of Tables 2.3.1 and 2.3.2 are graphed in Figure 2.3.1. The horizontal axis has two sets of labels, namely, τ (which holds for all n and all $\sigma>0$) and $\delta(\tau; \sigma/\sqrt{n})$ from (2.3.7) for $\sigma^2=2$, $n=9$. The vertical axis is labeled $P_{GLF}\{CS|\tau\}$. With τ on the horizontal axis this curve is the

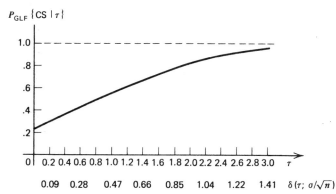

Figure 2.3.1   The probability of a correct selection as a function of τ for $k=4$ populations.

operating characteristic (OC) curve for the selection rule R with $k=4$ for any n and $\sigma > 0$. However, with $\delta(\tau; \sigma/\sqrt{n})$ on the horizontal axis the curve indicates how well the selection rule R is operating for $k=4$ and each possible true value of δ only for $n=9$ and $\sigma^2 = 2$; both axes assume we are in the generalized least favorable configuration.

Obviously, if $\delta = \delta^*$, then $P_{\mathrm{GLF}}\{CS|\tau\}$ reduces to $P_{\mathrm{LF}}\{CS|\tau_c\}$ where τ_c is computed from (2.3.4). (Interpretation of this quantity, and the corresponding confidence coefficient $\mathrm{CL}(\tau_c|R)$, was discussed in Section 2.3.2.) Therefore if $\delta = \delta^*$, Figure 2.3.1 with δ on the horizontal axis can be interpreted either as (1) the locus of all specified pairs (δ^*, P^*) for which the probability requirement is satisfied when $n=9$, $\sigma^2 = 2$, or as (2) the locus of all pairs $(\delta^*, P_{\mathrm{LF}}\{CS|\tau_c\})$ where $\tau_c = 3\delta^*/\sqrt{2}$.

2.3.4   Estimation of the True Probability of a Correct Selection

Thus far in this section all the analytical aspects of the selection problem could have taken place prior to the collection of the data, since the observations were used only to determine which population is selected. In this subsection we consider some data-analytic methods that make further use of the observations. We assume throughout that the sample size n is common, but its value was not necessarily determined by a preassigned pair (δ^*, P^*) because the investigator may have been unable to specify these numbers. Thus the n value could be based on considerations not under the control of the investigator (such as economic restrictions). In any case the selection rule R is defined (in one way or another) and we wish to estimate the probability of a correct selection attained by this procedure.

This probability depends on the true configuration of means, $\mu = (\mu_{[1]}, \mu_{[2]}, \ldots, \mu_{[k]})$ through the following $(k-1)$ differences between $\mu_{[k]}$ and every other mean,

$$\delta_1 = \mu_{[k]} - \mu_{[1]}, \quad \delta_2 = \mu_{[k]} - \mu_{[2]}, \ldots, \quad \delta_{k-1} = \mu_{[k]} - \mu_{[k-1]}. \quad (2.3.8)$$

We call these differences the δ values. Unfortunately we do not know any of the μ values, and hence cannot compute the δ values. However, we can estimate $\mu_{[j]}$ by the jth ordered sample mean $\bar{x}_{[j]}$, for $j = 1, 2, \ldots, k$, and thereby estimate the δ values by

$$\hat{\delta}_1 = \bar{x}_{[k]} - \bar{x}_{[1]}, \quad \hat{\delta}_2 = \bar{x}_{[k]} - \bar{x}_{[2]}, \ldots, \quad \hat{\delta}_{k-1} = \bar{x}_{[k]} - \bar{x}_{[k-1]}. \quad (2.3.9)$$

From these $\hat{\delta}$ values we can estimate the corresponding τ values by simply standardizing each $\hat{\delta}$ value as we did for δ^* in (2.3.4); that is,

$$\hat{\tau}_j = \frac{\hat{\delta}_j \sqrt{n}}{\sigma}, \quad (2.3.10)$$

where n is the fixed common sample size and σ is a known constant. Note that since the $\hat{\delta}$ values are ordered, the $\hat{\tau}$ values satisfy the same ordering relationship, namely, $\hat{\tau}_1 \geqslant \hat{\tau}_2 \geqslant \cdots \geqslant \hat{\tau}_{k-1}$.

We now show how these $\hat{\tau}$ values can be used to estimate the true probability of a correct selection. We first provide a lower bound P_L and an upper bound P_U for the true probability. When the lower and upper bounds are taken together, we can obtain an interval estimate of the true probability. Then we give a point estimate P_E of the true probability; this estimate always lies within the interval estimate.

We first describe each of these procedures in general, and then illustrate the methods by a numerical example.

To determine the lower bound P_L, we compute the $\hat{\tau}_j$ from (2.3.10) and consider each $\hat{\tau}_j$ separately to find the corresponding probability P_j of a correct selection between the jth and kth ordered populations, for each of $j = 1, 2, \ldots, k-1$. Then P_L is the product of these $(k-1)$ probabilities, or

$$P_L = P_1 P_2 \ldots P_{k-1}.$$

The probability P_j for each of $j = 1, 2, \ldots, k-1$ can be found by interpolating in Table A.2 in Appendix A with $k = 2$ for the corresponding $\hat{\tau}_j$. (We use $k = 2$ since each $\hat{\tau}_j$ is computed from exactly two sample means.)

Since the calculation of each P_j uses only the special case $k = 2$, the P_j can also be expressed in symbols as $P_j = \Phi(\hat{\tau}_j / \sqrt{2})$, where $\Phi(x)$ is the cumulative probability function of the standard normal distribution, given in the table in Appendix C. Therefore P_L can also be expressed in symbols as the following product:

$$P_L = \Phi\left(\frac{\hat{\tau}_1}{\sqrt{2}}\right) \Phi\left(\frac{\hat{\tau}_2}{\sqrt{2}}\right) \ldots \Phi\left(\frac{\hat{\tau}_{k-1}}{\sqrt{2}}\right). \qquad (2.3.11)$$

These two methods of calculating the P_j and therefore P_L always give the same result except for errors of rounding and/or interpolation.

The quantity P_L provides a good estimate of a lower bound for the true probability of a correct selection, as long as none of the $\hat{\tau}$ values is too close to zero. (The theory underlying this bound is based on well known, easily proved inequalities.)

We can also estimate an upper bound P_U for the true probability of a correct selection. The estimate of an upper-bound procedure, denoted by P_U, is found by first computing

$$\bar{\tau} = \frac{\sum_{j=1}^{k-1} \hat{\tau}_j}{k-1} = \frac{\hat{\tau}_1 + \hat{\tau}_2 + \cdots + \hat{\tau}_{k-1}}{k-1}, \qquad (2.3.12)$$

which is simply the arithmetic mean of the $\hat{\tau}$ values in (2.3.10). Then P_U is the entry in Table A.2 (with the original value of k) that corresponds to the value of $\bar{\tau}$ in (2.3.12).

The estimators P_L and P_U together provide an interval estimate $[P_L, P_U]$ of the true probability of a correct selection. If this interval estimate is very wide, it may be of little practical use. Hence it is desirable to describe also a method for finding a point estimate of the true probability of a correct selection. This point estimate, denoted by P_E, always lies between P_L and P_U. The method used to obtain P_E is actually a combination of the methods used to obtain P_L and P_U.

To find P_E the first step is to partition the $k-1$ values of $\hat{\tau}$ from (2.3.10) into subgroups or clusters that are of approximately equal size. Suppose we decide to form m clusters, with k_1-1, $k_2-1, \ldots, k_m-1$ values of $\hat{\tau}_j$ in the respective clusters. Since all $k-1$ of the $\hat{\tau}_j$ must be used, the sum of these cluster sizes must be $k-1$, that is, $\sum_{j=1}^m (k_j-1) = k-1$. These subgroups must be mutually exclusive; that is, each $\hat{\tau}_j$ can go into only one cluster. Moreover, if any two $\hat{\tau}$ values are in the same cluster, then any $\hat{\tau}$ value between them must also be in that same cluster. As a result, the easiest way to form the clusters is to partition the $\hat{\tau}$ values in their ordered arrangement, $\hat{\tau}_1 \geqslant \hat{\tau}_2 \geqslant \cdots \geqslant \hat{\tau}_{k-1}$. For example, we might put $\hat{\tau}_1$ and $\hat{\tau}_2$ in the first cluster, $\hat{\tau}_3$ in the second, $\hat{\tau}_4$, $\hat{\tau}_5$, and $\hat{\tau}_6$ in the third cluster, and so forth. Clearly there remains some arbitrariness in how the clusters are to be formed. The best results are obtained if the $\hat{\tau}$ values within each cluster are approximately equal in magnitude.

Let $\bar{\tau}_j$ be the arithmetic mean of the $\hat{\tau}$ values in the jth cluster. Then the jth cluster gives rise to a probability value P_j. The value of P_j is the entry in Table A.2 that corresponds to the values of $\bar{\tau}_j$ and k_j; k_j is of course one more than the number of $\hat{\tau}$ values in that cluster. The value of P_j is found in the same manner for each of $j = 1, 2, \ldots, m$. Then P_E is found by multiplying these m values together; that is,

$$P_E = P_1 P_2 \ldots P_m.$$

Example 2.3.2. We illustrate the calculation of each of the estimates explained earlier by the following sample means, calculated from data with $k=6$, $n=16$, $\sigma=2$:

$$\bar{x}_1 = 17.95, \quad \bar{x}_2 = 18.05, \quad \bar{x}_3 = 17.70, \quad \bar{x}_4 = 19.30, \quad \bar{x}_5 = 18.45, \quad \bar{x}_6 = 18.35.$$

Since the sample from population 4 produced the largest mean, we assert that population 4 has the true largest mean. How much confidence do we have that this is indeed a correct selection?

We first write the ordered sample means as

$$\bar{x}_{[1]} = 17.70, \quad \bar{x}_{[2]} = 17.95, \quad \bar{x}_{[3]} = 18.05,$$

$$\bar{x}_{[4]} = 18.35, \quad \bar{x}_{[5]} = 18.45, \quad \bar{x}_{[6]} = 19.30.$$

From (2.3.9) we calculate the $\hat{\delta}$ values, $\hat{\delta}_j = \bar{x}_{[k]} - \bar{x}_{[j]}, j = 1, 2, \ldots, k - 1$, as

$$\hat{\delta}_1 = 1.60, \quad \hat{\delta}_2 = 1.35, \quad \hat{\delta}_3 = 1.25, \quad \hat{\delta}_4 = 0.95, \quad \hat{\delta}_5 = 0.85,$$

and from (2.3.10) the $\hat{\tau}$ values are $\hat{\tau}_j = \hat{\delta}_j \sqrt{16}/2 = 2\hat{\delta}_j$, or

$$\hat{\tau}_1 = 3.20, \quad \hat{\tau}_2 = 2.70, \quad \hat{\tau}_3 = 2.50, \quad \hat{\tau}_4 = 1.90, \quad \hat{\tau}_5 = 1.70.$$

To find P_L we use Table A.2 with $k = 2$ and each of these $\hat{\tau}$ values to obtain the five values of P_j as

$$P_1 = .988, \quad P_2 = .972, \quad P_3 = .962, \quad P_4 = .911, \quad P_5 = .886.$$

Then P_L is the product

$$P_L = (.988)(.972)(.962)(.911)(.886) = .746.$$

Using the alternative method to calculate P_j, we first calculate $\hat{\tau}_j / \sqrt{2}$ (see the second column of Table 2.3.3), and find $\Phi(\hat{\tau}_j / \sqrt{2})$ from the table in Appendix C (see the third column), for each of the five values of j. Then, from (2.3.11), P_L is the product of the five numbers in the third column.

Table 2.3.3   Illustrative calculation of P_L

$\hat{\tau}$	$\hat{\tau}_j / \sqrt{2}$	$\Phi(\hat{\tau}_j / \sqrt{2})$
3.20	2.26	.988
2.70	1.91	.972
2.50	1.77	.962
1.90	1.34	.910
1.70	1.20	.885

$$P_L = (.988)(.972)(.962)(.910)(.885) = .744.$$

The difference in the two results for P_L, .746 and .744, is due to rounding and interpolation.

To calculate P_U we simply compute the mean of the five $\hat{\tau}$ values as

$$\bar{\tau} = \frac{3.20 + 2.70 + \cdots + 1.70}{5} = 2.40,$$

and find the corresponding probability of a correct selection from Table A.2 with $k = 6$. The result is $P_U = .848$.

Combining the lower and upper bounds, we obtain an interval estimate for the true probability of a correct selection, as $.746 \leqslant P\{CS\} \leqslant .848$ or $.744 \leqslant P\{CS\} \leqslant .848$. This interval should be reported along with the population selected, in this case population 4.

If we want a point estimate, instead of or in addition to the preceding interval estimate, the clustering procedure is used. Suppose we choose the $m = 3$ clusters shown in the first column of Table 2.3.4 and compute their respective means as in the third column.

Table 2.3.4   Illustrative calculation of P_E

Cluster j	Cluster size $k_j - 1$	Cluster mean $\bar{\tau}_j$	P_j
(3.20)	1	3.20	.988
(2.70, 2.50)	2	2.60	.941
(1.90, 1.70)	2	1.80	.830

The respective entries in the final column in Table 2.3.4 are found from Table A.2 for $\hat{\tau}_1 = 3.20$ with $k_1 = 2$, for $\bar{\tau}_2 = 2.60$ with $k_2 = 3$, and for $\bar{\tau}_3 = 1.80$ with $k_2 = 3$. The overall point estimate is the product of the three values of P_j, or $P_E = (.988)(.941)(.830) = .772$.

The value $P_E = .772$ should also be reported along with the population selected, because it estimates the true level of confidence associated with the selection. As explained in Remark 4 below, for $k = 6$ the true probability should be at least $\frac{7}{12} = .583$ in order to be useful in most practical situations we might encounter. In this case .772 is well above .583 and hence the selection is at an acceptable probability level.

Since the clustering used to find P_E is somewhat arbitrary, we also give here the estimates that would be found for the other possible clusterings with $m = 3$. The individual cluster sizes must then be some permutation of the integers 1, 2, 2, or some permutation of the integers 1, 1, 3. The results for each case are shown in Table 2.3.5.

Note that the values for P_E when $m = 3$ range from .759 to .798. The three values for the cluster sizes that are a permutation of 1, 2, 2 range between .767 and .772, whereas those for the cluster sizes that are a permutation of 1, 1, 3 have a much wider range. The clusters of sizes 1, 2, 2 are more uniform than those of sizes 1, 1, 3, and hence should normally give better results. Of the three values obtained here for P_E corresponding to cluster sizes 1, 2, 2, the result found in Table 2.3.4 seems to be the best

Table 2.3.5　*Calculations of P_E using all allowable partitions into three clusters*

Cluster j	$\bar{\tau}_k$	k_j	P_j	Cluster j	$\bar{\tau}_j$	k_j	P_j
(3.20, 2.70)	2.95	3	.966	(3.20, 2.70)	2.95	3	.966
(2.50, 1.90)	2.20	3	.896	(2.50)	2.50	2	.962
(1.70)	1.70	2	.886	(1.90, 1.70)	1.80	3	.830
$P_E = .767$				$P_E = .771$			
(3.20)	3.20	2	.988	(3.20)	3.20	2	.988
(2.70, 2.50)	2.60	3	.741	(2.70, 2.50, 1.90)	2.37	4	.889
(1.90, 1.70)	1.80	3	.830	(1.70)	1.70	2	.886
$P_E = .772$				$P_E = .778$			
(3.20)	3.20	2	.988	(3.20, 2.70, 2.50)	2.80	4	.940
(2.70)	2.70	2	.972	(1.90)	1.90	2	.911
(2.50, 1.90, 1.70)	2.03	4	.831	(1.70)	1.70	2	.886
$P_E = .798$				$P_E = .759$			

because that grouping places in the same cluster those $\hat{\tau}$ values that are closest together.

Other results could be obtained using $m = 2$ or $m = 4$ although we do not illustrate these calculations here. The only other possibilities are the extreme case $m = 5$, which makes $P_E = P_L$, and the extreme case $m = 1$, which makes $P_E = P_U$ (see Remark 3 below).

REMARK 1.　A more exact computation (using Hermite quadrature of degree 20) gives the result .802 for the point estimate of the true probability of a correct selection with the $\hat{\tau}$ values computed in this example. Thus the error in the point estimate $P_E = .772$, found using the three clusters of size $2, 2, 1$, respectively, is less than 5%. Although some of the values computed for P_E using other clusterings give a result closer to .802, we do not expect this to be true in general.　□

REMARK 2.　It would be most helpful if we could have some knowledge about the probability that the interval estimate $[P_L, P_U]$ covers the true probability of a correct selection so that it could be interpreted as a confidence interval estimate. Unfortunately this probability (or confidence coefficient) depends on the true configuration of μ values, and no results are available on how to estimate this unknown configuration.　□

REMARK 3.　Suppose that the $\hat{\tau}$ values are all different. In computing P_E the clustering may be extreme in either of two ways, namely, all $\hat{\tau}$ values are in a single cluster or each $\hat{\tau}$ value is in a different cluster. These extreme cases give $m = k - 1$ and $m = 1$, respectively, for the number of clusters; in the former case, $P_E = P_L$, and in the latter case, $P_E = P_U$. As a result, any sensible clustering of the $\hat{\tau}$ values will give a reasonable estimate and thus we do not mind leaving the division arbitrary. □

REMARK 4. In general, we know that for any k the true probability of a correct selection is never less than $1/k$. There is good reason to keep this probability greater than .5, and a rough rule of thumb is to make it at least $(.5 + .5/k)$ in the least favorable configuration. Hence if the computed estimate P_E is not at least $(.5 + .5/k)$, we can have little confidence in the selection made, in which case the only alternative is to take additional observations. The later calculations should include all the observations, although if this is done repeatedly the procedure should then be regarded as sequential. □

REMARK 5. The reader may wonder why we did not simply estimate a single δ value by $\hat{\delta} = \bar{x}_{[k]} - \bar{x}_{[k-1]}$ and use this $\hat{\delta}$ to compute $\hat{\tau} = \hat{\delta} \sqrt{n} / \sigma$ and then enter Table A.2 for the appropriate value of k to find the probability of a correct selection. We do not recommend this procedure, since the resulting estimate applies only under the least favorable configuration. With the methods of this section we use the data to estimate the true configuration and therefore make better use of the information available. □

REMARK 6. In finding P_E we are really clustering the values of the sample means, $\bar{x}_{[j]}$, for $j = 1, 2, \ldots, k-1$. However, since the $\bar{x}_{[j]}$ in each cluster are all to be subtracted from $\bar{x}_{[k]}$ to determine the $\hat{\tau}$ values, a specific clustering of the $\hat{\tau}$ values determines a specific clustering of the sample means. In effect, we are using the $\hat{\tau}$ values in order to partition the sample means into clusters. □

2.4 SAMPLE SIZES UNEQUAL

In many practical situations we do not have complete control over the sample sizes and they are not all equal. In Example 2.1.1 even though the experiment may begin with an equal number of cows on each type of feed, some of the cows may die during the period of the experiment. Then we would end up with unequal numbers of observations on milk yield for each feed. In Example 2.1.2 the baths may vary somewhat in size so that we end up with more observations at some temperatures than at others. In Example 2.1.3 we already mentioned that it is highly unlikely that the number of observations on traveling time is exactly ·the same for each route. The selection rule R for the problem of choosing the best one of k normal populations with a common known variance when the sample sizes are unequal was explained in Section 2.2 and illustrated by Example 2.2.2. However, the analytical methods of Section 2.3 apply only to the normal means problem with equal sample sizes. In particular, the tables in Appendix A and the figures in Appendix B are no longer applicable with unequal n_j.

An approximate solution for the selection problem with unequal sample sizes can be obtained by computing a certain generalized average sample size, denoted by n_0, by the *square-mean-root* (SMR) formula, given by

$$n_0 = \left(\frac{\sqrt{n_1} + \sqrt{n_2} + \cdots + \sqrt{n_k}}{k} \right)^2. \qquad (2.4.1)$$

Note that this formula makes $\sqrt{n_0}$ equal to the arithmetic mean of the square roots of the actual sample sizes. For example, suppose that we have samples for $k=4$ normal populations and the sample sizes are

$$n_1 = 14, \quad n_2 = 19, \quad n_3 = 24, \quad n_4 = 23.$$

Then from the SMR formula in (2.4.1), the average sample size is obtained as

$$n_0 = \left(\frac{\sqrt{14} + \sqrt{19} + \sqrt{24} + \sqrt{23}}{4} \right)^2 = (4.45)^2 = 19.80. \qquad (2.4.2)$$

REMARK. The result for n_0 obtained by the SMR formula is always greater than the geometric mean of the actual sample sizes, and always smaller than the arithmetic mean of the actual sample sizes. In the preceding example the geometric mean is 19.58 and the arithmetic mean is 20.00. In practice, the SMR result is rarely very different from the arithmetic mean; however, the SMR formula gives more accurate results most of the time and is therefore preferable to the arithmetic mean. □

The analytic methods explained in Section 2.3 for making confidence statements, calculating the operating characteristic curve, and estimating the true probability of a correct selection can now be carried out exactly as before except that the average sample size n_0 is substituted for n in any calculations. That is, we proceed as if each sample size were equal to n_0. Thus, for example, the relationship between δ^* and τ_c in (2.3.4) becomes

$$\tau_c = \frac{\delta^* \sqrt{n_0}}{\sigma} ; \qquad (2.4.3)$$

the confidence interval S in (2.3.5) changes to

$$S = \left[\mu_s, \mu_s + \frac{\tau_t \sigma}{\sqrt{n_0}} \right];$$
(2.4.4)

the relationship between the true value of τ and the true value of δ in (2.3.6) is now

$$\tau = \frac{\delta \sqrt{n_0}}{\sigma};$$
(2.4.5)

and the estimates of the τ values in (2.3.10) are computed as

$$\hat{\tau}_j = \frac{\hat{\delta}_j \sqrt{n_0}}{\sigma}.$$
(2.4.6)

Then Tables A.1 and A.2 and the figures in Appendix B are used as before, although the results are only approximate when the actual sample sizes are unequal.

Note that the effective sample size n_0 calculated from (2.4.1) will generally not be an integer. However, an integer is not needed for any of the calculations using n_0, and thus we need *not* round the value of n_0 to an integer.

The appropriate methods of analysis are illustrated in Example 2.4.1.

Example 2.4.1. Suppose that for $k = 4$ normal populations with a common variance $\sigma^2 = 2$, the sample sizes are

$$n_1 = 14, \quad n_2 = 19, \quad n_3 = 24, \quad n_4 = 23,$$

and the respective sample means are

$$\bar{x}_1 = 7.03, \quad \bar{x}_2 = 8.25, \quad \bar{x}_3 = 9.00, \quad \bar{x}_4 = 5.20.$$

Then the selection rule dictates that the population labeled 3 is asserted to be the one with the largest mean. In order to make approximate statements about this selection procedure, we use the average sample size $n_0 = 19.80$ found in (2.4.2) in equations (2.4.3) to (2.4.6) with the tables in Appendix A or figures in Appendix B for $k = 4$.

For example, from Table A.1 when $P^* = .90$ and $k = 4$, we find $\tau_t = 2.4516$. The width of the confidence interval S in (2.4.4) is then $(2.4516)\sqrt{2} / \sqrt{19.80} = 0.78$. Thus with confidence level (approximately) .90, we can say that the mean of population 3 is within 0.78 of the true largest mean, that is, $\mu_{[4]} - \mu_s \leqslant 0.78$. On the other hand, suppose that the investigator knows the value of δ^* that he wants as part of his confidence statement, namely, $\delta^* = 1.00$. Then from (2.4.3), the corresponding value of τ_c is $\tau_c = (1.00)\sqrt{19.80}/\sqrt{2} = 3.15$. From Appendix B, or from interpolation in Table A.2, the (approximate) confidence level is $CL(3.15|R) = .966$. As before, this confidence level can be interpreted as a point on the operating characteristic curve, or $P_{LF}\{CS|3.15\} = .966$. Thus .966 is an (approximate) lower bound for the probability of a correct selection over all configurations in the preference zone defined by $\delta^* \geqslant 1.00$. The entire operating characteristic curve could also be calculated in the usual way.

We may also want to make further use of the observations by incorporating them into the analysis as well as the selection procedure. With these sample means we use (2.3.9) to calculate the $\hat{\delta}$ values as $\hat{\delta}_1 = 3.80$, $\hat{\delta}_2 = 1.97$, $\hat{\delta}_3 = 0.75$, and from (2.4.6) the $\hat{\tau}$ values are

$$\hat{\tau}_1 = 11.96, \quad \hat{\tau}_2 = 6.20, \quad \hat{\tau}_3 = 2.36.$$

By interpolation in Table A.2 with $k = 2$ for each $\hat{\tau}$ value, we find $P_1 = 1.00$, $P_2 = 1.00$, $P_3 = .887$ so that $P_L = .887$. To find P_U we calculate $\bar{\tau} = 6.84$ and from Table A.2 for $k = 4$ we obtain $P_U = 1.00$. Since the $\hat{\tau}$ values in this example are so disparate, the only reasonable clustering would use $m = 3$, so that we obtain $P_E = P_L = .887$. Note that all of these estimates must be regarded as approximate answers, since they are based on an approximate common sample size, rather than on actual sample sizes.

REMARK.   Of course, we could also get an interval estimate for the case of unequal sample sizes by first assuming that the n values are all equal to the smallest n_j and then assuming that the n values are all equal to the largest n_j. The resulting estimators then form the desired interval. This will generally be much "rougher" (i.e., give larger intervals) than the method we have considered. $\square$

2.5   SELECTING THE POPULATION WITH THE SMALLEST MEAN

Suppose that for the normal distribution model of this chapter the goal is to select the population with the smallest μ value, $\mu_{[1]}$. Since the normal distribution is symmetric about its mean μ, and the sample mean $\bar{x}$ is a pure location statistic, the distribution of $\bar{x} - \mu$ does not depend on μ. In fact, the distribution is normal with mean zero, and hence is symmetric about zero. As a result, if each $\bar{x}$ is replaced by $-\bar{x}$, this is equivalent to replacing $\mu_{[j]}$ by $\mu_{[k-j+1]}$ for each $j = 1, 2, \ldots, k$. Further, $\bar{x}_{[j]}$ is replaced by $\bar{x}_{[k-j+1]}$ for $j = 1, 2, \ldots, k$. In particular, the smallest true mean becomes the largest true mean, and the smallest sample mean becomes the largest sample mean.

Tables A.1 and A.2 in Appendix A, and the figures in Appendix B, depend only on δ, and δ depends only on differences of μ values, which are unchanged by the replacement of $\mu_{[j]}$ by $\mu_{[k-j+1]}$ for $j = 1, 2, \ldots, k$. As a result, these tables and figures are equally applicable to the problem of selecting the population with the smallest mean $\mu_{[1]}$. The difference δ is then defined by

$$\delta = \mu_{[2]} - \mu_{[1]}. \tag{2.5.1}$$

In practice, we need not actually change the signs of the calculated values of $\bar{x}$ to carry out the selection procedure for this goal. Rather, we simply arrange the $\bar{x}$ values from smallest to largest, and the selection rule is to assert that the population giving rise to the smallest sample mean $\bar{x}_{[1]}$ is the one with the smallest population mean $\mu_{[1]}$.

All the analytical aspects of the problem of selecting the normal population with the largest true mean, as described in this chapter and the next, are equally applicable for the problem of selecting the normal population with the smallest mean as long as δ is defined as in (2.5.1). The reader is, however, cautioned not to assume that the same tables can be used for other goals, like selecting the second largest, or selecting the two largest with or without regard to order, and the like. He is also cautioned *not* to assume that when we get to other distributions, like the chi-square, multinomial, and so on, we can use the same table for the two problems of selecting the largest and the smallest. In fact, it is generally not true.

2.6 NOTES, REFERENCES, AND REMARKS

The theory for the early part of this chapter originated in Bechhofer (1954). Indeed, this model was the first to be considered in some depth. The tables needed to apply this theory were developed at an early stage, and the results of the calculation were included even in this first paper. (In fact, it is an interesting historical anecdote that the same normal ranking tables were independently computed by several people, for example, in Gupta's dissertation in 1954 (see Gupta (1956)), and independent computations were still going on even as late as 1968 (see Guttman and Milton (1969)).) Thus the theory was put in a form ready for usage from the beginning. Unfortunately it takes a long time before the general user of statistical methods becomes aware of the existence and utility of a new technique, and, as a result, the accumulation of papers applying these methods has been slow.

Later parts of this chapter, such as estimation of the true probability of a correct selection in Section 2.3.4, are quite new. The theory for the latter is in a recent technical report by Olkin, Sobel, and Tong (1976). In accordance with the work of Dudewicz and Tong (1971), the estimators $\hat{\delta}_j$ given here could be "shrunk" to obtain better estimators; this is not brought into our development. Bishop and Dudewicz (1976) address the problem of assessing the probability that our proposed interval estimate contains the true probability of a correct selection. The problem of estimating the probability of a correct selection when the sample sizes are unequal (Section 2.4) is discussed in Dudewicz and Dalal (1975).

The application of ranking and selection methods to normal distributions is considered first because it is by far the most important application. This is not because the normal distribution occurs more often in applications, but because the central limit theorem brings about an important relation between the normal distribution and almost every other distribution arising in practice, irrespective of whether the variable measured is continuous or discrete. Thus for many of the remaining applications, we return to the basic Tables A.1 and A.2 of Appendix A to find approximate answers for problems involving nonnormal distributions. In fact, most nonnormal tables are set up only for small sample sizes, and normal distribution theory is used for moderate and large sample sizes. This situation is exactly analogous to the corresponding situation in classical statistics when we are dealing with a sum of n independent, identically distributed random variables for moderate to large values of n. Under very mild restriction the distribution of this sum, when properly standardized, tends to a standard normal distribution as n becomes large; the normal theory is then used as an approximation if the fixed n is not too small.

Although the problem of selecting the population with the smallest mean presents no new difficulties (like the need for new tables) in the case of the symmetric normal distribution, the reader should not assume that the same result holds for ranking variances, where the symmetry is lacking. As we see later in Chapter 5, the problem of selecting the population with the smallest parameter is in general not equivalent to the problem of selecting the population with the largest parameter. The exception, of course, is the case $k = 2$ when the two problems are equivalent, regardless of the underlying distribution.

The goal considered in Guttman (1961) and in Guttman and Milton (1969) is to select a subset of populations that includes the one that ascribes the maximum probability to some fixed interval, say $(-\infty, a]$. This goal is not included in this book. However, we should point out that in many cases this formulation reduces to one of the problems previously studied; this fact is not clearly pointed out or made use of in these papers. In the simplest case of normal populations with a common known σ^2, their problem is identical with the Gupta (1956) formulation for selecting a subset containing the population with the smallest mean. However, the emphasis in their paper is on making the procedure free of parameters.

There is one important question that should be raised, namely, is there another procedure that can achieve the same probability requirement and yield a smaller sample size? If another procedure exists, then we should try to use it. Hall (1954, 1959) showed that no such other procedure exists, so that within the present formulation, we are doing the best we can. Hall also proved that this characteristic which he called *most economical* holds for many of the procedures considered in later chapters.

PROBLEMS

2.1 An experiment is carried out to determine which one of four teaching methods is most effective for learning statistics. The methods are: (1) IS—independent study, but with an instructor available to answer questions; (2) PT—formal instruction using a programmed text and a moderate class size (25 students); (3) FL—formal lecture with a large class size (80 students) but also problem sessions of small groups; and (4) S—seminar-style instruction with much opportunity for discussion and a small class size (15 students). The materials used for each method are considered of equal difficulty and clarity, and the professors are considered homogeneous in experience and effectiveness in instruction by these methods. Participants in this experiment are chosen randomly from a large group of college sophomores who are homogeneous in both intelligence and academic performance. They are then assigned randomly to the four groups. The group sizes are necessarily unequal, as indicated. At the end of the semester all students are given the same examination covering the material. The possible scores range from 0 to 100, with 100 a perfect score.

Assume that the scores for the populations represented by these four groups are normally distributed with variance $\sigma^2 = 9$. For the summary data shown below,

(a) Select the most effective teaching method.
(b) Make a confidence statement about the method selected, using confidence level .90.
(c) Calculate the operating characteristic curve.
(d) Estimate the true probability of a correct selection.
(e) Find lower and upper bounds for the true probability of a correct selection.

Group	IS	PT	FL	S
Sample size	20	25	80	15
Sample mean	70.3	78.6	75.8	83.2

2.2 An experiment is in the process of being designed to determine which one of seven named races of silkworm has the longest mean length in millimeters. The length for any race is known to be normally distributed with variance $\sigma^2 = 4$. How many cocoons of each race should be measured so that the probability of a correct selection is at least .90 whenever $\delta \geqslant \delta^* = 1.00$?

2.3 A constant source of unequal sample sizes in agricultural experimentation is unequal litter sizes. Obviously, the experimenter cannot control the litter sizes, and it would make no sense to eliminate certain animals in order to force equal sample sizes because the probability of a correct selection would then decrease. In a study to compare birth weights of pigs from five breeds of sows, one randomly chosen sow of each breed is impregnated and the birth weight is measured for each pig in the litter. For the weights (in pounds) shown in the following table, select the breed with the largest birth weight, assuming that the weights are normally distributed with common variance $\sigma^2 = 0.40$.

Birth Weights of Pigs in Litters for Five Breeds

Yorkshire	Poland China	Duroc-Jersey	Hampshire	Berkshire
2.6	2.8	3.5	2.9	3.1
3.1	3.3	2.8	3.1	2.9
3.2	4.4	3.2	4.2	2.6
2.9	3.6	3.1	2.5	3.0
2.0	3.3	3.0	3.3	
2.7	2.8	3.4		
2.5	1.1	2.8		
2.1	2.5			
	2.7			
	3.0			

2.4 Five varieties of corn are to be compared on the basis of yield in bushels per acre. The goal is to select the best variety. Seven acres are planted with each variety of corn and all fields are tended in the same way. Assume that the yields given below are normally distributed with common variance $\sigma^2 = 3$.

(a) Make a confidence statement at level $P^* = .90$ for $\delta^* = 0.15$.

(b) Estimate the true probability of a correct selection.

Yields of Corn in Bushels per Acre

Lancaster	Clark	Silver King	Osterland
4.6	5.3	12.1	7.8
5.2	6.1	15.9	7.5
3.9	7.2	9.2	8.3
5.7	4.7	10.5	9.1
6.3	5.2	11.4	6.9
6.8	6.3	8.6	7.7
4.8	8.1	10.5	8.1

2.5 In order to compare five machines, 25 test specimens are assigned randomly to five different machines that impose stress conditions. As this is a fatigue test, the time to failure in kilocycles is noted for each specimen. The goal is to select the machine with the largest mean time to failure. Assume the lifetimes are normally distributed with common variance $\sigma^2 = 100$.

Machine

A	B	C	D	E
206	201	198	200	215
209	216	202	202	219
214	238	218	210	226
231	257	229	214	230
249	263	243	236	245

(a) Make a selection, and give an appropriate confidence statement about your selection.

(b) Calculate the operating characteristic curve.

(c) Give a point estimate and an interval estimate of the probability of a correct selection.

2.6 Griffith, Westman, and Lloyd (1948) give a practical example of data collected to compare the speed of five different examiners in using a gauge to check the ovalness of ball bearing races. The gauge automatically records the extent of out-of-round condition. Since the gauge is part of a line inspection system, the speed of examiners in using this gauge is of great interest. Each examiner is timed on four different occasions. The following data are their gauging times in seconds.

Examiner				
1	2	3	4	5
13	14	14	15	13
14	14	15	14	16
12	13	13	14	14
14	14	15	14	14

Assume that the gauging time for each examiner is normally distributed with common variance $\sigma^2 = 0.90$.

(a) Select the examiner with greatest speed (smallest mean).

(b) Make an appropriate confidence statement at level .90 about your selection.

(c) Find an interval estimate of the true probability of a correct selection.

2.7 Originally, Peace Corps volunteers received training almost exclusively in the United States (denoted UST). The results of this training were not considered satisfactory. In the summer of 1966 the Peace Corps/India began in-country training programs (denoted ICT) to follow the UST programs. Jones and Burns (1970) reported a study designed to compare and evaluate the satisfaction of individual Peace Corps volunteers with the training program received. The volunteers were classified into the three groups:

I. Heavy ICT (6–9 weeks of UST and 4–6 weeks of ICT).

II. Light ICT (11 weeks of UST and 2 weeks of ICT).

III. No ICT (14 weeks of UST and no ICT).

The evaluation was based on responses to a lengthy questionnaire that investigated the individual's satisfaction with his training on each of six separate components (like language training) as well as in general. These separate scores were combined to give a single overall evaluation of satisfaction.

Suppose we assume that the populations of these scores for volunteers in each of Groups I, II, and III are normally distributed with a common variance σ^2, and that it is known that σ is about 2 for scores based on this type of questionnaire.

(*a*) Suppose you are designing an experiment to select the training group that produces the greatest satisfaction, as determined by the group with the largest mean combined score on this questionnaire, such that the probability of a correct selection is at least .95 whenever $\mu_{[3]} - \mu_{[2]} \geqslant 0.5$. How large a sample should be taken from each population?

(*b*) The study reported by Jones and Burns used unequal sample sizes, probably determined by extrastatistical considerations. The sample sizes and mean satisfaction scores (for males and females combined) were as follows:

Group	Mean	Sample Size
I	5.3	102
II	5.6	74
III	5.2	72 ·

What kind of statement could you make?

CHAPTER 3

Selecting the One Best Population for Other Normal Distribution Models

3.1 INTRODUCTION

This chapter is a continuation of Chapter 2 in that we are dealing with the model of k normal distributions where the mean is the parameter of interest. To avoid confusion we assume always that the goal is to select the population with the largest mean. (As explained in Section 2.5 of Chapter 2, this does not result in any loss of generality.)

For single-stage procedures we cover both the case where the variances are not known but are assumed equal to a common value, and the case where the variances are all known but are not necessarily all equal. In each case the selection rule is exactly as described in Section 2.2 of Chapter 2, but the other aspects of the procedure differ. In the former case the solution is approximate since we determine an estimate of the common variance σ^2 and treat it as if it were the true value. A sequential procedure for this same problem is given in Section 3.7. The case of unequal known variances is covered in Section 3.3.

In Section 3.5 we consider the related estimation problem of finding confidence intervals for the largest mean. A procedure for a confidence interval of random length is given for the case of common unknown variance, and a procedure for finding a confidence interval of fixed length is given for a common known variance.

For the case of unknown variances an exact solution can be obtained using a two-stage procedure. Section 3.4 covers the case where the variances are not known but are assumed equal, and Section 3.10 treats the problem where the variances are not known and are not necessarily equal.

Some other goals and models are also covered in this chapter. In Section 3.6 we treat the problem of selecting the best treatment in a two-way classification model of analysis of variance. Section 3.8 considers the goal of selecting the normal population whose mean value is closest to a known target constant. The goal in Section 3.9 is to select the one of k components that has the largest probability of surviving longer than a fixed known constant (called *reliability*) for an arbitrary distribution model; it is illustrated for the case of normal distributions with a common known variance.

3.2   VARIANCES COMMON BUT UNKNOWN

In the normal means problem of Chapter 2 we assumed that each normal distribution has the same variance and that this common value is known. Now we consider the model of normal distributions when the variances are homogeneous but the common value is unknown. That is, we assume it is known that $\sigma_1^2 = \sigma_2^2 = \cdots = \sigma_k^2 = \sigma^2$ say, but the value of σ^2 is not known.

This model occurs in many classical statistical problems, but usually it is associated with a test of the homogeneity hypothesis H_0: $\mu_1 = \mu_2 = \cdots = \mu_k$. The F test (or equivalently, the one-way analysis of variance test), which for $k = 2$ is equivalent to the t test, is indeed appropriate for the homogeneity hypothesis under the model assumed here. However, we repeat again that our selection problem is not to be compared with a test of homogeneity, as it is primarily concerned with identifying the population with the "best" mean rather than with determining whether there is any difference between the population means. Thus even for $k = 2$, where one might like to identify a one-sided (or two-sided) t test with the ranking and selection methodology, the analogy fails. The analogy does hold for the special case $k = 2$ if we use the t test for the null hypothesis $\mu_1 \leqslant \mu_2$ against the alternative $\mu_1 > \mu_2$, but this clearly is not a test of homogeneity. Further, there is no such simple analogy for $k \geqslant 3$.

As indicated in Chapter 2, σ^2 must be regarded as a "nuisance parameter" when we are interested in ranking the populations only according to their μ values. Without any information about σ^2, it has been shown that there is no common fixed sample size large enough so that the usual confidence statement about the selection procedure will hold for *all* possible values of σ^2. In this sense there exists no fixed sample size solution to the present problem. However, for a large total sample size N, we can obtain a good estimate of the common σ^2 by pooling the sample variances. If we treat this estimate as if it were the true value of σ^2, then an approximate fixed-sample-size solution can be obtained for the present problem.

REMARK.   Suppose the investigator is prepared to replace his specification of δ^* by the same specification in "standard units," that is, by specifying a value of $\Delta^* = \delta^*/\sigma$ such that the probability of a correct selection is at least P^* whenever $\Delta = \delta/\sigma \geqslant \Delta^*$ (rather than whenever $\delta \geqslant \delta^*$). Then the difficulty associated with not knowing the common variance σ^2 disappears and the problem of determining the common sample size needed reduces to a normal one-stage problem. We then use the specified values of k and P^* to enter Table A.1 of Appendix A and note the entry τ_t, equate this with $\Delta^*\sqrt{n}$, and solve for the required common sample size n as a function of Δ^* and P^*.

The question arises as to whether this is "cheating" or not, that is, whether it is allowable for the investigator to specify $\Delta^* = \delta^*/\sigma$ (in "standard units") rather than δ^* (in the units of the problem). First of all, it should be emphasized that there is no problem with the mathematical or probabilistic details in using a (Δ^*, P^*) requirement. The problem is that although most investigators are capable of expressing a (δ^*, P^*) requirement because δ^* is in the units of the problem, specification of $\Delta^* = \delta^*/\sigma$ in a meaningful way requires a much more sophisticated acquaintance with both the details of the application as well as the statistical analysis and its implications because Δ^* is unitless. Thus, although we wish to keep open the feasibility of this solution, we do agree that for most practical problems it is more appropriate to ask the investigator to specify a value of δ^*, which is in the original units of the problem.

Exact solutions can also be obtained by other sampling procedures. For example, a two-stage procedure is explained in Section 3.4. This procedure involves an initial sample, followed by a decision, followed by a second sample. $\square$

For the approximate fixed-sample-size solution based on a large total sample size $N = n_1 + n_2 + \cdots + n_k$, we estimate the common variance by the pooled sample variance s^2 computed by

$$s^2 = \frac{1}{N-k} \sum_{j=1}^{k} (n_j - 1)s_j^2, \qquad (3.2.1)$$

where for each $j = 1, 2, \ldots, k$, s_j^2 is the sample variance for the jth sample, given by

$$s_j^2 = \frac{1}{n_j - 1} \sum_{i=1}^{n_j} \left(x_{ij} - \bar{x}_j\right)^2. \qquad (3.2.2)$$

In this formula $\bar{x}_j$ is the mean of the jth sample, or

$$\bar{x}_j = \frac{\sum_{i=1}^{n_j} x_{ij}}{n_j}.$$

Then we use the methods explained in Section 2.3 or 2.4 of Chapter 2, according to whether the sample sizes are equal or unequal, respectively, in each case substituting s^2 for σ^2 in the analysis. (Of course, since we already have our sample data in this problem, the sample sizes are fixed and we do not consider the problem of determining the sample size.)

Example 3.2.1. Among the classrooms in the public schools of a large city, there are five different types of lighting. A new school is being built, and the city wants to use the type of lighting that can be expected to provide the greatest level of illumination. Random samples are taken from those classrooms with each type of lighting, and the illumination level for each is measured in foot-candles on the desk surface. The preliminary calculations on these data are shown below. The classrooms are known to be homogeneous as regards natural light and other factors. Assuming that each distribution is approximately normal and the variances are common but unknown, which lighting technique should be recommended?

	Lighting Techniques				
	1	2	3	4	5
Sample size n_j	15	12	21	16	16
Sample mean $\bar{x}_j$	37.2	42.6	35.1	26.5	31.7
Sample variance s_j^2	10.1	9.6	9.2	8.7	9.9

Technique 2 produced the largest sample mean of illumination, so we assert that technique 2 has the largest population mean. What confidence statements can we make about this assertion?

Because the total sample size $N = \Sigma n_j = 80$ is large, we can get reasonably good approximate solutions by estimating σ^2 by the pooled sample variance s^2. From (3.2.1) and (3.2.2) this estimate is calculated as

$$s^2 = \frac{14(10.1) + 11(9.6) + 20(9.2) + 15(8.7) + 15(9.9)}{75} = 9.47.$$

Since the sample sizes here are unequal, we must calculate the average sample size n_0 from the square-mean-root formula in (2.4.1) of Chapter 2 before we can use the methods of Section 2.4 of Chapter 2. The result is

$$n_0 = \left(\frac{\sqrt{15} + \sqrt{12} + \sqrt{21} + \sqrt{16} + \sqrt{16}}{5} \right)^2 = (3.984)^2 = 15.87.$$

Suppose the confidence level is specified as $P^* = .90$. For $k = 5$ populations Table A.1 gives the corresponding value of τ as $\tau_t = 2.5997$. From (2.4.4) of Chapter 2 the width of the confidence interval on $\mu_{[5]} - \mu_s$ is $2.5997\sqrt{9.47} / \sqrt{15.87} = 2.01$. As a result, with confidence level approximately .90, we can assert that the mean μ_s of the population selected, technique 2, differs from the true largest mean $\mu_{[5]}$ by no more than 2.01, that is, $\mu_{[5]} - \mu_s \leqslant 2.01$. On the other hand, suppose we want to make the confidence statement that the mean illumination for the population selected, technique 2, is within $\delta^* = 1.75$ of the true largest mean. Then we calculate τ_c from (2.4.3) of Chapter 2 as $\tau_c = 1.75\sqrt{15.87} / \sqrt{9.47} = 2.27$. From Table A.2 the confidence level for this statement is approximated by $CL(2.27 | R) = .845$.

It is important to recognize that the value of s^2 computed from (3.2.1) would change with each replication of the experiment, and any analysis that uses s^2 would change accordingly; for example, the confidence level in the selection procedure for a specified δ^* changes and the threshold value of δ for specified confidence P^* also changes. The change is not so great when N is large. However, for small values of N the pooled sample variance s^2 may not be a very accurate or reliable estimate of σ^2, and this could have a severe effect on the accuracy of the confidence level or on the threshold value of δ. Consequently, it is desirable to alter the preceding approach so that we can have some statistical faith in the accuracy of our procedure. The following plan is based on using s^2 to find an upper confidence limit, denoted by σ_U^2, for σ^2, and substituting σ_U^2 in place of s^2 in the calculations described earlier. The precise effect of this change in procedure on the level of confidence is difficult to assess. However, by using the upper confidence limit for σ^2, we are making an attempt to control the overall confidence we can have in the final selection of the best population. Suppose we use P_1^* as a confidence level associated with the construction of the upper confidence limit σ_U^2, and P_2^* as a confidence level for selecting the best population when the common σ^2 has the value σ_U^2. Then it seems reasonable to report either that the probability of a correct selection is at least P_2^* or that the overall confidence level associated with the final selection is the product $P^* = P_1^* P_2^*$, or to report these two P^* values separately.

For samples from k normal populations, a constant times the pooled sample variance s^2, namely $\nu s^2/\sigma^2$, has the chi-square distribution with degrees of freedom ν given by

$$\nu = \sum_{j=1}^{k} (n_j - 1) = \left(\sum_{j=1}^{k} n_j \right) - k = N - k, \qquad (3.2.3)$$

where N is the total sample size. Using this property of s^2, we obtain the upper confidence limit for σ with confidence coefficient P_1^* as the right-hand side of the inequality

$$\sigma < \sigma_u = \sqrt{\frac{\nu s^2}{c}}, \qquad (3.2.4)$$

where $c > 0$ is the $1 - P_1^*$ quantile of the chi-square distribution with ν degrees of freedom. Thus c is derived from the lower tail of the chi-square distribution, which is given in the table in Appendix D. For example, if $\nu = 10$ and the confidence level is $P_1^* = .95$, the value of c from Appendix D is $c = 3.94$.

As a numerical illustration, suppose that $k = 4$ and $s^2 = 3.26$ for $n_1 = 6$, $n_2 = 5$, $n_3 = 7$, $n_4 = 6$, so that $N = 24$. Then from (3.2.3), $\nu = 24 - 4 = 20$. If the confidence level for the upper confidence limit for σ is specified to be $P_1^* = .90$, the value of c from Appendix D with $\nu = 20$ is 12.443. Hence from (3.2.4) we obtain σ_U as $\sigma_U = \sqrt{20(3.26/12.443)} = \sqrt{5.24} = 2.29$. Now we use this value σ_U in place of σ in any further calculations. Since the n values are not all equal here, the appropriate confidence interval is given in (2.4.4) of Chapter 2. The value of n_0 is computed from (2.4.1) of Chapter 2 as $n_0 = [(\sqrt{6} + \sqrt{5} + \sqrt{7} + \sqrt{6})/4]^2 = 5.98$. Suppose the overall level for the statement in (2.4.4) of Chapter 2 is specified as $P^* = P_1^* P_2^* = .81$. Then we must use $P_2^* = .81/.90 = .90$ as the confidence level for selecting the best population, assuming that $\sigma^2 = \sigma_U^2 = 5.24$. From Table A.1 with $k = 4$ at $P_2^* = .90$ we find $\tau_t = 2.4516$. Thus the calculation needed for the width of the confidence interval in (2.4.4) of Chapter 2 is $\tau_t \sigma_U / \sqrt{n_0} = 2.4516(2.29)/2.45 = 2.29$. The confidence statement for the final selection at overall (approximate) level .81 is that $\mu_{[4]}$ lies in the interval $[\mu_s, \mu_s + 2.29]$.

We now illustrate the entire procedure recommended for a smaller common value n in Example 3.2.2. The sample sizes used are extremely small (for a practical application) to make it easier to follow the calculations.

Example 3.2.2. We return to the data given in Example 2.2.1 of Chapter 2, where the sample sizes are equal. However, we now assume that the variances are common but the common value is unknown. Recall that the Institutionalized Educable Mentally Handicapped (IEMH) group produced the largest sample mean score on the CMA (Children's Manifest Anxiety) scale. Hence we assert that the IEMH group has the largest population mean. How much confidence can we have in this assertion? The first step is to calculate each individual sample variance s_j^2 from (3.2.2). For example, for the Bright group we have

$$s_1^2 = \frac{\left[(19.0 - 12.9)^2 + (8.0 - 12.9)^2 + \cdots + (20.5 - 12.9)^2\right]}{8} = 35.5.$$

The results are summarized as follows.

Group	B	N	NEMH	IEMH
Sample size	9	9	9	9
Sample mean	12.9	21.2	22.4	30.1
Sample variance	35.5	54.9	10.5	19.6

The pooled sample variance is then found from (3.2.1) as

$$s^2 = \frac{8(35.5) + 8(54.9) + 8(10.5) + 8(19.6)}{32} = 27.6.$$

The next step is to use (3.2.3) to find ν and (3.2.4) to find σ_U. Since $k = 4$ and the common sample size is $n = 9$, the total sample size is $N = 36$ and the degrees of freedom is $\nu = 36 - 4 = 32$. Suppose we choose the confidence level $P_1^* = .95$ for this estimation procedure. Then from the table in Appendix D we find $c = 20.07$ and the upper confidence limit for σ is

$$\sigma_U = \sqrt{\frac{(27.6)(32)}{20.07}} = 6.63.$$

Now suppose we select $P_2^* = .95$ for the selection procedure based on $\sigma = \sigma_U = 6.63$. For $k = 4$ the corresponding value of τ from Table A.1 is $\tau_t = 2.9162$. Substituting $\tau_t = 2.9162$, $n = 9$, and $\sigma_U = 6.63$ for σ, in (2.3.5) of Chapter 2, we find that the width of the confidence interval is $\tau_t \sigma_U / \sqrt{n} = (2.916)(6.63)/\sqrt{9} = 6.44$. Thus with overall level approximatey $P^* = P_1^* P_2^* = (.95)^2 = .90$, we can state that $\mu_{[4]}$ lies in the interval $[\mu_s, \mu_s + 6.44]$, or equivalently, that $\mu_{[4]} - \mu_s \leqslant 6.44$ where μ_s is the mean of the population selected, that is, the mean CMA score of the IEMH group.

REMARK. The reader should note that for any specified overall probability of a correct selection $P^* = P_1^* P_2^*$, the choice of P_1^* is arbitrary except that P_1^* and P_2^* must both be between P^* and 1. As already noted, a fixed-sample-size solution does *not* exist for *all* values of the common unknown σ^2. For given total sample size N and given δ^*, define the deficiency $D(\sigma^2)$, as a function of the true σ^2, as the (nonnegative) amount by which the probability of a correct selection drops below P^*. It would be desirable to know when the deficiency $D(\sigma^2)$ equals zero under the approximate solution given in this section. It would also be desirable to know how to choose P_1^* so as to minimize the deficiency, even if the resulting method depends on the unknown value of σ^2. A Monte Carlo study could be used to confirm any conjectures made or asymptotic results found. Such studies have not been carried out in this case and in this sense the approximate solution here has not yet been tested for general applicability. Such testing is desirable for *any* approximate solution to a statistical problem.

In Example 3.2.2 we used equal values for P_1^* and P_2^* for the fixed value of P^*, that is, $P_1^* = P_2^* = \sqrt{P^*}$. The reader should not assume that this allocation is optimal. A cursory asymptotic analysis shows that it is better to take P_1^* close to one and P_2^* close to P^*. Thus for the data given in Example 3.2.2 we could take $P_1^* = .98$ (say) and then P_2^* would be $.90/.98 = .918$. This leads to the confidence interval $[\mu_s, \mu_s + 6.18]$, which is smaller than the one derived in the example. It appears that the optimal result (which is close to the answer 6.18 above) must be obtained by trial and error.

The same considerations arise if the n values are equal and we wish to determine the smallest $N = kn$ corresponding to a specified pair (δ^*, P^*); that is, the optimal allocation between P_1^* and P_2^* is to set P_1^* close to 1 and P_2^* close to P^*, using trial and error. □

3.3 VARIANCES UNEQUAL BUT KNOWN

In this section we discuss the normal means selection problem under the model of k normal populations with variances $\sigma_1^2, \sigma_2^2, \ldots, \sigma_k^2$, respectively, which are known but are not necessarily all equal.

Assuming that the sample sizes $n_1, n_2, \ldots, n_k$ are within the control of the investigator, an allocation that leads to a simple analysis (but may not be optimal) is the one for which the variances of the sample means are approximately equal. This means that, for some number n_0, the sample sizes satisfy the relationship

$$\frac{n_1}{\sigma_1^2} = \frac{n_2}{\sigma_2^2} = \cdots = \frac{n_k}{\sigma_k^2} = n_0. \tag{3.3.1}$$

To see what this means in terms of $n_1, n_2, \ldots, n_k$, we let N denote the preassigned fixed total number of observations in all k samples combined, that is,

$$N = n_1 + n_2 + \cdots + n_k, \tag{3.3.2}$$

and $\Sigma \sigma_i^2$ denote the sum of the k known variances. Using (3.3.1) and (3.3.2) together, we obtain the solution for n_0 as

$$n_0 = \frac{N}{\Sigma \sigma_i^2}. \tag{3.3.3}$$

Then from (3.3.1) the solution for each $n_j, j = 1, 2, \ldots, k$, becomes

$$n_j = n_0 \sigma_j^2 = \frac{N \sigma_j^2}{\Sigma \sigma_i^2}. \tag{3.3.4}$$

Any value computed from (3.3.4) that is not an integer must be rounded to an integer. For some of the n_j the rounding will have to be upward; for others it will have to be downward; the important point is that (3.3.2) must hold for the rounded values.

As a numerical illustration, suppose that we have $k = 4$ normal populations with variances $\sigma_1^2 = 4$, $\sigma_2^2 = \sigma_3^2 = 5$, $\sigma_4^2 = 10$. If $N = 100$, then (3.3.3) gives

$$n_0 = \frac{100}{24} = 4.17,$$

and the allocation by (3.3.4) is

$$n_1 = 16.68, \quad n_2 = n_3 = 20.85, \quad n_4 = 41.70.$$

These results must be rounded to integers in such a way that the sum is exactly 100, the preassigned total sample size. Rounding upward gives $n_1 = 17$, $n_2 = n_3 = 21$, and $n_4 = 42$, but these values add to 101. Hence we now have to reduce one sample size by one to make the sum exactly $N = 100$. For example, we might make $n_4 = 41$ rather than 42. With these sample sizes the selection rule, based on the sample means, is exactly as before.

The allocation determined from (3.3.4) is based on equalizing the variances of the sample means. The main advantage of this procedure is that the tables in Appendix A and the figures in Appendix B give exact results, that is, without any approximation. Further, in any calculations (except the sample means), we can use n_0 as if it were a common sample size (n_0 need not be rounded to an integer) and assume that there is a common variance $\sigma^2 = 1$. Then the methods of analysis described in Section 2.3 of Chapter 2 are carried out exactly as before with $n = n_0$ and $\sigma^2 = 1$.

REMARK. Note that the actual sample sizes, $n_1, n_2, \ldots, n_k$, are used to compute the sample means, and the n_0 computed from (3.3.3) is used only for determining the allocation and carrying out the analysis. $\square$

Example 3.3.1   We illustrate this procedure by considering the previous numerical example where $k=4$, $N=100$, and the variances are $\sigma_1^2=4$, $\sigma_2^2=\sigma_3^2=5$, $\sigma_4^2=10$, so that $n_0=4.17$. Suppose we round the allocation determined earlier to get $n_1=17$, $n_2=n_3=21$, $n_4=41$. The data are collected using these sample sizes; suppose the sample means are calculated (using the divisors n_j) to be

$$\bar{x}_1=122.4, \quad \bar{x}_2=136.2, \quad \bar{x}_3=130.1, \quad \bar{x}_4=125.6.$$

Then since $\bar{x}_2$ is the largest sample mean, we assert that the population labeled 2 is the one with the largest value of μ. With a confidence level .90, what can we say about this selection? From Table A.1 for $k=4$ we find that $\tau_t=2.4516$ for $P^*=.90$. Using $\sigma^2=1$ and $n_0=4.17$ we find $\tau_t\sigma/\sqrt{n_0}=2.4516(1)/\sqrt{4.17}=1.20$. Hence from (2.3.5) of Chapter 2 the confidence statement at level .90 is that $\mu_{[4]}$ is included in the interval $[\mu_s, \mu_s+1.20]$, or that the difference $\mu_{[4]}-\mu_s$ satisfies the inequality $\mu_{[4]}-\mu_s \leqslant 1.20$. Alternatively, suppose we want to make the confidence statement $\mu_{[4]}-\mu_s \leqslant 1.00$ for the selection rule R. What is the level of confidence? Substituting $\delta^*=1.00$, $n_0=4.17$, and $\sigma^2=1$ in Equation (2.3.4) of Chapter 2, we find $\tau_c=1.00\sqrt{4.17}/\sqrt{1}=2.04$. By interpolation in Table A.2 for $k=4$, the confidence level is $CL(2.04|R)=.831$.

REMARK.   The reader may be surprised that the common sample size solution n_0 used in the analysis for Example 3.3.1 is so much smaller than any of the actual sample sizes, thereby implying that we are losing information. The small value of n_0 is due to the fact that we set the common variance equal to 1. If the common variance had been chosen to be, say, 5, the common approximate sample size would be 20.85, which is more commensurate with the actual sample sizes. More generally, the common approximate sample size is determined from the ratio n_0/σ_0^2, where σ_0^2 is the value chosen for the common variance. Thus (3.3.3) in the general case could be replaced by

$$\frac{n_0}{\sigma_0^2} = \frac{N}{\Sigma\sigma_i^2} \quad \text{or} \quad n_0 = \frac{N\sigma_0^2}{\Sigma\sigma_i^2}. \tag{3.3.5}$$

For an arbitrary value of σ_0^2, n_0 is determined from (3.3.5) and substituted in (3.3.4) to determine the values of n_j. Since all the methods of analysis depend only on the ratio $\sigma_0/\sqrt{n_0}$, the results are always the same regardless of what value is taken for σ_0. (Note that n_0 computed from (3.3.5) is equivalent to n_0 computed from (3.3.1) only if $\sigma_0^2 = 1$.) □

3.4   A TWO-STAGE PROCEDURE FOR COMMON UNKNOWN VARIANCE

As we have noted, a single-stage procedure for selecting the normal population with the largest mean cannot satisfy the (δ^*, P^*) requirement for *all* σ^2 when σ^2 is unknown and δ^* is in the original (unstandardized) units.[†] This is the case because for any reasonable procedure that requires taking a single sample of n_j observations from the jth population, the probability of a correct selection will depend on the value of the unknown σ^2. If the true value of σ^2 is sufficiently large, then the probability of a correct selection will be arbitrarily close to $1/k$, which is smaller than any reasonable value of P^*.

If we use a two-stage procedure instead of a one-stage procedure, this difficulty can be avoided. In the first stage a sample of n observations is taken from each of the k populations. The pooled sample variance s^2 is then computed from

$$s^2 = \frac{(n-1)s_1^2 + \cdots + (n-1)s_k^2}{k(n-1)} = \frac{s_1^2 + \cdots + s_k^2}{k}, \qquad (3.4.1)$$

where s_j^2 is the sample variance for the jth sample, computed from (3.2.2). The number of degrees of freedom in s^2 is $\nu = k(n-1)$.

In the second stage we specify the (δ^*, P^*) values and take a second sample of size $N - n$ from each of the k populations. The value of N is obtained from the formula

$$N = \max\left(n, \left\{\frac{2s^2h^2}{\delta^{*2}}\right\}^+\right), \qquad (3.4.2)$$

where s^2 is defined in (3.4.1) and h is obtained from Table A.4 for the appropriate value of k and the degrees of freedom $\nu = k(n-1)$ and the specified P^*. The symbol $\{a\}^+$ means the smallest integer equal to or greater than a. Thus we have $\{3.3\}^+ = 4$, $\{3.6\}^+ = 4$, $\{3.0\}^+ = 3$. Note that

---

[†]We have noted in Section 3.2 that a single-stage solution does exist if we change the requirement to one on (Δ^*, P^*) where $\Delta^* = \delta^*/\sigma$. However, in the present discussion we are using a requirement on (δ^*, P^*).

if n is larger than $\{2s^2h^2/\delta^{*2}\}^+$, we have $N = n$ so that the initial sample size n is of adequate size and no second sample is needed.

The observations from both stages are then combined into a single sample of N observations from each population. The sample means for this pooled sample are computed and ordered as

$$\bar{x}_{[1]} \leqslant \bar{x}_{[2]} \leqslant \cdots \leqslant \bar{x}_{[k]}.$$

The proposed selection rule is to assert that the population corresponding to the sample mean $\bar{x}_{[k]}$ is the one with the largest μ value. With this procedure the usual probability requirement is satisfied. In other words, the probability of making a correct selection is at least P^* whenever $\delta = \mu_{[k]} - \mu_{[k-1]} \geqslant \delta^*$.

Example 3.4.1 (Example 3.2.2 continued). For the problem of Example 3.2.2 with $k = 4$ groups, suppose we are going to use a two-stage procedure and the data given previously represent an initial sample of size $n = 9$. The individual sample variances based on these nine observations were given as

$$35.5, \quad 54.9, \quad 10.5, \quad 19.6,$$

and the pooled estimate of σ^2 using (3.4.1) is

$$s^2 = 27.6$$

with $\nu = (8)(4) = 32$ degrees of freedom. We now need to determine whether or not additional observations should be taken. Suppose we specify $P^* = .95$ and $\delta^* = 5.0$. From Table A.4 we find $h = 2.15$. In order to obtain N from (3.4.2), we must first compute

$$\frac{2s^2h^2}{\delta^{*2}} = \frac{2(27.6)(2.15)^2}{(5.0)^2} = 10.21.$$

Since $\{10.21\}^+ = 11$, we obtain $N = \max(9, 11) = 11$. In order to have a total sample size of 11 per population, an additional two observations must be taken from each population. As indicated earlier, the final selection is based on the four overall means, each based on 11 observations.

REMARK. This two-stage procedure satisfies the (δ^*, P^*) requirement (it provides an exact solution) for any integer value of n in the first stage. Nevertheless it is desirable to make a "good" choice of n for several reasons. First of all, the expected total number of observations is convex as a function of n and has a minimum near the "point" where the two members on the right-hand side of (3.4.2) are equal. Of course, one of these members is unknown, and this "point" can only be estimated. Secondly, it is desirable to keep the sample size small at the second stage (without making n too large) since the additional information about s^2 that is available in the second stage is never used. Hence a large second stage is an indication of lower efficiency due to either a poor estimate of σ^2 or too small an initial sample size n (and usually both). Any prior information about σ^2 (or guessed value of σ^2) could easily be used to provide some idea of how large n should be. A suggested procedure is simply to replace s^2 by this guess, equate the members on the right-hand side of (3.4.2), and solve for n. No simple explicit method or formula is available for obtaining the best value of n exactly. Some recent work (cf., Section 3.11) recommended that the number of degrees of freedom $(n-1)$ associated with each population in the first stage should be at least 12, that is, $n \geqslant 13$.   □

3.5   CONFIDENCE INTERVALS FOR THE LARGEST MEAN

In the previous sections of this chapter we have described methods for *selecting* the normal population with the largest mean. In this section we consider a corresponding problem of *estimating* the largest mean, *without any selection*. That is, we are interested in determining a confidence interval estimate of the unknown largest mean without selecting the population with the largest mean. We consider only the model of k normal populations with common variance, and describe two different types of confidence interval procedures. In the first type a level of confidence is prescribed and the length of the confidence interval obtained depends on the outcome of the experiment. This is the usual approach to confidence interval estimation. In the second type of procedure the length of the confidence interval is prescribed and the confidence level attained depends on the outcome of the experiment.

3.5.1   Random-Length Confidence Interval Procedure

The following procedure gives the confidence interval for $\mu_{[k]}$ with a prescribed level of confidence P^* for the case of a common unknown variance σ^2. A sample of n_j observations is taken from the jth population, and the sample mean $\bar{x}_j$ and the sample variance s_j^2 are computed, for each $j = 1, 2, \ldots, k$. The estimate s^2 of the common variance σ^2 is the pooled sample variance based on $\nu = N - k$ degrees of freedom and computed as in (3.2.1) by

$$s^2 = \frac{\sum\limits_{j=1}^{k} (n_j - 1)s_j^2}{N - k}, \tag{3.5.1}$$

where $N = n_1 + n_2 + \cdots + n_k$. The interval

$$I = \left[\max_j \left(\bar{x}_j - d\frac{s}{\sqrt{n_j}} \right), \max_j \left(\bar{x}_j + d\frac{s}{\sqrt{n_j}} \right) \right] \qquad (3.5.2)$$

is a confidence interval for the value of the largest mean $\mu_{[k]}$ with exact confidence coefficient P^*. The positive constant d in (3.5.2), which depends on the degrees of freedom ν as well as on k and P^*, is obtained from Table A.5.

Example 3.5.1 (Example 3.2.1 continued).   In Example 3.2.1 we had $k = 5$ lighting techniques. The sample sizes were $n_1 = 15$, $n_2 = 12$, $n_3 = 21$, $n_4 = n_5 = 16$; the sample means were $\bar{x}_1 = 37.2$, $\bar{x}_2 = 42.6$, $\bar{x}_3 = 35.1$, $\bar{x}_4 = 26.5$, $\bar{x}_5 = 31.7$; and the pooled sample variance with degrees of freedom $\nu = 80 - 5 = 75$ was found to be $s^2 = 9.47$ so that the standard deviation is $s = \sqrt{9.47} = 3.077$. If we want to determine a 95% confidence interval for the mean value associated with the best lighting technique, that is, with the largest mean, then we need to compute $\bar{x}_j - ds/\sqrt{n_j}$ and $\bar{x}_j + ds/\sqrt{n_j}$ for each sample. For $k = 5$, $\nu = 75$, and $P^* = .95$, Table A.5 does not give the corresponding value of d explicitly. However, we can extrapolate (by drawing a curve with values of ν on the horizontal axis) to obtain $d = 2.37$. The computations needed to construct the confidence interval are shown in Table 3.5.1.

Equation (3.5.2) indicates that the two end points of the interval I are the largest numbers in the corresponding two last columns in Table 3.5.1.

Table 3.5.1   Computations for constructing a confidence interval I

Sample j	$\bar{x}_j - d\dfrac{s}{\sqrt{n_j}}$	$\bar{x}_j + d\dfrac{s}{\sqrt{n_j}}$
1	$37.2 - \dfrac{2.37(3.077)}{\sqrt{15}} = 35.3$	$37.2 + \dfrac{2.37(3.077)}{\sqrt{15}} = 39.1$
2	$42.6 - \dfrac{2.37(3.077)}{\sqrt{12}} = 40.5$	$42.6 + \dfrac{2.37(3.077)}{\sqrt{12}} = 44.7$
3	$35.1 - \dfrac{2.37(3.077)}{\sqrt{21}} = 33.5$	$35.1 + \dfrac{2.37(3.077)}{\sqrt{21}} = 36.7$
4	$26.5 - \dfrac{2.37(3.077)}{\sqrt{16}} = 24.7$	$26.5 + \dfrac{2.37(3.077)}{\sqrt{16}} = 28.3$
5	$31.7 - \dfrac{2.37(3.077)}{\sqrt{16}} = 29.9$	$31.7 + \dfrac{2.37(3.077)}{\sqrt{16}} = 33.5$

The largest in the left column is 40.5, and in the right is 44.7. Then the desired 95% confidence interval for $\mu_{[5]}$ is

$$I = [40.5, 44.7].$$

Since only one of the five sample means is contained in this interval, in this case $\bar{x}_2$, it is tempting to also assert that lighting technique number 2 is the best population with confidence level $P^* = .95$. However, in general the interval could contain several $\bar{x}$ values and then the theory for the preceding procedure does not include the control of any such assertions related to the selection problem.

REMARK. It is of some interest to note, and not difficult to show, that the final confidence interval in Section 3.5.1 must contain at least one of the sample means. This is not immediately apparent since the two different maxima on the lower and upper end of the confidence interval could come from different populations (i.e., from different values of j). On the other hand, it is possible, although not probable, that the optimum confidence interval of fixed length in Section 3.5.2 does not contain any of the observed sample means. In both cases the resulting confidence interval is not necessarily centered at the largest sample mean. In the former case the center of the final interval may be to the left or to the right of the largest sample mean; in the latter case it is always to the left of the largest sample mean. □

3.5.2 Fixed-Length Confidence Interval Procedure

We now present a procedure for finding a confidence interval for $\mu_{[k]}$ of fixed length L, regardless of the outcome of the experiment; the confidence level of this procedure is also determined. Only the case of k normal populations with common known variance σ^2 and common sample size n is considered.

The sample means are computed and ordered as

$$\bar{x}_{[1]} \leqslant \bar{x}_{[2]} \leqslant \cdots \leqslant \bar{x}_{[k]}.$$

The confidence interval of preassigned length $L > 0$ is given by

$$I = \left(\bar{x}_{[k]} - L + \frac{\sigma d}{\sqrt{n}}, \ \bar{x}_{[k]} + \frac{\sigma d}{\sqrt{n}} \right). \tag{3.5.3}$$

The constant d is determined by first computing

$$c = \frac{L\sqrt{n}}{\sigma}. \tag{3.5.4}$$

Then c and k are used to enter Table A.6 and find the corresponding value of d and the associated confidence level P.

Example 3.5.2 Suppose we have $k = 5$ normal populations with known variance $\sigma^2 = 100$ and we wish to determine a fixed-length confidence interval with length $L = 4$ for the largest mean. If we have a common sample size $n = 50$, then from (3.5.4) we compute

$$c = \frac{L\sqrt{n}}{\sigma} = \frac{4\sqrt{50}}{10} = 2.83.$$

From Table A.6 we use linear interpolation to obtain $d = 1.07$ and confidence coefficient $P = .82$.

Suppose the experiment is performed and the respective sample means are

$$80.4, \quad 86.2, \quad 75.4, \quad 92.1, \quad 80.7,$$

so that $\bar{x}_{[5]} = 92.1$. Then the confidence interval I from (3.5.3) is given by

$$\left(92.1 - 4 + \frac{(1.07)(10)}{\sqrt{50}}, \quad 92.1 + \frac{(1.07)(10)}{\sqrt{50}} \right) = (89.61, 93.61).$$

By construction, the length of this interval is, of course, 4, the prescribed value.

As mentioned earlier in this section, it is more usual in statistical procedures to start with a chosen confidence coefficient, with the help of which the confidence interval and its length are determined. We can also use Table A.6 to determine a confidence interval with a prescribed confidence coefficient P^* by entering Table A.6 with k and a desired confidence coefficient P^* for P. This determines corresponding values of d and c. Then we can either use that value of c to fix L and determine n by solving (3.5.4) explicitly for n in terms of L as

$$n = \left(\frac{c\sigma}{L} \right)^2, \tag{3.5.5}$$

or, alternatively, if the observations have already been taken or n is fixed by extrastatistical considerations, we can use n and c to find L by solving (3.5.4):

$$L = \frac{c\sigma}{\sqrt{n}}. \tag{3.5.6}$$

Example 3.5.3. Suppose we have $k = 5$ normal populations with common variance $\sigma^2 = 100$, and we wish to obtain a fixed-length confidence interval with fixed-confidence coefficient $P^* = .90$. From Table A.6 we obtain $d = 1.412$ and $c = 3.45$. Then if we wish to fix $L = 5$, we find the common sample size easily from (3.5.5) as

$$n = \left[\frac{(3.45)(10)}{5} \right]^2 = 47.6,$$

so that 48 observations per population are required. Alternatively, if we have already taken $n = 50$ observations per population, then from (3.5.6) we find that $L = (3.45)(10)/\sqrt{50} = 4.88$ is the length of the confidence interval for $k = 5$, $\sigma^2 = 100$, and $P^* = .90$.

3.6 ANALYSIS OF VARIANCE: CHOOSING THE BEST TREATMENT

One of the most widely used statistical procedures for unraveling different effects is the analysis of variance (ANOVA). In the standard two-way classification the data are usually presented in a $k \times c$ table like Table 3.6.1. The entry x_{ij} in cell (i,j) denotes the measurement obtained for categories A_i and $B_j (1 \le i \le k, 1 \le j \le c)$.

The standard fixed-effects model for the two-way classification states that this measurement x_{ij} is a composite of effects in the form

$$x_{ij} = \alpha_i + \beta_j + \gamma_{ij} + \varepsilon_{ij},$$

Table 3.6.1 Format for data presentation

	B_1	B_2	$\cdots$	B_j	$\cdots$	B_c
A_1						
A_2						
$\vdots$						
A_i				x_{ij}		
$\vdots$						
A_k						

where α_i denotes the (row) effect of the ith treatment with $\Sigma_i \alpha_i = 0, \beta_j$ denotes the (column) effect of the jth block with $\Sigma_j \beta_j = 0, \gamma_{ij}$ denotes the interaction between row A_i and column B_j with $\Sigma_i \gamma_{ij} = 0$ and $\Sigma_j \gamma_{ij} = 0$, and ε_{ij} denotes an error. These kc errors are assumed to be independently distributed, each having a normal distribution with mean zero and common variance σ^2.

Denote the ordered values of the parameter α representing the row (or treatment) effect by

$$\alpha_{[1]} \leqslant \alpha_{[2]} \leqslant \cdots \leqslant \alpha_{[k]}.$$

The present goal is to choose the A category that has the largest α value $\alpha_{[k]}$ with probability of a correct selection at least P^* whenever $\delta = \alpha_{[k]} - \alpha_{[k-1]} \geqslant \delta^*$. In the usual language of analysis of variance, the goal would be stated as selecting the row with the largest row (or treatment) effect. It should be noted that the analysis of variance technique provides a test of the null hypothesis that the α values are all equal but does not permit any assertion or conclusion (with controlled error) on the question of which treatment has the largest α value. Further, we generally believe at the outset that the α values are not all the same, and allow equality only for mathematical convenience.

We first note that the classical ANOVA models frequently assume that there is no interaction effect. A test for interaction can be achieved by taking a sample of $m \geqslant 2$ observations in each cell (which can be regarded as replicating the experiment) and carrying out an analysis of variance. For completeness, a typical analysis of variance (ANOVA) breakdown is displayed in Table 3.6.2.

Table 3.6.2   Analysis of variance (ANOVA) table for a two-way classification with m observations per cell

Source of variation	Sum of squares (SS)	Degrees of freedom (ν)	Mean square $= SS/\nu$
Rows	$SS_R = mc \sum_i (\bar{x}_{i\cdot} - \bar{x}_{\cdot\cdot})^2$	$k-1$	MS_R
Columns	$SS_C = mr \sum_j (\bar{x}_{\cdot j} - \bar{x}_{\cdot\cdot})^2$	$c-1$	MS_C
Interaction	$SS_I = m \sum_i \sum_j (\bar{x}_{ij} - \bar{x}_{i\cdot} - \bar{x}_{\cdot j} + \bar{x}_{\cdot\cdot})^2$	$(k-1)(c-1)$	MS_I
Within cells	$SS_E = \sum_i \sum_j \sum_l (\bar{x}_{ijl} - \bar{x}_{ij})^2$	$kc(m-1)$	MS_E
Total	$SS_T = \sum_i \sum_j \sum_l (\bar{x}_{ijl} - \bar{x}_{\cdot\cdot})^2$	$mkc-1$	

In this table x_{ijl} denotes the lth observation in the (i,j) cell, $\bar{x}_{ij}$ is the mean of the m observations in the (i,j) cell, $\bar{x}_{i.}$ is the mean of the mc observations in row i, $\bar{x}_{.j}$ is the mean of the mk observations in column j, and $\bar{x}_{..}$ is the overall mean of all the mkc observations taken. The entries in the "mean square" column are found by dividing each corresponding sum of squares by its own degrees of freedom; thus, for example, $MS_R = SS_R/(k-1)$.

To test for interaction we need $m \geqslant 2$; we then use the ratio of the interaction mean square to the within cells mean square. The hypothesis of no interaction is rejected with probability α of a Type 1 error if

$$\frac{SS_I/(k-1)(c-1)}{SS_E/kc(m-1)} = \frac{MS_I}{MS_E} > F_\alpha,$$

where F_α is the upper tail critical point (based on level α) of Fisher's F distribution with numerator degrees of freedom $(k-1)(c-1)$ and denominator degrees of freedom $kc(m-1)$. These critical values are given in Appendix O for $\alpha = .01$ and $\alpha = .05$.

In proceeding, we assume that for a particular set of data this hypothesis is not rejected so that we can act as if there are no interaction effects in our data. Then the γ_{ij} can be removed from the model and the measurement x_{ij} is expressed as

$$x_{ij} = \alpha_i + \beta_j + \varepsilon_{ij}.$$

Recall that in section 3.4 where the common variance σ^2 is unknown it is necessary to resort to a two-stage selection procedure. In the present analysis of variance problem the same type of procedure may be employed.

At the first stage, a sample of m observations is taken for each cell and the ANOVA table is computed in the form of Table 3.6.2. The variance σ^2 is usually estimated by

$$s^2 = MS_E = \frac{SS_E}{kc(m-1)} \qquad (3.6.2)$$

with degrees of freedom $\nu = kc(m-1)$. However, if there is strong evidence of no interactions, a pooled estimate of σ^2 can be found by combining the interaction sum of squares SS_I and the within cells sum of squares SS_E by the formula

$$s^2 = \frac{SS_I + SS_E}{(k-1)(c-1) + kc(m-1)} = \frac{SS_T - SS_R - SS_C}{mkc - k - c + 1}. \qquad (3.6.3)$$

This pooled sample variance has degrees of freedom

$$\nu = (k-1)(c-1) + kc(m-1) = mkc - k - c + 1.$$

Next we need to determine whether or not the size of the initial sample is adequate. This depends on the (δ^*, P^*) requirement through the quantity

$$M = \max\left(mc, \left\{ \frac{2s^2h^2}{\delta^{*2}} \right\}^+ \right), \tag{3.6.4}$$

where h is obtained from Table A.4 by entering the table with k, ν, and P^*. Recall that $\{a\}^+$ is the smallest integer greater than or equal to a.

If the maximum in (3.6.4) is mc, then additional observations are not needed. We simply order the treatment means as

$$\bar{x}_{[1.]} \leqslant \bar{x}_{[2.]} \leqslant \cdots \leqslant \bar{x}_{[k.]},$$

and assert that the treatment that gave rise to the largest mean is the one that has the largest α value.

If the maximum in (3.6.4) is $\{2s^2h^2/\delta^{*2}\}^+$, then $M - mc$ additional observations per treatment are needed. This number $M - mc$ and hence also M must be divisible by c because of the design. Thus if $M - mc$ is not divisible by c, M must be increased by just enough so that it is a multiple of c. Hence we assume that $M = m'c$. We then take an additional $m' - m = (M - mc)/c$ observations per cell, or equivalently, $M - mc$ additional observations per row. When the experiment has been completed, we compute the k treatment means using all the M observations in each row, order them from smallest to largest, and choose the treatment giving rise to the largest sample mean as the one with the largest α value.

REMARK 1. Clearly, if the treatments are displayed across the columns instead of down the rows (or if we are interested in selecting the column with the largest block effect), then the reader must interchange k and c in the preceding discussion and the treatment means become

$$\bar{x}_{[.1]} \leqslant \bar{x}_{[.2]} \leqslant \cdots \leqslant \bar{x}_{[.c]}. \quad \square$$

REMARK 2. If the hypothesis of no interaction is rejected, then it is usually necessary to consider more complicated models. $\square$

Example 3.6.1. Snedecor and Cochran (1967, p. 347) give data for an analysis of variance experiment to study the weight gain of male rats under six feeding treatments. The A categories were the *source of protein*, namely, beef, cereal, and pork, and the B categories were the *level of protein*, namely, high and low. Thus we have $k = 3$ and $c = 2$. Ten observations on weight gain in grams were made at each feeding treatment so that $m = 10$, and the fixed effects model was assumed. The data are given in Table

Table 3.6.3   Data on weight gain in grams

	High protein	Mean	Low protein	Mean	Row means $\bar{x}_{i.}$
Beef	73, 102, 118, 104, 81, 107, 100, 87, 117, 111	100.0	90, 76, 90, 64, 86, 51, 72, 90, 95, 78	79.2	89.6
Cereal	98, 74, 56, 111, 95, 88, 82, 77, 86, 92	85.9	107, 95, 97, 80, 98, 74, 74, 67, 89, 58	83.9	84.9
Pork	94, 79, 96, 98, 102, 102, 108, 91, 120, 105	99.5	49, 82, 73, 86, 81, 97, 106, 70, 61, 82	78.7	89.1
Column means $\bar{x}_{.j}$		95.1		80.6	

3.6.3. Suppose that the goal is to select the source of protein that produces the largest weight gain in male rats, and we want the probability of a correct selection to be at least $P^* = .95$ for any $\delta = \alpha_{[3]} - \alpha_{[2]} \geqslant 6.00 = \delta^*$.

The first computation for the data in Table 3.6.3 is the ANOVA table given in Table 3.6.4. By (3.6.1) we test for interaction using the ratio

$$\frac{589.1}{214.6} = 2.7,$$

which has numerator degrees of freedom 2 and denominator degrees of freedom 54. The table in Appendix O indicates that $F_{.01} = 4.20$ and $F_{.05} = 2.79$ so the hypothesis of no interaction is not rejected and we can proceed with our selection problem.

Table 3.6.4   ANOVA table for data in Table 3.6.3

Source of variation	Sum of squares	Degrees of freedom	Mean square
Row (protein source)	$SS_R = $ 266.5	2	133.2
Column (protein level)	$SS_C = $ 3,168.3	1	3,168.3
Interaction	$SS_I = $ 1,178.2	2	589.1
Error	$SS_E = 11,585.7$	54	214.6
Total	$SS_T = 16,198.7$	59	

As a first step, σ^2 is estimated by computing (3.6.3) or

$$s^2 = \frac{1,178.2 + 11,585.7}{56} = 227.9$$

with degrees of freedom $\nu = 56$. With $k = 3$, $\nu = 56$, $P^* = .95$, linear inter-polation on $1/\nu$ in Table A.4 gives $h = 1.95$. The right-hand member of (3.6.4) is then

$$\left\{ \frac{2(227.9)(1.95)^2}{(6.00)^2} \right\}^+ = \{48.14\}^+ = 49,$$

and the left-hand member of (3.6.4) is $10(2) = 20$. Thus

$$M = \max(20, 49) = 49,$$

and we need to take a total of

$$M - mc = 49 - 20 = 29$$

additional observations per treatment. Since 29 is not divisible by $c = 2$ we make M the next larger multiple of 2 or 50 so that $M - mc = 30$. Thus we need an additional $\frac{30}{2} = 15$ observations per cell.

After taking these additional observations, the row (treatment) means are computed from all 50 observations in each row and ordered. Recall that in this problem the large weight gain is desirable. Thus we identify the source of protein that produces the largest sample mean of weight gain and assert that this is the most effective source. The probability of a correct selection is at least .95 whenever the true protein source parameters satisfy $\alpha_{[3]} - \alpha_{[2]} \geqslant 6.00$.

REMARK. If it is not possible to take additional observations from the same blocks (for example, if the blocks were pig litters) to satisfy the (δ^*, P^*) requirement, the investigator can always increase δ^* or decrease P^* or both. Naturally the final assertion made about the probability of a correct selection depends on the sample sizes (and block sizes) actually used. □

3.7 A SEQUENTIAL PROCEDURE FOR COMMON UNKNOWN VARIANCE

When observations are expensive, we may wish to proceed sequentially, that is, by taking one observation at a time. This type of experimentation may be awkward or impossible to implement in some applications. However, sequential experimentation is quite practical in several areas of application, for example, with clinical trials, assembly lines, and simulation experiments.

A significant advantage (probably the most important advantage) in sequential experimentation is that if one population is clearly better than the others, it may show up as better after only a few observations have been taken. Such an outcome results in a considerable saving in the number of observations required especially if the experiment is to be repeated many times.

The distribution model here is k normal populations with a common unknown variance σ^2 and ordered means

$$\mu_{[1]} \leqslant \mu_{[2]} \leqslant \cdots \leqslant \mu_{[k]}.$$

The goal is to select the population with the largest mean $\mu_{[k]}$, and we wish this selection to be correct with probability at least P^* whenever $\mu_{[k]} - \mu_{[k-1]} \geqslant \delta^*$. However, in this section we devise a sequential procedure that satisfies the same (δ^*, P^*) requirement.

Any sequential procedure consists of three steps. First, we must have a "sampling rule" that indicates how the sampling is carried out. Second, we must have a "stopping rule" that indicates when the sampling is to be terminated. Finally, we must have a "decision rule" that indicates as a function of the observations what terminal decision to make; here the terminal decision is merely the selection of one population, which we assert to be best in some sense. Since we have k populations, a simple sampling scheme is to take observations in groups of k, where each group contains one observation from each population. At the rth stage, that is, when we have r observations from each population, the data can be presented in the format of Table 3.7.1. The individual sample variances are computed from (3.2.2) and are shown on the last line of this table.

Table 3.7.1   Format of data presentation

	Populations			
Stage	1	2	$\cdots$	k
1	x_{11}	x_{21}	$\cdots$	x_{k1}
$\vdots$	$\vdots$	$\vdots$		$\vdots$
r	x_{1r}	x_{2r}		x_{kr}
Sample variances	$s_{1,r}^2$	$s_{2,r}^2$		$s_{k,r}^2$

The additional subscript r is used with the sample variances as a reminder that these computations are based on r observations per population and that they change from stage to stage. The pooled estimate of the variance is then the average of the k sample variances, calculated as in

(3.4.1) by

$$s_r^2 = \frac{(r-1)s_{1,r}^2 + \cdots + (r-1)s_{k,r}^2}{k(r-1)} = \frac{s_{1,r}^2 + \cdots + s_{k,r}^2}{k}.$$

This estimate has degrees of freedom $\nu = k(r-1)$, but it is interesting to note that ν is not used in this procedure.

We assume that r is at least 5. In other words, the experiment is started with at least five observations per population, and no decision can be made before five observations are taken from each population. In the next stage an additional observation is taken from each of the k populations and similarly for each successive stage. The stopping rule is to continue taking observations until we reach a stage, say N, for which the inequality

$$s_N^2 \leqslant cN \qquad \text{or} \qquad \frac{s_N^2}{N} \leqslant c \qquad (3.7.1)$$

is satisfied; the constant c is to be determined so that the (δ^*, P^*) requirement is satisfied.

Since the conclusion about stopping is based on the value of s_N^2, this pooled variance must be computed at each stage. In other words, at the rth stage each sample variance is recomputed using all r observations in that sample. Their average s_r^2 is divided by r and compared with c. If it exceeds c, we go into stage $(r+1)$ and again recompute the sample variances and compare $s_{r+1}^2/(r+1)$ with c, and so forth. The letter N is used here to denote the total number of observations per population at the time of stopping.

At the time of stopping, the sample means based on the N observations are computed and ordered as

$$\bar{x}_{[1]} \leqslant \cdots \leqslant \bar{x}_{[k]}.$$

The decision rule is to designate the population yielding the largest sample mean $\bar{x}_{[k]}$ as the one with the largest μ value $\mu_{[k]}$.

Recall that a (δ^*, P^*) pair was specified for the selection procedure. The constant c depends on these values, on k, and on τ_t, which is the entry in Table A.1 of Appendix A that corresponds to the given values of k and P^*. Once τ_t is determined, c is calculated from

$$c = \frac{\delta^{*2}}{\tau_t^2}. \qquad (3.7.2)$$

REMARK. In any sequential procedure it is imperative that the stopping rule be such that with probability one we will *not* continue taking observations indefinitely. In the present case we see that as N increases, the ratio

s_N^2/N eventually decreases to zero and thus for any $c > 0$ the procedure will terminate at some finite stage with probability one. $\square$

Example 3.7.1. In attempting to assess the effect of certain competing sedative-type drugs, the drugs are studied extensively in many species of animals. These studies suggest that the drugs affect the limbic system of the brain, which is involved in emotional responses according to recent evidence (see, for example, Morgan, 1965). Suppose we have $k = 4$ drugs. The goal is to choose the drug that has the greatest effect on the limbic system of the brain. Each drug is administered to a group of animals and then readings are taken to measure the sedation effect. Thus large readings are clearly more desirable in this application. These readings are assumed to be normally distributed with a common unknown variance σ^2. We wish to select the drug with the largest population mean, that is, with the largest μ value. If we specify the probability of correct selection is to be at least $P^* = .95$ whenever $\mu_{[4]} - \mu_{[3]} \geq \delta^* = 0.50$ and for all values of the unknown σ^2, how shall we conduct a sequential experiment?

From Table A.1, with $k = 4$ and $P^* = .95$, we obtain $\tau_t = 2.9162$. From (3.7.2), then, the constant c is

$$c = \frac{\delta^{*2}}{\tau_t^2} = \frac{(0.50)^2}{(2.9162)^2} = 0.029.$$

Thus after the initial stage where five observations are taken on each drug, we continue to take observations until we reach the first value of N for which $s_N^2/N \leq 0.029$. Suppose the results of the first five observations are as given in Table 3.7.2. Then $s_5^2 = (0.33 + 0.25 + 0.15 + 0.04)/4 = 0.19$, and

$$\frac{s_5^2}{5} = \frac{0.19}{5} = 0.038.$$

Table 3.7.2   Data up to stage 5 (before any terminal decision can be made)

Stage	Drug			
	A	B	C	D
1	5.1	3.5	6.1	6.1
2	5.2	4.2	6.4	6.2
3	4.8	4.3	6.7	5.8
4	6.1	4.6	6.0	5.7
5	6.0	4.8	5.7	6.0
Totals	27.2	21.4	30.9	29.8
Sample means	5.44	4.28	6.18	5.96
Sample variances	0.33	0.25	0.15	0.04

Table 3.7.3   Data up to stage 6

Stage	Drug			
	A	B	C	D
1	5.1	3.5	6.1	6.1
2	5.2	4.2	6.4	6.2
3	4.8	4.3	6.7	5.8
4	6.1	4.6	6.0	5.7
5	6.0	4.8	5.7	6.0
6	6.1	4.6	5.9	6.1
Totals	33.3	26.0	36.8	35.9
Sample means	5.55	4.33	6.13	5.98
Sample variances	0.39	0.21	0.13	0.04

Since 0.038 is *not* smaller than $c = 0.029$, we take one more observation from each population. Suppose the results of the first six observations (from each population) are as in Table 3.7.3. Then $s_6^2 = (0.39 + 0.21 + 0.13 + 0.04)/4 = 0.19$, and

$$\frac{s_6^2}{6} = \frac{0.19}{6} = 0.032,$$

which is still not smaller than $c = 0.029$. Thus we continue sampling and take another observation from each population. Suppose the results of the first seven observations are as in Table 3.7.4. Now we compute $s_7^2 = (0.30 + 0.14 + 0.09 + 0.03)/4 = 0.14$, and

$$\frac{s_7^2}{7} = \frac{0.14}{7} = 0.020,$$

Table 3.7.4   Data up to stage 7

Stage	Drug			
	A	B	C	D
1	5.1	3.5	6.1	6.1
2	5.2	4.2	6.4	6.2
3	4.8	4.3	6.7	5.8
4	6.1	4.6	6.0	5.7
5	6.0	4.8	5.7	6.0
6	6.1	4.6	5.9	6.1
7	5.9	4.6	6.0	5.9
Totals	39.2	30.6	42.8	41.8
Sample means	5.60	4.37	6.11	5.97
Sample variances	0.30	0.14	0.09	0.03

which is smaller than $c = 0.029$. We then terminate the experiment at this seventh stage and assert that drug C is the best since it has the largest overall sample mean (or total). The probability of terminating this procedure with a correct selection is at least .95, regardless of the true value of the unknown σ^2, provided only that $\mu_{[4]} - \mu_{[3]} \geqslant 0.50 = \delta^*$.

REMARK. The number five of observations taken per population at the initial stage without making any decision in this sequential procedure can, of course, be replaced by any number greater than five without affecting the probability of a correct selection. If we are concerned with robustness because the underlying distribution is only approximately normal, it may be desirable to increase the number five to, say, 10 (perhaps even 20) in order to let the Central Limit Theorem start taking effect. □

3.8 SELECTING THE POPULATION WHOSE MEAN VALUE μ IS CLOSEST TO A SPECIFIED VALUE μ_0

There are a number of problems in which we have k normal populations with means $\mu_1, \ldots, \mu_k$, a common known variance σ^2, and a known target constant μ_0. Our concern is to choose that population whose μ value is closest to the target constant μ_0. In this context we are indifferent whether a μ_j is smaller or larger than μ_0; our only concern is how close μ_j is to μ_0, that is, the absolute difference. Thus we wish to rank the k populations according to the absolute values of the differences $|\mu_j - \mu_0|$ $(j = 1, 2, \ldots, k)$. This type of application arises in problems concerning the calibration of instruments, for example.

When $\mu_0 = 0$, we are ranking populations with respect to their $|\mu|$ values. If the variances of the k populations are common, this problem is equivalent to ranking the populations with respect to the values of $|\mu|/\sigma$.

Example 3.8.1. Consider a meter connected to a thermocouple of resistance R whose electromotive force E is to be measured. However, there is some noise in the system as Figure 3.8.1 shows. The reading of the meter is given, and we are required to infer the value of E. In the absence of noise this can be answered precisely. The signal-to-noise ratio is defined

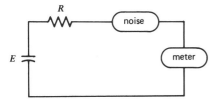

Figure 3.8.1 Diagram of a circuit with noise.

as the square of $|\mu|/\sigma$. If we have several different kinds of meters with a common known variance σ^2, then we wish to select the one that has the largest $|\mu|$ value.

Example 3.8.2. If several different programs are available for calculating the value of some particular function and each method has noticeable errors, we might want to identify the one that consistently gives more accuracy in the results. Assuming that the correct answers are available for some of the calculations and we are interested only in the absolute value of the error, the present model might apply for finding the program that has the smallest absolute error on the average (after a large sample of calculations). Here our goal is to select the population with the smallest value of $|\mu - \mu_0|$, where μ_0 is the correct answer.

We write $\theta_j = |\mu_j - \mu_0|$ for $j = 1, 2, \ldots, k$ and denote the ordered θ values by

$$0 \leqslant \theta_{[1]} \leqslant \cdots \leqslant \theta_{[k]}.$$

One goal is to select the population with the smallest θ value, $\theta_{[1]}$. Then for specified $\delta^* > 0$, we wish to make a correct selection whenever $\delta = \theta_{[2]} - \theta_{[1]} \geqslant \delta^*$ with probability at least P^*.

The procedure then is to take a sample of size n (to be determined) from each population and compute the sample means, $\bar{x}_1, \ldots, \bar{x}_k$. We then calculate $w_j = |\bar{x}_j - \mu_0|$ for $j = 1, 2, \ldots, k$ and order the w values as

$$0 \leqslant w_{[1]} \leqslant \cdots \leqslant w_{[k]}.$$

The selection rule is to identify the one population that gives rise to the smallest w value and assert that this population has the smallest θ value.

In some problems we wish to select the population with the largest θ value, $\theta_{[k]}$. Then the selection is to be correct with probability P^* whenever $\delta = \theta_{[k]} - \theta_{[k-1]} \geqslant \delta^*$. The rule here is to identify the population that gives rise to the largest w value and assert that it has the largest θ value.

The sample size n required for selecting the *smallest* θ value is determined from Table A.7. Table A.8 is used for the goal of choosing the *largest* θ value. Note that the entry in each of these tables is $\delta^* \sqrt{n}$. Thus the entry must be divided by δ^* and squared in order to find n. For example, suppose that we have $k = 4$ populations and wish to select the population with the smallest θ value such that the probability of a correct selection is at least $P^* = .95$ whenever $\delta = \theta_{[2]} - \theta_{[1]} \geqslant \delta^* = 0.40$. Table A.7 gives $\delta^* \sqrt{n} = 3.316$ so that

$$n = \left(\frac{3.316}{0.40} \right)^2 = (8.29)^2 = 68.7,$$

which is rounded upward to 69. Hence to carry out the procedure a sample of 69 observations is taken from each of the $k = 4$ populations. We then compute the sample means $\bar{x}_1$, $\bar{x}_2$, $\bar{x}_3$, $\bar{x}_4$ and corresponding w values $w_j = |\bar{x}_j - \mu_0|$ for the known target constant μ_0. The sample that produces the smallest w value, $w_{[1]}$, is asserted to be the one with the smallest θ value, $\theta_{[1]}$.

A statistic that is useful in a variety of contexts (in particular, in biostatistical analysis) is the coefficient of variation (CV), defined as $\sigma/|\mu|$. It should be noted that the coefficient of variation (CV) is unitless, and this is an important asset for many reasons; for example, we can compare CV's for different types of quantities (say, height and weight) without any further transformations. This concept of CV is frequently used when the data are all positive and hence μ is also, so we can safely assume that $\mu = |\mu|$. This fits into the previous model with $\mu_0 = 0$ and $w_j = |\bar{x}_j|$.

Suppose we have $k = 4$ normal populations with a common known variance σ^2 and unknown means, and we wish to select the one with the smallest CV in such a way that the probability of a correct selection is at least $P^* = .95$ whenever $|\mu_{[4]}| - |\mu_{[3]}| \geqslant \delta^* = 0.20$. (It should be noted that we are here measuring the "distance" between two CV's by taking the difference of the reciprocals of the CV's and disregarding the common known σ. It should also be noted that the smallest CV translates into the largest $|\mu|$ value.) From Table A.8 with $k = 4$ and $P^* = .95$ we obtain $\lambda = \delta^* \sqrt{n} = 3.258$, so that we need $n = (3.258/0.20)^2 = 265.4$ or (rounding upward) 266 observations from each of the $k = 4$ populations. To carry out the procedure we calculate the mean $\bar{x}$ for each of the $k = 4$ samples and the one that gives rise to the largest $|\bar{x}|$ is asserted to be the one with the smallest CV.

The reader should be cautioned about applying this technique when the μ values are close to zero. For $\mu = 0$ the CV is infinite, and this requires some special consideration. Since we want the population with the smallest CV, any populations with μ values close to zero would soon be out of contention and we would actually have a larger probability of a correct selection than we contracted for; that is, we would need fewer observations to satisfy the same (δ^*, P^*) requirement.

*3.9   SELECTING THE BEST POPULATION BASED ON SURVIVAL FUNCTIONS

The usual method of defining one population as being better than another is that some parameter is larger for one population than another. In a variety of examples the definition of better is based on a function of the parameters. This is particularly true in the context of the reliability of a

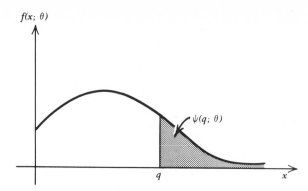

Figure 3.9.1   Reliability of a component $P(X > q)$.

system. For example, a component may have a density $f(x;\theta)$ that depends on a parameter θ. The probability that the component survives longer than a fixed, known threshold value c is called the *reliability* of the component. In symbols the reliability is defined as

$$\psi(q;\theta) = P\{X > q\} = \int_q^\infty f(x;\theta)\,dx, \qquad (3.9.1)$$

and is given by the shaded area in Figure 3.9.1. In many engineering examples, $f(x;\theta)$ may be the exponential distribution, or on occasion, the normal distribution. This same model also occurs in some social science contexts.

In this section we consider the situation where we have k components, where the jth component has parameter θ_j and reliability $\psi(q;\theta_j)$, and the goal is to select the component with the largest reliability. We first discuss what this means for a few different distribution models.

1. **Normal Distribution.**   For the normal density

$$f(x;\theta) = \frac{1}{\sigma\sqrt{2\pi}}\,\exp\left\{-\frac{1}{2}\left(\frac{x-\theta}{\sigma}\right)^2\right\},$$

the tail area or probability $\psi(q;\theta)$ is a monotone increasing function of θ. This means that as θ increases, $\psi(q;\theta)$ increases. Consequently, selecting the normal population with the largest reliability is equivalent to selecting the normal population with the largest mean θ. (In this discussion we assume that the variance σ^2 is known.)

2. **Exponential Distribution.**   Here the form of the density is

$$f(x;\theta) = \left(\frac{1}{\theta}\right)\exp\left\{-\frac{x}{\theta}\right\},$$

and the reliability or tail probability $\psi(q;\theta)$ is an increasing function of θ. Again, selecting the exponential population with the largest reliability is equivalent to selecting the exponential population with the largest θ.

Suppose we have k components and wish to choose the one with the largest reliability. Further, we assume that the underlying distributions are normal with different means and a common known variance σ^2. We need to specify differences in the reliabilities $\psi(q;\theta_j)$ for $j=1,2,\ldots,k$ as given by (3.9.1). This may be done by relating the differences between ψ values and differences between θ values. For the normal distribution we have

$$\psi(q;\theta)=1-\Phi\left(\frac{q-\theta}{\sigma}\right).$$

where $\Phi(z)$ denotes the cumulative standard normal distribution. Consequently,

$$\psi(q;\theta_2)-\psi(q;\theta_1)=\Phi\left(\frac{q-\theta_1}{\sigma}\right)-\Phi\left(\frac{q-\theta_2}{\sigma}\right).$$

Using the standard normal density $\phi(z)$, this latter difference may be approximated by

$$\frac{1}{\sigma}\phi\left(\frac{q-\frac{1}{2}(\theta_1+\theta_2)}{\sigma}\right)(\theta_2-\theta_1),$$

which in turn is bounded by $[1/(\sigma\sqrt{2\pi}\,)](\theta_2-\theta_1)$.

Consequently, we can relate the specified minimum difference Δ^* for $\psi(q;\theta_2)-\psi(q;\theta_1)$ to a corresponding minimum difference δ^* for $\theta_2-\theta_1$ by the relationship

$$\delta^* = \sigma\Delta^*\sqrt{2\pi}\,. \tag{3.9.2}$$

Then with this value of δ^* we proceed as in the normal means case for selecting the one population out of k that has the largest θ value; that is, we use Table A.1 for this problem.

Suppose that we have $k=4$ given normal populations with $\sigma=1$, a value of $q=2.0$, and we wish to select the population with the largest value of $\psi(q;\theta)$ such that the probability of a correct selection is at least $P^*=.90$ whenever one ψ value is at least .2 units more than each of the others. Then by (3.9.2), $\delta^*=.2\sqrt{2\pi}=.5013$ and using $\tau=\delta^*\sqrt{n}=2.4516$ we obtain $n=(2.4516/.5013)^2=23.9$, or rounding upward we need 24 observations from each of the $k=4$ populations.

3.10   VARIANCES UNKNOWN AND NOT NECESSARILY EQUAL

In this section we consider the goal of selecting the population with the largest mean $\mu_{[k]}$ for the model of k normal populations with unknown variances that are not necessarily equal. As in Section 3.4 when the variances are unknown but assumed common, in the present case we must use a two-stage procedure in order to satisfy the usual (δ^*, P^*) requirement that the probability of a correct selection is at least P^* whenever $\delta = \mu_{[k]} - \mu_{[k-1]} \geqslant \delta^*$, and we want this to hold uniformly for all values of the nuisance parameters $\sigma_1^2, \sigma_2^2, \ldots, \sigma_k^2$.

In the first stage a sample of common arbitrary size $n \geqslant 2$ is taken from each of the k populations. The sample means, $\bar{x}_1, \bar{x}_2, \ldots, \bar{x}_k$, where

$$\bar{x}_j = \frac{\sum\limits_{i=1}^{n} x_{ij}}{n},$$

are calculated and the individual sample variances s_j^2 are calculated from (3.2.2) for these n observations.

In the second stage we specify the (δ^*, P^*) requirement and take a second sample of size $N_j - n$ from the jth population, for $j = 1, 2, \ldots, k$. The value of N_j for $j = 1, 2, \ldots, k$ is obtained from the formula

$$N_j = \max\left[n+1, \left\{ \frac{h^2 s_j^2}{\delta^{*2}} \right\}^+ \right], \tag{3.10.1}$$

where the s_j^2 are defined in the preceding paragraph, the symbol $\{a\}^+$ means the smallest integer equal to or greater than a, and h is obtained from Table A.9 for the appropriate value of k, the specified value of P^*, and the size n of the first-stage sample. Note that if $n+1$ is larger than $\{h^2 s_j^2 / \delta^{*2}\}^+$ for some population j, only one additional observation need be taken from the jth population, but in any case at least one observation is taken from each population in the second stage; that is, $N_j - n \geqslant 1$ for each $j = 1, 2, \ldots, k$.

When the additional observations are taken from all populations, the means of the observations taken *in the second stage only* are calculated as

$$\bar{y}_j = \frac{1}{N_j - n} \sum_{i=n+1}^{N_j} x_{ij} \tag{3.10.2}$$

for $j = 1, 2, \ldots, k$. Then we calculate the weighted mean of the first and second stage means for the jth sample using the formula

$$\bar{z}_j = b_j \bar{x}_j + (1 - b_j) \bar{y}_j \tag{3.10.3}$$

where b_j is calculated from

$$b_j = \frac{n}{N_j}\left[1 - \sqrt{1 - \frac{N_j}{n}\left[1 - \frac{(N_j - n)\delta^{*2}}{h^2 s_j^2}\right]}\,\right] \qquad (3.10.4)$$

for $j = 1, 2, \ldots, k$. Note that the s_j^2 in (3.10.4) are the sample variances for the observations in the initial sample only, and the h value is that determined earlier from Table A.9.

The $\bar{z}$ values are ordered as

$$\bar{z}_{[1]} \leqslant \bar{z}_{[2]} \leqslant \cdots \leqslant \bar{z}_{[k]},$$

and the selection rule is to assert that the population that gives rise to the largest $\bar{z}$ value, $\bar{z}_{[k]}$, is the one with the largest true mean $\mu_{[k]}$.

It is interesting to note that the selection procedure here does *not* depend on the overall means of all N_j observations combined, that is, on

$$\bar{\bar{x}}_j = \frac{\displaystyle\sum_{i=1}^{N_j} x_{ij}}{N_j} \qquad (3.10.5)$$

for $j = 1, 2, \ldots, k$. The rule here based on the $\bar{z}$ values in (3.10.3) has an interesting performance characteristic, namely, the probability of a correct selection is at least P^* for all $\delta \geqslant \delta^*$, that are independent of the unknown variances $\sigma_1^2, \sigma_2^2, \ldots, \sigma_k^2$, but (curiously enough) the corresponding rule based on the overall means in (3.10.5) does not have this property. Thus the (δ^*, P^*) requirement is satisfied for the rule based on the $\bar{z}$ values, but it need not hold for the corresponding rule based on the $\bar{\bar{x}}$ values (which would appear to be more attractive).

As an example, suppose that we have $k = 3$ populations with unknown (possibly unequal) variances. An initial sample of common size $n = 20$ is taken from each population and the sample means and variances are given in Table 3.10.1.

Table 3.10.1   First-stage sample results

Population j	1	2	3
Sample mean $\bar{x}_j$	25.1	26.3	29.7
Sample variance s_j^2	39.3	54.6	50.8

We specify $P^* = .90$, $\delta^* = 3.0$. From Table A.9 with $k = 3$, $P^* = .90$, the h value is $h = 2.34$. The right-hand components of (3.10.1) are then calculated from $\{(2.34)^2 s_j^2 / (3.0)^2\}^+$, and the results are

$$23.9, \quad 33.2, \quad 30.9$$

for $j = 1$, 2, 3, respectively. Since each of these values exceeds $n + 1 = 21$, they are the N_j values. As usual, we round upward to obtain a conservative result. Then the N_j values are

$$N_1 = 24, \quad N_2 = 34, \quad N_3 = 31,$$

and the sizes of the second-stage samples are

$$N_1 - n = 4, \quad N_2 - n = 14, \quad N_3 - n = 11.$$

Suppose that these additional observations are taken and the sample means for the second stage *only* are

$$\bar{y}_1 = 22.7, \quad \bar{y}_2 = 33.6, \quad \bar{y}_3 = 25.2.$$

Then we use $\delta^* = 3.0$, $h = 2.34$, $n = 20$, the s_j^2 given in Table 3.10.1, and the N_j values given earlier to compute the b_j values from (3.10.4). For example, for $j = 1$ the result is

$$b_1 = \frac{20}{24}\left[1 + \sqrt{1 - \frac{24}{20}\left[1 - \frac{(24-20)(3.0)^2}{(2.34)^2(39.3)}\right]}\right] = 0.856.$$

Similarly, we find $b_2 = 0.664$, $b_3 = 0.671$. Using these b values, the $\bar{z}_j$ are found from (3.10.3) as

$$\bar{z}_1 = (0.856)(25.1) + (0.144)(22.7) = 24.75,$$

$$\bar{z}_2 = (0.664)(26.3) + (0.336)(33.6) = 28.75,$$

$$\bar{z}_3 = (0.671)(29.7) + (0.329)(25.2) = 28.22.$$

Since $\bar{z}_2$ is the largest z value, we assert that the population labeled 2 is the one with the largest mean $\mu_{[3]}$. This assertion is correct with probability at least $P^* = .90$ whenever $\delta = \mu_{[3]} - \mu_{[2]} \geqslant 3.0$.

Note that as n increases, the h values in Table A.9 approach the τ_t values in Table A.1 for the corresponding values of k and P^*. The entries in Table A.9 for $n = \infty$ were taken from Table A.1, rounding the third decimal upward in each case since the other entries in Table A.9 were rounded in the same manner in order to obtain conservative results.

REMARK 1. As in the two-stage procedure given in Section 3.4, the question of an optimal initial sample size n is not easily answered since it depends on the true values of the unknown parameters. In general, it is desirable to keep the second-stage sample size small for each population and at the same time minimize the total number of observations from each population in both stages combined. Thus any prior information about variances (or guessed values) could be used to try to equate the two quantities on the right-hand side of (3.10.1) by replacing s_j^2 in (3.10.1) by an a priori estimate of σ_j^2 and solving for a reasonable value of n for the jth population. However, we could then get several different values for n by considering different values of j; and some compromise would then be required. □

REMARK 2.   Using (3.10.1) and the fact that $n \geqslant 2$, it can be shown that the value of b_j in (3.10.4) is between zero and one (in particular, the quantity under the radical sign is never negative), so that the $\bar{z}_j$ in (3.10.3) are indeed weighted averages of the $\bar{x}_j$ and $\bar{y}_j$ for each $j = 1, 2, \ldots, k$.   □

3.11   NOTES, REFERENCES, AND REMARKS

In dealing with normal distributions with unknown variance in classical statistical theory, Student's t distribution must be used; the same phenomenon occurs in ranking and selection theory. However, here we have several populations and hence there are several cases to be considered. The most interesting case is that of a common unknown variance. The two-stage solution for this problem is based on two papers, Bechhofer, Dunnett, and Sobel (1954) and Dunnett and Sobel (1954). The former paper deals with the procedure and proves that it satisfies the desired P^* probability requirement, making use of a well-known two-stage sampling procedure of Stein (1945); the latter develops an exact analysis for the probability of a correct selection and involves a multivariate form of Student's t distribution. Dudewicz (1972) addresses the problem of assessing the level of confidence when the pooled sample variance is replaced by an upper confidence limit for σ^2.

Of course, there are many other cases to be considered. The model with known unequal variances in Section 3.3 has an exact solution when the n_j are such that the variances of the sample means are equal (see (3.3.1)); otherwise the solution is approximate. Some of this material is based on Bechhofer (1954).

The confidence interval procedure of Section 3.5.1 and the corresponding tables are due to Chen (1974). However, as the text indicates, the table values are so close to the (common) upper-tail percentage points (or equi-percentiles) of the one-sided multivariate t tables (see Krishnaiah and Armitage (1966)) that for practical purposes these two tables are interchangeable. The reason for this is that Chen required the probability $P\{-A < \max X < A\}$ for $A > 0$. If A is even moderately large this is approximated (and indeed bounded) by the probability $P\{\max X < A\}$.

The central idea of this paper has also been applied to the case of known variances with sample sizes not necessarily equal in Chen and Montgomery (1975). In this case the entries for $\nu = \infty$ in Table A.5 are used and this represents an independent check on these calculations.

The fixed-length confidence interval procedure is due to Dudewicz and Tong (1971). This paper is an outgrowth of previous work by the same authors separately (see Dudewicz (1970, 1972) and Tong (1971)).

Some preliminary ideas on the application of ranking and selection methods to the analysis of variance appear in Bechhofer (1954), but a two-stage procedure was not considered there. The latter approach is new with this book.

The sequential method of Section 3.7 is based on Robbins, Sobel, and Starr (1968). Although it involves a pooled sample variance computation at every stage, no new tables are required. Unlike the Stein two-stage procedure, it makes use of the information in the sample about σ^2 at each stage up to the time of stopping. Unfortunately we do not have numerical or Monte Carlo assessments or comparisons of the amount of saving that results when using the two-stage procedure versus the sequential procedure for this problem.

The idea of jointly controlling the probability of a correct selection and the level of confidence that a constructed interval contains the largest mean is a new feature in a paper of Rizvi and Saxena (1974); unfortunately the required tables for this procedure have not been calculated. The same idea is used by Alam and Thompson (1973) for ranking and estimation with Poisson processes.

The material in Section 3.8 on selecting the population with mean value closest to a known target constant and the tables used there are due to Rizvi (1971).

The problem of selecting the normal population with the largest mean for the case of unknown variances that are not necessarily common was considered in the paper by Dudewicz and Dalal (1975). This paper is the basis for Section 3.10; it deals with two-stage procedures, recommending $n \geqslant 13$ for the first stage sample, and has extensive tables. We note that Dudewicz and Dalal also give alternative procedures that use certain (nonintuitive) weighted averages of all the observations from both stages combined, but in the particular procedure described in Section 3.10 the $\bar{z}$ values are weighted averages of the means from the first and second stages. This makes the procedure given here intuitively more desirable for applications than any of the alternative procedures. Other two-stage procedures for this problem are considered by Ofosu (1972, 1974) and by Chiu (1974).

A three-stage elimination procedure for selecting the population with the largest mean for the case of variances common and unknown is in a recent paper by Tamhane (1976).

PROBLEMS

3.1 Black and Olson (1947) reported a study to compare dry shear strength of $k=6$ different resin glues for bonding yellow birch plywood. They obtained $n=10$ observations for each glue; each observation is the arithmetic mean of the measurements of five test pieces on dry shear strength in pounds per square inch -400. The data follow; the glues are numbered in increasing order of strength. Assume that the distributions for each glue are normal with common unknown variance.

(a) Select the glue with the largest dry shear strength.
(b) Make a confidence statement at level .95 about the glue selected, using s^2 as a point estimate of σ^2.
(c) Make a final confidence statement at overall level .810 about the glue selected, using an upper confidence limit for σ^2 that has level .90.

Shear strength of six types of glue

	Glue				
1	2	3	4	5	6
102	70	100	120	151	220
58	83	102	110	156	243
45	78	80	182	192	189
79	93	119	130	162	176
68	98	59	95	166	176
63	66	99	143	158	181
117	92	100	113	173	206
94	79	109	140	157	233
99	134	128	123	233	162
63	131	138	132	238	179

3.2 Suppose that the following summary data are for random samples of 200 persons taken from each of nine Southern states for a study concerning ages of women at the time of a decree for divorce or annulment. Assume that the data follow the normal distribution with common unknown variance.

(a) Select the state with the youngest average age at time of divorce or annulment.
(b) Make a confidence statement at level .90 about the state selected, using s^2 as a point estimate of σ^2.
(c) Make a final confidence statement at overall level near .95 about

the state selected, using an upper confidence limit for σ^2 that has level .90.

State j	$\bar{x}_j$	s_j^2	n_j
Alabama	27.7	36.2	200
Arkansas	26.4	39.4	200
Florida	35.6	46.1	200
Georgia	31.2	43.9	200
Louisiana	29.3	32.6	200
Mississippi	28.6	41.5	200
North Carolina	33.1	34.2	200
South Carolina	29.6	36.9	200
Tennessee	30.4	33.2	200

3.3   Fowler, Sullivan, and Ekstrand (1973) reported the results of two experiments concerning the effects of sleep on memory. One aspect of the study considered the effect of two conditions, namely, sleep and wakefulness, on loss of memory. Subjects were randomly divided into three groups, and all subjects learned two different paired-associate tasks, learning 15 pairs of common words as associates and learning 10 pairs of "nonsense" shapes. The first group learned during the daytime and were tested for recall after 3.5 hours of normal daytime activity. The second and third groups were awakened from sleep for the learning, one during the first half of the night and the other during the second half of the night. They then slept for 3.5 hours more and were awakened for the recall test. The data on percent loss of memory for the subjects as shown below for the two learning tasks are artificial, but they emulate the summary findings reported. Assume that the distributions are normal with a common unknown variance.

(a)   Select the best of the three conditions as regards average percent loss in memory on the verbal task and make an appropriate confidence statement about your selection.

(b)   Do (a) for the visual forms task.

Verbal task				Visual forms task		
Awake	Sleep 1	Sleep 2		Awake	Sleep 1	Sleep 2
48	19	36		32	26	22
45	16	31		35	14	25
51	14	24		41	10	36
42	22	39		26	31	10
55	25	43		19	40	8

3.4 Anderson (1962) reported the results of an experiment conducted to see whether (a) hope of success and (b) fear of failure in eighth-grade students was the same under three degrees of conditions of arousal, classified as high, medium, and low. The variables *hope* and *fear* were measured by projective tests; we consider only the *hope* variable here. Since the actual data are not available, we consider a similar experiment by supposing that 21 students are randomly divided into groups of sizes 6, 7, and 8. A teacher talks to each of the groups in such a way that the first group should become quite motivated to success, the second group less so, and the third group should have almost no arousal. The same projective test is given to all 21 students and their scores measure "hope of success." Assume that the distribution of scores is normal with common unknown variance, and the goal is to select the condition of arousal that produces the greatest "hope of success."

High arousal	Medium arousal	Low arousal
77	75	70
69	81	21
60	44	39
89	72	82
41	79	77
45	53	67
	62	48
		52

(a) Make a selection.

(b) Make a confidence statement about the selection with overall level approximately .95, using an upper confidence limit for σ with level .90.

3.5 Suppose that a two-stage procedure is planned for the situation described in problem 3.1, and the data given with $n = 10$ observations per sample are the results of the first stage. How many additional observations per sample should be taken in the second stage if we specify $\delta^* = 15.0$, $P^* = .95$?

3.6 For the data given in problem 3.1,

(a) find a random length confidence interval for the largest mean with confidence coefficient .95.

(b) find a confidence interval of length $L = 20$ for the largest mean, assuming that $\sigma^2 = 600$, and give the confidence coefficient.

(c)   Assuming again that $\sigma^2 = 600$, find L for a fixed-length confidence interval with confidence coefficient .95.

3.7   Suppose that a sequential procedure is planned for the situation described in problem 3.1, and the data given with $n = 10$ observations per sample are the results of the initial stage. For $\delta^* = 20.0$, $P^* = .90$,

(a)   have we taken enough observations?
(b)   Assume that the answer to (a) is no, and an additional observation is taken using each type of glue. For the following additional results, can we terminate the experiment?

1	2	3	4	5	6
84	95	96	107	175	183

3.8   Houston (1972) reported a study to investigate the benefit of unit pricing to the consumer who wants to purchase the most economical brand and size of any product. A sample of 53 housewives in a small Midwestern town comprised the subjects for the study. The subjects were randomly divided into three groups, one group for each of three grocery stores, A, B, and C. Each subject was instructed to go to her assigned store and purchase the one most economical brand and size from each of a list of 14 product classes, without regard to personal preference or perceived quality. Subjects were also told not to consider other product variations, such as color, ingredients, and so on. Every item in store A is marked according to unit price as well as by item price. Stores B and C do not give a unit price. Store B tends to favor "per item" pricing, and store C primarily uses multiple pricing (like two for 49 cents).

In this study the two hypotheses tested were

"(1)   the frequency with which the items chosen at store A do not correspond with the minimum unit-price items is significantly smaller than at stores B and C, and

(2)   the resulting monetary differences between the items chosen and the minimum unit-price items at store A are significantly smaller than those at stores B and C." (p. 53)

For hypothesis 1 the number of items chosen that did not correspond to the item with the smallest unit price in its product class was computed for each subject. For hypothesis 2 for each product the difference (in dollars) between the per unit price of the item chosen and the per unit price of the item with the smallest unit price in its product class was computed and multiplied by the quantity purchased; these were summed over all 14 products to give the dollar difference of the market basket chosen by each subject.

The summary data given are as follows:

Frequency of corresponding choices

Store A	Store B	Store C
$n_1 = 18$	$n_2 = 18$	$n_3 = 17$
$\bar{x}_1 = 4.71$	$\bar{x}_2 = 6.71$	$\bar{x}_3 = 6.35$
$s_1 = 1.32$	$s_2 = 1.90$	$s_3 = 1.73$

$$s^2 = 2.78$$

Price differences in dollars

Store A	Store B	Store C
$\bar{x}_1 = 0.36$	$\bar{x}_2 = 0.54$	$\bar{x}_3 = 0.84$
$s_1 = 0.140$	$s_2 = 0.288$	$s_3 = 0.249$

$$s^2 = 0.0548$$

In each case the data were analyzed using the one-way analysis of variance test and individual t tests to compare the means.

For each of the research hypotheses given earlier, the appropriate goal is to select the population with the smallest mean. For these data store A (with unit pricing) is the one selected. In the following problems assume that the distributions in each case are normally distributed with common unknown variance.

(*a*) What confidence statement can be made with confidence coefficient .90 about (i) the mean frequency for store A, and (ii) the mean price difference for store A?

(*b*) Find a confidence interval estimate for (i) the smallest mean frequency, and (ii) the smallest mean price difference, with confidence coefficient .90.

(*c*) If you were designing a similar experiment to investigate research hypothesis 2 and wanted to give a confidence interval at level .90 for $\mu_{[1]}$ with fixed length $L = .10$, how large a common sample size would you take? Assume that $\sigma^2 = 0.05$ to find your answer.

3.9 Locke and Whiting (1974) reported results of a survey on job attitudes of five classes of employees in the solid waste management industry. The number of persons in each class, their average total satisfaction score, and standard deviation are as follows:

Class j	n_j	$\bar{x}_j$	s_j
Unskilled	326	58.56	10.91
Skilled	303	60.19	9.25
Supervisory	146	64.90	7.69
Secretarial/clerical	75	65.84	8.83
Managerial	61	64.45	8.90

Assume that the true satisfaction scores for each group are normally distributed with a common unknown variance.

(a) Select the class with the largest true satisfaction score and make an appropriate confidence statement at level .90 about the selection.

(b) If you were designing such a study and wanted to choose a common sample size for each of the five populations, how large a sample would you choose? What assumptions are you making?

3.10   Kleijnen, Naylor, and Seaks (1972) present an example in which a firm that produces a final product from a process with independent stages is interested in selecting the one most profitable plan among $k = 5$ possible plans. They run simulation experiments with an initial sample of size $n = 50$ for each plan and assume that the profit using each plan is a random variable that is normally distributed with a common unknown variance. The data are as follows:

Plan j	Sample mean profit $\bar{x}_j$	Sample standard deviation s_j
1	2976.40	175.83
2	2992.30	202.20
3	2675.20	205.51
4	3265.30	221.81
5	3131.90	277.04

(a) Use a two-stage procedure to obtain the exact solution for $\delta^* = 100$, $P^* = .90$. If no second sample is required, make a selection.

(b) Do (a) for $\delta^* = 100$, $P^* = .95$.

(c) Assuming this is a one-stage procedure, find an upper bound for the common variance and make a confidence statement at level .95 about your selection.

(d) Find a random-length confidence interval with confidence coefficient .95 for the value of the largest mean profit.

CHAPTER 4

Selecting the One Best Population for Binomial (or Bernoulli) Distribution

4.1 INTRODUCTION

In this chapter we consider the problem of selecting the one best population out of k populations where each element in each population is classified into exactly one of two mutually exclusive (nonoverlapping) and exhaustive categories, such as male and female, good and defective, go and no-go. For convenience we always somewhat arbitrarily call these two categories "success" and "failure." In each of the k populations there is a certain probability that an element will be classified as a success. We denote these probabilities by $p_1, p_2, \ldots, p_k$. The *best* population may be defined as the one for which the probability of success is largest, or alternatively, as the one with the smallest probability of success, depending on the situation. A simple interchange of the labels *success* and *failure* reduces the latter problem to the former one; this interchange is discussed in Section 4.4.

The probability model that applies to independent dichotomous outcomes is either the Bernoulli or the binomial distribution; the relationship between these terminologies is explained later. Since both of these distributions involve the same parameter p, the problem under consideration here may be called either selection of the best Bernoulli distribution or selection of the best binomial distribution.

The binomial (or Bernoulli) distribution model is the prototype for many important practical problems in experimentation and investigation. In fact, it is safe to say that the binomial and normal distributions are the two most widely used distributions in statistics. In view of this we give a few specific illustrations of situations where the binomial (or Bernoulli) model applies and the basic problem of practical (as well as theoretical) interest is to select the best population.

Example 4.1.1 (Learning). In a learning experiment subjects study programmed text materials and are later tested on recall and comprehension. This test is scored as pass or fail. A total of k alternative sets of programmed materials are available; these cover the same subject matter at an equally advanced level. The problem is to determine which one of these k sets of materials is the most effective learning device. We define a success as a passing score on the test, so that p_j is the probability of a passing score for subjects exposed to the jth set of learning materials. Then the goal is to determine which one set has the largest p value. For this purpose a number of (homogeneous) subjects are divided randomly into k groups, with say n_j in the jth group. Each group studies a different set of programmed materials and then the test is administered. In each group the number of successes (passing scores) is counted; the selection procedure depends on these numbers, as well as the number k of populations and the number n_j of subjects in each group.

Example 4.1.2 (Drugs). Many different types of birth control pills are available by prescription. Any two types are regarded as different if they have different trade names, different manufacturers, or different proportions of various ingredients (primarily hormones). A physician wishes to determine which one among k types of pills is most effective in preventing conception when taken strictly according to instructions. As a result, if we define a success as *not* conceiving so that p_j is the probability of not conceiving when taking the jth type of pill, the goal is to identify the type of pill with the largest p value. A large number of women who engage in intercourse regularly and wish to prevent conception for a two-year period by using the pill are chosen as subjects. These women are divided into k groups, each of common size n (forming a total of kn subjects), and the same type of pill is given to all subjects within a group. Neither the subjects nor the physician is told which drug is given to which group (a double-blind experiment). For each group the number of women who do not conceive during the two-year period is counted; the selection procedure is based on n, k and the number of successes within each group.

Example 4.1.3 (Sales). A company that sells vacuum cleaners in a "door-to-door" manner is planning to conduct an extensive sales campaign in a certain metropolitan area. This company has not heretofore had salesmen call in this area, and they wish to establish the product quickly. However, only a skeleton sales force is available at present; hence the initial campaign must be on a smaller scale. Further, it is essential that the campaign be initiated in that section of the metropolitan area (sales district) that can be expected to yield the largest number of sales relative to number of calls. The profits from this campaign can then be used to hire additional salesmen to solicit the remainder of the area. As a result of these factors, the company wants to identify the best district to use in initiating the campaign. For this purpose they plan a sampling experiment as a pilot study of the districts. We assume that the metropolitan area divides itself naturally into k districts that are approximately homogeneous with respect to number and type of dwelling units, so that at the outset, each district is equally desirable from the company's point of view. However, if one district has a higher profit potential than the others, they wish to initiate the campaign there. For the pilot study n dwelling units are selected at random in each of the k districts, and calls are made at those units only. A success is defined as a sale, and p_j denotes the probability of a sale in the jth district. The number of sales is recorded for each district; the selection procedure depends on n, k, and the numbers of sales.

We now return to a discussion of the distribution model for these selection problems. An experiment or process for which each outcome is classified dichotomously, say as either success or failure, is called a *Bernoulli trial* and the *Bernoulli distribution* applies for each trial. If this kind of experiment is repeated a fixed number of times in such a way that the trials are independent and the probability of success is identical on every trial, the outcomes can be regarded as elements in a random sample from the Bernoulli distribution. The parameters in this distribution are p, the probability of a success on any one trial, and q, the probability of a failure. Since these parameters must sum to one for a complete (i.e., exhaustive) dichotomy, we always have $q = 1 - p$. As a result, only one parameter, say p, need be considered, and the general parameter θ in Chapter 1 is replaced by p in the Bernoulli problem. Since here we have k populations, we let p_j denote the probability of success for the jth Bernoulli population, for $j = 1, 2, \ldots, k$. As before, the ordered p values are denoted by

$$p_{[1]} \leqslant p_{[2]} \leqslant \cdots \leqslant p_{[k]}.$$

The goal is then to select the population with the largest p value, that is, the one that has parameter $p_{[k]}$. It should be emphasized that we are *not* trying to estimate the largest probability of success, but only to *identify* the population that has the highest probability.

REMARK.   Since equalities are allowed among the p_j's, the largest value of p is not necessarily unique. However, in those configurations of p values that enter into our probability calculations, the largest p value *is* uniquely defined.   □

For the purpose of selecting the population with the largest p value, we take one sample from each population in such a manner that the samples are mutually independent. For each sample we count the number of successes that occur. For the sample from the jth population, let n_j denote the number of observations in the sample and x_j the number of successes among these observations. Then for each sample, x_j (considered as a random variable) follows the binomial distribution with parameters n_j and p_j; the binomial distribution is denoted by $b(n_j, p_j)$. As a result, for any given values of n_j and p_j ($j = 1, 2, \ldots, k$), the model of independent samples from k Bernoulli populations is equivalent to the model of k independent binomial distributions, that is,

$$b(n_1, p_1), b(n_2, p_2), \ldots, b(n_k, p_k).$$

In the Bernoulli distribution the emphasis is on the outcomes of the individual trials, whereas in the binomial distribution, the emphasis is on the number of successes in n_j trials, for $j = 1, 2, \ldots, k$. Since p_j for the jth binomial distribution is the same as p_j for the jth Bernoulli population and the n_j are known constants, the goal can now be stated equivalently as selecting the binomial distribution with the largest p value. The selection rule appropriate for this problem is described in the next section.

Discussion of the analytical aspects of the problem of selecting the best Bernoulli population (or, equivalently, the best binomial population) begins in Section 4.3 for the situation where all k samples are of the same known size. For this case we can give a solution to the problem that is exact for all practical purposes. However, we must also consider the problem when the sample sizes are unequal, since in some instances the investigator has no control over the sample sizes. For instance, in Example 4.1.1 the k group sizes may be fixed (or limited) by practical considerations that result in unequal group sizes. An exact solution for unequal sample sizes is difficult, but an approximate solution is provided in Section 4.6.

Throughout this chapter the formulation is based on the use of the distance measure

$$\delta = p_{[k]} - p_{[k-1]}, \tag{4.1.1}$$

that is, the ordinary difference between the largest and next-to-largest p values. Unfortunately there is no one "natural" measure of distance in the binomial problem that has both practical and theoretical advantages. We restrict consideration to the simple difference in (4.1.1) for the following two reasons:

1. For the binomial distribution model, a solution to the problem of determination of sample size does exist for the measure of distance

$\delta = p_{[k]} - p_{[k-1]}$. In other words, for any specified values of δ^* and P^* (with $0 < \delta^* < 1$ and $1/k < P^* < 1$), we can find a common sample size n such that the probability of a correct selection is at least P^* for all $\delta \geqslant \delta^*$, that is, for all configurations of p values in the preference zone. This property does *not* hold for all measures of distance; that is, with some other distance measure, we may not be able to find a common sample size that guarantees the specified conditions unless we are willing to resort to more complicated sampling procedures. Indeed it can be shown that this is the case if the distance measure is the odds ratio δ_{OR} defined in (1.3.1) of Chapter 1.

2.   The measure of distance $\delta = p_{[k]} - p_{[k-1]}$ is simpler and perhaps more intuitive than any other measure. The investigator can therefore more easily grasp the meaning and consequences of the specified values δ^* and P^* in relation to the probability requirement.

If there are valid reasons for preferring some other measure of distance (such as the odds ratio, for example), the investigator must be willing to use more sophisticated solutions to the problem of finding the population with the largest p value. Two-stage procedures or fully sequential procedures may then be required. These methods are not discussed here since they are more complicated and require a new analysis and also new tables; in many cases both the analysis and the tables may not yet be available.

By using the distance measure $\delta = p_{[k]} - p_{[k-1]}$, we can restrict our consideration to fixed sample size (or one-stage sampling) procedures. In comparing the efficiency of our one-stage procedure with any other procedure, it makes a big difference whether or not the other procedures are also one-stage and utilize the very same formulation of the problem. In Section 4.5 we show that a large saving can be attained by bringing into consideration some prior information about the p values. A large saving could also be attained if we were to consider two-stage or sequential procedures. Naturally we cannot expect the usual efficiency criterion (based on expected sample size only) of a one-stage procedure to compare favorably with (for example) a fully sequential procedure that is designed to utilize the new information brought into the experiment with every observation taken.

4.2 SELECTION RULE

The procedure to select the Bernoulli population (or the binomial distribution) with the largest probability of success is quite simple and straightforward once the k sets of sample data are obtained. Since the parameters of interest are the probabilities of success in each population, it is natural to base our decision on the proportions of successes in each sample.

Suppose there are n_j observations in the jth sample and x_j is the number of successes observed in the jth sample. Let $\bar{x}_j$ denote the proportion of successes observed in the jth sample. Then $\bar{x}_j$ is calculated as

$$\bar{x}_j = \frac{x_j}{n_j}. \qquad (4.2.1)$$

These k sample proportions are then ordered from smallest to largest. Using the standard ordering notation, $\bar{x}_{[j]}$ is the jth smallest sample proportion, and the ordered sample proportions are

$$\bar{x}_{[1]} \leqslant \bar{x}_{[2]} \leqslant \cdots \leqslant \bar{x}_{[k]}.$$

The rule R (when there is no tie for first place) is simply to select the largest sample proportion $\bar{x}_{[k]}$ and assert that the population that produced that sample is the one that has the largest parameter value $p_{[k]}$.

REMARK. When the sample sizes are all equal, that is, when $n_1 = n_2 = \cdots = n_k = n$, say, each sample proportion in (4.2.1) has the same denominator. Therefore the ordering of the proportions $\bar{x}_j$ is exactly the same as the ordering of the integers x_j and the selection rule R can then be simplified as follows. The observed numbers of successes are ordered as

$$x_{[1]} \leqslant x_{[2]} \leqslant \cdots \leqslant x_{[k]}, \qquad (4.2.2)$$

and the rule R (when there is no tie for first place) is to assert that the population that gives rise to $x_{[k]}$ is the one with the largest probability of success $p_{[k]}$. $\square$

In the present case, as with any discrete distribution, we must allow for the possibility of ties between any two or more of the $\bar{x}_j$'s. This is especially true when all the sample sizes are equal. (The reason for this is that there may be ties in the numbers of successes; then when we divide by a common n, the ties remain, whereas when we divide by different n_j's, the resulting proportions will differ.) When ties occur for the largest value of $\bar{x}$, they should be "broken at random." More specifically, if r of the k values of the $\bar{x}_j$ are equal and the common value is larger than the other $k - r$ values of $\bar{x}_j$, one of the r samples tied for first place is selected at random. The population corresponding to the sample selected is then asserted to be the best one. The occurrence of ties and this randomization procedure to resolve them are taken into account in the derivation of the probability of a correct selection under the binomial or Bernoulli model. Hence, when ties occur, no modification is needed in the confidence statement made about the final selection. (Although it might be preferable to break such ties by taking some additional observations, the resulting procedure would not be a fixed sample size procedure; hence we do not consider this possibility here.)

The following two examples illustrate the selection rule for equal and unequal sample sizes respectively.

Example 4.2.1. In the situation described in Example 4.1.3, suppose that calls are made on 100 dwelling units in each of the three sales districts and the sales data are as given in Table 4.2.1. Defining a success as a sale, the problem is to select the district with the largest p value, where p is the probability that a sale is made in a randomly chosen house call. The procedure is to order the observed proportions of sales $\bar{x}_j$ and select the largest one. However, since $n_1 = n_2 = n_3 = 100$ here, we can simply order the observed numbers of sales x_j. The results are

$$x_{[1]} = 22 = x_3, \quad x_{[2]} = 27 = x_1, \quad x_{[3]} = 42 = x_2.$$

We then assert that the population designated as District 2 has the largest probability of a sale among the three populations. The sales campaign is therefore to be initiated in District 2.

*Table 4.2.1   **Data on sales in three sales districts***

District	Number of sales	Number of calls
1	27	100
2	42	100
3	22	100

Example 4.2.2. An organization plans to purchase a fleet of automobiles. The choice has been narrowed down to three different makes that are all in the same price range. The final decision is to be based on the expected maintenance cost for each of these three makes. The manager feels that the best indicator of low maintenance cost for each make is the proportion of cars not needing a single "early" major overhaul, where *early* is defined as within five years of the date of purchase. Thus he wants to select the one make that has the highest probability of not requiring an early major overhaul. (This defines the parameter p.) The data in Table 4.2.2 were collected from the records of distributors of these three makes, and the sample sizes are not equal. The entries in the last column indicate that the sample from make B has the highest proportion of cars not

Table 4.2.2   Data on early major overhauls (within five years of purchase date)

Car make	Number of cars not needing an early overhaul	Number of cars needing at least one early overhaul	Total number of cars	Proportion of cars not needing an early overhaul
A	286	107	393	$286/393 = .7277$
B	991	198	1189	$991/1189 = .8335$
C	340	85	425	$340/425 = .8000$

needing an early overhaul. Hence make B is asserted to be the one whose cars have the largest probability of not needing an early major overhaul, that is, the lowest maintenance cost.

4.3   SAMPLE SIZES EQUAL

In this section we consider some analytical aspects of the binomial selection problem when each sample is of the same size, that is, when $n_1 = n_2 = \cdots = n_k = n$, say. The problems covered are the determination of sample size and the calculation of the operating characteristic curve when n is fixed.

As explained in Section 4.1, a solution to these problems for any k can be found when n is common and the distance measure is $\delta = p_{[k]} - p_{[k-1]}$. The generalized least favorable configuration in the binomial case for all k and n is

$$p_{[1]} = p_{[2]} = \cdots = p_{[k-1]} \quad \text{and} \quad \delta = p_{[k]} - p_{[k-1]}. \quad (4.3.1)$$

Unfortunately the probability of a correct selection under this generalized least favorable configuration depends not only on the value of δ, but also on the value of $p_{[k]}$. However, for any particular k and n, the minimum probability can be computed for all configurations in the preference zone. Tables and figures based on exact computations are available for $n \leq 10$. A theoretical investigation has shown that the least favorable configuration for practical purposes is given by the single configuration

$$p_{[1]} = p_{[2]} = \cdots = p_{[k-1]} = .5 - \frac{\delta^*}{2} \quad \text{and} \quad p_{[k]} = .5 + \frac{\delta^*}{2}. \quad (4.3.2)$$

REMARK.   In general, the least favorable configuration depends on n because it is obtained by minimizing the probability of a correct selection for the given n and k. However, for $n \geq 10$ and any k, and for all n when $k = 2$, the least favorable configuration in the binomial problem is the configuration given by (4.3.2), which does not depend on n.   □

Note that in the configuration in (4.3.2), the values $p_{[k]}$ and $p_{[k-1]}$ are arranged symmetrically about .5. For $k=2$ this configuration is exactly the least favorable configuration for all n. For $k>2$, the error in using (4.3.2) was found to be too small to be of any practical significance even for moderate values of n. Moreover, for any fixed k, the error decreases to zero as n gets larger. The configuration in (4.3.2) has the nice property that it does not depend on n. As a result, for $n \geqslant 10$ and $k \geqslant 3$, the tables and graphs available have been computed using this one simple configuration (instead of the many exact least favorable configurations); for $k=2$, the integer solutions given are exact for all n appearing in the tables.

Before we can discuss the analytical aspects of the binomial selection problem, we must explain how to use the tables and graphs provided in this book. Table E.1 in Appendix E covers the cases of $k=2$, 3, 4, and 10 binomial populations. Each table entry is the common value of n which, for that k, guarantees the probability of a correct selection to be at least P^* whenever $\delta \geqslant \delta^*$, where $\delta = p_{[k]} - p_{[k-1]}$. These tables cover the selected values $P^* = .500, .600, .750, .800, .850, .900, .950, .990$ (listed across the top) and δ^* from .05 to .50 in increments of .05 (listed down the left side). The accompanying figures in Appendix F correspond to and supplement these tables, since they provide continuous curves that show the common n for the same selected P^* but for any possible value of δ^*, that is, $0 < \delta^* < 1$. Values read from the figures can also be found by interpolation in the tables. The recommended method of interpolation is explained and illustrated in Appendix E. Tables for the binomial problem are presently available only for $k=2$, 3, 4, and 10. Others could be developed, but the figures in Appendix F give approximate solutions for the common value of n for any k and for any specified pair (δ^*, P^*).

4.3.1 Determination of Sample Size

Suppose that we are in the process of designing an experiment to select the binomial population with the largest probability of success and the investigator has control over the sample sizes. The common sample size n is to be determined such that the probability of a correct selection is at least P^* whenever the distance measure $\delta = p_{[k]} - p_{[k-1]}$ is at least δ^*, where P^* and δ^* are specified values. The answer for any δ^* and P^* can be found from Appendix E or Appendix F. For any n and k the total number of observations is of course kn.

As a numerical illustration, suppose that $k=3$, as in the sales example of Example 4.1.3, and the specified values are $\delta^* = .10$ and $P^* = .90$. From either Table E.1 or the figure in Appendix F for $k=3$, we find the solution is $n=125$. If $k=3$, $P^* = .90$, and $\delta^* = .08$, Table E.1 does not give the answer explicitly. However, the figure for $k=3$ in Appendix F gives an n just slightly under 200. A more precise answer could be found by interpolation. As illustrated in Example 1*a* of the interpolation instructions, the interpolated result is $n = 195$, which compares favorably with the value "just under 200" read off the figure. If $k = 3$, $\delta^* = .12$, and $P^* = .82$, neither Table E.1 nor the figures in Appendix F explicitly cover this situation However, we can still interpolate in the table for $k = 3$ to find $n = 54$. (See Example 1*b* of the interpolation instructions.)

Now let us consider the following specific example.

Example 4.3.1.   A fire department has been asked to endorse one of four different types of fire control devices. The effectiveness of each device is measured in terms of whether it succeeds in putting out a fire, called a *hit*, or fails to put out a fire, called a *miss*. Define p as the probability of a hit. Then the appropriate goal is to select the type of device that has the largest p value. Suppose we specify that the probability of a correct selection must be at least $P^* = .99$ when $\delta = p_{[4]} - p_{[3]} \geqslant .10 = \delta^*$, that is, when the difference between the largest and next-to-largest p value is at least .10. How many independent observations should be made on each type of device, assuming a common number on each?

The requirements in this example are quite stringent; that is, P^* is very large and δ^* is fairly small. Hence we expect that the common sample size must be quite large. Using Table E.1 for $k = 4$ in Appendix E, with $P^* = .99$ and $\delta^* = .10$, we find $n = 360$. If we cannot afford to take four samples each of size 360 (a total of 1440 observations), we must specify either a smaller P^* or a larger δ^*, or both. For example, if $P^* = .90$ and $\delta^* = .15$, only 67 observations per sample are required when $k = 4$. In other words, if we make 67 independent tests on each type of device, then we can assert with probability at least .90 that the best model will be selected if the difference between the true p values of the best and second best types is at least .15. The important point here is that we can consult the tables before deciding on the (δ^*, P^*) pair to specify.

Once this method is used to determine n for the specified values (δ^*, P^*) we can make a confidence statement about the difference between $p_{[k]}$ and the parameter p_s, the p value of the population selected by the rule R. With confidence level P^* we can say either that

$$0 \leqslant p_{[k]} - p_s \leqslant \delta^*, \tag{4.3.3}$$

or equivalently, that $p_{[k]}$ lies in the interval

$$S = \left[p_s, p_s + \delta^* \right]. \tag{4.3.4}$$

Thus in Example 4.3.1, with a selection based on four samples each of size 67, we can say with 90% confidence that $p_{[4]} - p_s \leqslant .15$, or equivalently that the interval $[p_s, p_s + .15]$ covers $p_{[4]}$.

On the other hand, if the investigator knows the value of δ^* that he wants as part of the confidence statement in (4.3.3) or (4.3.4) he can determine the corresponding level of confidence in this assertion for any given values of k and n. The confidence level $CL(\delta^*|R)$ for the rule R depends on δ^*, k, and n, but not on the true configuration of parameters. The procedure is simply to interpolate in Table E.1 or the appropriate figure in Appendix F to find the probability that corresponds to this δ^*, k, and n. In Example 4.3.1 where $k=4$, suppose that the investigator specifies $\delta^*=.12$ and $n=50$. Then by interpolation in Table E.1 for $k=4$, the corresponding probability is .75, and hence the confidence level is $CL(.12|R)=.75$.

In any practical application, the true p values are unknown. Nevertheless suppose that in Example 4.3.1 the true unordered p values for the four types of devices are

$$p_1=.37, \quad p_2=.42, \quad p_3=.59, \quad p_4=.30.$$

Since $p_{[4]}=.59$ and $p_{[3]}=.42$, the true difference between the largest and second largest p values is $\delta=p_{[4]}-p_{[3]}=.17$. Then we know that the procedure based on 67 observations from each of the four types correctly selects the best type of device *more* than 90% of the time, since the preceding calculation indicated that $n=67$ is a large enough sample size to satisfy $\delta^*=.15$, $P^*=.90$. On the other hand, if the true difference δ were less than .15 and the set of p values is in the worst possible configuration, the selection would be wrong more than 10% of the time. For example, we might too often designate the second best as best; however, it should be noted that this would not be too critical an error. Roughly speaking, a specification of $\delta^*=.15$ indicates (implicitly) that the investigator is not too concerned about distinguishing between the best and second best population when the true difference is small, namely, less than .15.

REMARK 1.   Although it is correct to say that the probability of a correct selection for our procedure in the binomial (or any other) model approaches $1/k$ as the parameter values approach equality, this statement is not very meaningful from the point of view of loss, since the loss resulting from an incorrect selection gets smaller and smaller as the p values approach equality. If we define the loss because of an incorrect selection to be equal to one in the preference zone and zero elsewhere (i.e., in the indifference zone), then a decision-theoretic analysis of our selection procedure shows that it has an expected loss of at most $1-P^*$ for the entire parameter space. Alternatively, we might define the loss to be one if we select a population whose p value is less than $p_{[k]}-\delta^*$, and zero otherwise. Then by elementary decision theory our procedure again has an expected loss of at most $1-P^*$ throughout the parameter space. The result for the latter loss function shows that the procedure has desirable properties in both the preference zone and the indifference zone. $\square$

REMARK 2. It should be noted that the P^* value (or confidence level) of $1/k$ can always be attained by the procedure of selecting one of the k populations at random without taking even a single observation. This remark suggests that a minimal criterion for labeling any procedure as reasonable is that the probability of a correct selection should be at least $1/k$ for all points in the parameter space. Moreover, in general we should attempt to ensure a P^* value well above the level $1/k$, and indeed for most practical purposes we recommend keeping $P^* > .5 + (.5/k)$ for any k. In particular, note that we recommend keeping P^* greater than .5 for all k. $\square$

REMARK 3. Roughly speaking, if $P^* < .5$ the sequential procedure permits us to stop if the p values of two populations are about equal and much larger than all the other p values, since then we could toss a coin to decide between the two largest. However, this procedure is not commensurate with our stated goal to find the *best one* binomial population; it is more like selecting one of the best two populations. To avoid such "inconsistencies" we merely keep $P^* > .5$ for any k. The recommended expression $P^* > .5 + (.5/k)$ was not analytically derived. $\square$

4.3.2 Calculation of the Operating Characteristic Curve

Suppose now that the common sample size n is fixed and not under the control of the investigator, either because the data have already been obtained or because of practical limitations. Then we do not have a specified (δ^*, P^*) requirement, but the performance of the selection rule is still of concern. It is therefore of interest to calculate the operating characteristic curve.

As noted in the introduction to this section, the operating characteristic curve cannot be found (as a function of one variable) under the generalized least favorable configuration, since the calculation depends on $p_{[k]}$, and not just on δ. However, the curve can be found under the least favorable configuration. The tables and figures in Appendixes E and F do not give these probabilities directly as a function of δ^* for given k and n. Nevertheless we can interpolate in these tables to find the operating characteristic curve of the binomial selection procedure, that is, to find the probability $P_{LF}\{CS|\delta^*, n\}$, of a correct selection under the least favorable configuration, as a function of δ^* for given k and n. (Of course, for fixed k, n, and δ^*, this probability is equal to the confidence level $CL(\delta^*|R)$ for the statement in (4.3.3) or (4.3.4).) In other words, to obtain the OC curve, we

Table 4.3.1   *Corresponding values of $P_{LF}\{CS|\delta^*, n\}$ and δ^* for $n = 50$, $k = 4$*

$P_{LF}\{CS|\delta^*, n\}$	.50	.60	.75	.80	.90	.95	.99
δ^*	.060	.080	.120	.134	.176	.206	.266

force the specified *constants* (δ^*, P^*) to be functionally related *variables*, since n and k are now both fixed. However, we break our own rule here and keep the * notation for the variables δ^* and P^* (and keep n without the *) in order to emphasize the fact that δ^* and P^* do not denote true values.

As a numerical illustration, suppose that in Example 4.3.1, only $n = 50$ observations can be taken from each of the $k = 4$ populations. With this limitation on the common sample size, how well can we expect the selection rule of Section 4.2 to operate? For each P^* listed in Table E.1 for $k = 4$, we interpolate for $n = 50$ to find the corresponding value of δ^*. For example, when $P^* = .50$, Example $(2a)$ of the interpolation instructions for Table E.1 shows that the interpolated value is $\delta^* = .060$. The complete set of interpolated values for $n = 50$ and $k = 4$ is summarized in Table 4.3.1.

In Figure 4.3.1 the values given in Table 4.3.1 are plotted, with δ^* on the horizontal axis, and these points are connected by a smooth curve to give the operating characteristic curve. The interpretation of any particular

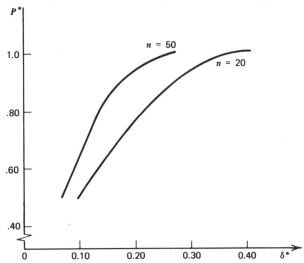

Figure 4.3.1   Operating characteristic curves for fixed $n = 20$, 50, $k = 4$.

point $(\delta^*, P_{LF}\{CS|\delta^*, n\})$ on the curve is that, for a sample of size $n = 50$ from each of $k = 4$ populations, the probability that a correct selection is made using the rule R is equal to $P_{LF}\{CS|\delta^*, n\}$ for the least favorable configuration, that is, when $\delta = \delta^*$, and is at least $P_{LF}\{CS|\delta^*, n\}$ for any configuration in the preference zone, that is, for any $\delta \geqslant \delta^*$.

Table 4.3.2 gives the corresponding values of $P_{LF}\{CS|\delta^*, n\}$ and δ^* for $n = 20$, $k = 4$ (computed by interpolation as in Table 4.3.1 but for $n = 20$). These values are also plotted in Figure 4.3.1. Note that for fixed δ^*, every $P_{LF}\{CS|\delta^*, n\}$ is smaller for $n = 20$ than for $n = 50$, and as δ^* increases the curve rises more slowly for $n = 20$ than for $n = 50$. This is true in general for any pair of n values. Thus when the common sample size is large, we not only have more information about the populations, but also (in some sense) the efficiency of our procedure increases faster as δ^* increases.

The operating characteristic curve is a contour of (δ^*, P^*) values for a fixed common value of n, and thus the curve indicates how well the selection procedure (for that fixed n) is operating at different levels of δ^*. For any n and k, the curve approaches the point $\delta^* = 0$, $P^* = 1/k$ (which is $P^* = .25$ in Figure 4.3.1), and the line $P^* = 1$ is a horizontal asymptote for δ^* increasing to unity.

Table 4.3.2   *Corresponding values of $P_{LF}\{CS|\delta^*, n\}$ and δ^* for $n = 20, k = 4$*

$P_{LF}\{CS|\delta^*, n\}$	.50	.60	.75	.80	.85	.90	.95	.99
δ^*	.095	.130	.190	.213	.238	.275	.322	.411

The operating characteristic curve gives an overall evaluation of how well the procedure can be expected to perform if the true δ is equal to δ^* and the least favorable configuration holds. However, the reader should remember that we usually do not know the true configuration, and hence we do not know δ. If we are interested in any particular value δ^* of δ and n is fixed, then the corresponding value P^* for $P_{LF}\{CS|\delta^*, n\}$ can be read from the OC curve or computed by interpolation in the tables of Appendix E; this is the true probability of a correct selection only for the least favorable configuration.

Operating characteristic curves may also be helpful for the problem of determining the sample size. If two or more common n values are under consideration, the OC curves based on each value of n could be plotted on the same graph (as was done for $n = 20$ and $n = 50$ in Figure 4.3.1) to compare the overall performance of the selection procedure for these n values under the same least favorable configuration.

REMARK. It is important to remember that these OC curves give the *minimum* probability of a correct selection over all points in the preference zone. Even if δ^* is the true value of δ and we are in the preference zone, the least favorable configuration still may not be the true configuration; that is, the $k-1$ smallest p values may not be equal. Hence in parts of the preference zone, the procedure will actually be operating better (and in some parts, much better) than the curve indicates. ☐

The ideas introduced in Section 2.3.4 of Chapter 2 for estimating the true probability of a correct selection in the normal means problem can be used in this chapter also. If in addition we are using asymptotic normal theory for the binomial problem, as explained later in Section 4.7 of this chapter, we can use not only similar ideas and methods but also the same tables. Thus in this model we could estimate p_j by $\bar{x}_j = x_j/n_j$ for each $j = 1, 2, \ldots, k$, use these values to estimate the δ values by $\hat{\delta}_j = \bar{x}_{[k]} - \bar{x}_{[j]}$ for each $j = 1, 2, \ldots, k-1$, and use these $\hat{\delta}$ values to estimate the true probability of a correct selection, by finding either an upper and lower bound or a point estimate.

4.4 SELECTING THE POPULATION WITH THE SMALLEST PROBABILITY OF SUCCESS

As mentioned in Section 4.1, it may be desirable to select the Bernoulli population (or the binomial distribution) with the smallest probability of success, that is, the one with p value equal to $p_{[1]}$. The population with the smallest p value must have the largest q value (where $q = 1 - p$). Therefore we might simply let the general parameter θ represent q and define the problem as selecting the population with q value equal to $q_{[k]}$. Since the sample estimate of q_j is $\bar{y}_j = 1 - \bar{x}_j = 1 - (x_j/n_j) = (n_j - x_j)/n_j$, the procedure now is to assert that the best population is the one that produced the sample with the largest value of $\bar{y}_j$. The distance measure is now interpreted as $\delta = q_{[k]} - q_{[k-1]}$, and the same tables can be used in the manner described in Section 4.3.

An equivalent possibility is to interchange the labels *success* and *failure*. (These labels are completely arbitrary in any case.) This has the effect of replacing $p_{[k]}$ by $(1 - p_{[1]})$, $p_{[k-1]}$ by $(1 - p_{[2]})$, and so on, and in general $p_{[j]}$ by $(1 - p_{[k-j+1]})$ for $j = 1, 2, \ldots, k$. Under this transformation the distance measure $p_{[k]} - p_{[k-1]}$ for the largest p value problem is replaced by $\delta = (1 - p_{[1]}) - (1 - p_{[2]}) = p_{[2]} - p_{[1]}$, which is the corresponding distance measure for selecting the population with the smallest p value. The determination of sample size and calculation of the operating characteristic curve remain unchanged under this transformation.

4.5   ALTERNATIVE FORMULATION: DETERMINATION OF SAMPLE SIZE USING SOME INFORMATION ABOUT THE PARAMETERS

In the formulation of the binomial selection procedures so far in this chapter (as in the original formulation), the least favorable configuration as given in (4.3.2) has the unknown p values centered about .5 and with difference δ^*. If the true configuration is far removed from this configuration, then the method of Section 4.3.1 for determination of sample size indicates a much larger sample size than is actually required. If some quantifiable information is available about some of the true p values, and therefore about the true configuration, it is quite important to take this information into account and have an alternative formulation, since such procedures can result in a large saving in the sample size required to achieve the same efficiency of the selection rule. We illustrate this situation and the alternative formulation by the following actual example.

Example 4.5.   Ski accidents are usually sprains or fractures, which require medical attention. When a skier falls the ski binding is supposed to separate from the ski. Failure of the binding to release is one of the major causes of ski accidents. A consulting service has been asked to determine which one of two popular types of ski bindings is the safest one to use, that is, the one for which the probability of failure is the smallest.

The plan is to obtain samples of n bindings of each type and give them to equipment rental agencies at ski resorts. These agencies have agreed to rent the bindings and to collect data on the accident rate. This rate is measured by the number of accidents per skier-day, that is, the total number of accidents divided by the number of "skier-days" (the number of ski days multiplied by the average number of skiers per ski day), for each binding given to these agencies. (Rental agencies keep records of accidents because of the possibility of legal suits for equipment failure.)

The particular problem now is to determine the sample size n that is needed so that the probability of a correct selection between these $k = 2$ types of bindings is at least $P^* = .90$ when $\delta \geqslant \delta^*$ for some realistic specified value δ^*. From extensive records of insurance companies it is known that the average rate of ski accidents is approximately .006, usually in the interval from .001 to .010. Thus a realistic value of δ^* could range only between .001 and .010, and the corresponding value of n for $P^* = .90$, determined by the method of Section 4.3.1 under the least favorable configuration, will be so large that for practical purposes the experiment is impossible. To make this point more clearly we have computed the n values required under the (unrestricted) least favorable configuration in

**Table 4.5.1   Common n value required with the
unrestricted configuration (4.3.2)**

δ^* \ P^*	.75	.90	.99
.001	227,468	821,184	1,352,765
.010	2,275	8,211	13,527

(4.3.2) for $\delta^* = .001$ and $\delta^* = .010$. These very large results are shown in Table 4.5.1.

The fact that δ^* ranges between .001 and .010 in this example means that $p_{[1]}$ and $p_{[2]}$ might be pairs such as (.001, .004), (.002, .004), and so on, but they cannot be very much more discrepant. As a result, the least favorable configuration, where the p values are centered about .5, cannot be even close to the true configuration, and we would expect that the n value determined with the least favorable configuration would be highly conservative.

Now suppose we make use of the information available. Assuming that $p_{[1]}$ and $p_{[2]}$ are both restricted to the interval [.001, .010], it would make sense to restrict the least favorable configuration to a comparable range. The least favorable configuration in this alternative formulation takes the form

$$p_{[2]} = .010, \quad p_{[1]} = .010 - \delta^*. \tag{4.5.1}$$

In other words, we are now restricted to the appropriate range of p values less than .010, rather than allowing the p values to be near .5, where the previous (unrestricted) least favorable configuration is located. The common sample size for $k = 2$ binomial populations for this restricted configuration in (4.5.1) can be computed for specified pairs (δ^*, P^*). The results are shown in Table 4.5.2.

**Table 4.5.2   Common n value required with the
restricted configuration in (4.5.1)**

δ^* \ P^*	.75	.90	.99
.001	8,562	30,908	50,916
.010	45	163	268

Comparing the results in Table 4.5.2 with the corresponding results in Table 4.5.1, we see that for $\delta^* = .001$ the n values differ by a factor of about 25, whereas for $\delta^* = .010$ they differ by a factor of about 50. This shows how very important it is to make use of any information available about the true p values.

To complete this example, suppose that we give 31 ski bindings of each of the two types to each of 10 rental agencies. Since there are approximately 100 ski days per season, this plan generates $n = (10)(31)(100) = 31,000$ observations on each type of binding in one season. The results given in Table 4.5.2 indicate that for $n = 31,000$, the probability of a correct selection is approximately (slightly over) .90 whenever $p_{[2]}$ is at most .010 and

$$p_{[2]} - p_{[1]} \geqslant \delta^* = .001.$$

4.6   SAMPLE SIZES UNEQUAL

In many practical problems it is not possible or not practical to have all the samples of the same size. The selection rule R for the binomial problem with unequal sample sizes was given in Section 4.2 and illustrated by Example 4.2.2. The rule is the same as for equal sample sizes when stated in terms of the $\bar{x}$ values. However, the analytical methods of Sections 4.3 and 4.5 apply only to the binomial problem when all the sample sizes are equal. In particular, the tables and figures in Appendix E and Appendix F are no longer applicable with unequal n_j.

An approximate solution to the problem can be obtained by computing a certain generalized average sample size n_0 and proceeding as if each sample size were equal to n_0. The average we use for computing n_0 is exactly the same as in the normal means problem (Section 2.4 of Chapter 2), that is, the *square-mean-root* (SMR) formula

$$n_0 = \left(\frac{\sqrt{n_1} + \sqrt{n_2} + \cdots + \sqrt{n_k}}{k} \right)^2. \qquad (4.6.1)$$

Using the data of Example 4.2.2 given in Table 4.2.2, we obtain

$$n_0 = \left(\frac{\sqrt{393} + \sqrt{1189} + \sqrt{425}}{3} \right)^2 = (24.97)^2 = 623.5.$$

(This result is smaller than the arithmetic mean, 669, and larger than the geometric mean, 583.4, as is always the case.) If we round 623.5 downward, we get the integer value $n_0 = 623$.

REMARK.   In general, if we are attempting to determine a common sample size for a specified pair of values (δ^*, P^*), as in Section 4.3.1, for example, we would round upward in order to obtain a conservative result. However, when we are averaging unequal values of n in order to determine an approximation to the confidence level or the probability of a correct selection, then rounding downward gives the conservative result, since P^* is an increasing function of n for fixed k and δ^*. Of course, it is not necessary to round n_0 to an integer for the calculations involving n_0. □

With this value n_0 in place of n, the analysis explained in Section 4.3 is carried out exactly as before, using the tables and figures in Appendix E and Appendix F. For example, consider Example 4.2.2 again where $k=3$, $n_0 = 623$, and suppose that we want to make the confidence statement $p_{[k]} - p_s \leqslant .05$. From Appendix E, with $k=3$, $n_0 = 623$, $\delta^* = .05$, we find the approximate confidence level is between .90 and .95. If we interpolate to get a more precise answer, the result is $CL(.05|R) = .933$. This same result can be interpreted as the approximate probability of a correct selection under the least favorable configuration with $\delta^* = .05$.

It is worth noting that the methodology of this section, with n_0 computed from (4.6.1), is consistent with the large sample theory and gives the best approximation when the values of n_j are not too spread out and no n_j is very small. Large sample approximations for the error associated with using (4.6.1) can be found, but these methods are omitted here since they are quite complicated.

4.7   LARGE SAMPLE APPROXIMATION

Although tables and figures are available for the solution to the binomial problem when the sample sizes are equal and not too large and when the difference measure is $\delta = p_{[k]} - p_{[k-1]}$, an approximate solution for larger sample sizes is useful. This approximate solution is based on normal theory and a variance-stabilizing transformation. The binomial problem is then reduced to a normal means problem with common known variance, and the tables in Appendix A can be used in the way explained in Section 2.3 of Chapter 2. With this approximation we frequently can avoid interpolation entirely, or at least can avoid the need for using special tables for the binomial problem. The approximate large sample solution is especially useful when we need to find n for some values of k, δ^*, and P^* that are not captions in the exact binomial tables (provided the resulting value of n is large), or when we want to see how the common sample size n varies as a function of δ^* (for δ^* close to zero) or as a function of P^* (for P^* close to one), or even when we want to see how n varies for a limited number of different values of k, provided the n values are all large. (The varying of k may be of interest in trying to find an efficient or optimal design for the experiment.)

Suppose we want to find the common sample size n required in order that the probability of a correct selection is at least P^* whenever $\delta \geqslant \delta^*$. The approximate value of n is estimated from one (or both) of the following formulas:

$$n = (1 - \delta^{*2})\left(\frac{\tau_t}{2\delta^*}\right)^2, \tag{4.7.1}$$

$$n = \left(\frac{\tau_t}{2\delta^*}\right)^2. \tag{4.7.2}$$

The value of τ_t in these formulas is the entry in Table A.1 in Appendix A that corresponds to k and the specified value of P^*. Note that the approximation given in (4.7.2) is a simplification of (4.7.1), based on the fact that δ^* is usually small enough so that the factor $(1 - \delta^{*2})$ is close to one and therefore has little effect in (4.7.1). Empirical studies have indicated that in many cases (4.7.1) and (4.7.2) give lower and upper bounds, respectively, on the exact value of n; as a result, it may be useful (at least as a check on arithmetic) to calculate both of these approximate values of n.

To illustrate the calculation of these approximations, suppose that for $k = 4$ populations the specified pair is $\delta^* = .05$ and $P^* = .90$. From Table A.1, the corresponding value of τ_t is 2.4516. The approximation in (4.7.1) yields

$$n = (1 - .0025)\left(\frac{2.4516}{.1000}\right)^2 = 599.53,$$

and (4.7.2) gives

$$n = \left(\frac{2.4516}{.1000}\right)^2 = 601.03.$$

These values round upward to 600 and 602, respectively. The exact sample size required is found from the table for $k = 4$ in Appendix E to be 601. Hence in this case the two approximations do bound the exact solution, and the error in either approximation is extremely small, namely, about 0.0017, or one-sixth of 1%.

4.8   NOTES, REFERENCES, AND REMARKS

Since the parameter p in the binomial distribution is not a pure location or scale parameter, there is some problem in finding the least favorable configuration. It is not enough to set $p_{[k]} - p_{[k-1]} = \delta^*$, where $p_{[k]} = \max p_j$ ($j = 1, 2, \ldots, k$) for the least favorable configuration; we must set $p_{[k]} = .5 + .5\delta^*$ and $p_{[1]} = \cdots = p_{[k-1]} = .5 - .5\delta^*$, as indicated in (4.3.2). The minimization of the probability of a correct selection can be regarded as a two-step process where we set $p_{[k]} - p_{[k-1]} = \delta^*$ in the first step and then minimize with respect to $p_{[k]}$. Thus the binomial case gives a preview of the analytic difficulties of finding the LF configuration in a general context. This work is based on the paper of Sobel and Huyett (1957) and is well documented with proofs, tables, and graphs, some of which appear here as Appendixes E and F. The calculation of the operating characteristic curve in Section 4.3.2 is new. The alternative formulation in Section 4.5 is also based on Sobel and Huyett (1957), but the impetus to include this here came from the ski accident problem, Example 4.5.1, which is based on an actual application of ranking and selection to a problem that was posed to one of the authors.

The square-mean-root (SMR) formula in (4.6.1) as applied in ranking and selection is new with this book. The case of unequal sample sizes was considered in connection with selecting a subset containing the best of k binomial populations in Gupta and Sobel (1960), but this problem appears here in Section 13.2.2 under "Subset Selection Procedures."

Many subsequent papers have used the tables of the Sobel and Huyett (1957) paper as a target and pointed out that considerable improvement is possible by utilizing some prior knowledge about the range of the p values or by using a two-stage or a fully sequential procedure. By putting all the p values near .5, the number of observations needed to satisfy a P^* condition is maximized. There are different ways of avoiding this difficulty. The alternative formulation in Section 4.5 is one approach; here we brought in a priori information and showed that it is compatible with the methods of ranking and selection. Other possibilities are to use a two-stage or a sequential procedure. If there is no a priori information and if we insist on a fixed sample size procedure, the tables and figures of Appendixes E and F give the best results that are possible under the indifference zone approach.

PROBLEMS

4.1 An agricultural experiment station needs to determine the depth of planting to recommend for a newly developed species of pine seedlings. They plan to plant a number n of 4 in. seedlings at four different depths (.5 in. apart) under comparable conditions in the winter and count the number of survivors in the fall. The present problem is to determine how many seedlings should be planted at each depth such that they can state their recommendation as follows: With confidence level $P^* = .95$, the selected depth of planting gives a survival probability that is at least $p_{[4]} - .10$, where $p_{[4]}$ is the depth that gives the true largest probability of survival. Find the number of seedlings needed if we "*know*" from past empirical studies that $p_{[4]} \leqslant .30$. (We take this to be an empirically established fact.)

4.2 Birth control is an essential feature of any humane program to control population expansion. International Conference on Family Planning Programs (1966) reported a study to compare the effectiveness of 14 different contraceptive methods. The research calculated the effectiveness of each method on the basis of 100 woman-years (the number of women per hundred who can expect to become pregnant after using that method for one year). These measures are the actual failure rates, irrespective of whether the failure was due to the fact that the method was ineffective or the method was used carelessly. The observed failure rate is 80 for women using no contraceptive; this means that 80% can expect to be pregnant by the end of a year. Consider now a similar experiment to compare the 10 most popular methods of birth control using samples of 200 women for each method. The data on percent failure rate are the same as the "high" results reported at the Conference. (The "high" results may be assumed to represent those women who had little medical supervision and were probably careless and not highly motivated.)

(a) Select the one most effective method, that is, the one with the lowest failure rate.

(b) Calculate the operating characteristic curve.

(c) Considering only the first eight methods listed (i.e., eliminating the pill and the IUD), select the best method.

(d) Calculate the operating characteristic curve for the selection procedure in (c).

Failure rates of contraceptive methods

	Failure rate	Sample size
Aerosol foam	32.0	200
Foam tablets	43.0	200
Jelly or cream	38.0	200
Douche	41.0	200
Diaphragm and jelly	35.0	200
Condom	28.0	200
Coitus interruptus	38.0	200
Rhythm	38.0	200
Steroid contraception (the "pill")	2.7	200
Intrauterine contraception (Lippes loop)	2.4	200

4.3 The graduate school at a major university has seven major divisions, Law (L), Medicine (M), Business Administration (BA), Engineering (EN), Arts and Sciences (AS), Education (ED), and Social Work (SW). All divisions receive many applications for graduate study. The following data give the number of applicants and the number of admissions by division for a sample of 600 applicants in 1974–75.

(a) Which division has the largest proportion of admissions?

(b) Calculate the operating characteristic curve.

Division	L	M	BA	EN	AS	ED	SW
Number of admissions	12	7	66	5	85	102	7
Number of applicants	85	74	102	49	122	136	32

4.4 In field tests of mine fuses, a sample of n fuses of each of five types is to be buried and run over by tanks. A "hit" occurs if the fuse goes off, and a "no hit" occurs otherwise. For each type of fuse the number of hits will be recorded. The goal is to select the type of fuse with the largest probability of a hit. How many fuses of each type should be tested in order to satisfy the requirement $\delta^* = .10$ when $P^* = .90$? What is the resulting confidence statement?

4.5 *Science News* (1974) reported an experiment at Duke University on sensitivity of taste buds of three groups of people—obese persons, normal weight persons, and elderly persons. Each member of each group was blindfolded and given strained bananas to taste. The percentages of correct identifications were

Obese persons	69%
Normal weight persons	41%
Elderly persons	24%

The medical psychologist performing the experiment claims that the low percentage for the elderly group is consistent with the theory that front taste buds, those sensitive to sweet and salty tastes, atrophy before those in back, which are sensitive to bitter and sour tastes. Suppose that the three subject groups in this experiment were random samples, each of size 500.

(a) With confidence level .90, make a confidence statement about the selection of the group with the highest taste sensitivity.

(b) Calculate the operating characteristic curve for the selection procedure.

4.6 "Don't-know" or "No-opinion" responses to items on a questionnaire are a problem in any experiment, especially in attitudinal studies. Berger and Sullivan (1970) reported a study designed to determine whether different methods of administering a questionnaire resulted in different numbers of "Don't-know" responses. The three methods under consideration were those commonly used to collect attitudinal data, namely, (1) face-to-face interview, (2) telephone interview, and (3) group (classroom) administration where responses were self-administered and anonymous. The experimenter drew a random sample of 150 college students and randomly assigned them into three groups of equal size. All the members of each group were asked to complete the same questionnaire, but a different method of administration was used for the three samples. The questionnaire consisted of 25 statements about academic policies of a university. The respondents were asked to describe their attitude toward each statement by checking one of these five responses—"Strongly agree," "Agree," "Disagree," "Strongly disagree," "Don't know."

Consider a similar experiment where the goal is to determine which of these three methods of administration produces the smallest proportion of an "excessive" number of "Don't-know" responses, where an excessive number is defined as a "don't-know" answer to 10 or more of the 25 statements.

(a) If you were designing such an experiment, how large a sample would you take? What factors did you consider in making this choice of sample size?

(b) Suppose that a sample of 180 students is randomly divided into three groups of equal size and the following hypothetical data represent the number of excessive "Don't-know" responses for each method of administration. Select the method that produces the smallest probability of excessive "don't-know" response and make a confidence statement at level .950.

Method	Sample size	Number of excessive "don't-knows"
1	60	10
2	60	20
3	60	16

CHAPTER 5

Selecting the One Normal Population with the Smallest Variance

5.1 INTRODUCTION

In this chapter we develop methods appropriate for selecting the one population with the smallest variance from a set of k independent normal populations. Hence the distribution model assumed in this chapter is the same as that in (2.1.1) of Chapter 2, or

$$N\left(\mu_1, \sigma_1^2\right), N\left(\mu_2, \sigma_2^2\right), \ldots, N\left(\mu_k, \sigma_k^2\right).$$

The values of $\mu_1, \mu_2, \ldots, \mu_k$ are assumed to be either all known or all unknown, and parameters of main interest are the σ^2 values. Using the usual ordering notation, the ordered σ^2 values are

$$\sigma_{[1]}^2 \leqslant \sigma_{[2]}^2 \leqslant \cdots \leqslant \sigma_{[k]}^2. \tag{5.1.1}$$

The goal is to select the one population that has the smallest value of σ^2, namely $\sigma_{[1]}^2$.

The reason we are concerned with the parameter σ^2 is that, in a single population, σ^2 represents the spread about the mean μ. If we have k populations, each of which satisfies a specification for a mean, then in many applications we prefer the one with the smallest spread.

Since the standard deviation σ is the positive square root of σ^2, the σ values occur with the same ordering as the σ^2 values; thus, in particular, if $\sigma_{[1]}^2 = \sigma_j^2$ for some j, then it is also true that $\sigma_{[1]} = \sigma_j$ for that same j. In other words, the ordering in (5.1.1) is equivalent to the ordering

$$\sigma_{[1]} \leqslant \sigma_{[2]} \leqslant \cdots \leqslant \sigma_{[k]}. \tag{5.1.2}$$

As a result, the goal for this model can also be stated as selecting the population with the smallest standard deviation $\sigma_{[1]}$.

As mentioned in Chapter 2, the normal distribution model is basic to many problems in both theoretical and applied statistics. When comparing the variances or standard deviations of k normal populations, the traditional concern is a test of homogeneity of the variances, that is, the null hypothesis

$$H_0: \quad \sigma_1^2 = \sigma_2^2 = \cdots = \sigma_k^2.$$

Bartlett's, Cochran's, or Hartley's test may be applied to this problem. Homogeneity of the variances is also of interest as a preliminary hypothesis when the ultimate hypothesis is homogeneity among means, because the classical procedure for this latter hypothesis is completely valid only under the assumption of homogeneity of variances. In fact, the test for means is known to be not very robust against violation of this assumption about variances. Thus in the classical procedure we generally would like to conclude that the variances are all equal so that we can proceed with the test for means. On the other hand, in the ranking and selection procedures of this chapter we assume at the outset that the variances are generally not all equal. Furthermore, our principal interest at present is only in variances, and we do not assume that this is a study preliminary to a later ranking of means. Hence the point of view is quite different here and we can expect to use different methodology.

Although the μ values are not of interest relative to our goal in this chapter, we need to distinguish the following two cases:

1. The values of $\mu_1, \mu_2, \ldots, \mu_k$ are all known; they may be common or arbitrary.

2. The values of $\mu_1, \mu_2, \ldots, \mu_k$ are all unknown; they may be assumed to be common or assumed to be arbitrary.

The ranking and selection procedures are somewhat different for these two cases. Both cases are covered in this chapter, but for case 2 the main interest is in the situation where $\mu_1, \mu_2, \ldots, \mu_k$ are all unknown and arbitrary.

Before explaining the statistical procedures for this goal and the two cases of this distribution model, we give a few illustrations of applications.

Example 5.1.1 (Thermostat Precision). Laboratory experiments are frequently carried out under controlled conditions. For example, in experiments designed to study growth of bacteria, it is very important that the test chamber be kept at a uniform temperature. The temperature is, of course, controlled by a thermostat. The heating or cooling unit is activated when the reading on the thermostat unit does not agree with the setting of the thermostat. As a result, the quality of a thermostat is frequently judged by the amount of variation between the actual temperature and the reading on the thermometer of the thermostat. The following experiment might be conducted to compare the quality of three different brands of thermostats. A test chamber is designed so that the temperature is controlled at 72°F uniformly throughout the chamber. A random sample of n thermostats of each of the three brands is placed in this chamber and the readings are taken from the thermometer of each thermostat. The goal is to select the best brand of thermostat, that is, the brand that has the smallest variance in readings. In this example the mean temperatures are all known to be equal to 72°F.

Example 5.1.2 (Laboratory Analysis). The precision of a method of chemical analysis is frequently defined as the reproducibility of results, since the analysis is seldom repeated for any particular specimen. Suppose that six different types of laboratory "kits," labeled $1, 2, \ldots, 6$, are available for measuring serum cholesterol in blood. A sample of blood is taken from each of six different persons. The sample from each person is split, that is, divided into n different smaller blood samples. Type 1 kit is used repeatedly on these "split samples" from the first person, type 2 kit on the second person, and so forth. In each case the level of serum cholesterol is measured. The goal of interest is to identify the kit type that has the smallest variation in measurements, that is, the smallest variance. Since six different persons are used as subjects, the mean measurement of serum cholesterol generally is different for each kit. If these six means are known, case 1 given earlier applies; if unknown, case 2 applies.

Example 5.1.3 (Drugs). Five supply houses are bidding for a contract to produce a certain drug. The common mean "strength" of the drug is specified chemically and all five suppliers meet this specification. Hence the decision about awarding the contract will be based on the magnitude of the variation of the strength around the specified mean for each supplier. The award will be made to the supply house producing the drug with the smallest variability in the strength of the final product, that is, the one with the smallest variance. Since the mean strength of the drug is known here, case 1 applies.

The selection rule for the goal of selecting the population with the smallest variance (or standard deviation) is presented in Section 5.2 for both cases 1 and 2. The analytical aspects of the problem are covered in Section 5.3 for equal sample sizes and in Section 5.4 for unequal sample sizes.

In formulating the problem for this model, we must first define a distance measure for the difference between the populations with parameters $\sigma_{[1]}$ and $\sigma_{[2]}$. One possibility is to use the ratio of the second smallest σ value to the smallest σ value; that is,

$$\delta = \frac{\sigma_{[2]}}{\sigma_{[1]}}. \tag{5.1.3}$$

In view of the ordering in (5.1.2), it is always true that $\delta \geq 1$, and $\delta = 1$ if the two smallest standard deviations are equal. As a result, the indifference zone must be defined as $\delta < \delta^*$ and the preference zone as $\delta \geq \delta^*$, for some $\delta^* > 1$. In other words, no separation between $\sigma_{[1]}$ and $\sigma_{[2]}$ corresponds to $\delta = 1$, and the numerical value of the degree of separation must be judged accordingly.

Recall that in Chapters 2, 3, and 4, the respective distance measures were equal to zero when the two extreme parameters were equal. To make this model analogous we could transform the δ in (5.1.3) either by subtracting one or by taking the logarithms. Either transformation would make the resulting distance measure equal to zero when $\sigma_{[1]} = \sigma_{[2]}$. However, when we subtract one from the distance measure in (5.1.3) we get $(\sigma_{[2]} - \sigma_{[1]})/\sigma_{[1]}$, which is a relative difference, specifically the difference relative to the size of $\sigma_{[1]}$. This does not happen with the logarithmic transformation, since (5.1.3) becomes $\log \sigma_{[2]} - \log \sigma_{[1]}$. The log transformation may be useful for obtaining large sample or asymptotic results, but it complicates the theory for exact (i.e., small sample) results.

Further, the ratio in (5.1.3) is a natural measure of distance because its magnitude is not affected by changes in scale. In the context of an invariant selection procedure, this means that if each observation in each sample is multiplied by the same constant, the problem and its solution remain the same. As a result, if one investigator makes measurements in feet while another uses meters, both arrive at the same selection and the same analysis.

REMARK. As we noted earlier, selecting the population with the smallest standard deviation is equivalent to selecting the population with the smallest variance. This equivalence obviously assumes that if the threshold value of δ in (5.1.3) is taken as δ^*, then the corresponding symbol that should be used to represent the threshold value for the ratio of the variances $\sigma_{[2]}^2 / \sigma_{[1]}^2$ is $(\delta^*)^2$. However, in the existing statistical literature the symbol δ^* is used to denote the ratio of the variances, rather than the standard deviations. The reader is cautioned to remember this difference in notation if he reviews the literature or checks with other sources on this problem. Since the standard deviations are in the units of the original data, we feel that it is easier for an investigator to specify the threshold value δ^* in the same units rather than in square units, which are much less intuitive and not generally used. Hence we believe that this deviation from tradition will improve the ease of use of the method. $\square$

Because δ in (5.1.3) ranges from 1 to infinity, it is sometimes preferable to use the reciprocal of δ as the distance measure. This reciprocal is denoted by Δ, and defined explicitly by

$$\Delta = \frac{\sigma_{[1]}}{\sigma_{[2]}} = \frac{1}{\delta}. \tag{5.1.4}$$

Thus if $\delta = 1.20$, then $\Delta = 0.83$. The slight advantage of Δ over δ is that the range of Δ is bounded in both directions, namely, $0 \leqslant \Delta \leqslant 1$. Of course, now the preference zone must be defined as $\Delta \leqslant \Delta^*$ and the indifference zone as $\Delta > \Delta^*$ for some $\Delta^* < 1$. Since it is a simple matter to change from δ to Δ or vice versa, we could use either one. However, the tables in this book are given in terms of Δ.

The generalized least favorable configuration in terms of the true value Δ is

$$\Delta = \frac{\sigma_{[1]}}{\sigma_{[2]}} \quad \text{and} \quad \sigma_{[2]} = \sigma_{[3]} = \cdots = \sigma_{[k]}, \tag{5.1.5}$$

or, in terms of the true value δ, it is

$$\delta = \frac{\sigma_{[2]}}{\sigma_{[1]}} \quad \text{and} \quad \sigma_{[2]} = \sigma_{[3]} = \cdots = \sigma_{[k]}. \tag{5.1.6}$$

The least favorable configuration for a specified Δ^* is the same as (5.1.5) with $\Delta = \Delta^*$, and for a specified δ^* is (5.1.6) with $\delta = \delta^*$.

In this chapter we define a procedure for selecting the normal population with parameter value $\sigma_{[1]}$ in such a way that the probability of a correct selection is at least P^* whenever $\Delta \leqslant \Delta^*$, where Δ^* and P^* are prespecified numbers such that $0 < \Delta^* < 1$ and $1/k < P^* < 1$. Note that the condition $\Delta \leqslant \Delta^*$ in (5.1.5) is equivalent to the condition $\delta \geqslant \delta^* = 1/\Delta^*$ in (5.1.6).

REMARK. The problem of selecting the population with the largest variance (or standard deviation) is not included in this chapter, partly because the exact (small sample) theory is not equivalent to the theory for selecting the population with the smallest variance and hence different tables are needed. The most important applications of the problem of selecting the population with the largest variance (scale parameter) lie in the area of life testing and reliability, where the usual model is the gamma distribution instead of the normal distribution. These topics are discussed in Chapter 14. $\square$

5.2 SELECTION RULE

The procedure R used to select the population with the smallest variance is quite simple to carry out, but it differs depending on whether the means are all known (case 1) or all unknown (case 2) because the statistics used to estimate the population variances are different. In particular, for any population, in case 1 the appropriate estimate of σ^2 is a constant times the sum of squares of the deviations of the observations around their true mean, whereas in case 2 the appropriate estimate of σ^2 is a constant times the sum of squares of the deviations of the observations around their sample mean. Although these two cases are quite distinct, the selection rule and analytical approaches to the problem are similar. Since there is little danger of confusion between the cases, we discuss them together in a parallel manner.

The notation used here is the same as in Chapter 2. Then x_{ij} denotes the ith observation in the jth sample of size n_j drawn from the jth population with distribution $N(\mu_j, \sigma_j^2)$. If the true means are known, the estimate of σ_j^2 based on n_j observations is denoted by V_j' and defined by

$$V_j' = \frac{\sum_{i=1}^{n_j} (x_{ij} - \mu_j)^2}{n_j} = \frac{T_j'}{n_j}. \tag{5.2.1}$$

Thus T_j' is the sum of squares of the deviations of the observations in the jth sample around their respective true mean. If the μ_j are all unknown, each μ_j is estimated as usual by the corresponding sample mean $\bar{x}_j$ for $j = 1, 2, \ldots, k$ [see (2.2.2) of Chapter 2]. The sample estimate of σ_j^2 is then denoted[†] by V_j and defined by

$$V_j = \frac{\sum_{i=1}^{n_j} (x_{ij} - \bar{x}_j)^2}{n_j - 1} = \frac{T_j}{n_j - 1}. \tag{5.2.2}$$

Thus T_j is the sum of squares of the deviations of the observations in the

---

[†]In this chapter we use the notation V_j for the sample estimate of σ_j^2, instead of the more traditional notation s_j^2, to avoid the necessity of having an exponent.

jth sample around the jth sample mean. In (5.2.2) we divide by $n_j - 1$ rather than n_j, since this makes V_j an unbiased estimator of σ_j^2 without affecting any large sample properties.

We compute V_j' or V_j for each sample and order them in the case of known means as

$$V_{[1]}' \leqslant V_{[2]}' \leqslant \cdots \leqslant V_{[k]}',$$

or respectively in the case of unknown means as

$$V_{[1]} \leqslant V_{[2]} \leqslant \cdots \leqslant V_{[k]}.$$

The selection rule R (when there are no ties for first place) in case 1 is to assert that the sample producing $V_{[1]}'$ is the one with the smallest population variance $\sigma_{[1]}^2$. Similarly, in case 2 we assert that the sample giving rise to $V_{[1]}$ comes from the population with variance $\sigma_{[1]}^2$. Since the estimates of the standard deviations, $\sqrt{V_j'}$ and $\sqrt{V_j}$ (in cases 1 and 2, respectively) have the same ordering as the V_j' and V_j (in cases 1 and 2, respectively), the respective rules can also be stated in terms of the $\sqrt{V_j'}$ and $\sqrt{V_j}$, respectively.

REMARK. When the sample sizes are all equal, that is, when $n_1 = n_2 = \cdots = n_k$, each sample estimate from (5.2.1) has the same denominator; the same applies to (5.2.2). Therefore the ordering of the V_j' or V_j is the same as the ordering of the T_j' or T_j, respectively. Then the selection rule R can be simplified to state that we order the T_j' for case 1 (T_j for case 2) and assert that the sample giving rise to $T_{[1]}'$ for case 1 ($T_{[1]}$ for case 2) has the smallest population variance. □

Since the normal distribution is continuous, the probability of a tie for the first place is zero and hence we need not take this into account in the probability analysis. However, ties can occur among the V_j', V_j, T_j, or T_j', because of rounding or taking too few digits in the original data. If there are ties, one sample should be chosen at random, as described in Section 2.2 of Chapter 2.

We illustrate the selection rule for cases 1 and 2 in terms of Examples 5.2.1 and 5.2.2, respectively. In each case the data are artificial and the sample sizes are quite small in order that the arithmetic calculations can be followed easily.

Example 5.2.1 (Diamond Cutting). Three different diamond cutting machines are each set to turn out diamonds that weigh 0.500 karats. Each machine is functioning properly, so there is no reason to think that the true mean weight is not 0.500 karats in each case. Variation in weight of the cut diamonds is of considerable interest to the producer of cut stones because a stone that is significantly underweight produces dissatisfied customers whereas a stone that is significantly overweight is very costly. Hence it is desired to select the one cutting machine that produces the smallest variation in the weight of the stone. If this variation is measured by the variance in the weight of the cut diamonds, the goal is to select the cutting machine with the smallest variance. For this purpose random samples of diamonds cut by each machine are collected and the weight in karats is measured accurately. The sample sizes are chosen proportionately to the total number of diamonds cut by these machines over the past year, and they are unequal; we assume that further data are not presently available so we must deal with unequal sample sizes. Suppose the data are as follows:

Data for weight (in karats) of diamonds cut by three machines

Machine number		
1	2	3
0.505	0.465	0.491
0.452	0.515	0.536
0.446	0.553	0.562
0.484	0.481	0.435
0.504	0.492	0.468
0.550		0.505
		0.431

Since the population means are known to be $\mu_1 = \mu_2 = \mu_3 = 0.500$, we calculate the sum of squares for the jth sample using $T'_j = \sum_{i=1}^{n_j} (x_{ij} - 0.500)^2$. Here the sample sizes are not equal and hence the V'_j must also be calculated. For example, when $j = 1$, we have from (5.2.1)

$$V'_1 = \frac{(0.505 - 0.500)^2 + (0.452 - 0.500)^2 + \cdots + (0.550 - 0.500)^2}{6}$$

$$= \frac{0.008017}{6} = 0.001336,$$

$$\sqrt{V'_1} = 0.037.$$

The results for all three sets of sample data are as follows.

Machine j	1	2	3
True mean μ_j	0.500	0.500	0.500
Sample size n_j	6	5	7
Sample mean $\bar{x}_j$	0.490	0.501	0.490
T_j'	0.008017	0.004684	0.015256
V_j'	0.001336	0.000937	0.002179
$\sqrt{V_j'}$	0.037	0.031	0.047

Since machine 2 has the smallest value of V_j' (and correspondingly $\sqrt{V_j'}$), we assert that machine 2 produces the diamonds with the most uniform weight.

Note that in this example it was not necessary to calculate the sample means since the μ_j are all known. The sample means do not enter into the calculation of the variance estimates. Naturally it may be worthwhile to look at the sample means before carrying out the ranking and selection procedure, but we do not consider that as part of the procedure under discussion here.

Example 5.2.2 (Tread Wear). Tread wear of tires is evaluated by measuring the depth of penetration at fixed points in the groove of the tire. In practice, the inspector usually makes only a single measurement on any one tire, and the accuracy using different instruments is substantially the same but the variances may differ. Hence it is desired to identify the instrument with the smallest variance in measurement of tread wear. In order to compare the variances of three different measuring instruments, an inspector makes five independent measurements with each instrument, all at the same fixed point on a single tire. The data are as follows:

Data for measurement of tread wear

	Instrument	
1	2	3
11.0	11.2	10.3
11.4	11.6	10.5
11.2	11.0	10.7
10.8	11.5	10.9
10.1	12.2	11.1

Since all measurements are made at the same point on the tire, it would be reasonable to assume that the tread wear of the tire is common and does not depend on the instrument used. The mean tread wears associated with the different instruments are not known. Hence we first estimate each population mean by the corresponding sample mean and then calculate T_j from (5.2.2) since the sample sizes are the same. The procedure then is simply to order the T_j and select the instrument that produces the smallest value of T_j. The results and an illustrative calculation are as follows. For instrument 1:

$$\bar{x}_1 = \frac{11.0 + 11.4 + \cdots + 10.1}{5} = 10.9,$$

$$T_1 = (11.0 - 10.9)^2 + (11.4 - 10.9)^2 + \cdots + (10.1 - 10.9)^2$$

$$= 1.00.$$

Instrument j	1	2	3
Sample size n_j	5	5	5
Sample mean $\bar{x}_j$	10.9	11.5	10.7
T_j	1.00	0.84	0.40
V_j	0.25	0.21	0.10
$\sqrt{V_j}$	0.50	0.46	0.32

Since T_j (or equivalently V_j or $\sqrt{V_j}$) is smallest for $j = 3$, we assert that instrument 3 is the one with the smallest variance in measurement.

5.3   SAMPLE SIZES EQUAL

In this section we consider some analytical aspects of the problem of selecting the normal population with the smallest variance, but only for the case where the sample sizes are equal, that is, $n_1 = n_2 = \cdots = n_k = n$. We can give an exact solution for this model when n is common, whether the true means are known or unknown. An approximate solution that may be used with equal sample sizes is discussed in Section 5.4.

5.3.1   Determination of Sample Size

In order to determine the sample size for the problem of selecting the population with the smallest variance, the first step is to choose a pair of constants (Δ^*, P^*) where $1/k < P^* < 1$ and $0 < \Delta^* < 1$. Here Δ^* is the threshold value of $\Delta = \sigma_{[1]}/\sigma_{[2]}$, that is, the quantity separating the indifference and preference zones. This choice is interpreted as meaning that the probability of a correct selection is to be at least P^* whenever $\Delta = \sigma_{[1]}/\sigma_{[2]} \leqslant \Delta^*$, and the problem is to find the smallest common sample size n that ensures this property.

For a fixed number k of populations and a specified pair (Δ^*, P^*), the common value of n can be found directly (or interpolated) from the tables in Appendix G. For $k = 2, 3, \ldots, 10$ these tables provide values of a quantity ν, called the *degrees of freedom*, for $\Delta^* = 0.25, 0.40, 0.50, 0.60, 0.70, 0.80, 0.85, 0.90, 0.95$ and for $P^* = .750, .900, .950, .975,$ and $.990$. If the specified Δ^* and/or P^* is not listed specifically in these tables, the required ν can be found by the method of interpolation explained in Appendix G

Using the quantity ν found from these tables, the common sample size n is then determined in the following manner. If the k population means, $\mu_1, \mu_2, \ldots, \mu_k$, are all known, then $n = \nu$. If these means are all unknown, then the required common sample size n is one more than ν, that is, $n = \nu + 1$.

REMARK. The quantity ν in the table of Appendix G is called the *degrees of freedom* because a function of the statistics used to determine the selection, that is, V_j' and V_j in (5.2.1) and (5.2.2), respectively, follows the chi-square distribution with parameter ν, customarily called the degrees of freedom (even though the physical significance of this term is not always apparent). In the case where the true means are all known so that V_j' applies, we have $\nu = n$. If the true means are all unknown and μ_j is estimated by $\bar{x}_j$ in order to calculate V_j, then $\nu = n - 1$. $\square$

Example 5.3.1 (Calibration). Suppose we have four scales and each scale can be calibrated; that is, we can place a weight of known magnitude on the scale under carefully controlled conditions and set the pointer at zero. If observations on the same weight are repeated, say, at different times and/or at various places, and/or under different temperatures for each of the scales, we expect to find discrepancies from the true weight, which is now zero. The primary source of these discrepancies is assumed to lie in the changing (uncontrolled) physical environment of the scale and not in either the observer or the scale. Thus we could arrange to have the *same* observer make every observation and still have mutually independent observations. Further, the discrepancies are quite likely to be normally distributed, and the true mean discrepancy for each observation is equal to zero because each scale has been calibrated under controlled conditions.

The ultimate goal is to select the scale for which the variance of the discrepancies in weight is the smallest. But first we wish to determine the required number of observations on each scale for the probability of a correct selection to be at least $P^* = .75$ whenever $\Delta = \sigma_{[1]}/\sigma_{[2]} \leqslant \Delta^* = 0.80$. The entry in Appendix G for $k = 4$ with $P^* = .75$ and $\Delta^* = 0.80$ is $\nu = 31$. Since the true means are known in this example, the required sample size is then $n = \nu = 31$ observations on each scale. When the observations are taken, the selection procedure R is to compute V_j' from (5.2.1) using $\mu_j = 0$ for $j = 1, 2, 3, 4$ (or simply T_j' since the sample sizes are equal), order these values, and assert that the scale for which the sample value V_j' (or T_j') is the smallest has the smallest true variance.

Once n is determined by this procedure for the chosen pair (Δ^*, P^*), we can make a confidence statement about the true variance of the population selected. Suppose that σ_s is the standard deviation of the population selected as best from the data, that is, the population that produced the sample statistic $V_{[1]}'$ or $T_{[1]}'$. Of course, we know that $\sigma_{[1]} \leqslant \sigma_s$, or equivalently, that $\sigma_{[1]}/\sigma_s \leqslant 1$, but we would also like to have a lower confidence bound for the ratio $\sigma_{[1]}/\sigma_s$. With confidence level P^* we can say that the ratio satisfies the inequality $\sigma_{[1]}/\sigma_s \geqslant \Delta^*$, or equivalently, that the ratio lies in the interval

$$\Delta^* \leqslant \frac{\sigma_{[1]}}{\sigma_s} \leqslant 1. \tag{5.3.1}$$

This is the same as saying that the true smallest standard deviation $\sigma_{[1]}$ is included in the (closed) interval

$$S = \left[\Delta^* \sigma_s, \sigma_s \right]. \tag{5.3.2}$$

Interpretation of the confidence statement may be facilitated by expressing it in terms of the relative difference in the following way. Since $\sigma_{[1]}/\sigma_s \geqslant \Delta^*$ if and only if $\sigma_s - \sigma_{[1]} \leqslant \sigma_{[1]}[(1/\Delta^*) - 1]$, we can also say with confidence level P^* that the standard deviation σ_s of the population selected is at most

$$\left(\frac{1}{\Delta^*} - 1 \right) 100\% \tag{5.3.3}$$

greater than the smallest standard deviation $\sigma_{[1]}$. Note that all these statements concern the correctness of the selection made, and not the value of any standard deviation σ.

Thus in Example 5.3.1 if we take 31 observations on each scale, we can say with confidence level $P^* = .75$ that the interval $[0.80\sigma_s, \sigma_s]$ includes $\sigma_{[1]}$, or that σ_s is at most $[(1/0.80)-1]100\% = 25\%$ greater than the smallest standard deviation.

The fact that Δ^* and P^* can ultimately be used to make confidence statements about the selection may be useful as a basis for specifying the pair of constants (Δ^*, P^*) that are used to determine the common sample size n. The investigator should feel free to consult the tables before deciding on the (Δ^*, P^*) pair to specify.

Note that in Appendix G, for any fixed k and P^*, the number of degrees of freedom ν increases as Δ^* increases. In fact, as Δ^* approaches one, the required value of ν (and therefore n) increases dramatically. The reason for this is that when Δ^* is close to one, $\sigma_{[1]}$ is very close to $\sigma_{[2]}$. Thus, roughly speaking, we need a very large common sample size to identify the best population when the ratio is very close to one. Consequently, the specification of Δ^*, especially when it is close to one, should be made only after careful consideration of the meaning and consequences of the specified value. Further, for any fixed k and Δ^*, the number of degrees of freedom (and therefore the sample size) increases as P^* increases. Hence the specification of a very large value of P^* (say, greater than .95) should be made only when either a very high level of confidence is absolutely essential to the problem at hand or the cost of taking observations is negligible.

Example 5.3.2. In the situation described in Example 5.1.2 regarding the precision of six laboratory kits, suppose that we are still in the process of designing the experiment and that the true mean serum cholesterol is not known for any of the six persons from whom blood samples are to be taken. We want to determine how many analyses to make with each type of kit if we specify that the probability of correctly selecting the best kit is to be at least $P^* = .90$ whenever $\Delta = \sigma_{[1]}/\sigma_{[2]} \leqslant \Delta^* = 0.70$. From Appendix G with $k = 6$, $P^* = .90$, and $\Delta^* = 0.70$, we find $\nu = 32$. Since the true means are unknown here, we need $n = \nu + 1 = 33$ analyses per kit to satisfy this (Δ^*, P^*) requirement. To illustrate the behavior of ν with different values of Δ^* or P^*, suppose we consider P^* fixed at .90 and see how ν varies as a function of Δ^*. If $\Delta^* = 0.75$, then $\nu = 48$ and $n = 49$, whereas if $\Delta^* = 0.80$, we find $\nu = 77$ and $n = 78$. Thus when Δ^* is changed from 0.70 to 0.80, we need a total of $6(78 - 33) = 270$ additional observations. On the other hand, suppose we consider Δ^* fixed at 0.70 and see how ν varies as a function of P^* in this example with $k = 6$. If $P^* = .95$, we find $\nu = 42$ and $n = 43$, and if $P^* = .975$, we find $\nu = 53$ and $n = 54$.

In general, if the sample size determined by the specified pair (Δ^*, P^*) cannot be met, the investigator may consider some other arbitrary value of n (or ν), and

1. See what happens to the confidence level for the same value of Δ^*, and/or

2. See what happens to the width of the confidence interval on $\sigma_{[1]}/\sigma_s$ for the same value of P^*.

In the first case we use the tables in Appendix G to find the probability of a correct selection under the least favorable configuration, $P_{LF}\{CS|\Delta^*,\nu\}$. This value, of course, can also be interpreted as the confidence level $CL(\Delta^*|R)$ for the statement made about the selection from (5.3.1), (5.3.2), or (5.3.3). In the second case, where P^* is fixed, we use Appendix G to see what happens to the width of the confidence statement in (5.3.1), (5.3.2), or (5.3.3) at the same level.

To illustrate these calculations, we consider the situation of Example 5.3.1 again where $k=4$ and the true means are known. Suppose now that the n value of interest is $n=40$ so that $\nu=40$, and the investigator wants to make the confidence statement $\sigma_{[1]}/\sigma_s \geqslant 0.75$. By interpolation in the table in Appendix G for $k=4, \Delta^*=0.75, \nu=40$, we find the corresponding probability is .91. Hence $CL(0.75|R)=P_{LF}\{CS|0.75,40\}=.91$. In terms of the probability of a correct selection this says that for the least favorable configuration, which from (5.1.5) is

$$\sigma_{[1]}=0.75\sigma_{[2]} \quad \text{and} \quad \sigma_{[2]}=\sigma_{[3]}=\sigma_{[4]},$$

the probability of a correct selection is equal to .91, and it is *at least* .91 for *all* configurations in the preference zone, that is, whenever $\sigma_{[1]} \leqslant 0.75\sigma_{[2]}$.

On the other hand, if the question of interest is what confidence statement can be made at level $P^*=.75$ when $k=4, \nu=40$, we interpolate in Appendix G to find the corresponding Δ^* value as 0.82. Hence the statement is $\sigma_{[1]}/\sigma_s \geqslant 0.82$. The least favorable configuration then changes to

$$\sigma_{[1]}=0.82\sigma_{[2]} \quad \text{and} \quad \sigma_{[2]}=\sigma_{[3]}=\sigma_{[4]},$$

and thus the preference zone now includes values of $\sigma_{[1]}$ that are much closer to $\sigma_{[2]}$; that is, the size of the preference zone has increased to include values of $\sigma_{[1]}$ that are between $0.75\sigma_{[2]}$ and $0.82\sigma_{[2]}$.

REMARK.   In general, it is well known that larger sample sizes are needed for problems dealing with variances (or standard deviations) than for problems dealing with means. With larger sample sizes (and a fixed P^*), we can make more meaningful statements in the sense that the corresponding Δ^* is then closer to 1. Also, with Δ^* close to 1 the value of $[(1/\Delta^*)-1]$ is closer to zero, so that σ_s is closer to $\sigma_{[1]}$. For example, if $[(1/\Delta^*)-1]=\varepsilon>0$ (say 0.10), then we can assert that the σ_s for the selected population is at most $100\varepsilon\%$ (say at most 10%) larger than the smallest standard deviation. In other words, if the investigator has difficulty deciding on a value to specify for Δ^*, he may want to set a value for ε and solve for Δ^*. The explicit result is $\Delta^* = [1/(\varepsilon + 1)]$. For example, with $\varepsilon = 0.10$ we find $\Delta^* = 1/1.1$ or about 0.9. If this Δ^* requires too large an n value for the given k, the investigator must then make his specifications less stringent. We repeat again that the tables in Appendix G can be studied *before* taking any observations. □

5.3.2   Calculation of the Operating Characteristic Curve

We now assume that the common sample size n is fixed permanently (either as determined by the method of the previous subsection or by other considerations), and we are going to take k samples of size n, one sample from each of k normal populations. The selection rule R for identifying the population with standard deviation $\sigma_{[1]}$ is determined. However, now we do not specify either of the values Δ^* and P^*. Nevertheless we can still compute the probability of a correct selection by procedure R under the generalized least favorable configuration as a function of the true value Δ for the fixed ν; that is, $P_{GLF}\{CS|\Delta, \nu\}$. This function is the operating characteristic (OC) curve of the selection procedure for this model and this goal. (Obviously if $\Delta = \Delta^*$, then $P_{GLF}\{CS|\Delta, \nu\}$ reduces to $P_{LF}\{CS|\Delta^*, \nu\}$, calculated in the last subsection.)

The procedure is to take arbitrary values of $P = P_{GLF}\{CS|\Delta, \nu\}$ and interpolate in the tables in Appendix G to find the corresponding value Δ for the appropriate values of k and ν, where $\nu = n$ if the true means are known and $\nu = n - 1$ if they are unknown. Then these pairs $(\Delta, P_{GLF}\{CS|\Delta, \nu\})$ are plotted on a graph (with Δ on the horizontal axis) to obtain the operating characteristic curve. The curve always approaches the point $P = 1/k$ as Δ approaches one, and approaches $P = 1$ as Δ approaches zero. For any given k and ν (or n), the OC curve indicates how well the selection rule is operating for different values of Δ.

To illustrate the procedure, we return to Example 5.3.1 where we have four scales calibrated in such a way that the true mean for each scale is known to be equal to zero. Suppose that the investigator has no control over the number of observations per scale, or that the observations have already been taken, and the number is $n = 40$. Since the means are known here, we also have $\nu = 40$. Consider $P = .75$. We go to the table for $k = 4$ in Appendix G and interpolate using the entries for Δ on both sides of $\nu = 40$ when $P = .75$, namely, $\Delta = 0.80$, $\nu = 31$, and $\Delta = 0.85$, $\nu = 56$. The interpolated result is $\Delta = 0.82$. Repeating this kind of interpolation in Appendix G for other values of P when $k = 4$, we obtain the pairs given in Table 5.3.1.

Table 5.3.1 $P_{GLF}\{CS|\Delta, \nu\}$ **for** $k = 4$, $n = \nu = 40$
and selected Δ values

Δ	1.00	0.82	0.75	0.71	0.68	0.65
$P_{GLF}\{CS|\Delta, \nu\}$	.250	.750	.900	.950	.975	.990

These points are plotted in Figure 5.3.1 and connected by a smooth curve. This OC curve indicates how well the selection rule R is operating for each possible true value of Δ when $n = \nu = 40$ and $k = 4$, and we are in the generalized least favorable configuration. Let us interpret two of these pairs, say $(\Delta, P) = (0.82, .75)$ and $(0.75, .90)$. Since Δ is the true value of $\sigma_{[1]}/\sigma_{[2]}$ in the generalized least favorable configuration, $\sigma_{[1]}$ is much closer to $\sigma_{[2]}$ when $\Delta = 0.82$ than when $\Delta = 0.75$; hence the probability of a correct selection must be smaller for 0.82 than for 0.75, as it is.

As mentioned earlier, if $\Delta = \Delta^*$, the $P_{GLF}\{CS|\Delta, \nu\}$ reduces to $P_{LF}\{CS|\Delta^*, \nu\}$. (Interpretation of this quantity, and the corresponding confidence coefficient $CL(\Delta^*|R)$, was discussed in Section 5.3.1.) Therefore, if $\Delta = \Delta^*$, Figure 5.3.1 with Δ^* on the horizontal axis can be interpreted either as (1) the locus of all specified pairs (Δ^*, P^*) for which the

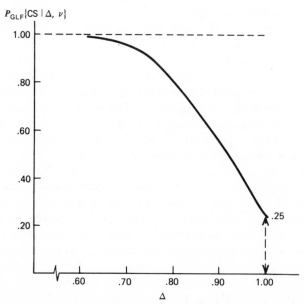

Figure 5.3.1 The probability of a correct selection as a function of Δ for $k = 4$ populations and $n = \nu = 40$.

probability requirement is satisfied when $n = \nu = 40$, or as (2) the locus of all pairs $(\Delta^*, P_{LF}\{CS|\Delta^*, \nu\})$ when $n = \nu = 40$.

5.3.3 Estimation of the True Probability of a Correct Selection

Up to this point in the chapter the observations have been used only to determine which population is selected. We now consider some data-analytic methods that make further use of the observations. We assume throughout that the sample size n is common, and that its value was determined by the method of Section 5.3.1 or by any other method. Nevertheless the selection rule R for identifying the normal population with the smallest variance is defined, and we wish to estimate the probability of a correct selection that is attained by this procedure.

This probability depends on the true configuration of standard deviations, $\sigma = (\sigma_{[1]}, \sigma_{[2]}, \ldots, \sigma_{[k]})$, through the following $(k-1)$ ratios between $\sigma_{[1]}$ and every other σ value,

$$\Delta_2 = \frac{\sigma_{[1]}}{\sigma_{[2]}}, \quad \Delta_3 = \frac{\sigma_{[1]}}{\sigma_{[3]}}, \quad \ldots, \quad \Delta_k = \frac{\sigma_{[1]}}{\sigma_{[k]}}.$$

These ratios, called the Δ values, are unknown since the σ values are unknown. However, we can estimate $\sigma_{[j]}$ by the jth ordered sample standard deviation and hence estimate the Δ values by the corresponding ratio of standard deviations. The sample standard deviations are computed from (5.2.1) or (5.2.2) for $j = 1, 2, \ldots, k$ as

$$s_j = \sqrt{V_j'} \quad \text{or} \quad s_j = \sqrt{V_j},$$

depending on whether the means are all known or all unknown, respectively. If $s_{[j]}$ denotes the jth ordered sample standard deviation for $j = 1, 2, \ldots, k$, the estimates of the Δ values are

$$\hat{\Delta}_2 = \frac{s_{[1]}}{s_{[2]}}, \quad \hat{\Delta}_3 = \frac{s_{[1]}}{s_{[3]}}, \quad \ldots, \quad \hat{\Delta}_k = \frac{s_{[1]}}{s_{[k]}},$$

or equivalently, for $j = 1, 2, \ldots, k$,

$$\hat{\Delta}_j = \sqrt{V_{[1]}' / V_{[j]}'} \quad \text{or} \quad \hat{\Delta}_j = \sqrt{V_{[1]}/ V_{[j]}}, \tag{5.3.4}$$

depending on whether the means are all known or all unknown respectively. Note that the $\hat{\Delta}$ values then satisfy the relationship $\hat{\Delta}_2 \geqslant \hat{\Delta}_3 \geqslant \cdots \geqslant \hat{\Delta}_k$.

We now show how these $\hat{\Delta}$ values can be used to obtain an approximate lower bound P_L and an approximate upper bound P_U on the true probability of a correct selection. The lower- and upper-bound approximations together provide an interval estimate of the true probability. We also show how to obtain a point estimate P_E of the true probability; this estimate usually lies within the interval estimate.

The approximate lower bound P_L can be computed as follows. We first interpolate in the table in Appendix G for $k=2$ with each $\hat{\Delta}_j$ separately to find the corresponding probability P_j of a correct selection between the jth ordered population and the best population, for each of $j=2,3,\ldots,k$. Then P_L is the product of these $(k-1)$ probabilities or

$$P_L = P_2 P_3 \cdots P_k.$$

(We use $k=2$ to determine each of the P_j since $\hat{\Delta}$ is computed from exactly two sample standard deviations.)

To find the approximate upper bound P_U, we first average the $\hat{\Delta}$ values by taking the geometric mean. This mean, denoted by Δ^{GM}, is defined as the $(k-1)$th root of the product of the $(k-1)$ values of $\hat{\Delta}$, or

$$\Delta^{GM} = \left(\hat{\Delta}_2 \hat{\Delta}_3 \cdots \hat{\Delta}_k\right)^{1/(k-1)}.$$

It is easier to first compute $\log \Delta^{GM}$, since it is the arithmetic mean of the $(k-1)$ logarithms of $\hat{\Delta}$ values or

$$\log \Delta^{GM} = \frac{\log \hat{\Delta}_2 + \log \hat{\Delta}_3 + \cdots + \log \hat{\Delta}_{k-1}}{k-1}. \tag{5.3.5}$$

Then Δ^{GM} is the antilogarithm of the result in (5.3.5). (The logarithms in (5.3.5) may be taken to any base, as long as the antilogarithm of the result is taken using the same base.) The approximate upper bound P_U is the probability given in the table in Appendix G for the original value of k and the value of Δ^{GM}.

The bounds P_L and P_U usually provide good lower and upper bounds for the true probability of a correct selection as long as the $\hat{\Delta}$ values are not too close to one.

The theoretical justification for these bounds is based on well-known inequalities, but rigorous proofs have been carried out only for the normal distribution when selecting means.

The estimators P_L and P_U together provide an interval estimate of the true probability of a correct selection, although the exact confidence level that can be used with this interval estimate has not been worked out. The interval may be very wide, and hence of little practical use. Therefore we also give a method of finding a point estimate P_E of the true probability of a correct selection. This method is in some sense a combination of the methods used to obtain P_L and P_U, and the result usually lies between P_L and P_U.

The first step in the method to find P_E is to group the $(k-1)$ values of $\hat{\Delta}$ from (5.3.4) into subgroups or clusters that are approximately equal in size and to compute the geometric mean of the $\hat{\Delta}$ values in each cluster. If there are m clusters, with $k_1-1, k_2-1, \ldots, k_m-1$ values of $\hat{\Delta}$ in the respective clusters, then the sum of these cluster sizes must be equal to $(k-1)$ since each $\hat{\Delta}_j$ can go into only one cluster and all the $\hat{\Delta}$ values must be used. In symbols this says that $\sum_{j=1}^{m} (k_j-1) = k-1$. Moreover, if any two $\hat{\Delta}$ values are in the same cluster, then any $\hat{\Delta}$ value between them must also be in that same cluster. Therefore the easiest way to form the separation into clusters is to partition the $\hat{\Delta}$ values in their ordered arrangement, $\hat{\Delta}_2 \geqslant \hat{\Delta}_3 \geqslant \cdots \geqslant \hat{\Delta}_k$. For example, the clusters and respective cluster means might be as follows.

$\hat{\Delta}$ values	$\hat{\Delta}_2, \ldots, \hat{\Delta}_{k_1}$	$\hat{\Delta}_{k_1+1}, \ldots, \hat{\Delta}_{k_1+k_2-1}$,	$\cdots$,	$\ldots, \hat{\Delta}_k$
Cluster	1	2	$\cdots$	m
Cluster geometric means	Δ_1^{GM}	Δ_2^{GM}	$\cdots$	Δ_m^{GM}

Note that Δ_j^{GM} denotes the geometric mean of the $\hat{\Delta}$ values in the jth cluster, and hence is the (k_j-1)th root of the product of the values of $\hat{\Delta}$ in the jth cluster; equivalently, Δ_j^{GM} is the antilogarithm of the arithmetic mean of the logarithms of the (k_j-1) values of $\hat{\Delta}$ in the jth cluster. Then the jth cluster gives rise to a probability value P_j that is found from the table in Appendix G for the computed value Δ_j^{GM} and k_j; of course, k_j is one greater than the number of $\hat{\Delta}$ values in the jth cluster. The value of P_j is found in the same manner for each of $j=1,2,\ldots,m$. Then P_E is found by multiplying these m values together; that is,

$$P_E = P_1 P_2 \cdots P_m.$$

Except for the restrictions mentioned earlier, there remains some arbitrariness in the value of m and where the partitions are inserted to form the clusters. The best result for P_E will be obtained if the $\hat{\Delta}$ values in each cluster are approximately equal in value. As a rule of thumb, we might choose the number of clusters m to be approximately $(k-1)/2$.

REMARK.   In finding P_U or P_E we are actually partitioning the values of the sample variances, $V_{[j]}$ (or $V'_{[j]}$), for $j = 2, 3, \ldots, k$ into clusters. However, since the $V_{[j]}$ (or $V'_{[j]}$) in each cluster are all to be divided into the same quantity $V_{[1]}$ to determine the $\hat{\Delta}$ values, a specific partition of the $\hat{\Delta}$ values determines a specific partition of the sample variances. Actually we are using the $\hat{\Delta}$ values in order to partition the sample variances into clusters.
□

Calculation of these estimates is illustrated in Example 5.3.3.

Example 5.3.3. Suppose we have $k = 5$ populations and the sample variances computed from (5.2.1), each based on $\nu = 40$ degrees of freedom, are

$$V_1 = 2.6, \quad V_2 = 3.9, \quad V_3 = 2.0, \quad V_4 = 4.7, \quad V_5 = 10.3.$$

Since the sample from population 3 produced the smallest variance, we assert that population 3 has the true smallest variance. We now show how to compute the interval and point estimates of the probability that this is indeed a correct selection.

The ordered values of V_j are

$$V_{[1]} = 2.0, \quad V_{[2]} = 2.6, \quad V_{[3]} = 3.9, \quad V_{[4]} = 4.7, \quad V_{[5]} = 10.3,$$

and thus the four $\hat{\Delta}$ values computed from (5.3.4) are

$$\hat{\Delta}_2 = \sqrt{\frac{2.0}{2.6}} = 0.8771, \quad \hat{\Delta}_3 = \sqrt{\frac{2.0}{3.9}} = 0.7161,$$

$$\hat{\Delta}_4 = \sqrt{\frac{2.0}{4.7}} = 0.6523, \quad \hat{\Delta}_5 = \sqrt{\frac{2.0}{10.3}} = 0.4407.$$

To find P_L we interpolate in the table in Appendix G for $k = 2$ and $\nu = 40$ to find P_j for each value of $\hat{\Delta}_j$ as

$$P_1 = .8216, \quad P_2 = .9825, \quad P_3 = 1.000, \quad P_4 = 1.000.$$

Although these results are not accurate to four significant figures, we use all four in the calculation of P_L and then round the result to two or three decimals. The result is

$$P_L = (.8216)(.9825)(1.000)(1.000) = .807.$$

Thus the true probability of a correct selection is estimated to be at least as large as .807.

To find P_U we first compute $\log \Delta^{GM}$ from (5.3.5) (using logarithms to the base e) as

$$\log \Delta^{GM} = \frac{-0.1311 - 0.3339 - 0.4273 - 0.8194}{4} = -0.4279.$$

Taking the antilogarithm we find $\Delta^{GM} = 0.6519$. Then we interpolate in the table for $k = 5$ in Appendix G, with $\Delta^{GM} = 0.6519$ and $\nu = 40$ to find the corresponding probability as .984. Thus the upper bound is $P_U = .984$, which means that the true probability of a correct selection is estimated to be no larger than .984.

Combining the results for P_L and P_U, we can say that the true probability of a correct selection is estimated as being between .807 and .984. Although this interval estimate may suffice for some applications, we may also want to provide a point estimate of the true probability of a correct selection. This point estimate is useful when the interval estimate is not considered sufficiently narrow for the problem at hand.

To find the point estimate P_E, since $(k-1)/2 = 2$ here, suppose we decide to use $m = 2$ clusters. Then there are only three allowable partitions, that is, $\hat{\Delta}_2$ and $\hat{\Delta}_3, \hat{\Delta}_4, \hat{\Delta}_5$, or $\hat{\Delta}_2, \hat{\Delta}_3$ and $\hat{\Delta}_4, \hat{\Delta}_5$, or $\hat{\Delta}_2, \hat{\Delta}_3, \hat{\Delta}_4$, and $\hat{\Delta}_5$. If we use the partition where $\hat{\Delta}_2, \hat{\Delta}_3, \hat{\Delta}_4$ are put into one cluster and $\hat{\Delta}_5$ into one cluster, then we compute P_1 and P_2 as follows. For the cluster of size $k_1 - 1 = 3$, we compute the logarithm of the geometric mean of $\hat{\Delta}_2 = 0.8771$, $\hat{\Delta}_3 = 0.7161$ and $\hat{\Delta}_4 = 0.6523$ from (5.3.5) as

$$\log \Delta_1^{GM} = \frac{-0.1311 - 0.3339 - 0.4273}{3} = -0.2974,$$

and the antilogarithm is $\Delta_1^{GM} = 0.7427$. Interpolating in the table for $k_1 = 4$ in Appendix G with $\nu = 40$ and $\Delta_1^{GM} = 0.7427$, we find $P_1 = .918$. For the cluster of size $k_2 - 1 = 1$, the geometric mean is the value $\hat{\Delta}_5$ itself, or $\Delta_2^{GM} = 0.4407$. Interpolation in the table for $k_2 = 2$ in Appendix G with $\nu = 40$ and $\Delta_2^{GM} = 0.4407$ gives $P_2 = 1.000$. Thus the point estimate P_E is the product of these two P values, or $P_E = (.918)(1.000) = .918$.

The point estimate P_E of the probability of a correct selection may be made using $m = 1, 2, 3,$ or 4 clusters; Table 5.3.2 provides the results for all allowable partitions.

Table 5.3.2 **Calculations of P_E using all allowable partitions into one, two, three, and four clusters†**

Cluster j	Partition	Δ^{GM}	k_j	P_j
1	(0.8771, 0.7161, 0.6523, 0.4407)	0.6519	5	.9842
	$P_E = .9842$			
	$= P_U$			
1	(0.8771)	0.8771	2	.8216
2	(0.7161, 0.6523, 0.4407)	0.5905	4	1.0000
	$P_E = .8216$			
1	(0.8771, 0.7161, 0.6523)	0.7427	4	.9180
2	(0.4407)	0.4407	2	1.0000
	$P_E = .9180$			
1	(0.8771, 0.7161)	0.7925	3	.8810
2	(0.6523, 0.4407)	0.5362	3	1.0000
	$P_E = .8810$			
1	(0.8771)	0.8771	2	.8216
2	(0.7161)	0.7161	2	.9825
3	(0.6523, 0.4407)	0.5362	3	1.0000
	$P_E = .8072$			
1	(0.8771)	0.8771	2	.8216
2	(0.7161, 0.6523)	0.6835	3	.9832
3	(0.4407)	0.4407	2	1.0000
	$P_E = .8078$			
1	(0.8771, 0.7161)	0.7925	3	.8810
2	(0.6523)	0.6523	2	1.0000
3	(0.4407)	0.4407	2	1.0000
	$P_E = .8810$			
1	(0.8771)	0.8771	2	.8216
2	(0.7161)	0.7161	2	.9825
3	(0.6523)	0.6523	2	1.0000
4	(0.4407)	0.4407	2	1.0000
	$P_E = .8072$			
	$= P_L$			

†P_L and P_U are the same for all the partitions

5.4   SAMPLE SIZES UNEQUAL

In some instances the sample sizes are fixed and are not all equal. This may occur because the data are collected before a statistician is consulted or because it is impossible or impractical to make the sample sizes equal. The selection rule R for this problem was explained in Section 5.2 and illustrated by Example 5.2.1. The methods of analysis described in Section 5.3 are no longer applicable because the tables in Appendix G apply only to equal sample sizes. Nevertheless we can give an approximate solution to the problem by calculating an average value for the degrees of freedom and substituting it for the common value used in Section 5.3.

Following methods similar to those used in Section 2.4 and 4.6, where we computed an average sample size, we compute here an average value for the degrees of freedom using the square-mean-root (SMR) formula given by

$$\nu_0 = \left(\frac{\sqrt{\nu_1} + \sqrt{\nu_2} + \cdots + \sqrt{\nu_k}}{k} \right)^2, \tag{5.4.1}$$

where ν_j is the degrees of freedom for the sample from the jth population. Thus if the true means μ_j are all known for $j = 1, 2, \ldots, k$, we have $\nu_j = n_j$ for $j = 1, 2, \ldots, k$, (and therefore $\nu_0 = n_0$), and if these means are all unknown, we have $\nu_j = n_j - 1$ for $j = 1, 2, \ldots, k$. Using this value ν_0 in place of ν, we can proceed as in Section 5.3 and use the tables in Appendix G. The value of (5.4.1) will generally not be an integer. However, an integer is not needed for interpolation in Appendix G, so there is no reason to round ν_0 to an integer. As can be expected, this approximation based on the SMR result gives good results when the sample sizes are not too disparate.

Example 5.4.1 (Example 5.2.1 continued).   Recall that in Example 5.2.1, we asserted that machine 2 is the one that cuts diamonds with the most uniform weight. Since the sample sizes are unequal, no analysis of this selection can be made before calculating the average number of degrees of freedom. Recall that $n_1 = 6$, $n_2 = 5$, and $n_3 = 7$ and the μ values are all known. Hence the corresponding numbers of degrees of freedom are $\nu_1 = 6$, $\nu_2 = 5$, and $\nu_3 = 7$, and ν_0 from (5.4.1) is

$$\nu_0 = \left(\frac{\sqrt{6} + \sqrt{5} + \sqrt{7}}{3} \right)^2 = (2.444)^2 = 5.97.$$

Thus we use $\nu = 5.97$ to interpolate in the tables in Appendix G. For $P^* = .75$ say, we find $\Delta^* = 0.63$.

The following statements can be made with a confidence level of approximately .75. From (5.3.2) the true smallest standard deviation $\sigma_{[1]}$ is included in the interval $[0.63\sigma_s, \sigma_s]$, and from (5.3.3) σ_s is at most $100[(1/0.63) - 1]\% = 58.7\%$ greater than the smallest standard deviation $\sigma_{[1]}$, where σ_s is the true standard deviation of machine 2. We can also say that the approximate probability that machine 2 is really the best one is at least .75 whenever $\Delta^* \leqslant 0.63$. These numerically weak results are due to our extremely small sample sizes. The adjective *approximate* is used in the above preceding sentence only because the exact sample sizes were replaced by the n_0 obtained from the SMR formula.

The following example is somewhat more realistic as regards sample sizes.

Example 5.4.2   (Example 5.1.2 continued). Suppose that the serum cholesterol study described in Example 5.1.2 was carried out before consulting a statistician and the sample sizes are unequal. The data are as follows:

Laboratory kit j	1	2	3	4	5	6
Sample size n_j	16	24	30	25	28	18
Sample mean $\bar{x}_j$	90.8	92.4	95.3	100.6	98.3	102.5
T_j	4828	4926	9296	5026	8124	6771
V_j	321.9	214.2	320.6	209.4	300.9	398.3
$\sqrt{V_j}$	17.9	14.6	17.9	14.5	17.3	20.0

The selection rule of Section 5.2 dictates that laboratory kit 4 should be asserted to be the one that makes serum cholesterol measurements with the smallest variance. Before any further analyses can be performed, we calculate the average number of degrees of freedom from (5.4.1) as

$$\nu_0 = \left(\frac{\sqrt{15} + \sqrt{23} + \sqrt{29} + \sqrt{24} + \sqrt{27} + \sqrt{17}}{6} \right)^2 = (4.712)^2 = 22.20.$$

This is the ν value that is used to interpolate in Appendix G. For example, if $P^* = .95$ we obtain $\Delta^* = 0.60$. Thus with a confidence level of approximately .95, we can say that the true standard deviation σ_s of kit 4 satisfies the inequality $\sigma_s \leqslant \sigma_{[1]}/0.60 = 1.67\sigma_{[1]}$, or that σ_s is at most 67% larger than $\sigma_{[1]}$.

5.5   LARGE SAMPLE APPROXIMATION

Although the tables in Appendix G give the solution for the common number of degrees of freedom ν, and hence the common sample size n, which corresponds to a specified (Δ^*, P^*) requirement, an approximate solution for larger sample sizes is also useful. This approximate solution is based on normal theory.

The approximate value of the degrees of freedom, which we denote by ν_0, is computed using the formula

$$\nu_0 = \frac{\tau_t^2}{2(\log \Delta^*)^2} + 2 \log (k-1), \qquad (5.5.1)$$

where τ_t is the entry in Table A.1 of Appendix A that corresponds to the specified P^* and the given k; the logarithms are taken to the base e (natural logarithms). The second term in (5.5.1) is a slight correction term based on empirical results; inclusion of this term tends to correct for the fact that the first term in (5.5.1) drifts below the true value of ν as k increases. Once ν_0 is computed from (5.5.1), the common sample size n_0 is approximated by $n_0 = \nu_0$ if the means are all known, and $n_0 = \nu_0 + 1$ if the means are all unknown. This approximation was used to obtain the larger ν values given in Appendix G. The error appears to be at most one unit throughout the table.

To illustrate the calculation of this approximation, suppose that for $k=4$ normal populations with known means and unknown variances, the investigator asserts that he wants a procedure R such that $P\{CS|R\} \geqslant .90$ whenever σ_s, the standard deviation of the population selected, is at most 25% larger than the smallest standard deviation. From (5.3.3) we see that this specifies $\Delta^* = 0.80$. For $P^* = .90$, $k=4$, Table A.1 gives $\tau_t = 2.4516$. Substituting these values in (5.5.1) we obtain

$$\nu_0 = \frac{(2.4516)^2}{2(\log 0.80)^2} + 2 \log 3 = 62.56,$$

and hence we estimate that $n_0 = \nu_0 = 63$ observations are required from each of the four populations. The reader may verify that the table for $k=4$ in Appendix G gives this same value, $n = \nu = 63$.

5.6   NOTES, REFERENCES, AND REMARKS

Strangely enough, the variance selection problem is fairly straightforward if all the means μ_j (which are now nuisance parameters) are known or if they are all unknown; however, if some means are known and some are unknown, complications arise in the theory. A practical solution is to solve the problem for the preceding two extreme cases and use the results as bounds for these more complicated cases.

The use of the distance $\Delta = \sigma_{[1]}/\sigma_{[2]}$ instead of its reciprocal δ or δ^2 is new with this book. This distance Δ is always less than 1 and hence easier to tabulate against; it also seems to be a more intuitive measure than δ^2 since it is a ratio in the *original* units of the problem. On the other hand, the Remark at the end of Section 5.3.1 relates Δ (as well as δ) to $\varepsilon = [(1/\Delta) - 1]$, which may be more useful than either Δ or δ in helping the investigator to specify the requirement.

It turns out that the problem of selecting the population with the smallest variance requires a table based on the same (integral) inequality that must be solved for the corresponding problem of selecting a subset of populations that *contains* the population with the smallest variance. Thus although the first paper on selecting the population with the smallest variance is by Bechhofer and Sobel (1954), a later paper by Gupta and Sobel (1962b) contains equivalent tables that can also be used for this problem. In addition, the papers by Gupta and Sobel (1962a, b) show that the exact solution also has a form that has been labeled *Dirichlet* because it involves some integrals studied by the nineteenth-century French mathematician G. P. Lejeune-Dirichlet; thus, as a third resource, the exact solution can be obtained from a table of Dirichlet integrals.

The idea of an operating characteristic curve for a ranking and selection problem and the section on estimation of the true probability of a correct selection are both new with this book. The fact that we could make a straightforward application to the variance problem illustrates how these basic ideas can apply to any ranking and selection problem. The same comment applies to the use of the square-mean-root (SMR) formula in Section 5.4 to handle the problem of unequal sample sizes.

As is shown in Chapter 14, the problem of selecting the normal population with the *smallest* variance is equivalent to the problem of selecting the gamma population with the smallest scale parameter, and hence the same tables are used. However, the problem of selecting the gamma population with the *smallest* scale parameter is *not* equivalent to that of selecting the gamma population with the *largest* scale parameter except when v is very large.

The problem of selecting the population with the largest variance is based on the same (integral) inequality that is used in Gupta (1963) for the corresponding problem on selecting a subset. Nevertheless because we use the new distance measure Δ rather than δ and because the entries in Gupta

(1963) do not cover the area of maximum practical value, we prepared new tables for the problem of selecting the population with the largest variance. Although we refer to this as the largest variance problem, the main applications are to problems involving the gamma distribution and in particular the exponential distribution, which is a central core in work dealing with reliability and life testing. Hence if the user does have the problem of finding the population with the largest variance, he will have to use the methods of Chapter 14, which deals with gamma distributions.

PROBLEMS

5.1 An experiment was carried out to determine which one of four different blends of gasoline gives the smallest variance of gasoline consumption as measured by miles per gallon. Four different drivers were assigned randomly to four different brand-new automobiles (of identical make and model) on each of 10 different runs over a specified course. The following data are the miles per gallon observed on each of the 10 runs.

(a) Select the blend that has the smallest variance of gasoline consumption.

(b) Calculate the operating characteristic curve.

(c) Find a point estimate and an interval estimate of the true probability of a correct selection.

	Blend		
1	2	3	4
8.7	7.8	8.8	9.3
8.9	8.2	10.2	8.6
10.1	10.5	9.1	8.5
11.3	15.1	9.6	10.5
9.7	11.5	10.4	10.8
9.9	10.8	10.3	9.2
10.6	9.5	9.5	11.3
10.2	11.9	9.2	11.5
9.3	13.5	8.9	9.7
11.1	7.7	9.7	8.9

5.2 A manufacturer of cloth is concerned about the variation in shrinkage of his polyester-cotton blend fabric. He suspects that the variation in percent shrinkage will be greater when washed at higher water temperatures. Tests are to be made by washing identical samples of the fabric in an ordinary machine, but at the three temperature settings of cold, warm, and hot water. How many fabric samples should be washed at each temperature so that the manufacturer can correctly select the water temperature that gives the smallest variation in shrinkage if he specifies $\Delta^* = 0.85$, $P^* = .90$?

5.3 Low variability of resistance in wire is very important because of its effect on durability. Four different kinds of wire are to be tested to determine the one with the smallest variability. How many observations should be taken on each wire in order to make the statement, "With confidence level .95, the standard deviation of the wire selected is at most 11% greater than the smallest standard deviation."

5.4 Barnett and Youden (1970) define the precision of a method of chemical analysis as the day-to-day reproducibility of results since patients are usually tested no more frequently than once a day. The precision of a "kit" is as important as its accuracy. Suppose that three different glucose kits have been established as accurate, so the best kit is defined as the one with the smallest variance in measurement around the true mean "assay" value, which is equal to 92 mg per 100 ml. The following data are the "assay" values in milligrams per 100 ml for 24 samples of a patient's serum that were assigned randomly to be analyzed by the three kits.

Kit A:	85,	97,	94,	86,	91,	88,	96,	87
Kit B:	81,	99,	100,	102,	79,	93,	90,	105
Kit C:	86,	94,	97,	101,	89,	94,	90,	91

(a) Select the kit with the best precision.

(b) Calculate the operating characteristic curve.

(c) Find a point estimate and an interval estimate of the true probability of a correct selection.

5.5 Wernimont (1947) reports a chemical laboratory experiment in which four new uncalibrated thermometers of specified quality but different brands were used to measure the melting point of a controlled sample of hydroquinine. For each thermometer the procedure was the same and the same heating bath was used. The data are given in degrees centigrade. Assume that the distribution of melting points for each thermometer is normal and that the goal is to select the brand of thermometer with the smallest variance in measurement of melting point.

(a) Select the best brand of thermometer.

(b) Make an appropriate confidence statement at level .90 about your selection.

Thermometer

A	B	C	D
174.0	173.0	171.5	173.5
173.5	173.5	172.5	173.5
173.0	172.0	171.0	171.0
173.0	173.0	172.0	172.0
173.5	173.0	173.0	172.5
173.0	173.0	173.0	173.0

5.6 In canning certain foods it is important to keep the variance of the "drained weight" under control. The drained weight is determined by opening a can that has been filled and sealed and weighing the solid contents. If a sample of such measurements indicates that the average weight and/or the variance in weights do not meet specifications, some action should be taken to restore control to the process (Grant and Leavenworth (1972), p. 40). Suppose a cannery has four six-hour shifts of workers each day, and an experiment has been designed to see which shift has the lowest variability in drained weight before processing but after filling size $2\frac{1}{2}$ cans (28 ounces undrained weight after processing) of standard grade tomatoes in puree. A random sample of 25 cans is taken from the production by each shift. The sample means and variances of drained weight in ounces are found to be

Shift j	$\bar{x}_j$	V_j
1	19.16	0.5531
2	21.36	0.5016
3	22.43	0.4732
4	20.78	0.5863

Assume that the drained weights are normally distributed with unknown means.

(a) Select the shift with the smallest variance in drained weight.

(b) Make an appropriate confidence statement about the selection with confidence level .975.

(c) Calculate the operating characteristic curve.

(d) Estimate the true probability of a correct selection.

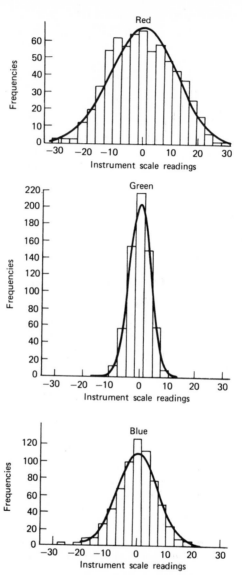

Figure for Problem 5.8 Figure showing histograms of actual frequency distribution of the red, green, and blue primaries. Reproduced from W. R. J. Brown, W. G. Howe, J. E. Jackson, and R. H. Morris (1956), Multivariate normality of the color-matching process, *Journal of the Optical Society of America, 46, 46–49, with permission.*

5.7 Preston (1967) states that although it is well known that the weather is hotter in summer than in winter in most parts of the world, it is not so well known that in some regions winter weather is more variable. This conclusion was based on data for average daily temperature at the Pittsburgh airport for each day in 1954. (Average daily temperature is computed by averaging the maximum and minimum of the 24 observations obtained by noting the temperature each hour.) Suppose we are designing a similar experiment to determine for the city of Honolulu which one of the four quarters (January–March, April–June, July–September, October–December) has the smallest variation in temperature. We consider the average daily temperatures in any one quarter because the quarters do not have the same number of days. A random sample of (common size) n of these temperatures is selected for each quarter. How large should n be if $P^* = .90$ and $\Delta^* = 0.70$?

5.8 A number of studies have been concerned with the sensitivity of the eye to small color differences. In one study Brown, Howe, Jackson, and Morris (1956) give the frequency distributions of data on the sensitivity to differences in the red, green, and blue primary. These distributions, which are shown in the figure on the facing page, are approximately normal, with the same mean but with different variances. Suppose we have three mixed colors and wish to choose the one with the smallest variance in sensitivity. How many observations should be taken so that the probability of a correct selection is at least .90 when $\Delta \leqslant \Delta^* = 0.85$,

(a) If the true means are unknown?

(b) If the true means are all equal to zero, as in the case of the primary colors?

CHAPTER 6

Selecting the One Best Category for the Multinomial Distribution

6.1 INTRODUCTION

In this chapter we consider the problem of selecting the one best category out of k mutually exclusive categories when each element in a single population is classified into one of these k categories. For each of the k categories there is a certain probability that any element will be classified into that category. We denote these probabilities by $p_1, p_2, \ldots, p_k$. The best category may be defined as the one with the largest probability, or alternatively, as the one with the smallest probability, depending on the situation. For the time being we consider only the case where *best* means the largest. The problem of selecting the category with the smallest probability is discussed in Section 6.5.

This model is a natural generalization of the Bernoulli or binomial distribution to the case where there are more than two possible outcomes (but only a finite number of possible outcomes) for each observation. The outcomes need not be labeled numerically or ordered in any way. The only requirement is that they be mutually exclusive and exhaustive so that every element can be classified into one and only one category. For example, consider an opinion poll where there are five possible answers by each respondent, namely, "Strongly agree," "Agree," "Indifferent," "Disagree," and "Strongly disagree." Here the five categories have an ordering (by amount of agreement), but this is not required. Similarly, we might be interested in classifying the people in some population according to eye color, say, blue, brown, hazel, and green. These categories have no ordering; they are simply labels.

The distribution model that applies to a sample from a population of this kind is called the multinomial distribution. The multinomial is one of the most useful discrete multivariate distributions, as it is the prototype for

158

many important practical problems in experimentation and investigation. In view of this we first give a few specific illustrations of situations where the multinomial distribution applies and the goal of practical as well as theoretical interest is to select the category that has the best probability.

Example 6.1.1 (Genetics).   Two types of corn (golden and green-striped) carry recessive genes. When these are crossed a first generation is consistently normal (neither golden nor green-striped). When this first generation is allowed to self-fertilize, four distinct types of plants are produced, normal, golden, green-striped, and golden-green-striped, although normal is predominant. Suppose a group of plants is allowed to self-fertilize until the fifth generation. It is desired to determine which of the four types of hybrid predominates after five such self-fertilizations. In symbols, we might denote the probability of producing a normal plant by p_1, a golden plant by p_2, a green-striped plant by p_3, and a golden-green-striped by p_4. The goal is to select the largest value among p_1, p_2, p_3, and p_4.

Example 6.1.2 (Blood Type).   A sample of blood is taken from each of a large number of Australian aborigines, and each person is classified as type A, type B, type AB, or type O. Since we wish to identify the blood type that is most rare, the goal is to determine which blood type occurs least frequently among these aborigines.

Example 6.1.3 (Voter Preference).   Five candidates are running for an office in a certain highly populated city. A random sample of registered voters is taken and each person is asked to name the one candidate he prefers. The goal is to determine which candidate has the greatest support among the potential voters.

Before discussing the multinomial ranking and selection problem in more detail, we first develop the multinomial distribution model. Consider an arbitrary experiment or process where each outcome is classified into one of k possible mutually exclusive possibilities, which we call categories (or cells). Let $p_j, 0 \leqslant p_j \leqslant 1$, denote the probability of an outcome in the jth category for $j = 1, 2, \ldots, k$. Since the k categories are mutually exclusive and exhaustive, the probabilities must sum to one, that is,

$$p_1 + p_2 + \cdots + p_k = 1. \qquad (6.1.1)$$

If an experiment of this type is repeated a fixed number of times, say n, then we can count the number of times an outcome is classified in each category. Let x_j denote the number of occurences for category j. Since each outcome must be in one of the k designated categories, we have

$$x_1 + x_2 + \cdots + x_k = n. \qquad (6.1.2)$$

The joint probability distribution of this set of outcomes is given by the multinomial distribution, specifically,

$$p(x_1, x_2, \ldots, x_k) = \frac{n!}{x_1! x_2! \ldots x_k!} \, p_1^{x_1} p_2^{x_2} \ldots p_k^{x_k}. \qquad (6.1.3)$$

For any specified nonnegative integer values of the components of the vector $(x_1, x_2, \ldots, x_k)$ and the nonnegative parameters $p_1, p_2, \ldots, p_k$, which sum to one, (6.1.3) gives the probability of exactly x_j outcomes in category j, jointly for all $j = 1, 2, \ldots, k$. Although it is not obvious, the sum of the probabilities in (6.1.3) over all sets of components of $(x_1, x_2, \ldots, x_k)$ that satisfy (6.1.2) is equal to 1.

A single observation in a multinomial model is regarded as a vector with k components, of which exactly one is equal to 1 and all the others are equal to 0. The position of the component with the 1 indicates which category describes the outcome of that observation. For example, if $k = 3$ and the observation vector is $(0, 0, 1)$ the outcome of this single observation is category 3. In general, if the jth component is 1, this indicates that the outcome is in the category labeled j. The value of x_j is then the sum over all n vectors of the number of 1's in the jth component.

The realization of a multinomial distribution can be regarded as equivalent to the rolling of a die with k (not necessarily balanced) sides n independent times and noting how many times each side appears, or the tossing of n balls randomly into k different cells and noting how many balls fall into each cell. In the former case x_j is the number of times the jth side appears or the frequency of the jth side, and in the latter situation x_j is the number of balls that fall into the jth cell or the frequency of the jth cell. Because these die-rolling and ball-tossing situations are so intuitive, it is customary to consider them as general prototypes for the realization of a multinomial distribution and to use the terminology of cell and category interchangeably.

The present goal in the ranking and selection problem for the multinomial distribution model is to select the one category with the largest probability of success. As usual, we denote the ordered parameters by

$$p_{[1]} \leqslant p_{[2]} \leqslant \cdots \leqslant p_{[k]}.$$

Clearly these ordered p values still sum to one, as in (6.1.1). The p values are all unknown, and the goal is to identify the category for which the p value is $p_{[k]}$. If more than one category has the same largest probability, we are content to find at least one. Note that we are *not* dealing with the problem of completely ordering the k categories according to their p values or with estimating any of the p values.

The selection rule is given in Section 6.3. As usual, the analytical aspects of the selection problem involve the probability of a correct selection for certain configurations of p values. This probability is equal to $1/k$ when the p values are all equal; as a result, if the probability requirement is $P^* \le 1/k$ the investigator need only perform an independent experiment with k equally likely outcomes. A probability requirement where $1/k < P^* < 1$ cannot be achieved for all possible true configurations of the p values, since the probability then depends on the unknown true parameter values. However, we can find a solution for certain configurations of concern, namely, those configurations in the preference zone for a correct selection.

In order to define the preference zone, we must first define a measure of distance δ. For the multinomial model, δ is defined as the ratio of the largest and next-to-largest p values, or

$$\delta = \frac{p_{[k]}}{p_{[k-1]}}. \qquad (6.1.4)$$

Note that $\delta \ge 1$ by definition, and also that $\delta = 1$ occurs only when two categories have the same largest probability. Since we are particularly concerned about a correct selection when δ is large, the preference zone is defined as $\delta \ge \delta^*$ for some specified number $\delta^* > 1$. We need to find a selection rule R such that the probability of a correct selection is at least P^* in all possible true configurations of p values for which $\delta \ge \delta^*$. In symbols, this says that

$$P\{CS|R\} \ge P^* \qquad \text{whenever} \quad \delta \ge \delta^*. \qquad (6.1.5)$$

We refer to (6.1.5) as the (δ^*, P^*) requirement, or, alternatively, as the PCS requirement.

Under the configuration $\delta \ge \delta^*$ where $\delta^* > 1$, the best category is uniquely defined (although we do not know which one it is), and hence the concept of $P\{CS|R\}$ is also well defined. The least favorable configuration for a general k in the multinomial model is given by

$$p_{[1]} = p_{[2]} = \cdots = p_{[k-1]} \qquad \text{and} \qquad p_{[k]} = \delta^* p_{[k-1]}. \qquad (6.1.6)$$

Since by (6.1.1) the sum of the p_j is equal to 1, (6.1.6) implies that

$$p_{[1]} = p_{[2]} = \cdots = p_{[k-1]} = \frac{1}{\delta^* + k - 1} \qquad \text{and} \qquad p_{[k]} = \frac{\delta^*}{\delta^* + k - 1}. \qquad (6.1.7)$$

The generalized least favorable configuration is obtained by replacing δ^* in (6.1.6) by δ. Note, however, that in this model *all* the p values necessarily change as δ or δ^* changes.

REMARK. It should be noted that for $k=2$ the multinomial problem reduces to the problem of selecting one of the two cells in a *single* binomial population, which is not the same as selecting the best one of k binomial populations for any k. Since the problems are different (as well as the distance measures), there is no reason to look for similarities between the two problems for the case $k=2$. $\square$

6.2 DISCUSSION OF THE DISTANCE MEASURE

In this section we discuss the distance measure given in (6.1.4) in more detail. Consider an example with $k=3$. Figure 6.2.4 shows the three regions that constitute the preference zone by a two-dimensional diagram using so-called barycentric coordinates. (In barycentric coordinates a point M in a given triangle ABC is determined by the three weights at the vertices A, B, and C that make M the center of gravity; the word barycentric means center of gravity.) The regions illustrated in Figure 6.2.4 represent the configurations of p values for which there is a strong preference for a correct selection and the probability requirement in (6.1.5) is met.

To construct Figure 6.2.4 we begin with an equilateral triangle ABC with altitude equal to one, that is, the vertical distance from each vertex to the midpoint of the opposite side is equal to one. This triangle is shown in Figure 6.2.1, with BC as the base. The unordered p values p_1, p_2, p_3, denote the distances from any point (p_1, p_2, p_3) to the sides BC, CA, and AB, respectively. The coordinates of the vertices are $A=(1,0,0)$, $B=(0,1,0)$, and $C=(0,0,1)$. The line AD represents the locus of points $p_3/p_2=\delta^*$, and the line CE represents the points where $p_1=p_2$. Consequently, the intersection of the lines AD and CE (point F) represents the point where $p_3=\delta^* p_2$ and $p_1=p_2$, that is, where $p_3=\delta^* p_2=\delta^* p_1$.

In Figure 6.2.2 we repeat Figure 6.2.1 but add a line that connects the points B and F and extends to intersect the side CA. This point of intersection is labeled D'. The lines AD and $D'B$ consist of all points where $\delta^*=p_3/p_2$ and $\delta^*=p_3/p_1$, respectively, for δ^* fixed at some value $\delta^*>1$. The first two coordinates of the point F are equal (since CE is an angle bisector that corresponds to $p_2/p_1=1$), and hence F corresponds to the point where $p_3/p_2=\delta^*>1$ and $p_3/p_1=\delta^*>1$. On AD between D and F we have $p_1 \leq p_2=p_3/\delta^*$, and hence the ordering $p_1 \leq p_2 \leq p_3$ holds on the line segment DF, and in the triangular region CFD we have $p_1 \leq p_2 \leq p_3/\delta^*$. Similarly, on $D'B$ between D' and F, that is, on the line segment $D'F$, we have $p_2 \leq p_1=p_3/\delta^*$ so that the ordering $p_2 \leq p_1 \leq p_3$ holds, and in the triangular region $CD'F$ we have $p_2 \leq p_1 \leq p_3/\delta^*$. These inequalities for CDF and $CD'F$, respectively, can be written equivalently as $p_3/p_1 \geq$

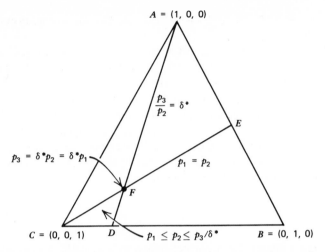

Figure 6.2.1   Barycentric coordinates showing point F where $p_3/p_2 = p_3/p_1 = \delta^*$.

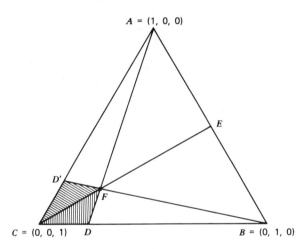

Figure 6.2.2   Representation in terms of barycentric coordinates of that part of the preference zone in the neighborhood of $p_3 = p_{[3]}$.

$p_3/p_2 \geqslant \delta^*$ and $p_3/p_2 \geqslant p_3/p_1 \geqslant \delta^*$. Combining both these triangular regions, we find that the quadrilateral region $CD'FD$ consists of all points (p_1, p_2, p_3) that satisfy both inequalities

$$\frac{p_3}{p_2} \geqslant \delta^* \qquad \text{and} \qquad \frac{p_3}{p_1} \geqslant \delta^*. \tag{6.2.1}$$

Thus if p_3 is the largest of p_1, p_2, and p_3, that is, if $p_{[3]} = p_3$, the quadrilateral region $CD'FD$, defined by the inequalities in (6.2.1), is the preference region for a correct selection of p_3.

We now find the corresponding preference region for selecting p_2 as the largest p value, and similarly for p_1. By symmetry, in Figure 6.2.1 we can obtain two other points in the equilateral triangle, one from the right-hand vertex and one from the top vertex. These points, F' and F'', respectively, are shown in Figure 6.2.3, along with F. The point F' is where $p_2/p_3 = \delta^*$ and $p_2/p_1 = \delta^*$, whereas at F'' we have $p_1/p_3 = \delta^*$ and $p_1/p_2 = \delta^*$.

Also by symmetry (see Figure 6.2.4), the preference region for a correct selection if p_2 is the largest p value is the quadrilateral region at the right-hand vertex, defined by the inequalities

$$\frac{p_2}{p_3} \geqslant \delta^* \qquad \text{and} \qquad \frac{p_2}{p_1} \geqslant \delta^*, \qquad (6.2.2)$$

and the preference region for a correct selection if p_1 is the largest p value is the quadrilateral region at the top vertex, defined by the inequalities

$$\frac{p_1}{p_2} \geqslant \delta^* \qquad \text{and} \qquad \frac{p_1}{p_3} \geqslant \delta^*. \qquad (6.2.3)$$

Since $k = 3$, we have three quadrilaterals, one corresponding to each of the possible decisions. The three regions defined by (6.2.1), (6.2.2), and (6.2.3) together constitute the preference zone for the multinomial problem when $k = 3$. The least favorable configuration for the multinomial problem when $k = 3$ is defined by the three points F, F', and F'' (see Figure 6.2.3).

REMARK. The use of barycentric coordinates for the parameter space permits a two-dimensional representation of the regions constituting the preference zone. If the parameter space were depicted in ordinary three-dimensional Euclidean space with axes p_1, p_2, and p_3, attention would be restricted to the first octant, where $p_j \geqslant 0$ for $j = 1, 2, 3$, and the equilateral triangle where $p_1 + p_2 + p_3 = 1$ (because of (6.1.1)). Then the geometrical picture is essentially the same as in Figure 6.2.3, except that in the preceding discussion the total diagram has been magnified so that all three altitudes (like CE in Figure 6.2.1) have length equal to 1. The reader may wish to draw a three-dimensional diagram to illustrate this preceding point. (If the geometrical discussion for $k = 3$ dimensions is replaced by one that uses only analytic methods in geometry, then there is no difficulty in extending these ideas to dimensions higher than three.   □

In order to give a more concrete illustration of the points (p_1, p_2, p_3) that constitute the preference zone when $k = 3$, consider the specific case where

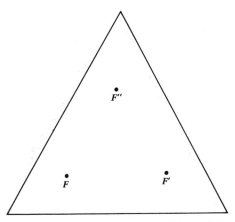

Figure 6.2.3 Barycentric coordinates showing LFC consisting of three points F: $p_3/p_2 = p_3/p_1 = \delta^*$, F': $p_2/p_3 = p_2/p_1 = \delta^*$, F'': $p_1/p_3 = p_1/p_2 = \delta^*$.

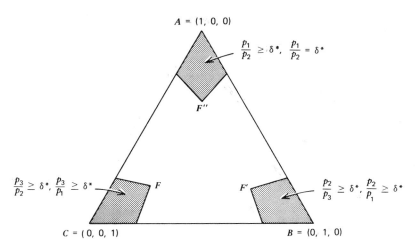

Figure 6.2.4 Total preference zone $p_{[3]}/p_{[2]} \geqslant \delta^*$ consists of the three shaded regions.

$\delta^* = 2.0$. The true largest p value may be p_1, p_2, or p_3. Which values of p_1, p_2, and p_3 lie in the preference zone for asserting that, say, p_3 is the true largest, that is, that $p_{[3]} = p_3$? A few of the infinite number of points in the preference zone are listed in Table 6.2.1.

The last line in Table 6.2.1 shows the point that is the least favorable configuration for $\delta^* = 2.0$. It should be noted that no point with $p_{[3]} < .50$ can be in the region of preference for a correct selection when $\delta^* = 2.0$. For

Table 6.2.1   Selected points in the region of preference for a correct selection of p_3 when $k = 3$ and $\delta^* = 2.0$

p_1	p_2	$p_3 = p_{[3]}$
.00	.00	1.00
.10	.00	.90
.05	.05	.90
.00	.10	.90
.20	.00	.80
.00	.20	.80
.10	.10	.80
.30	.10	.60
.10	.30	.60
.20	.20	.60
.25	.25	.50

example, the point $(.30, .30, .40)$ cannot be in this preference region; hence if this point is the true configuration, we cannot guarantee that the probability $P\{CS|R\}$ is at least P^*. Of course, the attained $P\{CS|R\}$ for such points can be computed for any fixed sample size n after the rule R is defined. For any reasonable procedure R, the result certainly is greater than $\frac{1}{3}$, since $k = 3$. For general $k > 2$ the probability is always greater than $1/k$ but is not necessarily greater than .5. For any k we suggest the rule of thumb that P^* be chosen such that $P^* \geqslant .5 + .5/k = .5(k+1)/k$; note that P^* is then always greater than .5.

From the preceding discussion it is clear that the choice $\delta^* = 2.0$ is fairly restrictive, and the minimum value of the difference $p_{[3]} - p_{[2]}$ in the preference region is $(\delta^* - 1)/(\delta^* + 2)$, which equals .25 for $\delta^* = 2.0$. If the value of δ^* is reduced to, say, $\delta^* = 1.2$, then this minimum is $\frac{1}{16}$, and many more configurations are included in the preference region. Some such additional points are shown in Table 6.2.2.

Table 6.2.2   Additional selected points in the region of preference for a correct selection of p_3 when $k = 3$ and $\delta^* = 1.2$

p_1	p_2	$p_3 = p_{[3]}$
.40	.00	.60
.00	.40	.60
.40	.10	.50
.10	.40	.50
.30	.20	.50
.20	.30	.50
.30	.30	.40
.3125	.3125	.375

A comparison of Table 6.2.2 with Table 6.2.1 shows that the region of preference for correctly selecting p_3 when $\delta^* = 1.2$ includes points for which $p_{[3]}$ is closer to $p_{[2]}$ than when $\delta^* = 2.0$. In particular, points such as $(p_1, p_2, p_3) = (.30, .30, .40)$ and $(.40, .10, .50)$ are included when $\delta^* = 1.2$, whereas they are not when $\delta^* = 2.0$.

Tables 6.2.1 and 6.2.2 are given here because a listing of some points that are in the preference region and some points that are not provides a better basis for specifying an appropriate value for δ^* and for understanding the meaning of the final confidence statement.

6.3 SELECTION RULE

Once the data are obtained, the procedure to select the category with the largest probability in the multinomial distribution is quite simple and straightforward. Suppose there are n observations and x_j of them are classified in category j for $j = 1, 2, \ldots, k$. These observed category (or cell) frequencies are then ordered from smallest to largest, as

$$x_{[1]} \leqslant x_{[2]} \leqslant \cdots \leqslant x_{[k]},$$

and the rule R (if there are no ties for first place) is to select the population that gives rise to the largest category frequency. Thus we assert that the category with the largest observed frequency $x_{[k]}$ has the largest probability of occurrence $p_{[k]}$ in the multinomial distribution.

Clearly, except in the cases where n is odd and $k = 2$ (and also in the trivial case where $n = 1$ for any k), some of the observed category frequencies x_j can be equal. When ties occur for the largest value of x_j, they can be broken in a random manner. More specifically, if r of the k values of x_j are equal and are larger than all the other x values, one of these r is selected at random, say by using a table of random numbers with probability $1/r$ for each, or by rolling a balanced cylinder with r marked sides. The category corresponding to the one selected is asserted to have the largest probability of occurrence. The probability that ties occur and the effect of this randomization to resolve them are taken into account in the derivation of the function that gives the probability of a correct selection, and hence no modification for ties is needed in any statement concerning the final selection. Of course, the probability of a tie approaches zero as n increases, regardless of the true configuration of p values. Hence if n is large, ties are unlikely to occur. In fact, a preponderance of ties for large n should make us suspicious about the appropriateness of the multinomial model that is being assumed.

REMARK. The values calculated for the probability of a correct selection are not conditional on the absence of ties. The conditional probability, given that there are no ties for first place, is asymptotically (as n becomes infinite) equivalent to the unconditional probability. For small values of n this conditional probability has not been calculated. It may also be of interest to consider some technique that requires additional observations to break ties, rather than our randomization technique, and to compare the properties of that procedure with ours. However, such a procedure would clearly not be a fixed sample size procedure and therefore we do not consider it at this point. □

Example 6.3.1   (Example 6.1.3 continued). In the voter preference problem of Example 6.1.3, suppose that we interview 250 persons selected randomly from the voter registration list of this highly populated city and ask each one his preference among the five candidates, A, B, C, D, E. The tallied data are as follows:

Candidate	A	B	C	D	E	Total
Frequency of Preference	39	63	12	78	58	250

Since candidate D is preferred more frequently than any other among the 250 persons interviewed, we assert that candidate D is the most preferred by the voting populace in this city, that is, candidate D is the one with probability of preference $p_{[5]}$. Note that for each j, the parameter p_j is the probability that a randomly selected voter in the city will prefer candidate j, or equivalently, the proportion of the voting population in this city that prefers candidate j. These proportions could be determined precisely only if a complete census of all the voters in this city were taken.

6.4   ANALYTICAL ASPECTS OF THE PROBLEM OF SELECTING THE BEST CATEGORY

In this section we consider the analytical problems of determining sample size and calculating the operating characteristic curve for the multinomial selection problem.

6.4.1   Determination of Sample Size

Determining sample size is tied up with the design of our experiment for selecting the category with the largest probability. Our present aim is to determine the smallest sample size n required to have a prescribed level of confidence that the selection rule R will correctly identify the best category among the k categories once the data are obtained. Recall that in the multinomial model, n denotes the total sample size.

The first step in determining the sample size is to specify a pair of values

(δ^*, P^*), where $1/k < P^* < 1$ and $\delta^* > 1$. This choice is interpreted as meaning that with this sample size, the probability of a correct selection is to be at least P^* in all (true) configurations for which the value of the distance measure δ in (6.1.4) is at least δ^*. Thus n is the smallest sample size for which (6.1.5) is satisfied.

The entries in Table H.1 of Appendix H give this required value of n for $P^* = .750, .900, .950, .975, .990$ and for δ^* from 1.2 to 3.0 (by jumps of size 0.2) for $k = 2, 3, \ldots, 10$. For other values of P^* and/or δ^*, n can be found by a method of interpolation explained in Appendix H.

Once n is determined by this method we can make a confidence statement about the magnitude of $p_{[k]}$ relative to the magnitude of the parameter p_s, the p value of the category selected as best by rule R. Of course, we know that $p_{[k]}/p_s \geqslant 1$, but we can also say, with confidence level P^*, that $p_{[k]}/p_s \leqslant \delta^*$. Combining these results, we obtain the confidence interval for $p_s/p_{[k]}$ as

$$\frac{1}{\delta^*} \leqslant \frac{p_s}{p_{[k]}} \leqslant 1 \quad \text{or} \quad \frac{p_{[k]}}{\delta^*} \leqslant p_s \leqslant p_{[k]}, \tag{6.4.1}$$

or, in terms of the difference,

$$0 \leqslant p_{[k]} - p_s \leqslant p_{[k]}\left(1 - \frac{1}{\delta^*}\right), \tag{6.4.2}$$

which means that p_s is at most $(1 - 1/\delta^*)100\%$ smaller than the true largest probability $p_{[k]}$. We can also assert with confidence level P^* that $p_{[k]}$ lies in the (closed) interval

$$S = \left[\, p_s, p_s\delta^* \,\right]. \tag{6.4.3}$$

It should be emphasized that these confidence statements are statements about the relationship between $p_{[k]}$ and p_s and not about the numerical value of any single parameter p.

As noted in Section 6.2, the problem of identifying the category with the largest probability is more difficult if δ^* is close to one (for fixed P^*) and if P^* is close to one (for fixed δ^*). (This is clear from Table H.1 since the values of n in the last column and first row are quite large for any k.) The first property is quite intuitive since if δ^* is close to one, then $p_{[k]}$ and $p_{[k-1]}$ are close together, and hence for a fixed confidence level a large sample size is needed to identify correctly the category with the largest p value. The second property, about P^*, is also intuitively clear, since a higher level of confidence also requires more observations.

REMARK. Note also from Table H.1 that the number of observations required increases as k, the number of categories, increases. It is certainly to be expected that, other things being equal, it is more difficult to identify correctly the one best category when there are more categories to choose from. At first sight, the proof of this property is not completely evident because all the p values must sum to one, and hence the least favorable configuration, and the p values determined by it, change with k, as (6.1.7) shows. Nevertheless we can consider the subspace where p_{k+1} is the smallest p value, $p_{[1]}, p_{[2]} = \cdots = p_{[k]}$, and $p_{[k+1]} = \delta^* p_{[k]}$. Then $p_{[k+1]}$ and $p_{[2]}$ are determined by p_{k+1}, since the p values sum to one. Keeping the ratios $p_{[k+1]}/p_{[2]} = p_{[k+1]}/p_{[k]}$ fixed at δ^*, we can lower the probability of a correct selection by increasing p_{k+1} until it equals $p_{[2]}$; the case of k populations corresponds to $p_{k+1} = 0$, and for $p_{k+1} > 0$ we are dealing with $k+1$ populations. Since we have lowered the probability requirement we now have to increase n to raise it back up to the level of the preassigned P^*. This proves that the number of observations required increases with k and the only property used in the proof is that we are dealing with the least favorable configuration. □

Example 6.4.1   (Examples 6.1.3 and 6.3.1 continued).   In the problem of voter preference among five candidates that was introduced in Example 6.1.3, suppose that the investigation is in the process of being designed. A random sample of n persons is to be interviewed and asked their (one) preference among candidates A, B, C, D, and E, which, for convenience, we now label as $1, 2, \ldots, 5$, respectively. As mentioned in Example 6.3.1, the parameter p_j is the probability that a randomly selected voter in this city will prefer candidate j. A correct selection then is the choice of the candidate with the largest p value. We wish to determine the smallest sample size n such that the probability of a correct selection is at least $P^* = .90$ whenever $\delta = p_{[5]}/p_{[4]} \geqslant \delta^* = 1.4$, say.

From (6.1.7), the p values in the least favorable configuration are

$$p_{[1]} = p_{[2]} = p_{[3]} = p_{[4]} = \frac{1}{5.4} \cong .185 \quad \text{and} \quad p_{[5]} = \frac{1.4}{5.4} \cong .259.$$

(Because of rounding, the sum of these five p values is only .999.) For this configuration, and hence also for any other configuration in the preference zone, we want to have $P\{CS|R\} \geqslant .90$. Table H.1 in Appendix H, for $k = 5$, $P^* = .90$, and $\delta^* = 1.4$, indicates that $n = 271$ observations are needed; that is, 271 persons must be chosen randomly and interviewed to satisfy the specified (δ^*, P^*) requirement. Then according to the selection rule R, the candidate with the largest frequency among these 271 observations is asserted to have the largest p value. If there are ties for first place, we randomize over these tied candidates and select one of them in such a way that each has the same probability of being selected. At level .90, from (6.4.2) the confidence statement we can make about this selection is that p_s is at most $(1-1/1.4)100\% = 28.6\%$ smaller than the largest p value, or equivalently, from (6.4.3) that $p_{[s]}$ lies in the interval $[p_s, 1.4p_s]$.

If the investigator regards n as large in this example, then the reason is that δ^* was specified as small and/or P^* was relatively close to one. Although 271 is not a large sample size in actual polls, for practical reasons he may not want to interview so many persons. Then perhaps a more realistic choice of δ^* is somewhere between 1.6 and 2.0, in which case n is between 58 and 134 for the same $P^* = .90$. If he specifies δ^* between 1.6 and 2.0, he can achieve $P^* = .75$ with n between 29 and 68. With the very mild requirement that $\delta^* = 2.6$ and $P^* = .75$, we need interview only 14 persons. The important point here is that the investigator can consult the tables before deciding on the (δ^*, P^*) pair that he considers appropriate. Of course, any terminal statement about the true δ and the associated confidence level is determined by the sample size actually used.

Example 6.4.2 (Tournament). Three persons are playing in a tournament of n games. Each game involves all three players and has exactly one winner, that is, no draw or tie (two-way or three-way) is possible in a single game. For the three players, A, B, and C, the probabilities of winning in any single game are p_A, p_B, and p_C, respectively, where $p_A + p_B + p_C = 1$. The problem is to determine the one player that is the best of these three, that is, the one with the largest p value. How large should n be so that the procedure R satisfies $P\{CS|R\} \geqslant P^* = .95$ whenever the ratio of the largest p value to the second largest p value is at least $\delta^* = 2.0$? The entry for $k = 3$ in Table H.1 indicates that a tournament of 42 games is needed to satisfy this (δ^*, P^*) requirement. If two or all three players tie for first place in the tournament, one of the contenders for first place is selected randomly, with equal probability for each such contender. Of course, we can still give prizes to all contenders for first place, but we assumed that one overall winner has to be chosen and that the important prize is to be given to the overall winner. Note from (6.1.7) that the least favorable configuration in this problem consists of the three probabilities $\frac{1}{2}$, $\frac{1}{4}$, and $\frac{1}{4}$. Even a three-person poker game could be an example of such a game if we order the values of the suits in such a way that no game could end in a tie. The only further alteration of the usual poker game is that the dealer should periodically change and the order of play (clockwise versus counterclockwise) should also change periodically. The reason for the latter change is that we want to have probabilities p_A, p_B, and p_C that do not depend on the order of the players. Perhaps a better example is the game of darts with three players; here we need only assume that the scoring (based on the distance from the center of the target) is accurate enough that the probability of a tie in any game (or round) of the tournament is negligible.

Sometimes the application is not quite as straightforward as in the previous two examples. The following example illustrates an application where some modification is needed.

Example 6.4.3   (Chess).   Two persons, A and B, play a tournament of n independent chess games to see which one is the better player. (The question of who gets the white pieces and therefore makes the first move is handled, say, by alternating, and hence we need *not* take such factors into account if n is moderately large.) The complicating factor is that chess games *can* end in a draw, and the probability of a draw, say, p_0, might even be larger than the maximum of p_A, the probability that A beats B in a single game, and p_B, the probability that B beats A in a single game; these probabilities are assumed to be constant from game to game. Note that here we are *not* interested in identifying the largest of the *three* probabilities p_A, p_B, p_0, which sum to 1, but only in identifying the larger of the *two* probabilities p_A and p_B. As a result, we can reduce the present problem with $k = 3$ to the problem of a single binomial, that is, to the multinomial problem with $k = 2$, by considering the conditional probabilities, r_A and r_B, of A winning and B winning, given that a draw does not occur. The n determined by the (δ^*, P^*) requirement is then only the number of "undrawn" games to be played; the total number of games N is a random variable that depends heavily on p_0. Once we determine the number of undrawn games required, n, we can compute the *expected* total number of games $E(N)$ as

$$E(N) = \frac{n}{1 - p_0}, \tag{6.4.4}$$

but we cannot determine the value of the random variable N. (The result in (6.4.4) is well known from inverse sampling theory.)

Let r_A and r_B be the conditional probabilities corresponding to p_A and p_B, respectively, that is, $r_A = p_A/(p_A + p_B)$ and $r_B = p_B/(p_A + p_B)$, so that $r_A + r_B = 1$. Let $r_{[1]} \leqslant r_{[2]}$ denote the corresponding ordered values of r_A and r_B. It should be noted that if $p_{[1]}$ and $p_{[2]}$ denote the ordered values of p_A and p_B, the ratio $r_{[2]}/r_{[1]}$ is equal to $p_{[2]}/p_{[1]}$, the distance measure in (6.1.4). Hence if we define the distance measure here as $\delta = r_{[2]}/r_{[1]}$, the intuitive meaning of δ and δ^* is similar in both the conditional and unconditional framework. Suppose we take $P^* = .99$ and $\delta^* = 2.0$; this means that we want a procedure R for which the probability of a correct selection is at least .99 whenever $\delta \geqslant 2.0$. From Table H.1 for $k = 2$ in Appendix H with $\delta^* = 2.0$ and $P^* = .99$, we find that $n = 47$ undrawn games are needed to satisfy this requirement.

From (6.1.7), for $\delta^* = 2.0$ and $k = 2$, the least favorable configuration in terms of the r values, that is, the conditional least favorable configuration, is $r_{[1]} = \frac{1}{3}, r_{[2]} = \frac{2}{3}$. Thus the least favorable configuration in terms of the (unconditional) p values is

$$p_{[1]} = \left(\frac{1}{3}\right)(p_A + p_B), \quad p_{[2]} = \left(\frac{2}{3}\right)(p_A + p_B), \quad p_0 = 1 - (p_A + p_B).$$

Note that $p_A + p_B = 1 - p_0$ is the expected proportion of undrawn games. From (6.4.4) the expected total number of games $E(N)$ required to achieve 47 undrawn games is given by $E(N) = 47/(1 - p_0)$. Since the denominator can take on any value between 0 and 1, the expected value of N can be substantially larger than 47. In fact, $E(N)$ approaches infinity as p_0 approaches 1, or equivalently, as the sum $p_A + p_B$ approaches zero. Thus if players A and B are close to being evenly matched then p_0 will usually be large (i.e., they may have many draws), and the tournament may require a large number of games.

For this reason it might be desirable to incorporate into the rules of the tournament an upper bound U on the total number of games to be played. If U is large enough (say three times the table value of n determined by the (δ^*, P^*) requirement), this modification often has little effect on the critical properties of our procedure R. Thus if $p_0 = \frac{1}{2}$ (which might be useful in the preceding chess example), the expected total number of games required is $47/(\frac{1}{2}) = 94$. If we take $U = 3(47) = 141$, then it can be shown that the probability of obtaining 47 undrawn games *before* reaching the upper bound U is approximately .99998. Hence the bound U would have almost no effect on the probability of a correct selection for our procedure, but it does protect us against long sequences of games, most of which end in a draw. However, the reader should note that this property was shown only for $p_0 = \frac{1}{2}$. In some extreme cases (for example, if p_0 is close to 1), the effect of U on the properties of R would have to be carefully investigated.

6.4.2　Calculation of the Operating Characteristic Curve

Suppose now that the total sample size n is fixed and not under the control of the investigator, say because of practical considerations. For example, an interviewer may enter a classroom with a fixed number of students. If this number were more than that required by the (δ^*, P^*) requirement, it would not make sense to remove some students just to get the exact n. If this number were less than that required, nothing could be done. On the other hand, data consisting of n observations may have already been obtained, and that n was fixed without any specified pair (δ^*, P^*). Nevertheless the performance of the selection rule R is still of interest, so we need to calculate the operating characteristic curve.

As noted in Section 6.1, the operating characteristic curve cannot be found (as a function of δ alone) under the generalized least favorable configuration, since the calculation depends on all the p values, and not just on δ. However, the curve can be found under the least favorable configuration. That is, we can give the operating characteristic curve for the multinomial selection procedure as $P_{LF}\{CS|\delta^*, n\}$, the probability of a correct selection under the least favorable configuration, as a function of δ^* for given k and n. (Of course, for fixed k, n, and δ^*, this probability is equal to the confidence level $CL(\delta^*|R)$ associated with the intervals in (6.4.1), (6.4.2), or (6.4.3).) In other words, to obtain the operating characteristic curve, we force the specified *constants* (δ^*, P^*) to be functionally related variables, since n and k are now both fixed, and, as in the binomial case, keep the * notation to emphasize the fact that they do not denote true values. Thus the operating characteristic curve represents the locus of all specified pairs (δ^*, P^*) that require the same value of n. Table H.1 of Appendix H does not give P^* directly as a function of δ^* for given n, but the values may be found by interpolation.

As a numerical illustration, suppose that $k = 3$. Some points on the operating characteristic curves for $n = 40$ and $n = 50$ are given in Table 6.4.1, and the two curves are shown in Figure 6.4.1. Each curve shows how much P^* increases for some increase in δ^*, and vice versa, for the fixed n.

Example 6.4.4   (Voting). A certain large community needs to decide whether to float a school bond issue. In order to assess the feeling of the population, n persons are chosen randomly and asked whether they are "in favor of," "indifferent to," or "opposed to" the bond issue. Let p_1, p_2, and p_3 be the probabilities associated with these three types of response, respectively, for the entire (large) population. We would like to make a correct selection of the most frequent response with probability at least $P^* = .90$ whenever $\delta = p_{[3]}/p_{[2]} \geq 2.0 = \delta^*$.

From Table H.1 of Appendix H for $k = 3$ we obtain the value $n = 29$. Consequently, we should take a random sample of size $n = 29$ and count the frequency of persons who respond "in favor," "indifferent," and "opposed". The category that has the largest frequency will be designated as the response with the largest probability.

For example, if we observe 15 persons opposed, 10 persons in favor, and 4 indifferent, then we would assert that p_1, the probability of "being

Table 6.4.1   *Corresponding values of* $P_{LF}\{CS|\delta^*,n\}$ *and* δ^* *for* $k = 3$, $n = 40$ *and* $n = 50$

$P_{LF}\{CS|\delta^*,n\}$	δ^* for $n = 40$	$n = 50$
.750	1.46	1.41
.900	1.80	1.70
.950	2.03	1.89
.975	2.27	2.08
.990	2.57	2.32

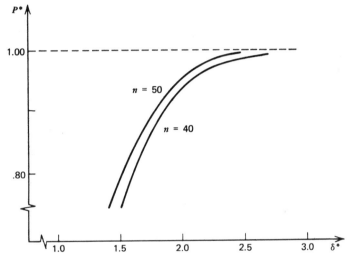

Figure 6.4.1   Values of (δ^*, P^*) for $k = 3$, $n = 40$ and $n = 50$.

opposed" is the largest in the population. Furthermore, the probability that this selection is correct is at least .90 if the true response probabilities p_1, p_2, p_3 have the property that $p_{[3]}/p_{[2]} \geqslant 2.0$.

But now suppose that the investigator feels that he could afford to take a sample of size 40 or 50 instead of 29. Then how would the (δ^*, P^*) values compare with the $(2.0, .90)$ result for $n = 29$? For $n = 40$ and $n = 50$, the pairs (δ^*, P^*) are shown in Table 6.4.1 and Figure 6.4.1. If the sample size is increased to 40, the P^* value increases to approximately .95, and for $n = 50$, P^* increases further to about .97, keeping $\delta^* = 2.0$ in both cases. Thus the investigator must decide whether this increase in P^* warrants the extra observations required.

In the following example the value of n is fixed by tradition, and it would require considerable re-education to attempt to change it.

Example 6.4.5 (World Series). In the World Series of baseball, there is a playoff between the two teams that win the pennant in their respective leagues. These teams, A and B, then play a series of seven games, at most. The team that wins four out of these seven games is declared the national champion. The question of interest is whether $n = 7$ games is a sufficiently large number in order to have a high probability that the better team is declared the champion. The fact that the series may be curtailed—that is, the full seven games may not be played because the series need not be continued after one team has four wins—is of no concern to us at present because the winner must have at least four wins whether or not the series is curtailed and the *probability* of winning the series is not affected by the curtailment. (The effect of curtailment is discussed more fully in Section 6.6.) The problem here is further simplified by the fact that a baseball game cannot normally end in a draw, that is, we can take the probability of a draw to be zero.

We now study this model and use our theory to answer the question about whether $n = 7$ games is sufficient. Let p_A (and $p_B = 1 - p_A$) denote the probability that A beats B in a *single* game (so that $p_B = 1 - p_A$ is the probability that B beats A in a single game), and let $p_{[1]} \leqslant p_{[2]}$ be the ordered values of these two probabilities.

Our goal is to select the better team, that is, the one with parameter $p_{[2]}$. For the two pennant-winning teams the ratio $\delta = p_{[2]}/p_{[1]}$ is within the range 1.00–2.00. If we took into account the proportion of games won by each team during that same season, a reasonable estimate of δ would probably be between 1.20 and 1.50. The specification $\delta^* = 1 p_{[1]} = \frac{5}{11} \cong .45$ and $p_{[2]} = \frac{6}{11} \cong .55$. Let us first see what n is needed for the pair $\delta^* = 1.20$ and $P^* = .90$, say. From Table H.1 of Appendix H for $k = 2$, we find $n = 199$. This means that the series would continue until one team wins 100 games out of a tournament of at most 199 games, and thus a series based on four out of seven games is an extremely poor procedure for selecting the better of two pennant-winning teams. In fact, a direct interpolation shows that for $\delta^* = 1.20, n = 7$, the corresponding value of P^* is .60. If we specified a value of δ^* smaller than 1.20, then the corresponding value of P^* would be even closer to .50 when $n = 7$. As a result (if the teams are close in ability) a World Series based on four games out of seven is not much better than the toss of a coin as a procedure for selecting the better team.

It could be argued that if approximately 199 games are needed to decide which one of the two teams is better, then perhaps each inning should be regarded as a separate game so that the seven-game series is magnified to $7(9) = 63$ inning-games. However, this argument is invalid since in about

75% of the innings the score would be tied (0–0, 1–1, or the like); hence by the argument given in the chess problem (Example 6.4.3) and the result in (6.4.4), the expected number of innings needed to obtain 199 *undrawn* inning-games is 4(199)=796, and this result corresponds to 796/9, or about 89 conventional games of nine innings each. This radical change in the game of baseball would help to reduce (from 199 to 89) the number of conventional games needed to determine the better team, but even the result of 89 games is still far from the traditional number of seven games.

There are various ways to estimate the true δ in this example and therefore find a reasonable value to specify for δ^*. One possibility is to estimate p_1 and p_2 by the ratio of games won by the two teams during that same season, and estimate δ by the ratio of these estimates. For example, in 1970, the two teams playing in the World Series were Cincinnati, representing the National League, and Baltimore, representing the American League. The data on games won and lost by these teams during the 1970 season are as follows:

	W	L	Percent won
Cincinnati	102	60	.630
Baltimore	108	54	.667

Thus we might specify δ^* as .667/.630 = 1.06; however, this would require a very large sample size. If we use $p_{[1]} = .45$, $p_{[2]} = .55$ we obtain $\delta^* = .55/.45 = 1.26$, whereas $p_{[1]} = .40$, $p_{[2]} = .60$ yields $\delta^* = 1.50$. Thus, a choice between 1.20 and 1.50 appears quite reasonable. With $P^* = .90$ and $\delta^* = 1.20$, the required n would be 199, and for $\delta^* = 1.50$ the required n is 42.

To summarize this example, if the teams are approximately equal in ability, then the number of games in the World Series should be magnified by a factor of about 28 for the conventional nine-inning game. Even if each inning is counted as a separate game, the traditional number of 7 conventional games (or 63 inning-games) is far from sufficient. A magnification of about 12.7 is desirable; this would give 800 inning-games, which corresponds to about 89 conventional games. This is a conservative estimate of the number of games needed to determine the better of two closely matched pennant-winning teams. Of course if the teams are highly discrepant in ability, then the probability of a correct selection may be ample under the present format of seven nine-inning games, but it is not likely that two pennant-winning teams will be so unevenly matched.

6.5   SELECTING THE CATEGORY WITH THE SMALLEST PROBABILITY

In some situations the goal of interest is to select the category with the smallest probability in the multinomial distribution, that is, the one with p value equal to $p_{[1]}$. The obvious selection rule is to identify the category that gives rise to the smallest cell frequency $x_{[1]}$ in the sample and assert that it is the best category. In spite of this clear analogy, it is interesting to note that this problem is completely unrelated to that of selecting the category with the largest cell probability. Not only are the two problems not equivalent for $k \geqslant 3$, but one is not reducible to, nor obtainable from, the other.

For $k \geqslant 3$, if the distance is defined by the *ratio* $p_{[2]}/p_{[1]}$ (or $p_{[1]}/p_{[2]}$), the ranking and selection problem has no solution. Hence it is necessary to use a different measure of distance. For the *difference* $\delta = p_{[2]} - p_{[1]}$, the problem does have a solution. However, if the difference is used, an entirely new set of tables based on a new set of formulas is needed. Conversely, the *difference* measure of distance introduces considerable difficulty into the problem of finding the category with the largest probability; this point is made clear in the Notes, References, and Remarks at the end of this chapter.

The entries in Table H.2 of Appendix H are the minimum sample size needed for selecting the category with the smallest probability such that a specified (δ^*, P^*) requirement is met. This requirement is that the probability of a correct selection is at least P^* whenever the configuration of parameters is in the preference zone defined by

$$\delta = p_{[2]} - p_{[1]} \geqslant \delta^*.$$

*6.6   CURTAILMENT

The idea of curtailment for $k = 2$ was introduced in Example 6.4.5. Recall that the two teams play each other until one team wins a total of four games. Although the maximum number of games ever played is seven, the champion team may be determined before seven games are played. For example, it may happen that one team wins four out of the first five games, including the fifth game. Then there is no need to play the full seven games to determine the champion; this outcome can be viewed as a saving of two games.

In general, curtailment is defined with respect to any rule as terminating the drawing of observations at a number smaller than n as soon as the final decision is determined; here n is the maximum number of observations that one is allowed to take. Thus curtailment is an "early stopping rule," and it yields a saving in the number of observations taken. Therefore we now discuss curtailment with respect to our sampling rule of looking for the cell with the highest frequency in n observations; we wish to evaluate the amount of saving that may result for various values of k and n.

The expected sample size needed for some specified performance is a key quantity for assessing the efficiency of a ranking and selection procedure (as in any statistical procedure). Let N denote the (random) number of observations required by the curtailment procedure. Any procedure, say R', that uses curtailment clearly has a smaller expected sample size $E(N|R')$ than the fixed sample size n needed for the corresponding procedure R without curtailment. Since both procedures satisfy the same probability requirement, the procedure R' is more efficient than the procedure R. Thus it is of considerable interest to compute $E(N|R')$. The difference between n and $E(N|R')$ is the expected saving due to curtailment. The expected percentage saved is

$$S_n = \left[\frac{n - E(N|R')}{n} \right] 100 = \left[1 - \frac{E(N|R')}{n} \right] 100. \qquad (6.6.1)$$

We discuss separately the small-sample and large-sample aspects of the expected saving.

The curtailment problem was entirely overlooked in the early journal literature on the multinomial selection problem. One could perhaps argue that the amount of saving due to curtailment is small and in fact tends to disappear for large n (with a fixed k and a fixed configuration of p values). This argument is answered in the following discussion. It should be emphasized that for any finite n there is always some positive expected saving; that is, the expected saving is never zero, and obviously, it is never negative. Hence it deserves some careful analysis.

We first consider the large sample aspects of the curtailment procedure. If n is very large the saving is positive for most configurations. The only exceptions are those configurations with $p_{[k]} = p_{[k-1]}$ (so that $\delta = 1$); here the asymptotic percent saved is equal to zero. In the generalized least favorable configuration where $p_{[1]} = p_{[2]} = \cdots = p_{[k-1]} = p'$, say, the percent saved due to curtailment approaches the limit

$$\frac{100(p_{[k]} - p')}{1 + p_{[k]} - p'} \qquad (6.6.2)$$

as n approaches infinity. This result holds for *any* k, and also for *any* configuration if p' in (6.6.2) is replaced by $p_{[k-1]}$. Although (6.6.2) does not involve k explicitly, the result does depend on k since the p values depend on k in the least favorable and generalized least favorable configurations. Thus for any specified set of values of $p_{[k]}$ and p', or of $p_{[k]}$ and $p_{[k-1]}$, we can compute the asymptotic percent saved by a simple substitution in (6.6.2). The maximum possible saving is always 50%. It is equal to this maximum of 50% for the configuration where $p_{[k]}=1$ (so that all other p values are equal to zero and $\delta = \infty$). As an example, suppose that $k=3$. Using (6.6.2), if $(p_{[1]},p_{[2]},p_{[3]})=(\frac{1}{4},\frac{1}{4},\frac{1}{2})$, we have a saving of approximately 20% for large values of n, and if $(p_{[1]},p_{[2]},p_{[3]})=(\frac{1}{8},\frac{2}{8},\frac{5}{8})$, we have a saving of approximately 27.3% for large values of n.

Of course, (6.6.2) cannot be computed if we do not know either both $p_{[k]}$ and $p_{[k-1]}$, or both $p_{[k]}$ and p'; knowledge of δ is not sufficient. However, the asymptotic saving due to curtailment can be computed as a function of δ when the true configuration is the generalized least favorable configuration (or equivalently, as a function of δ^* under the least favorable configuration). The percentages saved are given in Table I.2 of Appendix I for $k=2,3,4,5,10$, in each case for finite values of δ from 1.0 to 3.0 (by jumps of 0.2). Note that these results also depend on k. These percentages show a strict monotonicity, increasing as δ (or δ^*) increases for fixed k and decreasing as k increases for fixed δ (or δ^*). Although the percent saved is not very large in any case, it is worth noting that the saving is positive regardless of whether $p_{[k]} \geqslant 0.5$ or $p_{[k]} < 0.5$. As noted earlier, the only case in which the saving is negligible is when n is very large and $\delta = 1.0$, that is, when $p_{[k]}=p_{[k-1]}$; the expected saving is *never* zero.

We now consider the small sample aspects of curtailment. Table I.1 in Appendix I, for $k=2,3,4,5,10$, gives the exact values for the expected sample size N using curtailment, $E(N|R')$, as a function of δ (or δ^*) when the true parameter configuration is the generalized least favorable configuration (or the least favorable configuration), for various values of n. Here n, as usual, denotes the smallest integer that satisfies the (δ^*,P^*) requirement with the procedure R, and is thus given by Table H.1 in Appendix H. However, from the point of view of curtailment—that is, when the investigation may be terminated early—n is the *maximum* number of observations that might be needed. Thus the tables in Appendix I give the expected value of the number N of observations that will *actually* have to be taken to terminate the multinomial selection problem.

As an example, consider the situation where $k=3$, $P^*=.75$, and the configuration is $(p_{[1]},p_{[2]},p_{[3]})=(\frac{1}{4},\frac{1}{4},\frac{1}{2})$, so that $\delta=2.0$. From Table H.1 for $k=3$ in Appendix H, we obtain $n=12$. Then by Table I.1, for $k=3$ in Appendix I, again with $\delta=2.0$, the expected number of observations N

required using curtailment is only 10.63, which is a saving of $12 - 10.63 = 1.37$ observations, or 11.4%. Table I.2 shows that as n increases (say because P^* is increased), this saving increases steadily to the limiting value of 20% (which holds for $k = 3$ and $\delta = 2.0$ under the generalized least favorable configuration). As explained in the discussion of the asymptotic case, the same limiting value of 20% holds for arbitrary k under any (true) configuration for which $p_{[k]} - p_{[k-1]} = .25$.

REMARK. Although the tables in Appendix I show that the *expected sample size* using curtailment is a strictly increasing function of n for any fixed k and δ, the *percent saving* is *not* monotonic in n for all values of n. This is due to the number-theoretic aspects of the problem and is not the result of rounding. □

For any k consider the special case where $p_{[k]} = 1$ so that all other p values are equal to zero and δ is infinite. Then the expected sample size using curtailment is the integer part of $(n + 2)/2$ (i.e., the largest integer that does not exceed $(n + 2)/2$). Hence from (6.6.1) the percentage saved is $50[1 - (2/n)]\%$ for any n (which is approximately 50% when n is large). Thus when $\delta = \infty$ we obtain the maximum possible saving, and the same maximum holds for each value of k.

To investigate further the small-sample aspects of curtailment, we define $f_{[1]} \leqslant f_{[2]} \leqslant \cdots \leqslant f_{[k]}$ as the ordered frequencies of the k categories after m observations have been taken, where $1 \leqslant m \leqslant n$ and n is the maximum number that may be required. Let $\{x\}^+$ denote the smallest integer greater than or equal to x (thus if $x = 2.3$ or $x = 2.9$ or $x = 3$, then $\{x\}^+ = 3$). When the sampling is terminated (i.e., a winner is determined or n is reached with ties for first place), the largest frequency $f_{[k]}$ must lie in the interval:

$$\left\{ \frac{n}{k} \right\}^+ \leqslant f_{[k]} \leqslant \left\{ \frac{n+1}{2} \right\}^+. \tag{6.6.3}$$

If we stop before taking n observations because of curtailment, then the largest frequency $f_{[k]}$ must be equal to or greater than $(n - 3 + 2k)/k$. In symbols, this says that

$$f_{[k]} \geqslant \left\{ \frac{n - 3 + 2k}{k} \right\}^+. \tag{6.6.4}$$

For example, suppose $k = 3$ and $n = 10$. Then (6.6.3) indicates that in general the largest frequency at stopping time is always between four and six inclusive, and (6.6.4) indicates that under early curtailment the largest frequency must be at least five. (Clearly, f values such as $(4,3,2)$ and $(4,2,2)$, for example, are not possible early stopping points because a tie for first place could still result if 10 observations can be taken. Some examples of early stopping points are $(5,3,1)$, $(5,2,2)$ $(5,2,1)$, $(5,1,1)$, and $(6,0,0)$.)

The conditions in (6.6.3) and (6.6.4) are necessary, but not sufficient, for terminating the sampling. The basic test condition, which is both necessary and sufficient, is that

$$f_{[k]} - f_{[k-1]} > n - m, \tag{6.6.5}$$

where m is the number of observations already taken and n is the maximum number of observations that can be taken.

In the remainder of this section we give some exact and approximate analytic expressions for the expected value of the number of observations N needed under curtailment. For $k = 2$ the result is exact. For $k \geqslant 3$, separate approximations are given for n odd and n even; one of these approximations is also an upper bound.

For $k = 2$ the smallest integer value of n that satisfies the (δ^*, P^*) requirement is always an odd integer (see Table H.1 for $k = 2$), since the probability of a correct selection for every even integer is exactly the same as for the immediately preceding odd integer. Hence there can be no ties for first place when $k = 2$ and we curtail (that is, terminate the observations) as soon as either category has a frequency greater than or equal to $(n+1)/2$. The World Series baseball tournament is a well-known example, where a team needs four games out of $n = 7$ in order to win the series.

Consider a tournament consisting of at most n games for $k = 2$ teams with probabilities of success $p_{[1]} \leqslant p_{[2]}$ in a single game. The exact expected value of the number of games N played under the curtailment procedure R' for *odd* n (which are the only integers we need to consider by the preceding discussion) is

$$E(N|R') = \frac{n+1}{2} \left[\frac{1}{p_{[2]}} I_{p_{[2]}}\left(\frac{n+3}{2}, \frac{n+1}{2}\right) + \frac{1}{p_{[1]}} I_{p_{[1]}}\left(\frac{n+3}{2}, \frac{n+1}{2}\right) \right],$$

$$\tag{6.6.6}$$

where $I_x(a,b)$ is the well-known standard incomplete beta function. To carry out any computation, Table J.1 in Appendix J provides values of $I_x(w+1,w)$. Consequently, we may compute the expected number of games N with the help of a single table by using the entry pairs $(w, p_{[1]})$ and $(w, p_{[2]})$, where $w = (n+1)/2$. These are then combined as indicated in

(6.6.6). In the generalized least favorable configuration we set

$$p_{[2]} = \frac{\delta}{\delta+1}, \quad p_{[1]} = 1 - p_{[2]} = \frac{1}{\delta+1}, \tag{6.6.7}$$

and for the least favorable configuration we set $\delta = \delta^*$ in (6.6.7). The odd integer n is obtained from Table H.1 for $k=2$ in Appendix H as a function of δ^* and P^*. For $n=1$ it can be verified that the right-hand side of (6.6.6) is exactly 1, and for $n=3$ it is

$$2 + \left(1 - p_{[1]}^2 - p_{[2]}^2\right) = 2\left(1 + p_{[1]}p_{[2]}\right). \tag{6.6.8}$$

For $k \geqslant 3$ the problem is more complicated. The idea of stopping after n observations or as soon as any cell has $\{(n+1)/2\}^+$ observations in it is now only a sufficient condition for stopping and thus in general leads to an upper bound on $E(N|R')$. However, this upper bound is exact for certain small values of k and odd n (e.g., $k=3$ and $n=1,3,5,7$), whenever early curtailment (i.e., stopping before n observations) implies that some one cell has at least $\{(n+1)/2\}^+$ observations in it. As a result, we expect this upper bound to be close for many values of k and odd n.

For $k \geqslant 3$ and n odd, an expression for this upper bound can be given for the least favorable configuration. However, it is rather complicated to compute. Hence we give here only an *approximation* that can be calculated using Table J.1 of the incomplete beta function. The result is

$$E(N|R') \cong \left(\frac{n+1}{2}\right)(\delta^*+k-1)\left[\frac{1}{\delta^*}I_{\delta^*/(\delta^*+k-1)}\left(\frac{n+3}{2},\frac{n+1}{2}\right)\right.$$

$$\left. + (k-1)I_{1/(\delta^*+k-1)}\left(\frac{n+3}{2},\frac{n+1}{2}\right)\right]$$

$$+ nI_{(k-1)/(\delta^*+k-1)}\left(\frac{n+3}{2},\frac{n+1}{2}\right). \tag{6.6.9}$$

To illustrate the calculation from the formula in (6.6.9), suppose we consider $k=4$, $\delta^*=2.0$, and $n=15$ and we wish to approximate the expected number of observations needed under curtailment. Using Table J.1 with $w=(n+1)/2=8$, we obtain

$$E(N|R') = 8(5)\left[.5I_{.4}(9,8)+3I_{.2}(9,8)\right]+15I_{.6}(9,8)$$

$$= 20(.1423) + 120(.0015) + 15(.7161) = 13.77.$$

The exact result from Table I.1 is 13.78.

6.7 NOTES, REFERENCES, AND REMARKS

It is quite curious and nonintuitive that the problem of selecting the category with the largest probability is not equivalent (and not reducible) to that of selecting the category with the smallest probability. The former problem is treated in Bechhofer, Elmaghraby, and Morse (1959), and the latter is in a more recent paper by Alam and Thompson (1972). Although these two papers both use a standard type requirement for the probability of a correct selection, they actually require *different* measures of distance to obtain a solution. In the former paper the measure of distance is the *ratio* $p_{[k]}/p_{[k-1]}$ of (6.1.4), but in the latter the measure of distance is the difference $p_{[2]}-p_{[1]}$. Moreover, the latter paper shows that for $k \geqslant 3$ the ratio $p_{[2]}/p_{[1]}$ (or $p_{[1]}/p_{[2]}$) cannot be used as a measure of distance. The idea is simply that we can keep this ratio fixed and push both $p_{[1]}$ and $p_{[2]}$ close to zero by letting $p_{[k]}$ increase (and $p_{[k]}$ need not be equal to $p_{[2]}$ since $k \geqslant 3$). Then the categories with probabilities $p_{[1]}$ and $p_{[2]}$ cannot be distinguished from each other without taking larger and larger sample sizes, and thus the P^* requirement cannot be met in a practical application.

On the other hand, the analogous question arises as to whether we could use the difference $p_{[k]}-p_{[k-1]}$ as a measure of distance for the problem of selecting the category with the *largest* probability. Here the answer is again negative if we assume that the least favorable configuration is obtained by solving the equations:

$$p_{[k]}-p_{[k-1]}=\delta^*, \quad p_{[k-1]}=p_{[k-2]}=\cdots=p_{[1]}, \quad \sum_{j=1}^{k} p_{[j]}=1.$$

For example, suppose that $k=3$ and $\delta^*=.5$, so that

$$p_{[3]}=\tfrac{2}{3} \quad \text{and} \quad p_{[1]}=p_{[2]}=\tfrac{1}{6}=p. \qquad (I)$$

We wish to compare this configuration with the alternative one:

$$p_{[3]}=\tfrac{2}{3}, \quad p_{[2]}=\tfrac{1}{3}, \quad p_{[1]}=0. \qquad (II)$$

For $n=3$ the probability of a correct selection for each of the preceding two configurations (I) and (II), denoted by $P\{\text{CS}|(I)\}$ and $P\{\text{CS}|(II)\}$, respectively, can be calculated as follows:

$$P\{\text{CS}|(I)\}=\left(\tfrac{2}{3}\right)^3+3\cdot2\left(\tfrac{4}{9}\right)\left(\tfrac{1}{6}\right)+\tfrac{6}{3}\left(\tfrac{1}{6}\right)^2\left(\tfrac{2}{3}\right)=\tfrac{21}{27}=\tfrac{7}{9},$$

$$P\{\text{CS}|(II)\}=\left(\tfrac{2}{3}\right)^3+3\left(\tfrac{4}{9}\right)\left(\tfrac{1}{3}\right)=\tfrac{20}{27}<\tfrac{7}{9}.$$

Hence, at least for $n=3$, the configuration (I) is not LF for the proposed measure of distance that uses the difference $p_{[k]} - p_{[k-1]}$. Thus, although a deeper analysis may solve this problem, the present answer is that (I) is *not* the correct LF configuration.

The curtailment problem arises almost exclusively in the multinomial problem, since under our simple selection rule, the winning category is often determined before we complete all our sampling, as in the World Series of baseball. No mention of curtailment has previously been made and the development and tables dealing with the curtailment problem are new with this book.

PROBLEMS

6.1 Limestone beds in a stratigraphic unit can be classified according to texture as fine-grained, medium-grained, or coarse-grained. A geologist is interested in determining the textural classification that is predominant in a particular unit. He plans to take a random sample of n hand-specimen-sized pieces from this unit and classify them into these three categories. How large should n be such that he can state with confidence level $P^* = .90$ that p_s is at most 15% smaller than the true largest probability $p_{[3]}$, where p_s is the probability that a specimen is classified in the category selected as predominant? What is the least favorable configuration in this case?

6.2 In a certain state an insurance company classifies each automobile accident into one of the following four risk groups, according to which factor is designated as the major cause of the accident: Driving Under the Influence of Alcohol, Moving Violation, Under 21 Years of Age, and Over 65 Years of Age. The company wants to determine which one of these groups is the highest-risk group and hence should have the highest insurance rate. Records have been kept for six successive years, and hence a large amount of data is available. However, these data have not been summarized, and hence the decision must be based on sample data. How large a sample should they take so they can have $P^* = .95$ confidence in their selection when $\delta^* = 1.60$? What is the relevant confidence statement?

6.3 A management consultant wants to know whether there is any relationship between the sense of humor of a salesman of industrial chemicals and the volume of sales obtained. A random sample of 300 salesmen with outstanding sales volume over the past year are each classified into one of three categories—Poor, Average, and Excellent sense of humor. The data are as follows:

Poor	Average	Excellent	Total
73	135	92	300

(a) What kind of sense of humor is dominant among salesmen with outstanding sales records?

(b) Make an appropriate confidence statement for $P^* = .90$ and $\delta^* = 1.20$.

(c) Calculate the operating characteristic curve.

CHAPTER 7

Nonparametric Selection Procedures

7.1 INTRODUCTION

In each of the preceding chapters we had k values of some specific parameter θ, one representing each of k groups or populations, and the θ values were used to order the groups or populations. In most of these problems the parameter θ indexed a particular underlying distribution model $f(x; \theta)$ that was assumed to be the normal distribution, the binomial distribution, or the like. Then the best population among the k represented was defined to be the one with the largest (or correspondingly, the smallest) θ value.

Now we consider the situation where the underlying distribution model is not known and we are not willing to make the assumption that it belongs to a particular family. Such problems are usually called nonparametric, although the term *distribution-free* is sometimes used in order to emphasize the fact that the results are valid irrespective of the form of the underlying distribution model. Even though there is no specific distribution, we can still order the populations on the basis of the value of some positional measure representing each distribution. These measures are called quantiles or percentiles (to be defined specifically in Section 7.2). The most familiar quantile is known as the median or the 50th percentile point, but the 90th percentile point is also a quantile. Suppose we have k unspecified populations. Then the goal in this nonparametric problem might be stated as identifying the one population with the largest median, or the largest ninetieth percentile point, or the like.

Before discussing the problem and its solution in more detail, we give a few specific examples in which the goal is of practical interest.

Example 7.1.1. In a study of gasoline consumption by city buses, four blends of gasoline were considered. A specified route was laid out, and a specific time was selected. Then a single driver made 60 repeated runs over

187

this route with a single bus at the same time on weekdays, 15 times with each blend. By using the same driver, vehicle, and time for each run, it was hoped that only the difference in blends would affect gas mileage. The goal is to determine which blend has the largest median mileage per gallon.

Example 7.1.2.   A university is planning to conduct an extensive recruiting campaign for exceptionally bright graduate students, but the campaign will be limited to a single large metropolitan area. They are presently considering $k = 5$ different areas. Since the campaign is to be directed only at exceptionally bright students, they define the best area as the one for which the 75th quantile score on the Graduate Record Examination (GRE) is higher than that for any other area. A random sample of n prospective graduate students is taken from each of the five areas, and each student is given the GRE. The selection of an area for the campaign depends on these scores. We might be interested in determining the value of n needed to satisfy certain probability requirements.

Example 7.1.3.   A farmer who raises pigs and hogs for slaughter wants to determine which breed of swine has the largest median percentage of meat relative to body weight at the time of slaughter. Some of the primary breeds used for meat in the United States are the Poland China, the Duroc-Jersey, the Chester White, the Hampshire, and the Spotted Poland China. It is known that many brood sows are fattened to greatest profit after the second or third litter. Thus the farmer obtained a random sample of n sows from each of these $k = 5$ breeds and slaughtered them after the third litter. The percentage of meat relative to body weight is measured for each sow. The selection procedure depends on these percentages. We may wish to determine an n value to satisfy certain probability requirements.

When the median is the quantile of interest, the goal in this chapter may be considered analogous to the goal of selecting the population with the largest mean, because the mean and the median are both parameters that describe central tendency. In fact, the mean and median are equal for any symmetric distribution, which includes the normal distribution. However, the median is less affected by extreme values than the mean. As a result, for skewed distributions (for example, a distribution of personal incomes), the median may be a more "central" parameter than the mean. In such cases the investigator may be more interested in ranking the populations according to their medians than according to their means. Further, this nonparametric procedure for selecting the distribution with the largest median is valid for any continuous distribution; the validity of the procedures in Chapters 2 and 3 depends on the normal distribution model. In

Example 7.1.3 the data are measured as percentages, and it is known that percentages are rarely normally distributed without some transformation of the percentages. Hence the nonparametric procedure of this chapter may be a more appropriate one.

In general, the median procedure should be used instead of the mean procedure if the median is considered more representative of central tendency than the mean, if the population distributions are completely unknown, if the assumption of normal distributions seems untenable, or if the sample sizes are very small so that we cannot rely on the Central Limit Theorem to justify a procedure that depends on the assumption of normal distributions.

7.2 FORMULATION OF THE PROBLEM

Most people are familiar with the terms *percentile, median, quartile,* and the like, when used in relation to measurements, particularly in reports of test scores. These are points that divide the measurements into two parts, with a specified percentage on one side. Such points can also be defined for a probability distribution; then they are usually called quantiles. A quantile can be interpreted as a parameter of the probability distribution, whether the distribution is known or unknown. We now give a formal definition of a quantile.

Recall that a cumulative probability distribution function $F(x)$ can be depicted graphically as in Figure 7.2.1 or Figure 7.2.2, depending on whether the distribution is continuous or discrete, respectively. For our present problem we assume that the distribution $F(x)$ is continuous and is increasing (as in Figure 7.2.1, rather than Figure 7.2.2; see the following Remark), and is not of any particular specified form. Then a qth quantile,

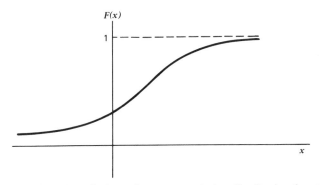

Figure 7.2.1   A typical continuous cumulative distribution function.

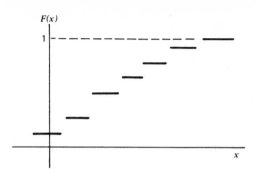

Figure 7.2.2   A typical discrete cumulative distribution function.

or a quantile of order q, for any $0 \leqslant q \leqslant 1$, is uniquely defined as the number ξ_q such that $F(\xi_q) = q$. This is equivalent to saying that for the random variable with distribution $F(x)$, the probability that the variable is less than or equal to the number ξ_q is equal to q, and, of course, that the probability that it is greater than ξ_q is equal to $1 - q$. The value of ξ_q can be determined graphically as shown in Figure 7.2.3. A horizontal line is drawn from the point q on the ordinate to the curve $F(x)$ and a perpendicular line is dropped to the abscissa. The point where this perpendicular line meets the abscissa is the number ξ_q. Note, then, that not only does q determine ξ_q uniquely, but also ξ_q determines q uniquely, for any given $F(x)$. Hence we may refer to either the ξ_q value or the q value.

REMARK.   The definition of a quantile in the preceding paragraph is fully satisfactory only in those cases where the point ξ_q exists and is a unique number. Obviously, for the discrete distribution depicted in Figure 7.2.2 (and in fact for any other discrete distribution), the qth quantile is not

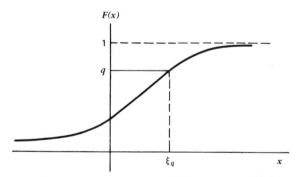

Figure 7.2.3   Determination of the qth quantile ξ_q.

unique for all q; the same problem arises for a continuous distribution if it is perfectly flat anywhere within the domain of x. Such complications are avoided if we assume that each of the distributions is continuous and strictly increasing. Further, this assumption makes these order quantiles simple to understand and interpret, and does not change the problem in any essential manner. □

Suppose that we want to order $k = 2$ cumulative distributions, $F_1(x)$ and $F_2(x)$, on the basis of the values of their respective qth quantiles for a given q, $0 \leq q \leq 1$. Then we say that

$$F_1(x) \precsim F_2(x) \quad \text{if} \quad \xi_{q1} \leq \xi_{q2},$$

where ξ_{q1} is the qth quantile for $F_1(x)$ and ξ_{q2} is the qth quantile for $F_2(x)$. This means that for the q in Figure 7.2.4, the ordering is $F_1(x) \precsim F_2(x)$, whereas for the same distributions but another q value as in Figure 7.2.5, the ordering may be reversed, that is, $F_2(x) \precsim F_1(x)$. As a result, we see that the ordering of the distributions is defined for each particular q value.

For an arbitrary k a set of k cumulative distributions $F_1(x), F_2(x)$, $\ldots, F_k(x)$ can be ordered on the basis of the values of their respective qth quantiles $\xi_{q1}, \xi_{q2}, \ldots, \xi_{qk}$, in the same way for a given q, $0 \leq q \leq 1$. Using our usual ordering notation, $F_{[j]}$ denotes the jth smallest among $F_1, F_2, \ldots, F_k$, and its qth quantile value is $\xi_{q[j]}$, which denotes the jth smallest among $\xi_{q1}, \xi_{q2}, \ldots, \xi_{qk}$. Thus the ordered population distributions are

$$F_{[1]} \precsim F_{[2]} \precsim \cdots \precsim F_{[k]},$$

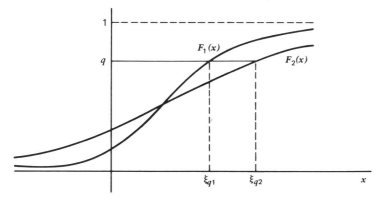

Figure 7.2.4 Comparison of the qth quantiles for two cumulative distributions $F_1(x)$ and $F_2(x)$, where $\xi_{q1} < \xi_{q2}$.

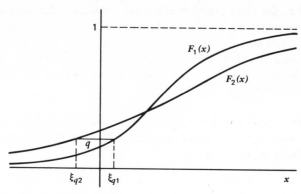

Figure 7.2.5 Comparison of qth quantiles for two cumulative distributions, $F_1(x)$ and $F_2(x)$, where $\xi_{q2} < \xi_{q1}$.

and the corresponding ordered qth quantile values are

$$\xi_{q[1]} \leqslant \xi_{q[2]} \leqslant \cdots \leqslant \xi_{q[k]}.$$

The goal is to identify the population $F_{[k]}$ for some particular value of q, or equivalently, the population with qth quantile value $\xi_{q[k]}$.

For example, with $k = 3$ and the q value indicated in Figure 7.2.6, we see that since $\xi_{q2} < \xi_{q1} < \xi_{q3}$, we have $F_2(x) \precsim F_1(x) \precsim F_3(x)$, or equivalently, $F_{[1]} = F_2(x)$, $F_{[2]} = F_1(x)$ and $F_{[3]} = F_3(x)$. Thus a correct selection occurs with this q value and these distributions if we select $F_3(x)$ and assert that it is $F_{[3]}$.

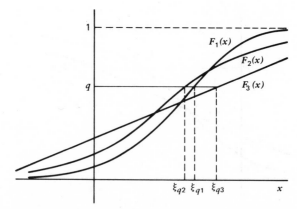

Figure 7.2.6 Comparison of three cumulative distributions that show the ordering $F_2(x) \precsim F_1(x) \precsim F_3(x)$, or equivalently, $\xi_{q2} \leqslant \xi_{q1} \leqslant \xi_{q3}$.

REMARK.   The ordering of the populations, and therefore the population $F_{[k]}$, may be different for different values of q (as is obvious from Figure 7.2.6). Hence we must be given a particular value of q in order to have a well-defined problem.   □

The next step in formulating our problem is to define a distance measure. Since we are dealing with distances between curves, or equivalently, distances between functions, rather than distances between parameter values, the discussion is a bit more involved than in the previous chapters. Consider two cumulative distributions $F_1(x)$ and $F_2(x)$, as in Figure 7.2.7. Note that in this figure, for $q < a$ we have $\xi_{q1} < \xi_{q2}$, so that $F_{[2]} = F_2(x)$, whereas for $q > a$ we have $\xi_{q1} > \xi_{q2}$, so that $F_{[2]} = F_1(x)$. To minimize confusion, suppose that the value for q is such that $\xi_{q1} < \xi_{q2}$ so that $F_{[2]} = F_2(x)$. We now specify a positive quantity d^* such that $d^* \leqslant q$ and $d^* \leqslant 1 - q$, and look at the points $q - d^*$ and $q + d^*$ on the ordinate, and the corresponding quantiles ξ_{q-d^*} and ξ_{q+d^*} on the abscissa for the distribution $F_{[2]}$ (see Figure 7.2.8). Let I denote the interval that has these

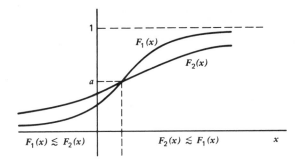

Figure 7.2.7   Indication of regions where $F_1(x) \lesssim F_2(x)$ and $F_2(x) \lesssim F_1(x)$.

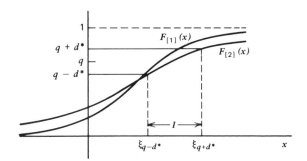

Figure 7.2.8   The indifference zone I for two distributions.

two quantiles as end points. For each x value in I, there is a (vertical) positive difference between $F_{[1]}(x)$ and $F_{[2]}(x)$. We look at all these differences and define the distance measure as the smallest. This minimum is denoted by d_{12}; in symbols, the distance measure is

$$d_{12} = \min_{x \in I} |F_{[1]}(x) - F_{[2]}(x)|, \tag{7.2.1}$$

where $x \in I$ means all x values in the interval I. In Figure 7.2.8, d_{12} appears at the left end of I.

In the general case with k populations, and for the goal of selecting the one population with the largest qth quantile, we let d_{ik} denote the smallest vertical distance between $F_{[i]}$ and $F_{[k]}$ for $i = 1, 2, \ldots, k-1$ in the interval I where the two values ξ_{q-d^*} and ξ_{q+d^*}, which are the end points of I, are determined by the $F_{[k]}$ curve. Then we have $k-1$ minima, namely, $d_{1k}, d_{2k}, \ldots, d_{k-1,k}$. The distance measure d is defined as the smallest of these pairwise minima. Hence, in symbols, for arbitrary k, the distance measure d is defined as

$$d = \min(d_{1k}, d_{2k}, \ldots, d_{k-1,k}), \tag{7.2.2}$$

where

$$d_{1k} = \min_{x \in I} |F_{[1]}(x) - F_{[k]}(x)|, \ldots, d_{k-1,k} = \min_{x \in I} |F_{[k-1]}(x) - F_{[k]}(x)|. \tag{7.2.3}$$

If $k = 2$, then clearly d defined in (7.2.2) is equal to d_{12} defined in (7.2.1).

With this distance measure we can define a procedure for selecting the population with distribution $F_{[k]}$, or equivalently, with qth quantile value $\xi_{q[k]}$, in such a way that the probability of a correct selection is at least P^* whenever $d \geqslant d^*$, where d^* and P^* are prespecified numbers such that $0 < d^* < \min(q, 1-q)$ and $1/k < P^* < 1$. These properties of the selection procedure hold for samples of common size n drawn from each of any k continuous distributions.

7.3 SELECTION RULE

Since the selection rule in this problem is based on the relative magnitude of k sample quantiles of order q, we must first define a sample quantile. We start with the sample median since that is the simplest case to understand. For a sample of n observations, say, $x_1, x_2, \ldots, x_n$, where n is odd, the sample median is defined as the $(n+1)/2$th smallest observation, or in our usual ordering notation, $x_{[(n+1)/2]}$. This observation has the property that the number of observations smaller than it is equal to the number of observations larger than it; both numbers are $(n-1)/2$. If n is even, the sample median can be defined as any number between $x_{[n/2]}$ and $x_{[(n+2)/2]}$, since each of these numbers has the property that $n/2$ observations are smaller and $n/2$ are larger than it. The common practice here is to say that the sample median is the midpoint (or equivalently, the arithmetic mean) of these two order statistics, that is, $(x_{[n/2]} + x_{[(n+1)/2]})/2$. (It would be more precise to say *a* sample median, rather than *the* sample median, since the median is not uniquely defined when n is even. However, with the preceding convention, we can use *the*.)

We now extend this reasoning to define a sample quantile of arbitrary order q, which we denote by y_q. Suppose first that $q(n+1)$ is an integer. Then the sample quantile of order q should be defined as the order statistic $x_{[q(n+1)]}$, since this value has the property that $[q(n+1)-1]$ observations are smaller than it and the remainder $n-[q(n+1)-1]-1=n-q(n+1)$ observations are larger than it. In other words, if $q(n+1)$ is an integer, we use the rth order statistic in the sample $x_{[r]}$, where $r=q(n+1)$ as the qth sample quantile y_q. If $q(n+1)$ is not an integer, the sample quantile lies between two order statistics, in particular between the rth and $(r+1)$th order statistics, $x_{[r]}$ and $x_{[r+1]}$, where r is the largest integer contained in $q(n+1)$. In order to have a single value, we define the qth sample quantile y_q as the following linear combination of $x_{[r]}$ and $x_{[r+1]}$:

$$y_q = [r+1-q(n+1)]x_{[r]} + [q(n+1)-r]x_{[r+1]}. \qquad (7.3.1)$$

Note that if $q = \frac{1}{2}$ so that we want the sample median, $r = n/2$ and y_q is the arithmetic mean of $x_{[n/2]}$ and $x_{[(n+2)/2]}$, as before. Note also that if $q(n+1)$ is an integer, then $r=q(n+1)$ and the right-hand side of (7.3.1) reduces to $x_{[r]}$. Hence (7.3.1) covers all cases and we can call y_q in (7.3.1) the sample estimate of the qth quantile for the population, where r is the largest integer contained in $q(n+1)$.

In our present problem we have k populations. Assume that we have n observations from each population. The sample data from the jth population are denoted by $x_{1j}, x_{2j}, \ldots, x_{nj}$, for $j=1,2,\ldots,k$. The n observations within the jth sample are ordered as $x_{[1]j} \leqslant x_{[2]j} \leqslant \cdots \leqslant x_{[n]j}$, and the sample estimate of the qth quantile for the jth population, denoted by y_{qj}, is the value of (7.3.1) for the jth sample. In symbols, if $q(n+1)$ is an integer, we set $r=q(n+1)$ and

$$y_{qj} = x_{[r]j}, \qquad (7.3.2)$$

whereas if $q(n+1)$ is not an integer, we set r equal to the largest integer contained in $q(n+1)$ and compute y_{qj} from

$$y_{qj} = [r+1-q(n+1)]x_{[r]j} + [q(n+1)-r]x_{[r+1]j}. \qquad (7.3.3)$$

Then this procedure is repeated for each sample, so that we obtain the k estimates $y_{q1}, y_{q2}, \ldots, y_{qk}$. These k values are then ordered from smallest to largest, say as $y_{q[1]} \leqslant y_{q[2]} \leqslant \cdots \leqslant y_{q[k]}$, and the selection rule is simply to assert that the population with the largest quantile $\xi_{q[k]}$ is the one for which the sample quantile is largest, that is, the one that gives rise to $y_{q[k]}$.

To make the selection rule complete, we must consider the possibility of ties occurring. Within the jth sample if two or more observations are equal to $x_{[r]}$ or $x_{[r+1]}$, then y_{qj} is still uniquely defined. If there are ties for first place between samples, that is, among the $y_{q1}, y_{q2}, \ldots, y_{qk}$, then we will be content to select one of the contenders by some random mechanism and assert that it has the largest population quantile $\xi_{q[k]}$.

As a numerical illustration, suppose that $q = .50$ so that we are interested in selecting the one population with the largest median among $k = 3$ populations. If $n = 55$ then $q(n+1) = 28$, an integer, so that $r = 28$. The sample statistics to be compared are the values of $x_{[28]j}$ for $j = 1, 2, 3$, that is, the 28th-order statistics (the sample medians) for each of the three samples. The population that produces the largest sample median is asserted to have the largest population median. Again with $q = .50$, if $n = 56$ then $q(n+1) = 28.5$ so that (7.3.3) must be used with $r = 28$. The result is $y_{qj} = .5x_{[28]j} + .5x_{[29]j}$. On the other hand, suppose that $q = .90$ and $n = 50$. Then $q(n+1) = 45.9$ so that (7.3.3) must be used with $r = 45$. The result is $y_{qj} = .1x_{[45]j} + .9x_{[46]j}$.

Example 7.3.1 illustrates completely how to carry out the selection procedure.

Example 7.3.1   (Agriculture).   Cochran and Cox (1957, pp. 96–97) describe an experiment to investigate the best means of applying sulfur to the soil to reduce scab disease of potatoes. Sulfur is considered an effective additive since it increases the acidity of the soil, and a very acid soil inhibits the growth of scab. The methods of sulfur application under consideration were three different amounts, 300, 600, and 1200 lb. to be spread per acre, and two seasons, fall (F) and spring (S). Thus there are six treatments, denoted here for convenience by F3, S3, F6, S6, F12, and S12. Four observations of a "no scab" index are obtained for each treatment, as shown below. Each "no scab" index is computed by averaging the results obtained by taking 100 potatoes at random from that treatment and grading each potato according to the percentage of the surface area of the potato that is not infected. The goal is to select the treatment that is most effective; that is, the one for which the median "no scab" index is greatest. Since the sample sizes are small here and the data are percentages, the assumption of normal distributions is questionable.

Table 7.3.1   *"No scab" index for six methods of sulfur application*

Treatment					
F3	S3	F6	S6	F12	S12
91	70	84	82	90	83
91	93	90	76	96	93
84	79	82	88	96	84
96	91	82	81	95	83

Hence we use the nonparametric procedures of this chapter to select the treatment with the largest median "no scab" index.

For each treatment the sample median $y_{.5j}$ is the simple average of the second- and third-order statistics. These follow:

Treatment j	F3	S3	F6	S6	F12	S12
$y_{.5j}$	91.0	85.0	83.0	81.5	95.5	83.5

Since the largest value $y_{.5j}$ is 95.5, we assert that the method of applying in the fall 1200 lb. of sulfur per acre will produce potatoes with the largest median "no scab" index, and hence with the least median scab disease.

If the sample sizes are unequal, say the sample from the jth population is of size n_j, the sample estimate y_{qj} is computed from (7.3.2) or (7.3.3) with n replaced by n_j, for each $j = 1, 2, \ldots, k$. Example 7.3.2 illustrates the selection procedure for unequal sample sizes.

Example 7.3.2   (Linguistics).   Studies of literary style and vocabulary have been used frequently to compare authors, to distinguish one author from another, and even for identification in cases where authorship is a matter of dispute. A number of quantitative indexes have been found useful in such studies, for example, word length, sentence length, frequency of usage of certain words, to name only a few. One of the first such studies used sentence length as the literary characteristic of interest. This study, reported by Yule (1939), gave a frequency distribution of sentence length in the works of three authors, Bacon, Coleridge, and Macaulay; these data are shown in Table 7.3.2. Determine which one of these three authors has the largest median sentence length.

The respective sample sizes are $n_1 = 936$, $n_2 = 1207$, and $n_3 = 1251$, where Bacon is sample 1, Coleridge is sample 2, and Macaulay is sample 3. Then since $q = .5$, $q(n_1 + 1) = 468.5$, $q(n_2 + 1) = 604$, and $q(n_3 + 1) = 626$. From

Table 7.3.2   Frequency distribution of sentence length of three authors

Number of words per sentence	Author			Number of words per sentence	Author		
	Bacon	Coleridge	Macaulay		Bacon	Coleridge	Macaulay
1– 5	3	11	46	91– 95	15	13	1
6–10	16	58	204	96–100	13	8	2
11–15	49	90	252	101–105	10	10	1
16–20	45	95	200	106–110	12	4	0
21–25	99	131	186	111–115	5	2	1
26–30	85	120	108	116–120	6	6	0
31–35	112	112	61	121–125	7	5	1
36–40	75	103	68	126–130	5	2	0
41–45	62	101	38	131–135	3	2	0
46–50	56	76	24	136–140	3	0	0
51–55	51	53	20	141–145	5	2	0
56–60	46	45	12	146–150	1	3	0
61–65	36	39	8	151–155	3	1	0
66–70	25	37	2	156–160	0	1	0
71–75	27	29	4	161–165	0	1	0
76–80	18	16	8	166–170	1	0	0
81–85	23	15	2	171–175	0	1	0
86–90	13	14	2	Over 175	6	1	0
				Total	936	1207	1251

(7.3.2) and (7.3.3), the respective sample medians are $.5x_{[468]1} + .5x_{[469]1}$, $x_{[604]2}$, and $x_{[626]3}$. However, the values of these order statistics cannot be found exactly from the data given. The data in Table 7.3.2 are reported in the form of a frequency distribution, and the individual order statistics for each sample are not available. In other words, Bacon is reported to have three sentences of length 1–5. From this information we cannot tell whether $x_{[1]1} = 1$, 2, 3, 4, or 5, and similarly for $x_{[2]1}$ and $x_{[3]1}$.

Therefore we must use a different method for computing the medians here. In a frequency distribution of n_j observations the qth quantile is defined as the value below which exactly qn_j percent of the observations fall. The first step in locating the interval in which the median lies, called the *median class*, is to construct a cumulative frequency distribution for each sample up to the point where the cumulative total is equal to or greater than qn_j. For the data of Table 7.3.2 with $q = .5$, we have $qn_1 = 468$, $qn_2 = 603.5$, and $qn_3 = 625.5$. The cumulative distributions are given in Table 7.3.3.

Table 7.3.3 **Cumulative distribution for data in Table 7.3.2**

Number of words	Author		
per sentence	Bacon	Coleridge	Macaulay
5 or less	3	11	46
10 or less	19	69	250
15 or less	68	159	502
20 or less	113	254	702
25 or less	212	385	
30 or less	297	505	
35 or less	409	617	
40 or less	484		

From Table 7.3.3 we see that the median number of words per sentence is between 35 and 40 for Bacon, between 30 and 35 for Coleridge, and between 15 and 20 for Macaulay. To calculate the value of the median for each sample, we use linear interpolation within the median class. This interpolation procedure assumes that the observations within any median class, which here is of width 5, are evenly distributed over those 5 units. For Bacon the median is $(468-409)/(484-409) = \frac{59}{75}$ of the way up the interval, which includes all numbers over 35 but not larger than 40. We take 35.5 as the lower limit of this class. Hence the median for Bacon is

$$35.5 + 5\left(\frac{59}{75}\right) = 39.4.$$

For Coleridge and Macaulay, the corresponding calculations are

$$30.5 + 5\left(\frac{603.5 - 505}{617 - 505}\right) = 34.9,$$

$$15.5 + 5\left(\frac{625.5 - 502}{702 - 502}\right) = 18.6.$$

The sample median for Bacon is the largest of the three, so that we assert that works by Bacon have a longer median sentence length than works by Coleridge or Macaulay.

Since data for large sample sizes are often given in the form of a frequency distribution, we now give a formula that can be used with such data to determine the qth quantile y_q, for any q, of a sample of size n. Let

Iq denote the interval in which the qth quantile falls. Then the formula is

$$y_q = x_L + c_{Iq}\left(\frac{qn - \Sigma f_{bIq}}{f_{Iq}}\right),$$

where x_L = the true lower limit of the interval Iq,

$\quad c_{Iq}$ = the width of the interval Iq,

$\quad \Sigma f_{bIq}$ = the sum of the frequencies in all intervals before Iq, and

$\quad f_{Iq}$ = the frequency in the interval Iq.

This formula always gives a unique value for y_q unless Iq is an interval with frequency equal to zero. If Iq has frequency zero, the usual convention is to say that the qth quantile is the midpoint of Iq.

7.4 ANALYTICAL ASPECTS OF THE SELECTION PROBLEM

We assume that the experiment to select the population with distribution $F_{[k]}$, or equivalently with qth quantile value $\xi_{q[k]}$, is to be based on a common number n of observations from each of the k populations, and the selection rule R of Section 7.3 is to be used to make the selection. In this section we consider the problem of determining the common sample size required to satisfy a specific requirement. Since tables have been constructed only for $q = .50$, we consider here the analysis only for the problem of selecting the population with the largest median. (At the time of writing this book, tables are being prepared for the problem of selecting the population with the largest 75th percentile.) However, the methods of analysis are identical regardless of the order of the quantile, and tables could be constructed to deal with a quantile of any order.

7.4.1 Determination of Sample Size

The first step is to specify a pair of constants (d^*, P^*) such that $0 < d^* < \min(q, 1-q)$ and $1/k < P^* < 1$. This choice is interpreted as meaning that the probability of a correct selection using the rule R of Section 7.3 is at least P^* whenever $d \geq d^*$, where d is defined in (7.2.2). The problem then is to determine the smallest common sample size n such that this (d^*, P^*) requirement is satisfied. Table K.1 in Appendix K gives these values of n for $q = .50$ for $k = 2, 3, \ldots, 10$, $P^* = .750, .900, .950, .975, .990$, and $d^* = .05, .10, .15, .20$. For example, suppose we want to select the population with the largest median. If we have $k = 3$ populations and specify $d^* = .15$, $P^* = .90$, then Table K.1 indicates that we need to take $n = 55$ observations from each of the three populations.

The investigator needs to develop some experience and intuition to guide him in the specification of an appropriate value for d^* in order to use these nonparametric methods efficiently. The d values are distances between the cumulative distributions $F_{[k]}$ and $F_{[1]}, \ldots, F_{[k-1]}$. In other chapters, the δ values are distances between the parameters $\theta_{[k]}$ and $\theta_{[1]}, \ldots, \theta_{[k-1]}$. The δ values (and δ^*) are related to the units of the problem, but the d values (and d^*) are not; consequently, d^* is a less intuitive quantity than δ^*.

*7.5 THE SECRETARY PROBLEM

Not all nonparametric selection problems fall into the framework of the previous sections of this chapter. A very interesting nonparametric selection problem is the so-called *Secretary Problem* or *Marriage Problem*; in some situations a variation of this problem may be highly practical as well as interesting.

To make the problem more specific, suppose we have a certain position available and n candidates apply for it. We want to fill the position with the best candidate. We assume that a unique ranking exists among these candidates; that is, if all n candidates could be observed together, then we could rank them (without ties) from rank 1 (best) to rank n (worst). These ranks are called *absolute ranks*. The goal is to select the candidate with the smallest absolute rank, since the smaller the absolute rank the better the candidate.

The problem stems from two facts: (1) We do not or cannot observe the candidates all together, but rather we observe them one at a time and in a random order, over which we have no control, and (2) once a candidate is rejected, he or she cannot be recalled. As a result, when the kth candidate appears for an interview, we have the *relative ranks* of these k candidates, that is, their ranks relative to each other. Thus we know the number of predecessors who are better than the present candidate. However, we do not know the ranks of the remaining $n - k$ candidates that have not yet appeared.

At any stage we may select the candidate being observed, in which case the process ends, or we may reject the candidate and go on to the next candidate. The last candidate must be selected if no previous candidate was accepted. The selection question then becomes, when should we stop sampling and what criterion shall we use for stopping?

The proposed stopping rule R is carried out as follows. We are given n integers, $s_1, s_2, \ldots, s_n$, such that

$$s_1 \leqslant s_2 \leqslant \cdots \leqslant s_{n-1} \leqslant s_n = n.$$

At each stage i we can find the relative rank y_i of the ith candidate relative to the $(i-1)$ predecessors. Of course, $y_1 = 1$. If $y_1 \leqslant s_1$ (and this occurs if and only if $s_1 \geqslant 1$), we choose candidate 1 and stop sampling. If $y_1 > s_1$ (and this occurs if and only if $s_1 = 0$), we go into stage 2 and interview candidate 2. If $y_2 \leqslant s_2$, we choose candidate 2, and if $y_2 > s_2$, we go into stage 3. In general, we continue sampling until the first time that $y_i \leqslant s_i$ and then stop and select the ith candidate. By setting $s_n = n$ we are forced to accept the last candidate if we have not already stopped before. The determination of the best constants s_i under such a procedure is complicated; for the proposed procedure R, these values are already computed and given in Table 7.5.1 for $n = 2(1)10$.

Table 7.5.1 **Table of critical constants s_i for selecting the best among**
n candidates to be used with the stopping rule—stop with i
observations if the relative rank of the ith candidate is
less than or equal to s_i

Stage

n	1	2	3	4	5	6	7	8	9	10
	s_1	s_2	s_3	s_4	s_5	s_6	s_7	s_8	s_9	s_{10}
2	1^a	2								
3	0	1^a	3							
4	0	1	2^a	4						
5	0	1	1	2^a	5					
6	0	0	1	2	3^a	6				
7	0	0	1	1	2	3^a	7			
8	0	0	1	1	2	2	4^a	8		
9	0	0	1	1	1	2	3	4^a	9	
10	0	0	0	1	1	2	2	3	5^a	10

[a] It is interesting to note that $s_{n-1} = [(n-1)/2]$ for all n, where $[x]$ is the largest integer not greater than x.

For example, if $n = 4$, we find $s_1 = 0$, $s_2 = 1$, $s_3 = 2$, $s_4 = 4$. Suppose the candidates are A, B, C, D, and their respective absolute ranks are 1, 2, 3, 4. Suppose further that the candidates appear in the order B, A, C, D. The relative ranks are then $y_1 = 1$, because the relative rank of B alone is 1; $y_2 = 1$, because the rank of A among A and B is 1; $y_3 = 3$ because the rank of C among A, B, C is 3; $y_4 = 4$ because the rank of D is 4. According to our stopping rule, we first determine whether $y_1 \leq s_1 = 0$. Since it is not, we check whether $y_2 \leq s_2 = 1$. This inequality is verified, so we stop at stage 2 and choose candidate A. Then y_3 and y_4 are not observed.

On the other hand, suppose that the candidates appear in the order C, B, D, A. Then $y_1 = 1$, $y_2 = 1$, $y_3 = 3$, $y_4 = 1$. We would again stop at stage 2, but now candidate B is selected.

As these two examples for $n = 4$ indicate, the best candidate will not always be selected under procedure R (or under any procedure). The

Table 7.5.2   Enumeration of all orders of appearance for $n = 4$ candidates and stopping stage for each permutation under the optimal rule

$$R: (s_1 = 0, s_2 = 1, s_3 = 2, s_4 = 4)$$

Order of appearance of candidates	Relative ranks				Results of using procedure R		
	y_1	y_2	y_3	y_4	Stage at stopping	Candidate selected	Absolute rank S of candidate selected
A B C D	1	2	3	4	4	D	4
A B D C	1	2	3	3	4	C	3
A C B D	1	2	2	4	3	B	2
A C D B	1	2	3	2	4	B	2
A D B C	1	2	2	3	3	B	2
A D C B	1	2	2	2	3	C	3
B A C D	1	1	3	4	2	A	1
B A D C	1	1	3	3	2	A	1
B C A D	1	2	1	4	3	A	1
B C D A	1	2	3	1	4	A	1
B D A C	1	2	1	3	3	A	1
B D C A	1	2	2	1	3	C	3
C A B D	1	1	2	4	2	A	1
C A D B	1	1	3	2	2	A	1
C B A D	1	1	1	4	2	B	2
C B D A	1	1	3	1	2	B	2
C D A B	1	2	1	2	3	A	1
C D B A	1	2	1	1	3	B	2
D B C A	1	1	2	3	2	A	1
D A C B	1	1	2	2	2	A	1
D B C A	1	1	2	1	2	B	2
D B A C	1	1	1	3	2	B	2
D C A B	1	1	1	2	2	C	3
D C B A	1	1	1	1	2	C	3

selection depends on the order of appearance as well as on the values of s_i given in Table 7.5.1 for procedure R. For $n=4$ there are $4!=24$ different orders of appearance. Table 7.5.2 enumerates each of these 24 orderings and also indicates which candidate is selected under procedure R for each ordering. The final column gives the absolute rank (denoted by S) of the candidate selected, assuming that the absolute ranks of A, B, C, D are $S = 1, 2, 3, 4$, respectively.

Now consider another example where we have $n=7$ candidates, A, B, C, D, E, F, G, and procedure R is used. Here A is best, B next best, and so on, so that their absolute ranks are $S = 1, 2, 3, 4, 5, 6, 7$, respectively. The rule for procedure R is to stop sampling as soon as the relative ranks satisfy $y_i \leqslant s_i$ where the s_i values for $n=7$, given in Table 7.5.1, are

$$s_1 = 0, \quad s_2 = 0, \quad s_3 = 1, \quad s_4 = 1, \quad s_5 = 2, \quad s_6 = 3, \quad s_7 = 7. \quad (7.5.1)$$

Suppose the candidates appear in the order D, B, C, E, A, G, F. Then the relative ranks are easily found to be

$$y_1 = 1, \quad y_2 = 1, \quad y_3 = 2, \quad y_4 = 4, \quad y_5 = 1, \quad y_6 = 6, \quad y_7 = 6. \quad (7.5.2)$$

The first time a y value in (7.5.2) is less than or equal to its corresponding s value in (7.5.1), we stop sampling and select that candidate. This occurs here at stage 5, when A appears, so we select candidate A.

Again with $n=7$, suppose the order of appearance is D, C, B, E, A, G, F. Then the relative ranks are

$$y_1 = 1, \quad y_2 = 1, \quad y_3 = 1, \quad y_4 = 4, \quad y_5 = 1, \quad y_6 = 6, \quad y_7 = 6. \quad (7.5.3)$$

Now comparing the y values in (7.5.3) with the s values in (7.5.1), we find $y_i \leqslant s_i$ for the first time at $i=3$, so we stop at stage 3. Thus candidate B is selected under procedure R. Even though we did not select the best candidate with this order of appearance, we are able to terminate the interviews under procedure R after seeing *only three candidates*.

Two interesting questions arise for this type of problem. One is how well will we do in our selection, or, in other words, what is the expected absolute rank of the candidate selected? The other question is, At what stage can we expect to terminate the interviewing process by making a selection; that is, how many candidates can we expect to interview? These two expected values can be easily calculated for small values of n once the $n!$ orders of appearance are enumerated and the corresponding selection determined. That is, we need to prepare a table like Table 7.5.2 for the appropriate small value of n, using procedure R.

To illustrate these calculations, we return to the case $n = 4$. From Table 7.5.2 we can tally the number of times the candidate with absolute rank S is selected and present the results in the form of a frequency distribution, as shown in Table 7.5.3. Note that candidate A is selected with probability $\frac{10}{24} = .417$. The expected absolute rank of the candidate selected under procedure R is the sum of the products of the absolute ranks (column 2) and the corresponding relative frequency of selection (column 3) or

$$E(S|R) = 1\left(\tfrac{10}{24}\right) + 2\left(\tfrac{8}{24}\right) + 3\left(\tfrac{5}{24}\right) + 4\left(\tfrac{1}{24}\right) = 1.875.$$

Table 7.5.3   Frequency distribution of selection when $n = 4$ and procedure R is used

Candidate	Absolute rank (S)	Frequency of selection
A	1	10
B	2	8
C	3	5
D	4	1
		24

To compute the expected value of the stopping stage (or equivalently, the expected number of candidates interviewed), we tally the stopping stages in Table 7.5.2 to get the frequency distribution in Table 7.5.4.

Table 7.5.4   Frequency distribution of stopping stage when $n = 4$ and procedure R is used

Stage	Frequency
1	0
2	12
3	8
4	4

If T is the number of candidates interviewed, then the expected value of T under procedure R is given by the calculation

$$E(T|R) = 2\left(\tfrac{12}{24}\right) + 3\left(\tfrac{8}{24}\right) + 4\left(\tfrac{4}{24}\right) = 2.667.$$

For a large number of candidates it can be shown that $E(S|R)$ approaches 3.8695. In other words, using the preceding optimal procedure R we expect on the average to select a candidate with absolute rank approximately 4 (or slightly smaller) for all large values of n.

For large values of n the proposed procedure R is approximated by the following procedure R'. Let n/e of the candidates (or the closest integer thereto) go by without making any selection, where $e = 2.71828\ldots$ is the base of the natural logarithm. After that, select the first candidate that is better than all his predecessors; as with procedure R, the last candidate must be selected if we have not already stopped. It is interesting to note (and we omit the details) that if we apply procedure R' to the case $n = 4$ we obtain the results

$$E(S|R') = 1.875 = E(S|R),$$

and

$$E(T|R') = 2.833 > E(T|R).$$

Since procedure R' needs more interviews (on the average) to accomplish the same expected absolute rank as procedure R, this tends to confirm the optimal nature of procedure R and indeed to explain what is meant by optimal in this context.

REMARK.   It is interesting to point out that for $n = 4$ under procedure R' the probability of selecting candidate A is $\frac{11}{24}$ whereas it is only $\frac{10}{24}$ under procedure R. Hence if our criterion for optimality of a procedure were simply to maximize the probability of selecting candidate A (the best candidate) without regard to the value of $E(T)$, then we would have to select procedure R' in preference to procedure R. We note this property of procedure R' for $n = 4$, but the same property holds for larger n. $\square$

Instead of considering the expected value of the *number* of candidates interviewed we may wish to look at the expected *proportion* of candidates interviewed. Under procedure R the expected proportion values are $\frac{5}{6}$ $= .833$ for $n = 3$, $\frac{2}{3} = .667$ for $n = 4$, $\frac{89}{150} = .593$ for $n = 5$, and so forth. Apparently this sequence will continue to decrease; at least, we conjecture that this is the case. Then the limiting value of the sequence is of interest. It appears that as n gets large this limit is positive and is greater than $1/e = 0.368\ldots$, in view of the preceding remark about procedure R' being close to procedure R. A very rough conclusion based on the earlier calculation is that the expected proportion for both procedures R and R' will not be far from .5 for a long sequence of moderate values of n.

7.6 NOTES, REFERENCES, AND REMARKS

The basic ideas for this nonparametric treatment of ranking and selection are developed in Sobel (1967). Unfortunately the tables given there cover only the case of the median or 50th percentile, and we have not developed tables for other quantiles.

The problem of selecting a subset containing the population with the largest α quantile was treated by Rizvi and Sobel (1967). It is interesting to note that when dealing with stochastically ordered families, the problem of ordering k such populations is equivalent to ordering them according to any fixed quantile, say the median. Clearly, there must be more efficient ways of ordering stochastically ordered populations than just to order their medians. However, some early attempts at this problem led to difficult theoretical problems on what constitutes the least favorable configuration (see Lehmann (1963), Puri and Puri (1968, 1969), Rizvi and Woodworth (1970)). This problem of ranking stochastically ordered populations was also considered by Alam and Thompson (1971).

An adaptive procedure for the problem of selecting the population with the largest location parameter is provided in Randles, Ramberg, and Hogg (1974); that procedure is not covered in this book.

Rizvi and Saxena (1972) developed distribution-free interval estimates of the largest α quantile.

The main source for the "Secretary" or "Marriage" problem is Chow, Moriguti, Robbins, and Samuels (1964), but all the calculations carried out to illustrate the method and its efficiency are new with this book. The stopping rule R is due to Robbins (see Chow, Robbins, and Siegmund (1971) for a discussion).

PROBLEMS

7.1 Williams (1940) reported the following frequency distribution of the lengths of sentences from works of Chesterton, Shaw, and Wells. Select the author with the longest median sentence length, and make a confidence statement at level .90 about your selection.

Number of words	Author		
per sentence	Chesterton	Shaw	Wells
1– 5	3	19	11
6–10	27	62	66
11–15	71	68	107
16–20	112	77	121
21–25	108	66	75
26–30	109	64	61
31–35	64	57	52
36–40	41	30	27
41–45	28	28	29
46–50	19	24	17
51–55	9	36	12
56–60	6	16	8
61–65	1	9	5
66–70	0	6	4
71–75	0	7	3
76–80	1	7	1
81–85	0	9	0
86–90	0	5	0
Over 90	1	10	1
Total	600	600	600

7.2 The U.S. Bureau of the Census (1960) reported the following data on median age at first marriage for white men aged 45–54 who were foreign born: United Kingdom, 26.7; Ireland, 29.4; Germany, 27.5; Poland, 28.1; U.S.S.R., 27.0; Italy, 26.1; Canada, 26.2 years. Since these data are from a census, the following statement is assumed to be true (and accordingly to have 100% confidence): "Of all white men aged 45–54 alive in the United States in 1960 who were born in either the United Kingdom, Ireland, Germany, Poland, U.S.S.R., Italy, or Canada, men born in Ireland marry at a larger median age than do corresponding men born in the other six countries." Suppose that a follow-up study is to be carried out to see whether the pattern of age at first marriage has changed among European immigrants, and because the cost of another census is prohibitive the conclusions must be based on a random sample of n white men aged 45–54 and born in each of these seven countries. How large should n be so that we can say with 95% confidence that the country selected as having the largest median age at first marriage is correct when $d^* = .10$?

7.3   The U.S. Bureau of the Census (1960) reported the following data on the median difference in age (husband – wife) for all couples married once in the United States in 1960, by type of residence.

Urban areas:
Central cities	2.5
Urban fringe	2.3
Other urban	2.5
Rural nonfarm	2.8
Rural farm	3.0

A follow-up study is being planned to see whether there has been a change in the type of residence that has the largest median excess in age of husband over wife. How large a sample should be taken from each type of residence if we specify $P^* = .95$, $d^* = .10$.

CHAPTER 8

Selection Procedures for a Design
with Paired Comparisons

8.1 INTRODUCTION

In this chapter we consider the goal of selecting the best object when several, say k, different objects are compared using some specific design with paired comparisons. That is, only two objects are compared at any one time. The term *object* should be interpreted in a broad generic sense—it may refer to a food item, a medical treatment, a psychological stimulus, a teaching method, or the like.

In one specific design, called the *completely balanced paired-comparison design*, every possible pair of different objects is presented to each of some number of judges. The term *judge* should also be interpreted in a broad generic sense—it may refer to an observer, a rater, a scorekeeper, or the like. For each pair presented, each judge states which object is preferred among the two. In our formulation and analysis the judge is obliged to select one of them; that is, he cannot declare a tie.

The reader will no doubt recognize the similarity of this design format to the familiar round robin (RR) tournament used in many sports. In such a tournament, each of k players (or teams) plays every other player some fixed number of times. Thus the players in the tournament assume the roles of the judges in that they determine the winner in each game of two players, but they are also the objects. The total number of times any given pair of players play together is the same as the number of judges to whom any given pair of objects is presented. Several other paired-comparison designs also have a tournament analogy. When it is convenient we may resort to the more familiar game terminology so that we can discuss the different designs for selecting the best object in terms of the different strategies for selecting the best player in a tournament.

Suppose there are k objects and r judges, or equivalently, k players and each pair of players play exactly r times in a round robin tournament. In either case, r is referred to as the *number of replications* in the experiment. Note that r is not the same as either n, the sample size per population, or N, the total number of observations. Since each replication gives rise to a total of $\binom{k}{2} = k(k-1)/2$ comparisons between pairs of objects, the total number of observations is $N = rk(k-1)/2$. The value of N does not play an important role in our problem, since we are primarily interested in determining the number of replications r needed to obtain statistically reliable comparisons; the value of r uniquely determines N since k is fixed.

The r replications are assumed to be independent. In order to justify this assumption it is desirable that each of the r complete replications be done by a different judge. When this is not possible and a single judge is used, there are methods for trying to ensure that the judge's responses from replication to replication are independent; for example, we could randomize the order of presentation between pairs and within pairs.

A paired-comparison design is used frequently in sensitivity studies and preference studies, especially when the objects to be compared can only be judged subjectively because it is impossible or impractical to make an objective measurement that will determine the preferred object. In the ranking and selection formulation for this problem, the goal is to select the best object, that is, the one with the most "merit." However, it should be clear that the true merit is not represented by some parameter θ that indexes a specific underlying distribution model $f(x; \theta)$. Hence we cannot state our goal in the usual manner. Nevertheless the goal here is a realistic one and we assume that the true merit of any object is measurable on some ordinal scale and that our sampling of the preferences of judges will reflect the ordering preferences in some population represented by the judges. Thus, although it is not emphasized below, it is also important to use statistical methods carefully to select the judges so that they can be considered representative of the particular population under study.

Before proceeding with the formulation and development of the selection problem, we give some specific examples in which this goal is appropriate and a paired-comparison design is a practical, effective means of obtaining information on relative merit within a group of objects.

Example 8.1.1 (Taste Tests). In taste-testing experiments it is very difficult for even a professional taster simultaneously to compare the taste of (and state a preference among) more than two different edible or potable products. Hence if there are $k > 2$ products to be compared, a common procedure is to present the products to each taster in pairs so that he needs to compare only two products at any one time. These pairwise comparisons are repeated until all possible pairwise comparisons are completed. This procedure is replicated (with randomization) with each of

r different tasters. For example, suppose we have $k = 5$ different blends of tea and the goal is to select the most preferred blend. Then if all the $\binom{5}{2} = \dfrac{5(4)}{2} = 10$ pairs of blends are presented to each of $r = 6$ tasters, we have a total of $6(10) = 60$ statements of preference on which to base our selection.

Example 8.1.2 (Tournament).   Consider a sport in which only two teams (or players) can play against each other at any one time and each game between two teams is continued until there is a winner (i.e., no ties are possible). A round robin tournament is frequently used to determine the best of k teams in such sports. In this kind of tournament each pair of teams plays together and such sets of $k(k-1)/2$ games can be replicated r times. Suppose that 10 teams are vying for the championship in such a tournament where each team plays exactly 30 games with every other team. Then a total of $9(30) = 270$ games are played by any one team and a total of $30\binom{10}{2} = 1350$ games are played altogether. Once the tournament is completed, we have data on the winners for each of the 1350 games. The goal is to select the best team. Note that in this example the teams play the role of the objects in the general paired-comparison experiment. Hence $k = 10$ here and $r = 30$; in tournament terminology a *round* consists of one game between every pair of players and is analogous with our use of the term *replication*.

Most of our following discussion centers on the completely balanced design, or equivalently, the round robin strategy, since it is one that has been widely used. If we insist that *every* pair of objects must be compared by each judge (and, moreover, that each pair be compared the same number λ of times by each judge and that the value of λ be the same for all judges and all pairs), then the completely balanced design is a good one in the sense of being "efficient." (At the present time we do not give a precise definition of efficiency; the idea is illustrated later with examples.) However, it should be pointed out that the requirement that every pair must be compared necessitates a very large number of comparisons and in fact is overly rigid for the purpose of identifying the one best object (or player). If we drop this requirement and use some other (specific) design, we can expect substantial improvements in efficiency.

As a result, in Section 8.6 of this chapter we also consider an alternative specific paired-comparison design, known in game terminology as the *knock-out* (KO) tournament. For the purpose of finding the best object among $k = 8$ objects, we show later that using the knock-out strategy as opposed to the round robin strategy reduces the required number of comparisons by more than $33\frac{1}{3}\%$, for the same probability of a correct selection prespecified at the nominal value $P^* = .90$. These remarks about increased efficiency do not apply if the goal is to obtain a complete ranking of all the k objects, since in this problem it is clearly more useful to make comparisons between all possible pairs.

8.2 THEORETICAL DEVELOPMENT

In the general case, suppose we want to compare k objects, which we denote by $T_1, T_2, \ldots, T_k$. These objects are to be judged only in pairs on some specific characteristic. For any one pair, say T_i and T_j, either T_i is preferred to T_j or vice versa. Let π_{ij} denote the *preference probability* for T_i over T_j for any $i = 1, 2, \ldots, k, j = 1, 2, \ldots, k$ $(i \neq j)$; that is,

$$\pi_{ij} = P\left\{ T_i \text{ is preferred to } T_j \right\}. \tag{8.2.1}$$

Suppose further that for each $i = 1, 2, \ldots, k$, object T_i has a true "merit" ν_i, when judged on the basis of the specified characteristic. These true merit values, $\nu_1, \nu_2, \ldots, \nu_k$, can be represented by k points on an ordinal scale; that is, if T_i has more merit than T_j, then $\nu_i > \nu_j$. Of course, the observed merits, as estimated by the observed preferences, vary from experiment to experiment as well as from judge to judge. Hence we may view the observed merit of object T_i as arising from a continuous random variable Y_i that varies on the same scale and in the same direction as the ν_i. Then in a paired comparison of the objects T_i and T_j, a correct preference occurs if T_i is preferred to T_j when $Y_i > Y_j$, and also T_j is preferred to T_i when $Y_i < Y_j$. Equivalently we can express the preference probabilities in (8.2.1) as

$$\pi_{ij} = P\left\{ Y_i > Y_j \right\} = P\left\{ Y_i - Y_j > 0 \right\} = 1 - \pi_{ji}. \tag{8.2.2}$$

If a merit scale can be constructed such that for all i and j, the probabilities in (8.2.1) "satisfy a linear model" (defined in the following Remark), we can find a solution to our problem of selecting the object with the highest true merit. Theory and applications have shown that this idea of satisfying a linear model is a reasonable assumption in some practical situations. In other cases we may be forcing an essentially multivariate problem into a linear, univariate model, and the value of this would have to be assessed in each case separately.

REMARK.   The preference probabilities π_{ij} in (8.2.1) are said to satisfy a linear model if they can be written as some function of $\nu_i - \nu_j$, say $H(\nu_i - \nu_j)$, which (1) increases monotonically from 0 to 1, and (2) satisfies the relation $H(-x) = 1 - H(x)$. Intuitively, we want $H(\nu_i - \nu_j)$ to increase with the difference $\nu_i - \nu_j$, since this measures the difference in true merits. Also, we want $\pi_{ij} = 1 - \pi_{ji}$, as (2) states. □

We denote the best object by $T_{[k]}$, where $T_{[k]}$ is defined as the object with the largest true merit $\nu_{[k]}$. The selection rule is described in the next section. However, as part of the theoretical development of the problem, we must define a distance measure and a preference zone, and discuss the various configurations of the π_{ij} so that some of the usual analytical questions about the selection rule can be answered.

The relevant parameters here are the π_{ij} for all $i \neq j$. Under the assumption of a linear model, the $k(k-1)$ parameters π_{ij} can be expressed as functions of only the $k-1$ differences $\nu_i - \nu_j$, and the relationship between the π_{ij} and the $\nu_i - \nu_j$ satisfies the following:

$$\pi_{ij} > .5 \quad \text{if and only if} \quad \nu_i > \nu_j,$$

$$\pi_{ij} = .5 \quad \text{if and only if} \quad \nu_i = \nu_j, \qquad (8.2.3)$$

$$\pi_{ij} < .5 \quad \text{if and only if} \quad \nu_i < \nu_j.$$

The use of these ν values greatly simplifies the discussion of parameter configurations, since we need to deal with only k parameters on a linear scale.

Unfortunately the theory at present has not resolved the question of which configuration of the π values is least favorable. However, there is a particular configuration that suffices for many practical purposes, even though it is not the one that is least favorable. In this configuration, some particular set consisting of $k-1$ of the objects are "equal" in merit in that they have equal ν values; and the other object, $T_{[k]}$, has more merit in that it has a larger ν value than all the others. In view of (8.2.3) this means that within the set of $k-1$ objects, the probability of preferring one to another is exactly equal to .5 for each pair, and that the probability that $T_{[k]}$ is preferred to any of the others is greater than .5. In terms of the parameters π_{ij}, for this configuration we have

$$\pi_{ij} = \pi > .5 \quad \text{for} \quad i = [k] \text{ and } j \neq [k],$$

$$\pi_{ij} = .5 \quad \text{for} \quad i \neq [k], j \neq [k], \text{ and } i \neq j. \qquad (8.2.4)$$

(Here π is used as a symbol for a probability and is not to be confused with the usual usage as the constant $3.14159\ldots$.)

For this configuration the distance measure δ could be defined as the excess of π over .5; that is,

$$\delta = \pi - .5. \tag{8.2.5}$$

This is interpreted as the common amount, measured by the excess in probability over .5, to which object $T_{[k]}$ is preferred over each of the other objects. Note that δ ranges from 0 to .5, and is close to zero when π is only slightly larger than .5, that is, when the best object has only a little more merit than all the others. Hence the preference zone for a correct selection is $\delta \geqslant \delta^*$ where $0 < \delta^* < .5$. Of course, in a practical problem we expect to have δ^* close to zero.

Since specification of a threshold value for δ in (8.2.5) as δ^* where $0 < \delta^* < .5$ is equivalent to the specification of a threshold value for π in (8.2.4) as π^* where $.5 < \pi^* < 1$, we can use π itself as the distance measure. A configuration is then regarded as being in the preference zone if it satisfies both

$$\pi_{ij} \geqslant \pi^* \quad \text{for} \quad i = [k] \text{ and } j \neq [k]$$

$$\pi_{ij} = .5 \quad \text{for} \quad i \neq [k], j \neq [k], \text{ and } i \neq j. \tag{8.2.6}$$

Note that the configuration (8.2.6) is the same as (8.2.4) with $\pi = \pi^*$. Because all the π values in (8.2.6) are equal except when $i = [k]$, in which case π_{ij} has "slipped" to the right, we call the configuration in (8.2.6) a *least favorable slippage* (LFS) configuration. Accordingly, for general π where $.5 < \pi < 1$, the configuration in (8.2.4) can be regarded as a *generalized least favorable slippage* configuration.

REMARK. It should be noted that the LFS configuration is not the same as the LF configuration, and hence is not in agreement with the general approach of this book. However, we may interpret this LFS solution as based on the LF configuration within the class of slippage configurations. In this respect it provides a reasonable approach and solution to the balanced paired-comparison design for the applications being considered.

Normally, our approach is to look for a least favorable configuration and then find the solution for that state of nature, whereas the LFS solution carried out for any nominal P^* and a specific state of nature S does not continue to satisfy the P^* requirement for other states of nature that are apparently more favorable than S. Thus if the number of players is increased (with π^* held fixed), it might appear natural at first sight to expect that the probability of a correct selection should decrease. However, this also (rapidly) increases the number of observations on the better player, which could be in short supply and are badly needed to *increase* the probability of a correct selection. Thus this probability (of a correct selection) decreases for small k, reaches a minimum, and then increases with larger k. The number of replications required then does just the reverse; it increases with increasing k when k is small, reaches a maximum, and decreases with k when k is large. Thus the least favorable configuration could be obtained by setting some merits equal to zero since this effectively reduces the value of k but increases the number of observations on the better player.

As a result, it is necessary for us to put the present solution in proper perspective with respect to our general plan of finding a solution based on the least favorable configuration. Later, in Section 8.5, we do consider an alternative solution to the balanced paired-comparison design, using a different configuration that may be more appropriate for certain problems.
□

8.3 SELECTION RULE

Recall that in the balanced paired-comparison design with k objects and r replications, we have a total of $N = rk(k-1)/2$ observations of preference statements. The data collected then might be recorded as in Table 8.3.1.

Table 8.3.1 Recorded data on preferences

Replication	Object i	Object j 1	2	3	...	k
1	1	—	1	1	...	0
	2	0	—	1	...	1
	.	.	.			.
	.	.	.			.·
	.	.	.			
	k	1	0	1	...	—
2	1	—	0	1	...	1
	2	1	—	0	...	0
	.	.	.			.
	.	.	.			.
	k	0	1	1	...	—
.	.			.		
.	.			.		
.	.			.		
r	1	—	1	0	...	1
	2	0	—	1	...	0
	.	.	.			.
	.	.	.			.
	k	0	1	0	...	—

In Table 8.3.1 a unit in the ith row and jth column, that is, in the (i,j) position, for any replication indicates that in that replication, object i (the row label) is preferred to object j (the column label), whereas a zero in the (i,j) position indicates that object j is preferred to object i. For example, in replication 1, the unit in the first row and second column indicates that object 1 is preferred to object 2. Of course, this means there must be a zero in the second row and first column, because that represents the same pair of objects. Thus, in general, a unit in the (i,j) position implies there will be a zero in the (j,i) position for that same replication. Further, there is never any entry in the diagonal (i.e., the (i,i) position) for any replication since *different* objects are presented in each pair.

Although Table 8.3.1 reflects all the information present in the record of preferences, some of this information is not pertinent for the goals and procedures we have in mind. The information we intend to use is the total number of times each object is preferred to each other object, irrespective of which replication produced those preferences. This total for object i over object j is the number of units in row i and column j summed over all the r replications.

A convenient matrix form for presentation of these totals is a preference table, illustrated in Table 8.3.2. As in Table 8.3.1, the same objects are listed both across the rows and down the columns. The (i,j) entry here, which we denote by f_{ij}, indicates the total number of times object i is preferred to object j in all the r replications. Note that for any i and j $(i \neq j)$, the sum of the (i,j) entry and the (j,i) entry must equal r, the number of replications. In symbols this says that for all pairs (i,j) with $i \neq j$,

$$f_{ij} + f_{ji} = r. \tag{8.3.1}$$

As a result of this relationship, only one half of the table is actually needed; the other half can be regarded an an arithmetic check.

The total of the ith row,

$$f_{i.} = \sum_{j=1}^{k} f_{ij}, \tag{8.3.2}$$

indicates the total number of times object i is preferred over any other object, and the total of the jth column,

$$f_{.j} = \sum_{i=1}^{k} f_{ij}, \tag{8.3.3}$$

indicates the total number of times object j is not preferred when com-

Table 8.3.2　*Preference table: Number of times object i is preferred to object j*

		Object j			Row total
		1	2　　　...	k	
Object i	1	$-$	f_{12}　...	f_{1k}	$f_{1.}$
	2	f_{21}	$-$　　...	f_{2k}	$f_{2.}$
	$\vdots$	$\vdots$	$\vdots$	$\vdots$	$\vdots$
	k	f_{k1}	f_{k2}　...	$-$	$f_{k.}$
Column total		$f_{.1}$	$f_{.2}$　...	$f_{.k}$	$N = \dfrac{rk(k-1)}{2}$

pared with any other object. Hence the sum of the ith row total and the ith column total must equal $(k-1)r$, the total number of comparisons of object i with any other object. In symbols this says that

$$f_{i.} + f_{.i} = (k-1)r. \tag{8.3.4}$$

As a mnemonic for remembering this notation (which is standard for two-way tables), the position of the dot (.) indicates the symbol that has been summed over.

Example 8.3.1 is given to clarify the format for data presentation.

Example 8.3.1 (Food Preference Experiment).　Gulliksen (1956, p. 130) presents data from an experiment designed to determine preferences between five types of meat: tongue, pork, lamb, beef, and steak. Ninety-two college students serve as subjects, and each subject is asked to indicate a preference between all possible pairs of these meats. Since there are $\binom{5}{2} = 10$ pairs of meats, the total number of observations is $10(92) = 920$. The preference table is given in Table 8.3.3.

In Table 8.3.3 the entry 24 in row 1, column 2, for example, indicates that tongue is preferred over pork by 24 of the 92 subjects. Note that the sum of corresponding row and column totals is $4(92) = 368$ in each case, as (8.3.4) indicates, and that the sum of any two entries symmetric about the diagonal is equal to 92, as (8.3.1) indicates.

Before discussing the procedure for selecting the most preferred type of meat, we return to the general discussion. The total in the ith row of the last column of Table 8.3.2, that is, the $f_{i.}$ defined by (8.3.2), indicates the

Table 8.3.3   ***Preference table: Number of times the meat in the left column is preferred to the meat in the top row***

	Tongue	Pork	Lamb	Beef	Steak	Row total
Tongue	–	24	13	1	0	38
Pork	68	–	21	4	0	93
Lamb	79	71	–	13	6	169
Beef	91	88	79	–	19	277
Steak	92	92	86	73	–	343
Column total	330	275	199	91	25	920

number of times object i is preferred over any other object. Hence $f_{i.}/n$ is the appropriate estimate of ν_i, where ν_i is the true relative merit of the ith object (on a scale ranging between 0 and 1), and the selection rule is to select the object with the largest value of $f_{i.}/n$ for $i = 1, 2, \ldots, k$, or equivalently, the object with the largest value of $f_{i.}$ among $f_{1.}, f_{2.}, \ldots, f_{k.}$. Using our usual ordering notation, this largest row sum is expressed as $f_{[k].}$. Note that if the ith object has the largest row sum, it necessarily follows from (8.3.4) that the ith object has the smallest column sum, that is, $f_{[k].} = f_{.[1]}$. Thus we assert that the object that produces $f_{[k].}$ (or equivalently, $f_{.[1]}$) has the most merit.

Of course, it can happen that two or more objects are tied for first place. If this occurs, we are content to select any one of these contenders as the best one. We recommend that one be selected at random by the methods described in previous chapters. Many other possibilities exist for breaking ties but we do not discuss them here.

Example 8.3.1 (continued). According to the selection rule stated earlier, we assert that steak is the meat with the greatest merit, that is, the most preferred type of meat.

Example 8.3.2 (Baseball League Championship). The baseball records for the 1948 season in the United States show that there were 8 teams in the American League and each team played every other team 22 times (except for minor fluctuations). Ignoring these fluctuations, we can say that each team participated in $7(22) = 154$ games, and the total number of games played in the American League was $\binom{8}{2} 22 = 28(22) = 616$. Mosteller (1951, p. 210) gave the records in the form shown in Table 8.3.4. This matrix form is similar to Tables 8.3.2 and 8.3.3, except that here the (i, j)

Table 8.3.4 *Proportion of 22 games won by the team in the left column when playing the team in the top row in 1948*

	Clev.	Bost.	N.Y.	Phil.	Det.	St. L.	Wash.	Chic.	Prop. won
Clev.	–	.522	.455	.727	.591	.636	.727	.727	.626
Bost.	.478	–	.636	.545	.682	.682	.682	.636	.620
N.Y.	.545	.364	–	.545	.591	.727	.773	.727	.610
Phil.	.273	.455	.455	–	.455	.818	.636	.727	.546
Det.	.409	.318	.409	.545	–	.500	.727	.636	.506
St. L.	.364	.318	.273	.182	.500	–	.455	.619	.387
Wash.	.273	.318	.227	.364	.273	.545	–	.571	.367
Chic.	.273	.364	.273	.273	.364	.381	.429	–	.337
Prop. lost	.374	.380	.390	.454	.494	.613	.633	.663	.500

element represents the *proportion* x_{ij} of games in which the ith team defeated the jth team. Thus the (i,j) element here is $x_{ij} = f_{ij}/22$, and in this case the (i,j) and (j,i) elements sum to 1. The last column gives the proportion of games won by the ith team, that is, $x_{i.} = f_{i.}/154 = \sum_{j=1}^{k} f_{ij}/(22)(7)$, and the last row gives the proportion lost by the jth team, that is, $f_{.j}/154$. We show .500 as the overall proportion won (or lost), that is, $\sum_{i=1}^{k} x_{i.}/8 = .500$ since every game had a winner and a loser (i.e., ties were not possible).

Even though the entries in Table 8.3.4 are given as proportions, rather than frequencies, the selection rule is applied as before since the team that won the largest proportion of games also won the largest number of games. From the data we see that Cleveland should be asserted to be the best team in the American League in 1948.

Example 8.3.3 (Cereal Preference). A paired-comparison experiment was performed to determine the preference of enlisted men in the U.S. Army among seven well-known ready-to-eat breakfast cereals. The subjects did not actually taste any cereal; they were asked to indicate their preference when presented with each possible pair of the standard 1 oz. boxes of these cereals. Hence this cannot be regarded as a taste-preference experiment, since the appeal of the label (among other things) may also influence the preferences. Each of 50 subjects made the $7(6)/2 = 21$ comparative judgments, and the data reported in Jones and Bock (1957, p. 153) are shown in Table 8.3.5. Since cereal 5 has the largest number of stated preferences, we assert that it is the most preferred kind.

Table 8.3.5　Preference table: Paired preferences among seven ready-to-eat breakfast cereals

	1	2	3	4	5	6	7	Total
1	–	23	24	16	20	32	24	139
2	27	–	23	21	13	35	26	145
3	26	27	–	24	19	31	25	152
4	34	29	26	–	26	39	28	182
5	30	37	31	24	–	37	28	187
6	18	15	19	11	13	–	18	94
7	26	24	25	22	22	32	–	151
Total	161	155	148	118	113	206	149	1050

8.4 ANALYTICAL ASPECTS OF THE PROBLEM OF SELECTING THE BEST OBJECT

In this section we consider the problem of determining the number of replications needed for a balanced paired-comparison experiment with a specified triple of values $(\pi^*, P^*, .5)$. We write the specified values as a triple, since (8.2.6) specifies two different values for the parameter π, namely, $\pi_{[k]j} = \pi^*$ and $\pi_{ij} = .5$ otherwise. We also consider the problem of calculating the operating characteristic curve for a fixed number of replications. In each case we find a solution only for the *least favorable slippage* configuration given in (8.2.6).

It can be shown that the least favorable slippage configuration is not in general the least favorable configuration. However, if we have some basis for insisting that all objects, except for the best one, have equal merit, then the least favorable slippage configuration is least favorable. Some discussion of what may happen when the true configuration is different from (8.2.4) is given later in Section 8.5.

8.4.1 Determination of Number of Replications

Suppose that for a particular k, we are in the process of designing a paired-comparison experiment to select the object with the greatest merit, and we wish to determine the smallest number of replications r needed to satisfy a specified $(\pi^*, P^*, .5)$ requirement where $\pi^* > .5$ and $1/k < P^* < 1$. These values have been computed under the least favorable slippage (LFS) configuration, and the integer solution for r is given in Table L.1 of Appendix L for $k = 2, 3, \ldots, 10, 12, 14, 16, 18,$ and 20. The values covered for π^* are .55 to .95 in steps of .05, and $P^* = .75, .90, .95, .99$. For values not listed specifically in these tables (but within their range), the method of interpolation explained in Appendix L can be used. It is important to remember that these tables give the number of replications r, and not the sample size per population or the total sample size required. The total sample size is $N = rk(k-1)/2$; for any fixed k, N increases quite rapidly as r increases.

Example 8.4.1 (Example 8.3.2 continued). In the American League baseball season of 1948, recall that there were $k = 8$ teams. Suppose that this same number of teams compete in some future season, and the problem is to determine the number of times each team should play every other team, that is, the value of r such that the probability that our selection rule will lead to a correct selection under the least favorable slippage configuration is at least $P^* = .90$ when $\pi^* = .55$ (i.e., when the probability that the best team will defeat any one of the others in a single game is only .55 and the others all have equal chance of beating one another). This may seem like a very stringent specification for π^* (and it may not seem highly consistent with the data in Table 8.3.4), but frequently the teams in each league are relatively evenly matched and the second best team is almost as good as the best team. From Table L.1 with $k = 8$, we find that $r = 103$, and hence the number of games that must be played in the season is $7(103) = 721$ games per team, or a total of $N = 103(28) = 2884$ games. From this same table we see that the 22 games played in 1948 are not enough to guarantee even a P^* value of .75 assuming a value of π^* as small as .55. Since $\pi^* = .55$ requires such a large number of replications, let us see what happens for a larger π^*. If $\pi^* = .60$, for example, 22 games are satisfactory under the least favorable slippage configuration for some P^* between .75 and .90. By interpolation in this table , we find that 22 games would be satisfactory if we were to specify $P^* = .87$. The important point here is that we can consult the tables *before* deciding on the values to specify for π^* and P^*.

In general, as would be expected, whenever π^* is close to .50 for any k, the number of replications required under the least favorable slippage configuration is very large. Further, as P^* increases for any k and π^*, the value of r is nondecreasing. It is especially interesting to note that for fixed π^* and P^*, a few of the values of r increase as k increases (namely, some of those for $k \leqslant 4$ and $P^* \leqslant .90$), but eventually r decreases as k increases. Thus even though it is more difficult to identify one object as better than all the others when k is large, each new replication brings in $k(k-1)/2$ additional observations so that r need not necessarily increase when k increases.

In the situation of Example 8.3.1, where we had $k = 5$ types of meat, suppose we are in the process of designing an experiment to select the most preferred type. If we set $P^* = .90$ and $\pi^* = .60$, then under the least favorable slippage configuration we need $r = 33$ replications to guarantee this $(\pi^*, P^*, .5)$ condition. If we raise P^* to .95, the replication number increases to 46. Also, if π^* is increased, the discrepancy between the best and second best meat becomes greater, so that the required value of r is nonincreasing under the least favorable slippage configuration.

REMARK. For the special case $k = 2$, this model again reduces to that of a single binomial distribution (or a multinomial distribution with $k = 2$) with the probability $p_{[2]} = \pi$ that one player wins a single game and probability $p_{[1]} = 1 - \pi$ for the other player. If we let δ^* denote the specified lower bound for the ratio $\delta = p_{[2]}/p_{[1]}$ and let π^* denote the specified lower bound for π, then δ^* and π^* are related by the expression $\delta^* = \pi^*/(1 - \pi^*)$, or equivalently, by $\pi^* = \delta^*/(1 + \delta^*)$. Except for this interchange of the measure of distance, the two problems and the solutions given are identical for the special case $k = 2$. For example, if $k = 2$, the solution obtained using Table L.1 with $\pi^* = .75$, $P^* = .90$, indicates that $r = 7$ games are required. The corresponding value of δ^* for the multinomial problem is $\delta^* = .75/.25 = 3$ and the solution obtained using Table H.1 with this δ^* and $P^* = .90$ is again 7. In fact, all the entries for $k = 2$ in Table H.1 are equal to the entries for $k = 2$ in Table L.1 if $\delta^* = \pi^*/(1 - \pi^*)$ and hence these tables supplement each other.

Since the special case $k = 2$ is equivalent for the two formulations, the whole discussion in Example 6.4.5 about the problem of the World Series in baseball, where $k = 2$, is also appropriate to this chapter. Conversely, for the special case $k = 2$ the preceding discussion on the problem of designing a paired-comparison experiment ties in with the corresponding problem of Example 6.4.5. $\square$

8.4.2 Calculation of the Operating Characteristic Curve

In this subsection we assume that the number of replications is fixed and known, either because the data have already been collected or because the number of replications is (or was) determined by considerations other than statistical ones. The question to be answered is, What kind of overall performance can we expect from the selection rule of Section 8.3? Alternatively, the value of r may have been determined by the methods of Section 8.4.1, but we may still be interested in knowing the P value for other π^*, or the π value for other P^*. These questions can be answered by calculating the operating characteristic curve.

As noted in Section 8.2, we do not know the least favorable configuration for the selection problem with a paired-comparison design, and hence we cannot calculate the operating characteristic curve under that configuration. However, the curve can be calculated under the least favorable slippage configuration given in (8.2.6). This function gives the locus of all pairs (π^*, P^*) in the triple $(\pi^*, P^*, .5)$ that are satisfied for a given r, and hence it can be interpreted as either the probability of a correct selection under the least favorable slippage configuration for any value π^*, that is, $P_{LFS}\{CS|\pi^*, r\}$, or as the confidence level $CL(\pi^*|R)$ associated with our selection rule R for any π^*. For any fixed r and k we must interpolate in the appropriate table in Appendix L to find the corresponding (π^*, P^*) pair. These pairs are plotted on a graph to determine the operating characteristic curve. This calculation is illustrated by the following example.

Example 8.4.2 (Effect of a Flavor Enhancer). Monosodium glutamate (MSG) is frequently used in different concentrations as a taste stimulant or flavor enhancer. An experiment to determine which concentration produces the preferred flavor of dehydrated apple slices was conducted at the Virginia Agricultural Experiment Station by L. L. Davis and the data are reported in Bradley (1954, p. 389). Apple slices containing four different concentrations, labeled as A, B, C, and D, were presented in pairs to seven different judges. Here A, B, and C represent increasing concentrations of MSG whereas D has no MSG and acts as a check or control. The preference table is given as Table 8.4.2, where the (i,j) entry indicates the number of preferences for concentration i over concentration j.

Table 8.4.2   *Preference table: Preferences of the flavor of dehydrated apple slices with different concentrations of MSG*

	A	B	C	D	Total
A	–	4	4	0	8
B	3	–	5	2	10
C	3	2	–	3	8
D	7	5	4	–	16
Total	13	11	13	5	42

Since concentration D was chosen the most frequently in these paired comparisons, we assert that concentration D produces the best flavor with dehydrated apple slices. (Recall that D has no MSG.) To assess the performance of our selection procedure we interpolate in Table L.1 for $k = 4$ using $r = 7$ to obtain the pairs of points (π^*, P^*). The method of interpolation is explained and illustrated in Appendix L. The results for $r = 7$ are as follows:

π^*	.66	.73	.76	.82
P^*	.75	.90	.95	.99

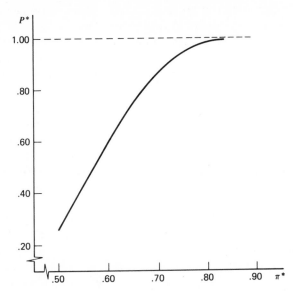

Figure 8.4.1 Operating characteristic curve under the least favorable slippage configuration for $k = 4$ and $r = 7$.

These values are plotted in Figure 8.4.1 to give the operating characteristic curve for $r = 7$. (Strictly speaking, we should now drop the * in π^* since only the $r = 7$ and the rounded P^* values are specified, but no confusion should arise.)

*8.5   CONJECTURED LEAST FAVORABLE CONFIGURATION

As remarked earlier, the entries for r in Table L.1 of Appendix L, computed under the least favorable slippage configuration given in (8.2.6), are not always increasing as k increases. For example, for the pair $(\pi^*, P^*) = (.55, .75)$, we have $r = 71$ for $k = 4$ and $r = 68$ for $k = 5$. In fact, for this (π^*, P^*) and any $k \geqslant 5$, we have $r \leqslant 68$. This is equivalent to saying that for r fixed, the probability of a correct selection increases as k increases for certain values of π^*. This fact indicates that the least favorable slippage configuration cannot be the overall least favorable configuration. As a result, in this section we give a new conjecture for the overall least favorable configuration; we call this the *conjectured least favorable* (CLF) configuration.

The result given here is not proposed as an alternative to the formulation and solution that was described in previous sections. However, we do think that this new solution is of some interest, and we recommend that it be used in a round robin tournament where there may be evidence of "collusion." Collusion occurs if two players, say A and B, are in close competition and some other player or players decide to lose to, say, player A so that A can gain a higher total score than player B. If the collusion is serious, then even the CLF approach may not solve the problem. (Such alleged collusions have been widely publicized in recent years in international chess competitions.)

The solution based on the conjectured least favorable configuration can be found from Table L.1 in the following way. We now let k_0 denote the particular value of k that applies to this tournament, that is, the fixed total number of players. Then we regard $\tilde{k}$ as a variable over the integers $\tilde{k} = 2, 3, \ldots, k_0$, and determine r under the conjectured least favorable configuration for a specified (π^*, P^*) requirement as the maximum value of r, over $\tilde{k} = 2, 3, \ldots, k_0$, under the least favorable slippage configuration for the $(\pi^*, P^*, .5)$ requirement with the same values of π^* and P^*. In other words, for any fixed values of π^* and P^*, we check the tables in Appendix L.1 for all $\tilde{k}$ less than or equal to k_0, and see which $\tilde{k}$ value gives the maximum value of r. This maximum value is the number of replications required under the conjectured least favorable configuration.

As a numerical illustration, suppose we have $k_0 = 8$ players in competition and we specify $\pi^* = .65$ and $P^* = .90$. For these values the tables in Appendix L.1 for each integer $\tilde{k} = 2, 3, \ldots, 8$ give the following respective values of r:

$\tilde{k}$	2	3	4	5	6	7	8
r	17	18	16	15	13	12	11

The conjectured least favorable solution is $r = 18$, which is attained for $\tilde{k} = 3$, and the total sample size required is $18\binom{8}{2} = 18(28) = 504$. On the other hand, the least favorable slippage solution, attained for $k = \tilde{k} = 8$, is $r = 11$, or a total of $11\binom{8}{2} = 11(28) = 308$ comparisons. Thus in this case the conjectured least favorable solution requires a total sample size increase of about 70% over the least favorable slippage solution, to attain the same P^* level with the same value of π^*.

The tables in Appendix L.2 have been prepared as an aid to the investigator in determining the conjectured least favorable solution. For each specified pair (π^*, P^*) these tables give the value of $\tilde{k}$, for $2 \leq \tilde{k} \leq k_0$ $(k_0 \geq 3)$, which has the maximum r under the least favorable slippage configuration. Once $\tilde{k}$ is determined from these tables, the corresponding value of r is found from the appropriate table in Appendix L.1. (Obviously, if $k_0 = 2$, the value of r is found directly from Table L.1 for $k = 2$.) It is interesting to note that the results in Appendix L.2 are almost independent of k_0. They are completely independent for $P^* = .90, .95$, and $.99$ and all π^* listed when $k_0 \geqslant 3$, and also for all P^* and all π^* listed when $k_0 \geqslant 4$.

Since the conjectured least favorable solution gives fairly large values of r in a balanced paired-comparison (round robin) design, we recommend that it be used only when there is evidence of collusion. Of course, collusion cannot occur in say a food-tasting experiment, and therefore we do *not* recommend this solution for such problems. Rather, if we are committed to a round robin design in an experiment for which there is not any evidence for, or cannot be, any collusion, we do recommend the least favorable slippage solution described in Section 8.4.

REMARK. The conjectured least favorable solution given here is based on the following reasoning. If some proper subset of the k_0 players have no merit (i.e., each of these has a corresponding ν value that is equal to zero), then each of these players is defeated by any player with positive merit. In particular, the effect of setting say $k_0 - j$ merits (or ν values) equal to zero is to reduce the value of k_0 to j in the calculation of the probability of a correct selection under the least favorable slippage configuration. However, the value of $k = k_0$ in the total sample size remains unchanged, that is, $N = rk_0(k_0 - 1)/2$, and the total sample size varies with j only through r. Hence if j is the value of $\tilde{k}$ for which r is a maximum for some specified $(\pi^*, P^*, .5)$ requirement, then the conjectured least favorable configuration is attained by setting $k_0 - j$ merits equal to zero and putting the remaining merits in the usual least favorable slippage configuration.

The reasoning in the preceding paragraph is based on the assumption that the least favorable situation will not occur with π values in the range between 0 and $1 - \pi^*$. Since this assumption has not been proved, we call this the *conjectured* least favorable configuration. Nevertheless this method does show that the conjectured least favorable solution is attainable, and hence this solution forms a lower bound for the one associated with the overall least favorable configuration. $\square$

8.6   AN ALTERNATIVE MORE EFFICIENT DESIGN

As noted earlier, the completely balanced paired-comparison (or round robin) design gives fairly large answers for r. Thus we should consider some other designs that are more efficient in identifying the object with the most merit. In this section we look at an alternative paired-comparison design in which the balanced aspect is sacrificed in order to increase efficiency.

The most common alternative design drops the requirement that every possible pair of objects be compared by each judge. The best-known version of this setup is called the *knock-out* (*KO*) *tournament* in sports, and we use the tournament terminology in this section.

The knock-out tournament, again with say k players (or teams), consists of R rounds, where k and R satisfy the relationship $k = 2^R$, or equivalently, $R = \log_2 k$. (Note that $\log_2 k = 1.4427 \log_e k$, where $\log_e k$ is the natural logarithm.) In the first round the players are paired off, and each of these pairs plays a series of games that leads to the declaration of a winner within that pair (i.e., the series cannot end in a draw). The loser in any pair is then knocked out of the tournament, so that after the first round the number of players is reduced by a factor of 2, to 2^{R-1}. These remaining players are then paired off for a second round of the same nature. Thus, with each round played, one-half of the players are eliminated, and a champion is determined after R rounds. Since each pairing gives rise to a player that is knocked out, the total number of pairings in the tournament is $k - 1$. For example, if we start with $k = 16$ players, then we need $R = \log_2 16 = 4$ rounds and a total of 15 pairings are made—eight in the first round, four in the second round, two in the third round, and one in the final (fourth) round.

Note then that a winner *could* be declared after R rounds or a total of $k - 1$ games in a knock-out tournament, as compared to $k(k-1)/2$ games in a round robin tournament. However, this winner is unlikely to be the best player if, in the KO tournament, each round consists of only one game, and if, in the RR tournament, each team plays every other team only one time (only one replication). As a result, in order to compare the efficiency of the two types of tournaments for determining the best player, we must consider the total number of games required for a specific kind of selection rule. Of course, we already know that for specified values $(\pi^*, P^*, .5)$, the tables in Appendix L can be used to obtain the required number r of times that each player must play every other player in an RR tournament under either the least favorable slippage configuration or the conjectured least favorable configuration; thus the total number of games required is easily calculated as $N = rk(k-1)/2$. For an efficiency comparison we must also compute the expected total number of games required in a knock-out tournament, denoted by $E(N|\text{KO})$, to satisfy the same (or an equivalent) requirement with specified values for π^* and P^*.

For the purpose of computing $E(N|\text{KO})$, we must first describe the contest between any pair of players in the knock-out tournament. Suppose they play a series of independent games and stop as soon as one of them wins w games, where w is computed at the outset as a function of π^* and P^*, and, as before, π^* denotes the probability that the best player will defeat any opponent. We compute w by setting the probability that the best player will win in any one round equal to $(P^*)^{1/R}$. Then, assuming that the successive rounds are independent, the probability that the best player will win in all R rounds of the tournament is at least P^*, as desired. For each of the $k-1$ pairings, the number of games (or comparisons) needed to get w wins is a chance variable denoted by G. The expected value of G, $E(G)$, can be denoted by $E_{\pi^*}(G)$ and $E_{.5}(G)$, depending on whether the best player is included in the pair or is not included. Clearly, we are assuming the state of nature described in the configuration (8.2.6), with $\pi_{ij} = \pi^*$ for $i = [k]$, $j \neq [k]$, and $\pi_{ij} = .5$ otherwise.

The value of w is defined as the smallest integer such that

$$I_{\pi^*}(w,w) \geqslant (P^*)^{1/R}, \tag{8.6.1}$$

where R is the smallest integer equal to or greater than $\log_2 k$, and $I_{\pi^*}(w,w)$ denotes the incomplete beta function and can therefore be read from Table J.2 in Appendix J. With this integer value of w, we can now compute the expected value of the total number of comparisons, $E(N|KO)$. Since R of the $k-1$ pairings contain the best player and $k-1-R$ pairings match players of equal merit, $E(N|KO)$ is calculated as the weighted average

$$E(N|KO) = RE_{\pi^*}(G) + (k-1-R)E_{.5}(G). \tag{8.6.2}$$

In general, if there are two players, say A and B, and p is the probability that A beats B and $q = 1 - p$ is the probability that B beats A in a single play (or comparison), then the value of $E_p(G)$ for given w is

$$E_p(G) = \frac{w}{p}I_p(w+1,w) + \frac{w}{q}I_q(w+1,w), \tag{8.6.3}$$

where $I_x(w+1,w)$ denotes the incomplete beta function and can be read from Table J.1 in Appendix J. Thus (8.6.3) can be used to obtain the two expected values needed to compute (8.6.2).

To illustrate these computations, we return to the example considered in Section 4 where we had $k = 8$ objects, $\pi^* = .65$, and $P^* = .90$. Then $R = \log_2 8 = 3$ and $(P^*)^{1/R} = (.90)^{1/3} = .9655$. Using (8.6.1) and Table J.2 in Appendix J, we find that $w = 18$ gives $I_{.65}(18,18) = .9664$, and $w = 17$ gives $I_{.65}(17,17) = .9623$, and hence we use $w = 18$ to satisfy (8.6.1). (Note that this happens to coincide with the r value determined for the conjectured least favorable solution in Section 8.5.) Substituting $w = 18$, $p = \pi^* = .65$, and the value of $I_{.65}(19,18)$ from Table J.1 in Appendix J in (8.6.3), we obtain

$$E_{.65}(G) = \frac{18}{.65}(.9543) + \frac{18}{.35}(.0215) = 27.53.$$

Then we compute $E_{.5}(G)$ by substituting $w=18$, $p=.5$, and the value of $I_{.50}(19,18)$ from Table J.1 in Appendix J in (8.6.3) to obtain

$$E_{.50}(G) = \frac{18}{.50}(.4340) + \frac{18}{.50}(.4340) = 31.25.$$

Substituting these results in (8.6.2) with $k=8$ and $R=3$, we obtain

$$E(N|KO) = 3(27.53) + 4(31.25) = 207.59.$$

This is the KO solution for the expected total number of games required for $\pi^* = .65$ and $P^* = .90$. Randomizing on w between 17 and 18 has a negligible effect.

From the table for $k=8$ in Appendix L.1 the least favorable slippage solution is $r=11$ for $\pi^* = .65$, $P^* = .90$, so that $N = 11(8)(7)/2 = 308$, and thus the average saving using the KO solution is 32.6%. (We found in Section 8.5 that the conjectured least favorable solution is $r=18$ so that $N = 18(8)(7)/2 = 504$; thus the average saving here using the KO solution over the CLF solution is 59%.) The average saving for the KO solution over the LFS solution was found to be of the same order of magnitude, namely, about one-third, for different values of π^* and P^*.

These results show that if we drop the rigid requirement that all pairs of players must meet each other, or equivalently, that all pairs of objects have to be compared, there exist designs that are more efficient and satisfy the same probability requirements specified by π^* and P^*. The knock-out tournament format is employed more in tennis competition, such as Wimbledon, and the round robin tournament format is used more in league competitions such as baseball and football.

REMARK. One could also argue that further improvements are possible, for example, by keeping track only of the total number of games won in the knock-out tournament. However, this could bring in biases if by chance one player was pitted only against players with very low merit. In addition, the necessary theory and tables for such improvements are not available for general k. $\square$

8.7 SELECTING THE WORST OBJECT

In some applications the goal of interest may be to select the *worst* object among k objects in a balanced paired-comparison design. For example, in a drug-screening study the investigator may wish to identify that one drug that is least effective so that he can eliminate it from further experimentation. Alternatively, if a restaurant owner decides that he must eliminate one of the entrees on the regular menu, he wants to determine which entree is the least popular among his regular customers.

Such problems are easily handled by the methods of this chapter, since the word *best* can be defined in whatever manner is appropriate to the problem. For example, if we want to identify the worst player in a round robin tournament we can say that player i is preferred to player j if player i loses to player j. The entry x_{ij} in the preference table (Table 8.3.1) is equal to 1 if player i loses to player j and equal to zero if player i beats player j. Then the selection rule and methodology analysis can be applied to this problem in exactly the same way as for the previous goal of identifying the *best* player, that is, the one with the highest probability of losing.

8.8   NOTES, REFERENCES, AND REMARKS

The main source for the original ideas on paired comparisons is David (1963). Moreover, the discussion in Section 8.2 and the analyses in Section 8.4 follow the formulation of David (1963). However, it should be pointed out that David does not consider this problem from the point of view that is used here. The examples, the orientation toward solving for an appropriately sized design, the idea of a least favorable slippage configuration, and the conjectured least favorable configuration are all new with this book. This also applies to our efficiency comparisons between different types of tournaments in Section 8.6. In essence, we are applying our background and structure to the problem that David treated from another point of view. Some other important references in this area are Bühlmann and Huber (1963), Huber (1963a, b), Trawinski and David (1963).

The general area of inverse sampling (i.e., sampling until a preassigned result is obtained) and other sampling methods has been treated in many papers but is not included in this book. One application of inverse sampling is in ranking, say, three players in a tournament. For example, if only two can play against one another (as in tennis), one can keep dropping out the loser. Such a procedure was considered by Sobel and Weiss (1970) in which one player has to be called best. In a variant of this Bechhofer (1970) allows a fourth decision that no one of the three is best. An inverse sampling procedure was used for selecting the best cell (i.e., the one with the largest cell probability) in a multinomial distribution by Cacoullos and Sobel (1966). An inverse sampling procedure for selecting the best binomial population using the play-the-winner rule was studied extensively by Sobel and Weiss (1972a, b), and a survey of many of the procedures appears in Sobel and Weiss (1972b). A survey of adaptive sampling for clinical trials is given by Hoel, Sobel, and Weiss (1972).

PROBLEMS

8.1 The Chronicle of Higher Education (1976, p. 14) reported some results of a survey of some 3500 college teachers by two political scientists, Seyour Martin Lipset of Stanford University and Everett C. Ladd of the University of Connecticut (The Ladd-Lipset Survey). In an effort to determine the self-identification of faculty members, respondents were asked to designate which of five terms—*intellectual, scholar, scientist, teacher,* or *professional*— best described themselves. The summary results were as follows:

Identification classification	Percentage
Intellectual	11
Scholar	12
Scientist	11
Teacher	32
Professional	44
Total	110

The total added to more than 100% because of some multiple responses. Given that some respondents could not decide on one of the five, perhaps a paired-comparison design should have been considered. Then each respondent would be asked to choose one identification classification over another for each of the 10 possible pairs of classifications.

(*a*) If you were designing such a paired-comparison experiment for the purpose of identifying the most preferred identification classification, how many respondents would be required to satisfy the requirement $\pi^* = .55$, $P^* = .90$ under the least favorable slippage configuration?

(*b*) What is the answer to (*a*) if $\pi^* = .75$, $P^* = .90$?

(*c*) Interpret the meaning of π^* in terms of this problem and explain why the increase in π^* leads to such a decrease in the required number of respondents.

(*d*) Suppose the paired-comparison experiment is carried out with $r = 20$ respondents. Calculate the operating characteristic curve for the selection rule.

8.2 A director of placement services at a major university has a grant to carry out a study of the views of various groups who recruit and employ new college graduates. One aspect of the study concerns the relative importance to employers of various items included on an applicant's written resume in determining whether to request a personal interview of that applicant. The items on a resume are to be grouped into the following six categories:

1. PD—personal data
2. ED—education experience (major, minor, institution, etc.)
3. EM—employment experience
4. OW—other work experience (campus activities, publications, etc.)
5. HA—honors and awards
6. RF—references

One way of investigating the relative importance of these criteria would be a paired-comparison experiment.

(a) How many employers should be asked to make the $6(5)/2 = 15$ comparative judgments if we specify $P^* = .95$, $\pi^* = .60$?

(b) Suppose the experiment is performed using the r value obtained in (a), but only 30 employers completed the questionnaire. The preference table follows.

   (i) Which category is the most important to these employers?
   (ii) Calculate the operating characteristic curve.

Number of times the category in the left column is preferred to the category in the top row

	PD	ED	EM	OW	HA	RF	Row total
PD	–	1	1	8	2	3	15
ED	29	–	18	20	28	16	111
EM	29	12	–	16	24	18	99
OW	22	10	14	–	15	12	73
HA	28	2	6	15	–	7	58
RF	27	14	12	18	23	–	94
Column total	135	39	51	77	92	56	450

CHAPTER 9

Selecting the Normal Population
with the Best Regression Value

9.1 INTRODUCTION

In this chapter we return to the basic model introduced in Section 2.2, where we have k normal populations with a common, known variance σ^2 and unknown means $\mu_1, \mu_2, \ldots, \mu_k$, that is

$$N\left(\mu_{[1]}, \sigma^2\right), N\left(\mu_2, \sigma^2\right), \ldots, N\left(\mu_k, \sigma^2\right).$$

The goal again is to select the single population that has the largest mean value μ, but we now have a different experimental setting and sampling procedure.

Consider the broad class of models in which the μ values can be considered average "treatment effects." The treatments can be fertilizers, medicines, tests, learning experiences, or the like. One possible sampling procedure is to assign experimental units randomly to the k treatment groups and another is to assign treatments randomly to the experimental units. In either case only random factors determine which unit receives any particular treatment. This type of experimental design (called a *completely randomized design*) is a means of reducing sampling bias, that is, eliminating the effects of differences between the experimental units that are not attributable to the effects of the treatments. Then when the units are subjected to treatments, any differences in μ values can be attributed primarily to differences in treatment effects.

However, in many experimental situations this kind of design is not possible or practical. For example, if the treatments are k different teaching methods, it may be necessary to use k different schools and implement a different teaching method at each school. There may be differences between students at the different schools, for example, neighborhood income levels or level of educational achievement of parents, and such

234

factors may affect the results considerably. If the effect of these factors could be eliminated from the measures of effectiveness of teaching methods, more reliable comparisons would be possible. We can eliminate these effects by using techniques of regression analysis.

9.2　SELECTION PROCEDURE

We explain the selection procedure for this model in terms of an example.

Example 9.2.1 (Learning)　A "wordlike" item is defined as a group of letters that do *not* form a real English word or a familiar foreign word, for example, *nept*. A nonsense string is a sequence of wordlike items, which may or may not also include real English suffixes, prefixes, or even words to show grammatical relationships (called *function words*, *form words*, or *functors*), such as prepositions, conjunctions, or the like. An example without functors is *nept et voor mux prut*, and an example with functors is *the nept et vor mux prutly*, which has the functors *the* and *ly*. Since 1957, when the linguist C. E. Osgood made the suggestion, nonsense strings have been used frequently in experiments to study verbal learnings, recall, and psycholinguistics (see for example, O'Connell(1970)).In such experiments subjects "learn" one or more nonsense strings until they attain some stated "criterion of retention or recall." The method of presentation of these strings is one of the many factors that may have a significant effect on learning or recall or both. Suppose the problem is to study the following four different written methods of presentation of a 12-item string. The wordlike items are listed on the page as: (1) horizontally with instructions to read from left to right and denoted here by LR; (2) horizontally with instructions to read from right to left and denoted by RL; (3) vertically with instructions to read from top to bottom, denoted by TB; and (4) vertically with instructions to read from bottom to top, denoted by BT. The subjects are 36 eighth-grade students, nine drawn randomly from each of four different schools. The four methods of presentation are assigned randomly to the four schools, one method per school. Method BT is assigned to school 1, TB to school 2, LR to school 3, and RL to school 4. At each school each subject studies the 12-item string presented in the assigned method for eight seconds and is then given 30 seconds to write it. This process of studying and writing is repeated with the same nonsense string until the subject can reproduce the string perfectly. Twenty-four hours later these same students are given a test of recall. The data given in Table 9.2.1 are hypothetical scores on such a recall test; the scores range from 0 to 10, where 10 is a perfect score; the larger the score the greater the recall.

Table 9.2.1   Recall scores of students at four different schools using four forms of written presentation

	BT	TB	LR	RL
	5.2	7.9	10.0	6.5
	7.1	6.7	7.7	4.3
	6.9	8.1	10.0	3.8
Individual	9.2	9.2	8.6	7.4
student scores	6.3	8.3	9.8	3.6
	4.1	7.6	7.9	7.9
	5.3	8.9	9.6	6.2
	6.9	9.4	8.2	5.4
	7.3	7.5	8.4	4.4
Total	58.3	73.6	80.2	49.5
Sample size	9	9	9	9
Sample mean	6.48	8.18	8.91	5.50

Suppose that the goal is to identify the best method of presentation, that is, the one with the largest mean recall. Since method LR has the largest sample mean, we would assert that horizontal left to right is the best written presentaion to facilitate recall.

However, note that each method of presentation is used at a different school. Might there then be a "learning ability or intelligence effect"? If one school has students who have a greater learning ability on the average

Table 9.2.2   Recall and grade score of student at four different schools using four forms of written presentation

	BT		TB		LR		RL	
	Grade	Recall	Grade	Recall	Grade	Recall	Grade	Recall
	v	w	v	w	v	w	v	w
	2.4	5.2	3.0	7.9	3.7	10.0	2.6	6.5
	2.8	7.1	2.6	6.7	3.1	7.7	2.4	4.3
	3.0	6.9	3.1	8.1	3.5	10.0	2.0	3.8
Individual	3.4	9.2	3.2	9.2	3.4	8.6	2.8	7.4
student	2.6	6.3	2.9	8.3	3.6	9.8	2.2	3.6
scores	2.3	4.1	2.8	7.6	3.2	7.9	2.7	7.9
	2.9	5.3	3.3	8.9	3.5	9.6	2.7	6.2
	2.7	6.9	3.4	9.4	3.3	8.2	2.3	5.4
	3.1	7.3	2.9	7.5	3.3	8.4	2.1	4.4
Total	25.2	58.3	27.2	73.6	30.6	80.2	21.8	49.5
Sample size	9	9	9	9	9	9	9	9
Sample mean	2.80	6.48	3.02	8.18	3.40	8.91	2.42	5.50

than students at the other schools, that group may exhibit greater recall ability with nonsense strings irrespective of what method of presentation is used. If we had some average measure of learning ability for each student at every school (in addition to the scores on the recall test), we could to some extent eliminate the "learning ability effect." Suppose we consider the student's grade point average, (GPA) (denoted by v in the Table 9.2.2) as a measure of learning ability (although it may not be the best measure available). The data then appear in the form of Table 9.2.2.

In the previous analysis of the recall data, we did not use any information concerning the grade point average of each student. Now we describe a method that permits us to take this additional information into account in making a selection.

We first look at the following ordered sample means of grade and recall (ordered separately):

Grade	Method		Recall	Method
2.42	RL		5.50	RL
2.80	BT		6.48	BT
3.02	TB		8.18	TB
3.40	LR		8.91	LR

Note that the largest sample mean of recall occurred at the school in which method LR was used, and that the students in the same school also had the largest sample grade point average. This suggests that their higher learning ability might account for their superior recall ability more than does the method itself.

As a first step toward taking a possible learning ability effect into account, we plot the data pairs for each sample in a scatter diagram and calculate the sample correlation coefficients. The nine pairs of points (grade, recall) $= (v, w)$ for each of the four methods are shown on four different graphs in Figure 9.2.1. Note that in each case the points lie approximately on a straight line, and the correlation coefficients are relatively large. In general, such learning curves would not be linear over the whole range of GPA values, since the curve has to flatten out for large GPA values. However, within the middle range a linear approximation is not unreasonable.

As a result, we assume that for each method the mean recall (the mean of w) is dependent on the grade and the dependency relationship can be described by a straight line as

$$\mu_{w|v} = E(W|v) = \alpha + \beta v.$$

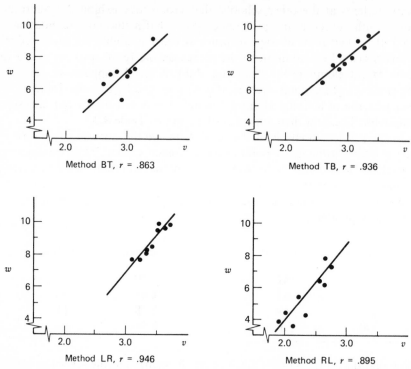

Figure 9.2.1   Scatter diagrams and calculated regression lines for data in Table 9.2.1.

Since we have four different methods of presentation here, we have four different regression lines as

$$E_j(W|v) = \alpha_j + \beta_j v_j \qquad \text{for } j = 1, 2, 3, 4. \tag{9.2.1}$$

Here $E_j(W|v)$ denotes the true mean recall value conditional on the grade v for the jth population, and α_j and β_j are unknown regression parameters. Such a linear relationship as in (9.2.1) may hold between $\mu_{w|v}$ and v regardless of the distribution, but it always holds if w and v have a joint bivariate normal distribution, in which case the conditional distribution of w given v is also a normal distribution.

Given a set of data that consist of n observations on the pairs (v, w) for each of the k populations, we may compute each of the k sample regression lines. The data appear in the format of Table 9.2.3, where v_{ij} and w_{ij} denote the ith observation in the jth sample.

Table 9.2.3   Data presentation for pairs (v, w)

1		2		...	k	
v_{11}	w_{11}	v_{12}	w_{12}	...	v_{1k}	w_{1k}
v_{21}	w_{21}	v_{22}	w_{22}	...	v_{2k}	w_{2k}
.	.	.	.		.	.
.	.	.	.		.	.
v_{n1}	w_{n1}	v_{n2}	w_{n2}	...	v_{nk}	w_{nk}

The general equation of the line for the sample from population j is given by $w_j = a_j + b_j v_j$, where a_j and b_j are estimates of α_j and β_j, and are found from the standard regression formulas

$$b_j = \frac{\sum\limits_{i=1}^{n} (w_{ij} - \bar{w}_j)(v_{ij} - \bar{v}_j)}{\sum\limits_{i=1}^{n} (v_{ij} - \bar{v}_j)^2}, \qquad a_j = \bar{w}_j - b_j \bar{v}_j. \qquad (9.2.2)$$

The numerical results of (9.2.2) for the data of Table 9.2.2 are as follows:

$$\text{Method BT: } w_1 = -3.85 + 3.69 v_1,$$

$$\text{Method TB: } w_2 = -1.58 + 3.23 v_2,$$

$$\text{Method LR: } w_3 = -6.62 + 4.57 v_3,$$

$$\text{Method RL: } w_4 = -6.30 + 4.87 v_4.$$

Each of these regression lines, along with the correlation coefficient, r, is shown in the respective scatter diagram in Figure 9.2.1. The four lines are shown together in Figure 9.2.2.

Using each equation, we can determine for each method of presentation the predicted regression value of recall, that is, the predicted mean recall value (w) as a function of the grade (v). For example, when $v = 3.0$, the mean predicted recall values are $7.22, 8.11, 7.09, 8.31$ for methods BT, TB, LR, RL, respectively; when $v = 2.0$, the respective mean predicted recall values are $3.53, 4.88, 2.52, 3.44$.

Thus for $v = 3.0$, method RL has the largest predicted regression value, and for $v = 2.0$, method TB has the largest sample regression value. In general, the decision as to which method is best in the sense of having the

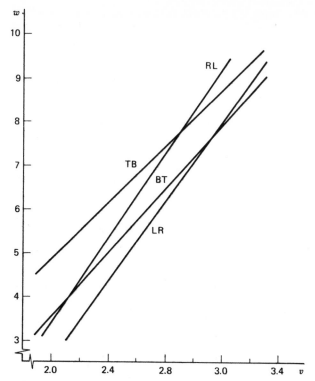

Figure 9.2.2   Calculated regression lines for the data in Table 9.2.1.

largest true regression value $E(W|v)$ may depend heavily on which v value
is used to make the comparison of predicted regression values.

REMARK.   The reader should note that selecting the population with the
largest regression value at some particular value of v is not exactly the
same as selecting the population with the largest regression parameter β
(although these two problems may be highly interrelated). Although b_j is
an estimate of β_j and we could easily order the b_j and select the largest b
value $b_{[k]}$; that is *not* the problem we are discussing here. The b_j are
calculated only to find the equations of the sample regression lines and
thereby determine the predicted regression values of w.   □

The choice of a value of v as a point of comparison of predicted
regression values is not necessarily a statistical question. It may well
depend on what type of students attend these schools. These data suggest
that different methods of presentation may be better for different segments
of the population, which is an eminently reasonable supposition.

In order to give a specific illustration of the procedure, we choose the mean of all 36 values of v for a point of comparison. This value is found as

$$\frac{25.2 + 27.2 + 30.6 + 21.8}{36} = 2.91.$$

Then we substitute $v = 2.91$ in each of the sample regression equations to determine the predicted regression value of recall. The results are as follows:

Method	BT	TB	LR	RL
Recall values obtained from regression equations	6.89	7.82	6.68	7.87

These regression values are the predicted values of mean recall for each population when $v = 2.91$. Therefore we choose the method with the largest sample regression value and assert that it has the largest population recall value at the specific grade 2.91. Here we select method RL. For any $v \geqslant 2.91$, we would still select method RL as best. However, for $v = 2.42$ (where method RL has its mean grade value), we have changed our selection from method RL to method TB by taking the v values into account.

The remainder of the statistical analysis is the same as in Section 2.3. Methods RL and TB have sample means that differ only slightly. Consequently, we may worry about whether method RL is really the best one when $v = 2.91$. With some prescribed confidence, what can we say about how close method RL is to the true best method? From Table A.1 in Appendix A with $P^* = .90$, say, and assuming that $\sigma^2 = 2$ we find for $k = 4$ that $\tau_t = 2.4516$. Thus from (2.3.4) of Chapter 2 we can say with confidence .90 that we have chosen a population whose mean recall, given $v = 2.91$, is within

$$\mu_{[4]} - \mu_s = \frac{\tau_t \sigma}{\sqrt{n}} = \frac{(2.4516)\sigma}{\sqrt{9}} = 0.82\sigma = 1.16$$

of the true largest mean. Thus even if method RL does not have the largest true mean, we have a fairly high confidence (.90) that the mean of method RL, μ_s, is within $0.82\sigma = 1.16$ of the true largest mean $\mu_{[4]}$.

9.3 NOTES, REFERENCES, AND REMARKS

The application of ranking and selection ideas to the problem of selecting the normal population with the best (i.e., largest) regression value appears to be new with this book.

PROBLEMS

9.1 DiPaola (1945) reported a study designed to aid in estimating the average number of oversize rivet holes (w) from the number of minor repairs (v) on a unit of airplane fuselage frame sections. The rivets are used to join aluminum sheets and frames. Suppose that the following data represent the (v, w) values for random samples of units from three different manufacturers.

Manufacturer 1		Manufacturer 2		Manufacturer 3	
v	w	v	w	v	w
22	45	30	52	17	38
26	52	37	70	36	56
21	49	33	59	29	65
28	60	13	40	17	43
33	67	12	32	25	41
32	61	18	41	27	61
33	70	18	48	40	72
25	54	24	49	14	34
34	52	23	55	41	69
35	67	36	63	30	60

(a) Calculate the linear regression lines of w on v for each of the three manufacturers.

(b) Select the manufacturer with the smallest regression value of number of oversize rivet holes for $v = 20$ and for $v = 25$.

9.2 The yield of a chemical process (w) depends on the kind of catalyst used in reaction. However, the estimate of the yield may be improved by considering the amount (v) of the catalyst added to the reaction. Suppose that the following are observations on yield (in percent) taken for five different quantities of catalyst (in pounds) using each of four different kinds of catalyst.

Catalyst 1		Catalyst 2		Catalyst 3	
Amount	Yield	Amount	Yield	Amount	Yield
0.8	60.22	0.8	65.34	0.8	62.46
1.0	63.72	1.0	66.91	1.0	65.78
1.2	62.39	1.2	68.76	1.2	67.21
1.4	65.41	1.4	70.66	1.4	66.73
1.6	71.26	1.6	79.83	1.6	70.24

(a) Calculate the linear regression lines to estimate average yield from amount for each of the three catalysts.

(b) Select the catalyst that produces the largest regression value of yield for a catalyst of 1.2 lb.

CHAPTER 10

Selecting Normal Populations Better than a Control

10.1 INTRODUCTION

A certain type of experiment known as a treatment-control experiment is a useful design in many important problems of investigation and experimentation. This design applies to the situation in which we want to compare each of k treatments with a single "control" treatment.

The distribution model for this type of experiment is k treatment populations

$$f(x;\theta_1), f(x;\theta_2), \ldots, f(x;\theta_k), \tag{10.1.1}$$

and a single control population

$$f(x;\theta_0).$$

Each of these $k+1$ distributions is of the same form; they differ only by the value of a single parameter θ. The treatment parameters are $\theta_1, \theta_2, \ldots, \theta_k$ and the control parameter is θ_0. We assume that the θ values measure the goodness of the respective populations and that larger θ values correspond to better populations. The word *better* can be defined in whatever manner is appropriate to the problem. (For example, in some instances *better* might mean a smaller number of fatalities in a drug experiment. In other instances we may be looking for drugs that induce tumors and then *better* could be taken to mean the drug that induces *more* tumors.) The values of $\theta_1, \theta_2, \ldots, \theta_k$ are all unknown. The value of θ_0 may be either known or unknown.

If θ_0 is known, θ_0 is referred to as a *standard* (rather than a control), and no observations are taken from the standard population. In fact, θ_0 may be simply a constant, interpreted as a "standard" value of θ. Then the problem may be called one of selecting all populations better than a standard, and samples are taken only from the k populations in (10.1.1). Here it is understood that we wish to select all those better and *only* those; that is, we are *not* selecting a subset *containing* all populations better than a standard. Since the latter type of problem has also been treated in the literature and is covered here in Chapter 13, it is important to distinguish between these two different types of goals.

Whether θ_0 is known or unknown, the goal in this problem is to select from the k populations in (10.1.1) the group of exactly those that have θ values larger than θ_0. Of course, this means that the populations not selected are the ones that are asserted to have θ values smaller than θ_0. As a result, this goal could equivalently be stated as one of selecting all populations worse than a control (or standard). Since θ values in the vicinity of θ_0 can be changed from "better than θ_0" to "worse than θ_0" (and vice versa) by small changes in magnitude, a realistic *indifference zone* for this goal must include θ values on both sides of θ_0, that is, θ values that we can assert to be either better or worse than the control population (or standard) and still regard the decision as correct under our definition.

Before proceeding formally to describe the indifference and preference zones and other aspects of this problem, we give some specific examples of practical situations where this goal is the one of interest. These examples should also give the reader a better idea of what is meant by the terms *control* population and *standard*.

Example 10.1.1. In cancer research, drugs are compared on the basis of many factors, but in this example we assume that the factor of interest is survival time, as measured by mean survival time in, say, days. Suppose that a large group of mice are inoculated with carcinogens that produce a certain type of tumor. These mice are then divided randomly into $k = 5$ groups; each of the five groups is given a different drug, A, B, C, D, and E, respectively, at some safe dosage level. Another drug F that has been under experimentation for some time is known to produce a mean survival time of 6.3 days. The goal is to determine which of the new drugs, A, B, C, D, and E, produce a mean survival time greater than 6.3 days, the current "standard" survival time for this experiment. Although we are treating the drug F as a standard in this example, another possibility (which is usually desirable in drug experiments) is to treat F as a control; then we would take observations on F also. The latter procedure would be especially appropriate if some conditions of the present experiment are different from corresponding conditions of previous experiments or are not controllable to the same extent.

Example 10.1.2. Approximately 19 million Americans suffer from depression. The term *depression* covers a broad range of feelings ranging from moodiness to a desire to end one's life. This experiment is designed to study the effectiveness of various methods of treating persons with a minor degree of depression. The generic names of four groups of mood-elevating drugs are meprobromate, diazepam, chlordiazepoxide-hydrochloride, and thioridazine. A group of $5n$ persons diagnosed as suffering from minor depression are randomly divided into five groups, each of size n. A drug of the first type is given to each member of the first group, and similarly for the second, third, and fourth groups. Since these four drugs are quite mild, it is difficult to determine whether a patient's report of feeling better is due to the effect of the drug or due to a psychosomatic effect, stemming from the person's belief that the drug is beneficial. Therefore the fifth group of size n is given a placebo; that is, they experience the psychological effect of having taken some medication, but the medication itself has no physiological value. All subjects are treated in the same clinic with the same instructions for the same period of time, and none of the subjects is informed of which drug (or placebo) he is receiving. The same quantitative criterion is used to measure the degree of depression of each subject both before and after the "treatment." Differences between these measures indicate the effectiveness of the treatments in dispelling mild feelings of depression. The goal is to determine which of the four types of drugs has a mean effectiveness that is larger than the mean effectiveness of the placebo in the control group. Note that $k = 4$ in this example, since the control group is not included in the value of k.

The only particular distribution model that we treat in this chapter is the one in which all the populations are normal. For selection problems where the θ values are the population means, we assume that the k treatment populations have a common variance σ_1^2. Thus the k normal distributions associated with the k treatment effects are

$$N(\mu_1, \sigma_1^2), N(\mu_2, \sigma_1^2), \ldots, N(\mu_k, \sigma_1^2),$$

and the control population has a normal distribution with variance σ_0^2.

$$N(\mu_0, \sigma_0^2).$$

Our goal is to select exactly that subset of the k treatment populations whose μ values are larger than μ_0, which is the μ value for the control population. Since the θ values in the general discussion are the μ values here, the variances must be regarded as nuisance parameters, especially when they are unknown. For the case where $\sigma_1^2 = \sigma_0^2 = \sigma^2$, we consider the following two cases in this chapter:

1. The common variance σ^2 is known.

2. The common variance σ^2 is unknown.

In either case the value of μ_0 may be known (standard) or unknown (control). We also consider briefly the case where $\sigma_1^2 \neq \sigma_0^2$ but both are known and μ_0 is unknown. Then we treat the corresponding estimation problem of making a joint confidence statement about the k differences between each treatment population mean and the unknown control mean in the case of a common unknown variance.

In the final section of this chapter the θ values are the population variances σ_1^2 and thus the goal is to select those treatment populations for which the variance is smaller than a known standard variance σ_0^2.

10.2   FORMULATION OF THE PROBLEM FOR MEANS

To formulate the problem for means more explicitly, we first designate the indifference and preference zones appropriate to this goal and this model. As mentioned earlier, the indifference zone, which we denote by the region R_I, should consist of those μ values that are close to μ_0 on either side. For some two positive specified constants, say δ_1^* and δ_2^*, R_I is then defined as all μ values that lie in the interval

$$R_I: \quad \mu_0 - \delta_1^* < \mu < \mu_1 + \delta_2^*.$$

This interval in the parameter space is illustrated in Figure 10.2.1. Note that since the entire parameter space consists of μ values that are all real numbers, this indifference zone leaves *two* other disjoint intervals in the parameter space; these intervals are labeled R_S (S for smaller) and R_L (L for larger) in Figure 10.2.1. In the interval R_S the μ values are much smaller than μ_0, and the corresponding treatment populations are worse than the control population (or standard). In the interval R_L the μ values are much larger than μ_0 and the corresponding populations are better than the control. In symbols, for the positive constants δ_1^* and δ_2^*, the parameter space has now been partitioned into the following three intervals:

$$
\begin{aligned}
R_S: & \quad \mu \leqslant \mu_0 - \delta_1^*, \\
R_I: & \quad \mu_0 - \delta_1^* < \mu < \mu_0 + \delta_2^*, \\
R_L: & \quad \mu \geqslant \mu_0 + \delta_2^*.
\end{aligned}
\qquad (10.2.1)
$$

Since R_S and R_L are not in the indifference zone, they must both be considered part of the preference zone. In other words, even though our goal is to identify those treatment populations whose μ values lie in R_L, we

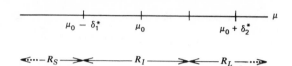

Figure 10.2.1   Indifference and preference zones for population means.

are also concerned with a precise definition of R_S since we do not want any population with μ value in R_S to be classified as belonging to R_L.

As a result, this problem is reduced to the problem of classifying the k values of μ either as in R_S or as in R_L. A population is misclassified if we say it is in R_L when it is really in R_S, or if we say it is in R_S when it is really in R_L. A population cannot be misclassified if its true μ value is in R_I. Thus a correct decision (CD) about a single population is defined as one that classifies that population as worse when it is in R_S or in R_I; it is also a correct decision to classify a population as better when it is in R_L or in R_I. An overall correct decision (OCD) is then defined as meaning that a correct decision is made for *each* of the k treatment populations; that is, *every* population is classified correctly.

REMARK.   The labeling of better or worse depends on the context of the problem. In some instances smaller is better; in others larger is better.   □

With this formulation of the indifference and preference zones, we can say that our problem is to find a decision procedure R based on samples of common size n such that the probability of an overall correct decision $P\{\text{OCD}|\mu\}$ is at least some specified constant P^*; that is,

$$P\{\text{OCD}|\mu\} \geqslant P^*, \tag{10.2.2}$$

for all configurations of μ values in the preference zone. Here and below μ is also used as a vector of population means. The value of P^* must be between $(.5)^k$ and 1, that is, $(.5)^k < P^* < 1$, since the smallest probability that a random assignment (without using any observations) of all k populations to either R_S or R_L gives an overall correct decision is $(.5)^k$; this occurs when none of the k populations is in R_I.

The probability of an overall correct decision depends on the true configuration of μ values $\mu = (\mu_0, \mu_1, \ldots, \mu_k)$, as well as on k and n. However, for any particular k and n we can compute the minimum probability over all configurations in the preference zone. This minimum occurs at the least favorable configuration μ_{LF}. Thus if we adopt the conservative approach of defining n as the smallest integer that satisfies

$$P\{\text{OCD}|\mu_{LF}\} \geqslant P^*,$$

then we know that

$$P\{\text{OCD}|\mu\} \geqslant P^*$$

for all configurations of μ values in the preference zone.

For the control and standard problem the least favorable configuration for any specified positive constants δ_1^* and δ_2^* is obtained by setting roughly half of the μ values at $\mu_0 - \delta_1^*$ and the remaining μ values at $\mu_0 + \delta_2^*$. More precisely, define $k' = k/2$ and $k'' = (k+1)/2$; if k is even, we set

$$\mu_{[1]} = \cdots = \mu_{[k']} = \mu_0 - \delta_1^*, \qquad \mu_{[k'+1]} = \cdots = \mu_{[k]} = \mu_0 + \delta_2^*, \quad (10.2.3\text{a})$$

and if k is odd, we set

$$\mu_{[1]} = \cdots = \mu_{[k'']} = \mu_0 - \delta_1^*, \qquad \mu_{[k''+1]} = \cdots = \mu_{[k]} = \mu_0 + \delta_2^*. \quad (10.2.3\text{b})$$

Here in the odd case (10.2.3b), after setting $k'' - 1 = (k-1)/2$ of the μ values at each of the extreme points, $\mu_0 - \delta_1^*$ and $\mu_0 + \delta_2^*$, we have set the single remaining μ value at the left extreme, but it could be placed on either side. It may seem strange that for the LF configuration the allocation of the k population means to the two extreme points does not depend on δ_1^* or δ_2^*. This is partly explained by the fact that δ_1^* and δ_2^* enter the decision rule defined in the next section.

For the standard problem we need to determine the smallest value of the probability of a correct selection over all configurations of the form

$$\mu_{[1]} = \cdots = \mu_{[r]} = \mu_0 - \delta_1^*, \qquad \mu_{[r+1]} = \cdots = \mu_{[k]} = \mu_0 + \delta_2^*$$

for different values of r. It turns out that the minimum does not depend on r, that is, on how many μ's are set equal to $\mu_0 - \delta_1^*$. The minimum of the probability of a correct selection is given by $\Phi^k(\sqrt{n}\,(\delta_1^* + \delta_2^*)/2\sigma)$, which is then set equal to P^*. (This is the origin of (10.4.2) defined later in Section 10.4.1.)

10.3 DECISION PROCEDURE

In this section we describe the decision procedure R for selecting all normal populations with μ values greater than μ_0 when all the populations have a common variance. The decision rule for procedure R is the same whether this common variance is known or unknown, although the sampling part of the procedure is different. However, the decision rule differs according to whether μ_0 is known or unknown. The case of μ_0 unknown (i.e., the control case) is considered first.

Table 10.3.1   Format for data presentation

Population	0	1	...	k
Sample observations	x_{10}	x_{11}	...	x_{1k}
	x_{20}	x_{21}	...	x_{2k}
	.	.		.
	.	.		.
	.	.		.
	x_{n0}	x_{n1}	...	x_{nk}
Sample mean	$\bar{x}_0$	$\bar{x}_1$	...	$\bar{x}_k$

Suppose we take a random sample of size n from each of the k treatment populations, and another random sample of size n from the control population, in such a way that independence holds between all samples as well as within each sample. Let x_{ij} denote the ith observation from the jth population, for $i = 1, 2, \ldots, n$ and $j = 0, 1, \ldots, k$. The data can then be presented using the format in Table 10.3.1. The sample means are calculated in the usual way as $\bar{x}_j = \sum_{i=1}^{n} x_{ij} / n$ for $j = 0, 1, \ldots, k$.

For two positive constants δ_1^* and δ_2^*, the decision procedure R (for any such data) is to classify population j as better than the control, that is, in R_L ($L = $ larger), if the corresponding sample mean $\bar{x}_j$ satisfies the inequality

$$\bar{x}_j > \bar{x}_0 + .5(\delta_2^* - \delta_1^*), \tag{10.3.1}$$

and to classify it as worse than the control, that is, in R_S ($S = $ smaller), if

$$\bar{x}_j < \bar{x}_0 + .5(\delta_2^* - \delta_1^*), \tag{10.3.2}$$

for each of $j = 1, 2, \ldots, k$. Note that $\delta_2^* - \delta_1^*$ may be either positive or negative or zero.

With this procedure, every one of the k treatment populations is classified as either better than or worse than the control unless equality holds in (10.3.1) or (10.3.2) for some one or more of $j = 1, 2, \ldots, k$. This equality is very unlikely but if it should happen, we will be content to toss a balanced coin for each of these samples and classify the populations as "better" or "worse" than the control according to whether a head or a tail appears, respectively.

If the mean of the control population μ_0 is known, we refer to the control as a standard and there is no reason to take a sample from the standard population. In fact, if μ_0 is a known constant, like a standard value, then there may not be any corresponding population. As a result, in the situation where μ_0 is known the data consist of only the last k columns in Table 10.3.1 (i.e., the column for the population labeled zero is nonexistent). As a result, we have only k sample means, $\bar{x}_1, \bar{x}_2, \ldots, \bar{x}_k$, and we replace $\bar{x}_0$ in (10.3.1) and (10.3.2) by μ_0. Then for two positive constants δ_1^* and δ_2^*, the decision procedure R is to classify population j as better than the control (or standard), that is, in R_L, if the corresponding sample mean $\bar{x}_j$ satisfies the inequality

$$\bar{x}_j > \mu_0 + .5(\delta_2^* - \delta_1^*), \tag{10.3.3}$$

and to classify it as worse than the control (or standard), that is, in R_S, if

$$\bar{x}_j < \mu_0 + .5(\delta_2^* - \delta_1^*), \tag{10.3.4}$$

for each of $j = 1, 2, \ldots, k$. The preceding remarks about the decision procedure and the method of dealing with ties in the previous case for μ_0 unknown apply equally to this procedure for μ_0 known (except that here a tie occurs if $\bar{x}_j = \mu_0 + .5(\delta_2^* - \delta_1^*)$ for some one or more of $j = 1, 2, \ldots, k$). Note that in both the control and standard cases the common variance does not enter into the decision rule R.

We illustrate these decision procedures by the following example.

Example 10.3.1. The pesticide DDT is an environmental pollutant that has been in mass use since late in World War II. Its concentration in human fat is often as high as 12 ppm (parts per million). DDT enters the human body in all sorts of ways, through the air we breathe, drinking water, fruits and vegetables, in addition to exposure to any substance sprayed with DDT. Excessive concentration of DDT in human fat has very ominous results, which include a higher incidence of certain types of cancer, excess production of estrogen, hypertension, cerebral hemorrhage, and so on. Ehrlich and Ehrlich (1970) report that the mean DDT concentration in the United States is in the range 7–12 ppm. Suppose that a scientific committee concerned with ecology states that a DDT concentration of 10 ppm in human body fat is the border line between safe and unsafe levels. The following data give the average values of the DDT concentration for samples of size $n = 100$ from each of $k = 9$ different populations during the 1960s. For the specified constants $\delta_1^* = 1.00 = \delta_2^*$, exactly which of the groups represented have an average DDT concentration larger than the standard of $\mu_0 = 10$ ppm?

We assume that the populations from which each of these samples were drawn are normally distributed with common variance. Then the decision rule from (10.3.3) is to classify the jth population as "better" than this standard (here *better* is being used to mean larger concentration values of DDT) if

$$\bar{x}_j > 10,$$

Table 10.3.2   Mean concentration of DDT in human body fat

Population j	Mean DDT $\bar{x}_j$ (in ppm)
U.S. (white, over 6 years)	8.4
U.S. (nonwhite, over 6 years)	16.7
Alaskan Eskimo	3.0
Canada	3.8
United Kingdom	3.3
Hungary	12.4
France	5.2
Israel	19.2
India	26.2

and from (10.3.4) to classify it as "worse" than this standard if

$$\bar{x}_j < 10.$$

According to this rule the populations asserted to be "better" than the standard value, that is, those with unacceptable levels of DDT in body fat, are United States (nonwhite, over 6 years), Hungary, Israel, and India. Thus out of the $k = 9$ groups considered here, exactly these four are asserted to have average DDT concentrations above the acceptable level.

10.4   ANALYTICAL ASPECTS

Now that the decision procedure has been defined, we can consider some analytical aspects of the problem of selecting exactly those normal populations with common variance whose μ values are better than a control or better than a standard value μ_0. The problem of interest here is determining the number of observations that should be taken from each population. In particular, we suppose that the experiment is in the process of being designed, and we want to choose the smallest common sample size n such that we can make the statement that the probability of an overall correct decision is at least some specified value P^*; that is, the probability satisfies (10.2.2) for all μ values in the preference zone. As explained in Section 10.2, we can find a solution to this problem under the least favorable configuration of μ values given in (10.2.3) for any specified values of δ_1^*, δ_2^*, and P^*, and the same solution satisfies the probability requirement in (10.2.2). As a result, the first step is to specify the values for δ_1^*, δ_2^*, and P^*, where $\delta_1^* > 0$, $\delta_2^* > 0$, and $(.5)^k < P^* < 1$. With these values we can determine n, whether the common variance is known or unknown. Since the solution is quite different in these two cases, we consider them separately. Of course, for each of these we still have to consider whether μ_0 is known or unknown.

10.4.1 Common Variance Known

When the common variance σ^2 is known, the common sample size n that is required to satisfy the specified values δ_1^*, δ_2^*, and P^* depends on whether μ_0 is known or unknown.

If μ_0 is unknown, n is found by computing the quantity

$$n = \frac{8\sigma^2 b^2}{(\delta_1^* + \delta_2^*)^2}, \qquad (10.4.1)$$

where b is the entry in Table M.1 of Appendix M that corresponds to the value of P^* for the given k. If the result in (10.4.1) is not an integer, it should be rounded upward to obtain a conservative result.

If μ_0 is known, n is found as the solution to the equation

$$\Phi\left(\frac{\sqrt{n}\,(\delta_1^* + \delta_2^*)}{2\sigma}\right) = (P^*)^{1/k}, \qquad (10.4.2)$$

where $\Phi(x)$ is the standard cumulative normal distribution, which is given in the table in Appendix C. To aid in the solution of equation (10.4.2), Table M.2 in Appendix M gives the value of B that is the solution to the equation $\Phi(B) = (P^*)^{1/k}$ for selected values of k and P^*. Once B is found from this table, the required value of n can be found from the simple expression

$$n = \frac{4\sigma^2 B^2}{(\delta_1^* + \delta_2^*)^2}. \qquad (10.4.3)$$

If the result from (10.4.3) is not an integer, it should be rounded upward to obtain a conservative result.

Note that in the special case where $\delta_1^* = \delta_2^* = \delta^*$, say, (10.4.1) simplifies to

$$n = \frac{2\sigma^2 b^2}{\delta^{*2}} \qquad (10.4.4)$$

for the case of μ_0 unknown, and (10.4.3) simplifies to

$$n = \frac{\sigma^2 B^2}{\delta^{*2}} \qquad (10.4.5)$$

for the case of μ_0 known.

We now give examples to illustrate these procedures for finding n.

We use the background of Example 10.1.2 to illustrate the case in which μ_0 is unknown. There are $k = 4$ types of drugs, but a total sample of $5n$

persons is needed since observations must also be taken on the control (placebo) group. Suppose that $\sigma^2 = 4.00$ and we specify $P^* = .95$, $\delta_1^* = 1.00$, $\delta_2^* = 1.20$. Then we find $b = 2.212$ from Table M.1. We substitute these values in (10.4.1) to obtain $n = 8(4.00)(2.212)^2/(2.20)^2 = 32.4$. This result is rounded upward to 33. Hence a total of 165 persons are needed, 33 for each of the four treatment groups and 33 for the control group. If the resources are such that this common sample size is too large, a smaller value must be specified for P^*, or larger values must be specified for δ_1^* and δ_2^*. For example, if $P^* = .90$, the reader may verify that $n = 25$ for the same $\delta_1^* = 1.00$, $\delta_2^* = 1.20$, and hence a total of only 125 persons are needed.

In the situation of Example 10.1.1 where the "standard" survival time is stated as $\mu_0 = 6.3$ days, suppose we want to determine the common sample size n needed for the $k = 5$ groups of mice inoculated with different drugs if the common variance is $\sigma^2 = 1.20$. If we specify the value $P^* = .95$, then Table M.2 in Appendix M indicates that the solution to the equation $\Phi(B) = (.95)^{1/5}$ is $B = 2.319$. Then from (10.4.3), if $\delta_1^* = 0.50$ and $\delta_2^* = 0.60$, the required value of n is $n = 4(1.20)(2.319)^2/(1.10)^2 = 21.33$. Rounding this result upward, we need a sample of 22 mice for each drug group, or a total of $5(22) = 110$ mice. The reader may verify that if we keep $\delta_1^* = 0.50$, $\delta_2^* = 0.60$, but specify the larger value $P^* = .975$ (so that $B = 2.573$), then the required n value increases to 27 mice per drug.

Example 10.4.1. The *New York Post* (January 23, 1976, p. 4) reported that consumers in the city were cheated in the order of $25 to $30 million in 1975 because of short weight and short count merchandise. The Bureau of Consumer Affairs issued 55,975 violations that year, primarily to markets. The source of a shortweight violation may be the food manufacturer or the market. The market is held responsible, but the fine is usually paid by the manufacturer if the market did not package the item. The biggest violation in New York history was 1200 shortweight flour packages shipped to the city by three major manufacturers. Of course, the manufacturer is concerned about overweight as well as short weight.

Suppose that we need to design an experiment to investigate shortweight in "5 lb" flour packages. We have k normal populations (brands) and wish to choose those better than the standard $\mu_0 = 5$. How many observations should we take from each population?

As a first step we define the regions

$$R_S: \quad \mu \leqslant \mu_0 - \delta_1^*,$$
$$R_I: \quad \mu_0 - \delta_1^* < \mu < \mu_0 + \delta_2^*,$$
$$R_L: \quad \mu \geqslant \mu_0 + \delta_2^*.$$

The choice of δ_1^* and δ_2^* leads to interesting alternatives.

In this type of problem we have to distinguish carefully between (1) routine classification (or quality control) for the purpose of protecting the consumer and (2) identifying shortweight packages for the purpose of fining the manufacturer or marketer responsible. In case (1) we will take $0 \le \delta_1^* \le \delta_2^*$ so that the critical constant on the right-hand side of (10.3.3) and (10.3.4) is greater than or equal to μ_0. In case (2) we will take $0 \le \delta_2^* \le \delta_1^*$ so that the critical constant on the right-hand side of (10.3.3) and (10.3.4) is less than or equal to μ_0. A consumer organization would sponsor case (1), and a government regulation agency would tend to want to use case (2). In particular, the agency might wish to show its neutrality by taking $\delta_1^* = \delta_2^*$ but the common value still must be specified. Since the newspaper article stressed the finding of shortweight packages, we start our illustrations with case (2).

Suppose that the weights of "5 lb" bags of flour are normally distributed with mean $\mu_0 = 5.0$ lb and standard deviation $\sigma = 0.2$ lb, regardless of the brand. For specified $(\delta_1^*, \delta_2^*) = (0.050, 0.025)$ as in case (2), for $k = 3$ brands, and for given $P^* = .90$, say, we can solve for n in (10.4.2) or use the explicit expression in (10.4.3). Using (10.4.2) we have

$$\Phi\left(\frac{\sqrt{n}\,(0.050 + 0.025)}{2(0.2)} \right) = \Phi(0.1875\sqrt{n}\,) = (.90)^{1/3} = .9655.$$

The table in Appendix C shows that $\Phi(1.819) = .9655$. Equating the two arguments of Φ gives

$$0.1875\sqrt{n} = 1.819$$

and hence

$$n = 94.12.$$

As a result, 95 observations are needed from each of the $k = 3$ brands being investigated. The selection rule (10.3.3) and (10.3.4) for this case is to classify brand j as being shortweight if $\bar{x}_j$ satisfies

$$\bar{x}_j < \mu_0 + .5(\delta_2^* - \delta_1^*) = 5.0 + .5(-0.025) = 4.9875 \cong 4.99,$$

and to classify it as satisfactory (i.e., weight at least 5) if $\bar{x}_j > 4.99$.

To illustrate the middle ground or common part of cases (1) and (2), we take $\delta_1^* = 0.050 = \delta_2^*$, and again use $k = 3$ and $P^* = .90$. Then we obtain

$$\Phi\left(\frac{\sqrt{n}\,2(.050)}{2(0.2)} \right) = \Phi(0.25\sqrt{n}\,) = (.90)^{1/3} = .9655,$$

$$0.25\sqrt{n} = 1.819 \text{ or } n = 52.94.$$

Thus in this example we would need 53 observations from each of the $k=3$ populations. The selection rule (10.3.3) and (10.3.4) for this case is to classify brand j as being shortweight if $\bar{x}_j < 5.0$ and as satisfactory (i.e., weight at least 5) if $\bar{x}_j > 5.0$.

To illustrate case (1) we need only interchange the values of δ_1^* and δ_2^* in the first illustration, that is, we take $(\delta_1^*, \delta_2^*) = (0.025, 0.050)$ for the same k and the same P^*. Then the required $n = 95$ for each of the $k=3$ populations is also the same. However the rule now is to classify brand j as shortweight if and only if $\bar{x}_j < 5.01$.

Now suppose the experiment has already been performed using $n = 50$ observations from each of the $k=3$ brands. Then what can we say about the probability of an overall correct decision $P\{\text{OCD}\}$ if $\delta_1^* = 0.050$ and $\delta_2^* = 0.025$? Substituting these values in the left-hand side of (10.4.2) gives

$$\Phi\left(\frac{\sqrt{50} \; (0.050 + 0.025)}{2(0.2)}\right) = \Phi(1.33) = .908.$$

This gives $(P^*)^{1/3} = .908$ so that $P^* = (.908)^3 = .75$, and the probability of an overall correct decision is $P\{\text{OCD}\} \geqslant .75$. Clearly the same result holds if we interchange δ_1^* and δ_2^* for our illustration of case (1).

On the other hand, with $\delta_1^* = \delta_2^* = 0.050$, the same k and the same n, we obtain

$$P\{\text{OCD}\} \geqslant \Phi^3\left(\frac{\sqrt{50} \; (0.050 + 0.050)}{2(0.2)}\right) = \Phi^3(1.77) = (.962)^3 = .89.$$

These numerical results illustrate the variations and constancies in the properties of the selection procedure as a result of different choices of δ_1^* and δ_2^*.

10.4.2   Common Variance Unknown

In many practical situations the value of the common variance σ^2 may not be known to the investigator. In this event we cannot find a single-stage sampling procedure to solve the problem (without redefining the distance function or specifying an upper bound for σ^2). That is, we cannot find a common n and a single-stage procedure such that the probability of an overall correct decision is at least P^* for all positive values of the nuisance parameter σ^2. We consider here only the case where μ_0 is unknown.

A solution can be found using a two-stage sampling procedure in this case. At the first stage we take an initial sample of size n from each population, including the control, and use these data to estimate σ^2. The

*Table 10.4.1   **Sample values based on the initial sample of size** n*

Population j	0	1	...	k
Sample observations	x_{10} . . . x_{n0}	x_{11} . . . x_{n1}	... ...	x_{1k} . . . x_{nk}
Sample sums of squares	T_0	T_1	...	T_k

initial sample data are presented in the format of Table 10.4.1; the sums of squares of the deviations of each sample around its sample mean, denoted by T_j for $j = 0, 1, \ldots, k$, are shown in the last row of Table 10.4.1. These quantities T_j are calculated for each j using the formula

$$T_j = \sum_{i=1}^{n} \left(x_{ij} - \bar{x}_j \right)^2.$$

For each j, the sample variance $s_j^2 = T_j/(n-1)$ is an estimate of the common population variance σ^2. A better estimate of σ^2 is obtained by pooling these estimates. This pooled estimated, denoted by s^2, is computed by the following formula:

$$s^2 = \frac{T_0 + T_1 + \cdots + T_k}{(k+1)(n-1)}. \qquad (10.4.6)$$

The total number of degrees of freedom associated with s^2 is $\nu = (k+1)(n-1)$.

In the second stage we take an additional sample of size $N - n$ from each population. This gives a *total* sample of N observations from each population. Now the question of interest is what is the minimum integer value of N required such that the probability of an overall correct decision is at least P^* for specified values of δ_1^* and δ_2^*. For any specified values of δ_1^*, δ_2^*, and P^*, N is the smallest integer that satisfies

$$N \geqslant \text{maximum} \left\{ n, \frac{8s^2}{\left(\delta_1^* + \delta_2^* \right)^2} h_\nu^2 \right\}, \qquad (10.4.7)$$

where h_ν (which depends on ν given earlier) is obtained from Table M.3 of Appendix M, and s^2 is the value computed from (10.4.6).

We have not discussed the question of how to choose the initial sample size n. In some instances we may not have any control over this choice.

For example, we may have classrooms of $n = 30$ students, say, or a fixed number n of patients who can be observed within a given time period. If we do have some control over the value of n, a reasonable approach is as follows. We first guess a reasonable value for σ^2. Then we act as if this guessed value is the true value of σ^2 and determine the value of n required for specified values δ_1^*, δ_2^*, and P^*, using the procedure described in Section 10.4.1 for σ^2 known. Then we take an initial sample of size n from each of the populations, compute s^2 from (10.4.6), and determine N from (10.4.7), exactly as before. The better the guess of σ^2, the less will be the sum of the expected total sample size and the variance of the total sample size.

10.5 ALLOCATION OF OBSERVATIONS WITH UNEQUAL SAMPLE SIZES

In the previous sections we assumed the samples were all of a common size n. In some situations it is impossible or impractical to choose the same number of observations from each population. Now suppose that we take n_0 observations from the control population $f(x; \theta_0)$ and n_j observations from the treatment population $f(x; \theta_j)$, for each $j = 1, 2, \ldots, k$. For simplicity, we assume that $n_j = n_1$ for each $j = 1, 2, \ldots, k$. When a control population is to be compared with several different *sets* of treatment populations, we may want to choose n_0 larger than n_1. The present question is, how should n_0 and n_1 be chosen?

We assume the normal distribution model with unknown means and a common variance for the treatment populations only. That is, $f(x; \theta_0)$, represents the control population with distribution $N(\mu_0, \sigma_0^2)$, and the k treatment populations have distributions $N(\mu_1, \sigma_1^2), \ldots, N(\mu_k, \sigma_1^2)$, respectively; we do not assume that $\sigma_0^2 = \sigma_1^2$.

There are several criteria we could use to choose the sample size allocation for n_0 and n_1. One criterion is to minimize the expected number of populations that are misclassified by the selection procedure. A second criterion is to maximize the probability of a correct selection. The solution is approximately the same using either criterion. This indicates that the sample sizes should be chosen such that their ratio is proportional to the ratio of the standard deviations as follows:

$$\frac{n_0}{n_1} \cong \frac{\sigma_0 \sqrt{k}}{\sigma_1}. \tag{10.5.1}$$

For example, if $\sigma_0 = 3$, $\sigma_1 = 5$, $k = 4$, then from (10.5.1) we have $n_0/n_1 \cong \frac{6}{5}$, so that $n_1 = 5n_0/6$. For any n_0 and n_1 the total number of observations N is then

$$N = n_0 + kn_1 = n_0 + 4n_1 = n_0 + 4\left(\frac{5n_0}{6}\right) = \frac{13n_0}{3}.$$

Hence $n_0 = 3N/13$ and $n_1 = 5N/26$. Thus if we wish to take a total of $N = 100$ observations, we should have $n_0 \cong 23$ and $n_1 \cong 19$. Note that because n_0 and n_1 must both be integer-valued, we may not be able to fulfill exactly the condition in (10.5.1). In this example the total is $N = 99$ observations. If we insist on having $N = 100$ exactly, we simply take $n_0 = 24$ and $n_1 = 19$.

An explicit solution for n_0 and n_1 can be given by substituting $N = n_0 + kn_1$ into (10.5.1). The result is

$$n_1 = \frac{N\sigma_1}{\sigma_0\sqrt{k} + \sigma_1 k}, \qquad n_0 = N - kn_1. \qquad (10.5.2)$$

In the preceding numerical example, (10.5.2) gives $n_1 = 19.23$. If we round this value upward, the result is $n_1 = 20$, $n_0 = 20$; if we round downward, the result is $n_1 = 19$, $n_0 = 24$, as suggested earlier. Since 19.23 is much closer to 19 than to 20, the latter is preferable.

Example 10.5.1.   (Drug Screening).   Dunnett (1972) described a screening experiment for anticancer drugs. Cancer cells are implanted in a group of mice so that tumors will grow. The mice are then randomly divided into a control group and k treatment groups. The control group receives no drugs; k different anticancer drugs are assigned randomly to the k treatment groups. After a fixed period of time, the tumors are removed and weighed for each mouse. The magnitude of the weight of a tumor is considered indicative of the effectiveness of the treatment administered in curing, arresting, or reducing the cancer. The control sample should be larger than the treatment samples because it is very important that the control mean be precise when it is to be compared with many different drugs. Suppose that there are $k = 6$ treatment groups to compare with the control group, the tumor weights for all groups are normally distributed, the variance of the control group is $\sigma_0^2 = 0.40$, the variance of all the treatment groups is $\sigma_1^2 = 0.36$, and a total of $N = 160$ observations can be taken. How should the observations be allocated?

From (10.5.2) we compute

$$n_1 = \frac{160\sqrt{0.36}}{\sqrt{0.40}\ \sqrt{6} + \sqrt{0.36}\ (6)} = 18.6.$$

If we round upward to get $n_1 = 19$, then $n_0 = 160 - 6(19) = 46$.

In this example, it seems reasonable that the allocation $(18, 18, 19, 19, 19, 19)$ with unequal sample sizes for the treatment groups, and 48 for the control might yield better results, but the best result with equal n_1 values is the one given here, that is, $n_1 = 19, n_0 = 46$.

10.6   JOINT CONFIDENCE STATEMENTS FOR MEANS

In an experiment comparing k populations (or treatments) with a control or an unknown standard, it frequently is desirable to make a joint confidence statement about all the k differences of the unknown population means and the control, $\mu_1 - \mu_0, \mu_2 - \mu_0, \ldots, \mu_k - \mu_0$. The joint confidence statement will be based on $\bar{x}_j - \bar{x}_0$ ($j = 1, \ldots, k$), where $\bar{x}_j$ is the mean of the sample from the jth population and $\bar{x}_0$ is the mean of the sample from the control population. Assume that these sample means are all based on a common sample size n and that all populations are normal with a common unknown variance σ^2.

It makes a substantial difference whether we treat these k confidence intervals for $\mu_j - \mu_0$ as separate statements and control the error rate per comparison, or whether we control the confidence level that the k statements hold simultaneously. We prefer the latter approach for reasons that will be stated, but it should be noted that both are in common usage. First of all, it is usually more comfortable in the sense of involving less risk to work with a specified error rate per experiment rather than per comparison. Thus the use of a 95% joint confidence set would ensure no errors in 95% of such experiments. On the other hand, using a 95% confidence interval for each comparison, we would not be able to pull out even a single pair of such comparisons and make a probability statement about their being jointly correct. The reason for this is that we cannot assume independence between comparisons, and the error rate per comparison tells us nothing about the dependence between two comparisons.

10.6.1   Joint Confidence Statement for Comparisons with a Control

We treat here only the case where μ_0 is *unknown*. As before, the model assumes that we have $k + 1$ normal populations with a common unknown variance σ^2, and samples of common size n are drawn from each of the k populations. Since the populations have a common variance σ^2, the best estimate of σ^2 is the pooled sample variance s^2, computed from

$$s^2 = \frac{s_0^2 + s_1^2 + \cdots + s_k^2}{k+1}, \qquad (10.6.1)$$

where s_j^2 is the sample variance for the data from the jth population. This estimate has degrees of freedom v computed from

$$v = (k+1)(n-1). \tag{10.6.2}$$

The joint confidence statements about the differences $\mu_j - \mu_0$ for $j = 1, 2, \ldots, k$ may be either one-sided upper (or lower) or two-sided. The one-sided *upper* confidence statement is that for all $j = 1, 2, \ldots, k$, the inequalities

$$\mu_j - \mu_0 \leqslant \bar{x}_j - \bar{x}_0 + d's\sqrt{\frac{2}{n}} \tag{10.6.3}$$

hold simultaneously with probability at least P^*, that is, with confidence level P^*. If a one-sided *lower* confidence statement with confidence level P^* is required, (10.6.3) becomes

$$\mu_j - \mu_0 \geqslant \bar{x}_j - \bar{x}_0 - d's\sqrt{\frac{2}{n}}. \tag{10.6.4}$$

The two-sided confidence statement is that for $j = 1, 2, \ldots, k$, the inequalities

$$\bar{x}_j - \bar{x}_0 - ds\sqrt{\frac{2}{n}} \leqslant \mu_j - \mu_0 \leqslant \bar{x}_j - \bar{x}_0 + ds\sqrt{\frac{2}{n}} \tag{10.6.5}$$

hold simultaneously with probability at least P^*.

The constant d' in (10.6.3) or (10.6.4) is given in Table A.4 of Appendix A for $P^* = .950$ and $P^* = .990$. The constant d in (10.6.5) is given in Table M.4 of Appendix M for $P^* = .950$ and $P^* = .990$. In either case the table is entered with the value of v, the degrees of freedom in s^2 computed from (10.6.2), and a value determined by k. Table M.4 is entered with k, the number of populations *excluding* the control. In Table A.4 the heading k refers to the *total* number of populations, so this table is entered with the value of $k + 1$ (even though the heading says k).

REMARK. The values of d' and d given in Tables A.4 and M.4, respectively, are exact for common sample sizes and the normal distribution model with common unknown variance. If the sample sizes are unequal, the same tables may still be used for d' and d, respectively, if $\sqrt{2/n}$ in (10.6.3), (10.6.4), and (10.6.5) is replaced by $\sqrt{(1/n_j) + (1/n_0)}$, but the associated level of confidence is then only approximate (the number of degrees of freedom is $v = \sum_{j=0}^{k}(n_j - 1)$). Note that with unequal sample sizes, s^2 must be computed as the weighted average of the sample variances, or

$$s^2 = \frac{(n_0-1)s_0^2 + (n_1-1)s_1^2 + \cdots + (n_k-1)s_k^2}{(n_0-1) + (n_1-1) + \cdots + (n_k-1)}. \qquad \square$$

Example 10.6.1. As a numerical example, consider again the situation of Example 10.5.1, but now suppose that samples of size $n=20$ are taken from each of the $k=6$ treatment groups as well as the control group, and the summary data are as shown below:

Group j	$\bar{x}_j$	n	s_j^2
0 (Control)	2.36	20	0.35
1	2.12	20	0.40
2	1.83	20	0.32
3	0.96	20	0.20
4	1.10	20	0.26
5	0.74	20	0.22
6	1.21	20	0.36

We assume here that the tumor weights for all groups are normally distributed with a *common* unknown variance. It is desired to make a one-sided upper joint confidence statement at level $P^* = .99$ about the difference between each of the six treatment means and the control mean tumor weight.

The first step is to calculate the pooled sample variance from (10.6.1). The result is

$$s^2 = \frac{(0.35) + (0.40) + \cdots + (0.36)}{7} = 0.3014,$$

with degrees of freedom $v = 133$. From Table A.4 of Appendix A with $k+1=7$, $v=133$, $P^*=.99$, by interpolating with $1/v$ we obtain $d'=2.935$. Substituting these results into the last term in the right-hand side of (10.6.3), we obtain

$$d's\sqrt{2/n} = 0.5095.$$

The differences between each sample mean and the control sample mean

$\bar{x}_0 = 2.36$ are as follows:

$$\bar{x}_1 - \bar{x}_0 = -0.24, \qquad \bar{x}_4 - \bar{x}_0 = -1.26,$$

$$\bar{x}_2 - \bar{x}_0 = -0.53, \qquad \bar{x}_5 - \bar{x}_0 = -1.62,$$

$$\bar{x}_3 - \bar{x}_0 = -1.40, \qquad \bar{x}_6 - \bar{x}_0 = -1.15.$$

As a result, the six statements that can be made with joint (i.e., overall) confidence level .99 are as follows:

$$\mu_1 - \mu_0 \leqslant 0.27, \qquad \mu_4 - \mu_0 \leqslant -0.75,$$
$$\mu_2 - \mu_0 \leqslant -0.02, \qquad \mu_5 - \mu_0 \leqslant -1.11,$$
$$\mu_3 - \mu_0 \leqslant -0.89, \qquad \mu_6 - \mu_0 \leqslant -0.64.$$

Example 10.6.2. The example that follows was given in Dunnett (1955) as an adaptation of an example first given by Villars (1951). An experiment was conducted to compare the effectiveness of three different chemical treatment processes on fabric manufactured by a standard method. The variable of interest is the breaking strength of the fabric measured in pounds, and large values represent stronger fabric. A piece of cloth manufactured by the standard method is divided into 12 identical pieces; these pieces are assigned randomly into four groups, each of size three. The first group is untreated (the control). Each of the remaining three groups is treated with a different chemical process. Then breaking strength is measured for each piece of cloth. For the following data develop a one-sided lower joint confidence statement for $\mu_j - \mu_0$ at level .95 and also a two-sided joint confidence statement at level .95, assuming that all the populations are normal with a common unknown variance.

Breaking Strength (in Pounds)

	Control	Process 1	Process 2	Process 3
	55	55	55	50
	47	64	49	44
	48	64	52	41
Totals	150	183	156	135
$\bar{x}_j$	50	61	52	45
s_j^2	19	27	9	21

We illustrate the calculation of s_j^2 for $j = 1$ as follows:

$$s_1^2 = \frac{(55-61)^2 + (64-61)^2 + (64-61)^2}{2} = 27.00.$$

The pooled sample variance from (10.6.1) is $s^2 = 19.00$ with $\nu = 8$. For the one-sided confidence limits we find $d' = 2.42$ from Table A.4, and for the two-sided confidence limits we find $d = 2.88$ from Table M.4, in each case using $\nu = 8$, $P^* = .95$, but entering with $k + 1 = 4$ in Table A.4 and with $k = 3$ in Table M.4. Then the absolute value of the last term in (10.6.4) (i.e., the allowance) is $2.42\sqrt{19}\ \sqrt{2/3} = 8.61$ and the joint one-sided lower confidence statement is as follows:

$$\mu_1 - \mu_0 \geqslant 2.39, \quad \mu_2 - \mu_0 \geqslant -6.61, \quad \mu_3 - \mu_0 \geqslant -13.61.$$

The investigator can then make the following three statements jointly at level .95:

1. The true breaking strength of process 1 exceeds the standard breaking strength by at least 2.39 lb.
2. The true breaking strength of the standard exceeds that of process 2 by at most 6.61 lb.
3. The true breaking strength of the standard exceeds that of process 3 by at most 13.61 lb.

As a result, at level .95 we can conclude that only process 1 is significantly better than the standard.

The allowance for the two-sided confidence limits in (10.6.5) is $2.88\sqrt{19}\ \sqrt{2/3} = 10.25$, and the joint confidence statement is as follows:

$$0.75 \leqslant \mu_1 - \mu_0 \leqslant 21.25, \quad -8.25 \leqslant \mu_2 - \mu_0 \leqslant 12.25, \quad -15.25 \leqslant \mu_3 - \mu_0 \leqslant 5.25.$$

These inequalities have a joint confidence level equal to .95.

10.6.2 Joint Confidence Statement Versus a Collection of Individual Confidence Statements

In the previous subsection we presented a procedure for making k joint confidence statements on the k differences $\mu_j - \mu_0$ with overall level of confidence P^*. Another possible approach is to make k separate confidence statements on these k differences, each at some specified level of confidence, say P^*. In this subsection we develop a table to show explicitly how much we would lose or gain by making these k separate confidence statements each at level P^*, instead of a single joint statement at overall level P^*.

The one-sided joint confidence statements are given by (10.6.3) and (10.6.4) where d' is obtained from Table A.4. The two-sided joint statements are given by (10.6.5) where d is found in Table M.4. The respective separate confidence statements are of the same form but the constants differ and are found from Student's t distribution with $\nu = (k+1)(n-1)$ degrees of freedom. In particular, the constant d' in (10.6.3) and (10.6.4) is replaced by the upper-tailed critical value $t_{\nu,\alpha}$ of a one-sided t test statistic at level $\alpha = 1 - P^*$. The constant d in (10.6.5) is replaced by the critical value $t_{\nu,\alpha/2}$ of a two-sided t test statistic at level $\alpha = 1 - P^*$.

The entries in Table 10.6.1 indicate the extent of the errors involved if one were to use a confidence level of $P^* = .95$ or $P^* = .99$ on each comparison separately, neglect to report this, and treat this P^* as if it were

Table 10.6.1 *Probability that each of k confidence intervals will simultaneously contain its respective parameter*

Here the caption P^* is the nominal level of each separate confidence interval and the table entry is the associated joint confidence level.

Total number of populations[a] $k+1$	Degrees of freedom ν	One-sided P^*		Two-sided P^*	
		.95	.99	.95	.99
5	20	.86	.97	.85	.97
	30	.86	.97	.85	.97
	∞	.86	.97	.84	.97
10	20	.77	.94	.75	.94
	30	.77	.94	.75	.94
	∞	.77	.94	.74	.92
15	20	.73	.92	.68	.92
	30	.73	.92	.67	.91
	∞	.71	.92	.65	.91

Note. In utilizing any table outside of this chapter for the problem of comparing k populations with a control it should be noted that the total number of populations there is usually k but here it is $k + 1$. Hence it may be necessary to replace k by $k + 1$ before utilizing any table *outside of this chapter* for comparisons with a control. We have attempted to retain the standard, usual terminology in both cases. As a result, the tables in this chapter start with $k = 1$ but in other chapters they start with $k = 2$.

the overall confidence level. The table entries are the achieved probabilities that each of the k separate confidence intervals will simultaneously contain its respective parameters for selected values of k and the degrees of freedom ν. Both one-sided and two-sided confidence intervals are considered. It is clear from this table that for moderate values of P^*, the actual probability achieved may be quite small. Note that the results for fixed k are stable with respect to changes in ν. Although this table includes three values of ν ($\nu = 20, 30, \infty$), our main interest here is $\nu = \infty$ since in this case s^2 can be treated as a constant equal to σ^2.

REMARK.   It is interesting to note that for the same P^* the results for the one-sided and two-sided cases are quite similar, and are more similar for small ν than for $\nu = \infty$. Of course, all these entries have to be consistent with the Bonferroni and Boole inequalities. This means that for any fixed k and ν (in both the one-sided and two-sided cases), we have by Boole's inequality for any fixed P^*,

$$1 - P\{k \text{ correct comparisons}\} \leqslant kP\{\text{an error in any single comparison}\}$$

$$= k(1 - P^*).$$

Hence

$$P\{k \text{ correct comparisons}\} \geqslant 1 - k(1 - P^*). \tag{10.6.6}$$

For example, if $k + 1 = 5$ (or $k = 4$), we obtain .80 for the right-hand side of (10.6.6) when $P^* = .95$, and .96 when $P^* = .99$; if $k + 1 = 10$, we obtain .55 when $P^* = .95$ and .91 when $P^* = .99$; and if $k + 1 = 15$, we obtain .30 when $P^* = .95$ and .86 when $P^* = .99$. All the entries in Table 10.6.1 satisfy these respective lower limits.   □

10.7   SELECTING NORMAL POPULATIONS WITH RESPECT TO VARIANCES THAT ARE BETTER THAN A STANDARD

In the previous sections of this chapter we have seen how to compare each of a set of k normal populations to a standard, when the comparison is made with respect to the means. In many problems we are not concerned with means, but rather with variances. This may occur, for example, when we have k products and wish to guarantee that there is not too much variability about the mean.

Example 10.7.1. Suppose we have k alloys, each of which satisfies certain specifications for mean tensile strength. We now are given a specification for the variability and wish to compare each alloy with the standard.

More specifically, we have k normal populations with variances $\sigma_1^2, \sigma_2^2, \ldots, \sigma_k^2$,

$$N\left(\mu_1, \sigma_1^2\right), N\left(\mu_2, \sigma_2^2\right), \ldots, N\left(\mu_k, \sigma_k^2\right),$$

and the standard variance σ_0^2, which is known. We wish to define a procedure for selecting exactly those populations for which $\sigma_i^2 \leqslant \sigma_0^2$, and this implies not selecting any of the others. (Note that in problems dealing with variances, we almost always wish to choose populations with small, rather than large, variances, so that a better population now means a smaller variance.)

If m denotes the (unknown) number of populations that are better than a standard, the generalized least favorable configuration is

$$\sigma_{[1]}^2 = \cdots = \sigma_{[m]}^2 = a\sigma_0^2, \qquad \sigma_{[m+1]}^2 = \cdots = \sigma_{[k]}^2 = c\sigma_0^2, \qquad (10.7.1)$$

where $c > 1 > a$. Then we wish to guarantee that the probability of a correct selection is at least some specified value P^* when such a configuration prevails for specified c and a.

The quantity $(c - a)\sigma_0^2$ separates those populations that are better than a standard from those worse than that standard. A simplification is achieved by reducing the number of constants; suppose we let $a = 1/c$. Then the generalized least favorable configuration in (10.7.1) becomes

$$\sigma_{[1]}^2 = \cdots = \sigma_{[m]}^2 = \frac{\sigma_0^2}{c}, \qquad \sigma_{[m+1]}^2 = \cdots = \sigma_{[k]}^2 = c\sigma_0^2, \qquad (10.7.2)$$

and we obtain the least favorable configuration from (10.7.2) by merely setting c equal to the specified value c^*.

The procedure is as follows. Given P^* and k, we first compute $(P^*)^{1/k} = Q$ using Table M.5. Then we enter Table M.6 with Q and c^* to find a value for the degrees of freedom ν. The common sample size n is then $n = \nu + 1$. Now n observations are taken from each of the k populations. The sample variances are computed and ordered as

$$s_{[1]}^2 \leqslant s_{[2]}^2 \leqslant \cdots \leqslant s_{[k]}^2.$$

At this stage we enter Table M.6 again with c^* and ν to obtain a value $B < 1$; this B is the second entry in the cell corresponding to (c^*, ν). The procedure is to select all populations for which the sample variance is less than or equal to $B\sigma_0^2$, and assert that the variances of these populations are less than or equal to σ_0^2. Diagrammatically this selection rule may be represented as follows:

$s_{[1]}^2 s_{[2]}^2 \cdots s_{[m]}^2$	$B\sigma_0^2$	$s_{[m+1]}^2 \cdots s_{[k]}^2$
populations better than the standard σ_0^2		populations worse than the standard σ_0^2

Example 10.7.2. Five brands of thermometer have been calibrated according to means, and we wish to choose exactly those brands for which the variance is smaller than that of a known standard for which $\sigma_0^2 = 0.6$. How many observations n shall we take on each brand so that the probability of a correct selection is at least $P^* = .90$ whenever c in (10.7.2) satisfies $c > c^* = 1.4$.

We first use Table M.5 to compute $(P^*)^{1/5} = (.90)^{1/5} = .979$. Now using Table M.6 with $c^* = 1.4$ and $Q = .979$, we find by linear interpolation that the degrees of freedom is $\nu = 74.7$, which is rounded upward to 75, so that $n = \nu + 1 = 76$. To complete the procedure, we take a sample of 76 observations from each of the five brands. Suppose the results for the five brands are

Brand	A	B	C	D	E
Sample variances	0.23	0.75	0.80	0.46	0.55

Now entering Table M.6 with $c^* = 1.4$ and $\nu = 75$, we obtain $B = .973$. Then $B\sigma_0^2 = (.973)(0.6) = 0.58$. Since the sample variances for brands A, D, and E are all less than 0.58, we assert that brands A, D, and E are better than the standard and the two others are not.

10.8   NOTES, REFERENCES, AND REMARKS

The basic source for this model for selecting populations better than a control is in a technical report by Tong (1967), later published in Tong

(1969). This paper treats the case of a control but omits the easy problem that arises in the case of a known standard; both cases are included here. The single-sample procedure for a common known variance and the two-stage procedure for a common unknown variance are both in Tong (1969). The allocation results in Section 10.5 come from Sobel and Tong (1971).

The original discussion on joint confidence statements is in Dunnett (1955), and some of his tables are revised in Dunnett (1964); this is the origin of the material in Section 10.6.1. Dudewicz, Ramberg, and Chen (1975) give two-stage procedures and tables for joint confidence statements in the case of unknown (not necessarily equal) variances. The material in Section 10.6.2 on comparing the probability level of a joint confidence statement versus a collection of individual confidence statements is new with this book.

Schafer (1976) treats the problem of selecting exactly those populations better than a standard (or control); this paper deals with variances and is the basis for Section 10.7.

An earlier paper that selects a subset containing all populations better than a control or standard is Gupta and Sobel (1958). The subject of nonparametric ranking procedures for comparison with a control was treated by Rizvi, Sobel, and Woodworth (1968).

PROBLEMS

10.1 A manufacturing plant is considering buying a new machine in hope of improving its production level. Seven different machine companies have submitted bids and all these bids are acceptable. The plant wishes to determine which of these seven machines have a higher average production level than the machine presently in use. Then they will choose the company with the lowest bid out of those selected as better than the one presently in use. Assume that the production level for each machine is normally distributed with common variance.

(a) What is the decision rule for $\delta_1^* = \delta_2^* = 10$?

(b) Assume that the common variance is known to be $\sigma^2 = 900$ and the average production level for the machine presently in use is known to be $\mu_0 = 126.7$. How many machines should be sampled from each of the seven new types for the probability of an overall correct decision to be at least $P^* = .90$ when $\delta_1^* = \delta_2^* = 10$?

(c)   Assume that the common variance is known to be $\sigma^2 = 900$ but the average production level for the machine presently in use is unknown. How many machines from each of the eight types should be included in the sample if $P^* = .90$ when $\delta_1^* = \delta_2^* = 10$?

(d)   Answer (a), (b), and (c) if $\delta_1^* = 8$, $\delta_2^* = 12$.

(e)   Assume that both the common variance and the average production level for the machine presently in use are unknown. An initial sample of size $n = 15$ is taken from each of the seven machines and the data give the pooled estimate $s^2 = 987.2$. An additional sample of $N - n$ observations is to be taken. How large should N be so that $P^* = .90$, $\delta_1^* = 10$, $\delta_2^* = 10$?

(f)   Answer (e) for $\delta_1^* = 8$, $\delta_2^* = 12$.

10.2   A medical school wants to make a recommendation to undergraduates on what majors in addition to the so-called premed major (the control group) will best prepare them for successful study in medical school. The majors under consideration are classified into the following $k = 5$ groups: biological sciences, physical sciences, mathematical sciences, social sciences, and humanities. Suppose that for the present study the potential for success of applicants to medical school is measured only by a quantitative score that is computed by a fixed formula combining undergraduate grade point average and score on the Medical School Admissions Test. The plan is to take a sample of n graduates of the medical school that had undergraduate majors in each group and determine their score from the records. We assume that these scores are normally distributed with a common variance and that the mean score (μ_0) for premed majors is unknown.

(a)   What is the decision rule for $\delta_1^* = 0.7$, $\delta_2^* = 0.3$?

(b)   If the common variance is known to be $\sigma^2 = 1.4$, how large should n be so that $P^* = .95$, $\delta_1^* = 0.7$, $\delta_2^* = 0.3$?

(c)   If the common variance is unknown, how large should N be if $n = 10$, $s^2 = 1.8$, $P^* = .95$, $\delta_1^* = 0.7$, $\delta_2^* = 0.3$?

10.3   Four new types of amplifier circuits are to be investigated for noise in comparison with the standard type of circuit now in use, which is known to have a mean noise of 2.6 and a variance $\sigma^2 = 1.00$. The investigator specified $\delta_1^* = \delta_2^* = 0.50$, $P^* = .90$.

(a)   How many observations should be taken on each type of circuit?

(b)   What is the decision rule?

10.4   Fichter (1976) states that one critical measure of the output of a nonelectrical detonator is the velocity of the detonator fragments, because higher velocity fragments mean greater output as long as the fragments have relatively uniform size and weight. A filter is installed between the detonator and the instrument that senses velocity. This filter prevents fragments smaller than a standard size from passing through so that velocity is measured for only the larger fragments. Thus the purpose of the filter is to ensure velocity measurements only for fragments that are uniform in size and weight. Fichter reported an experiment to compare two filter sizes, 16 mil and 30 mil, with the standard 20 mil size. With the 16 mil and 30 mil filter, the velocities of 10 fragments were measured in millimeters per microsecond; 94 fragments were measured with the 20 mil filter. The summary data from his experiment are shown below:

Group j	n_j	$\bar{x}_j$	s_j
20 mil (standard)	94	3.813	0.16689
16 mil	10	3.896	0.13769
30 mil	10	3.520	0.35261

Make two-sided joint confidence statements at level approximately .99 about the differences between the true means of the two filters under consideration and the 20 mil filter.

10.5   Dunnett (1964) gives data on an experiment performed at Lederle Laboratories to determine the effect of three drugs on the fat content of the breast muscle in cockerels. A group of 80 cockerels were divided randomly into four groups of 20 each. The first group, A, received no drug and hence acted as a control. The other three groups, B, C, and D, were given, respectively, stilbesterol and two levels of acetyl enheptin. At four different specified times, five cockerels from each group were sacrificed so that (among other things) the fat content of the breast muscle could be measured. One of the goals was to determine which of the three treatments produce a percentage fat of fresh tissue in cockerels that is significantly lower than the corresponding percentage for the control group of cockerels. The means of the percentage fat for the four samples were given as

$$\bar{x}_A = 2.493, \quad \bar{x}_B = 2.398, \quad \bar{x}_C = 2.240, \quad \bar{x}_D = 2.494,$$

each based on 20 observations. The ANOVA table (see Section 3.6 of Chapter 3) was given as follows.

Source of variation	Sum of squares (SS)	Degrees of freedom (v)	Mean squares = SS/v
Treatments	0.8602	3	0.2867
Sacrifice times	0.7574	3	0.2525
Interaction	1.1911	9	0.1323
Within cells (error)	6.9492	64	0.1086

Since this experiment used a fixed-effects model and the interaction was found to be nonsignificant, the appropriate estimate of the common variance is found from (3.6.3).

Find a one-sided upper joint confidence statement at level .95 for comparison of the three treatments with the control and make appropriate conclusions for this experiment.

CHAPTER 11

Selecting the t Best
Out of k Populations

11.1 INTRODUCTION

In the previous chapters we have dealt, at some length, with the goal of selecting the single best population from a set of k populations. There are situations, however, where we may wish to choose the two best populations, or the three best populations without ordering them. As an example, consider a clinic that dispenses drugs in large quantities, and no single manufacturer can fulfill all the needs of the clinic. As a result, several brands of drugs must be chosen, and the clinic wants to choose the best two brands, or the best three brands, and so on. In this example we would not want to select the brands in an ordered fashion, that is, we would not specify one as the best, some other one as the next best, and so on. In other situations the ordering *is* important. For example, the clinic might be interested in selecting the best brands of drugs in an ordered manner if they want to purchase the majority of drugs from the manufacturer that has the best brand. Thus there are two different cases here, one where the selection is to be made without regard to order, and one where the selection is ordered.

In this chapter we consider the goal of selecting a fixed number t of best populations out a group of k populations, with emphasis on the case of selecting them in an unordered manner. In either case the present goal is a straightforward generalization of the goal for $t = 1$. Whether the selection is made in an ordered manner or in an unordered manner, the formulation uses the indifference zone approach, and the primary emphasis is on determining the smallest common sample size needed to satisfy a specified (δ^*, P^*) requirement. Tables are available for some specific distribution models. However, for many distributions, the appropriate table for one or both of these cases may not have been calculated.

This goal is treated very briefly in this chapter; we consider only the model of normal distributions with common known variance, and only for a selection made without regard to order. However, the same treatment can easily be applied to other distribution models.

The distribution model is k normal populations with a common known variance σ^2, or

$$N\left(\mu_1,\sigma^2\right), N\left(\mu_2,\sigma^2\right),\ldots, N\left(\mu_k,\sigma^2\right).$$

The ordered μ values are denoted as usual by

$$\mu_{[1]} \leqslant \mu_{[2]} \leqslant \cdots \leqslant \mu_{[k]}.$$

The goal is to choose the t best populations, that is, the populations with μ values

$$\mu_{[k]}, \mu_{[k-1]},\ldots, \mu_{[k-t+1]},$$

in an unordered manner. The distance measure δ is defined as the difference between the smallest mean of the t best populations and the largest mean of the remaining $k - t$ populations, or

$$\delta = \mu_{[k-t+1]} - \mu_{[k-t]}.$$

In a sense this distance function measures the minimum separation of the means between the two groups composed respectively of the t best populations and the $k - t$ worst populations.

The generalized least favorable configuration in terms of the true value of the distance measure δ is

$$\mu_{[1]} = \mu_{[2]} = \cdots = \mu_{[k-t]}, \qquad \mu_{[k-t+1]} = \cdots = \mu_{[k]},$$

$$\delta = \mu_{[k-t+1]} - \mu_{[k-t]}. \tag{11.1.1}$$

The least favorable configuration in terms of a specified valued δ^* is the same as (11.1.1) with $\delta = \delta^*$.

With this formulation, the details discussed in Chapter 2 where $t = 1$ carry over to the present case where $t \geqslant 2$. We wish to define a selection procedure R such that the probability of a correct selection is greater than or equal to a specified value P^* whenever the distance δ is greater than or equal to a specified value δ^*.

A sample is taken from each of the k populations, and the sample means are computed. These k sample means are then ordered, and the ordered values are denoted by

$$\bar{x}_{[1]} \leqslant \bar{x}_{[2]} \leqslant \cdots \leqslant \bar{x}_{[k]}.$$

The selection rule is simply to assert that the populations that produced the t largest sample means, $\bar{x}_{[k]}, \bar{x}_{[k-1]}, \ldots, \bar{x}_{[k-t+1]}$, are the t best populations.

11.2   SAMPLE SIZES EQUAL

Now suppose that the experiment to select the t best populations is in the process of being designed, and the present task is to determine the appropriate common sample size n for a specified pair (δ^*, P^*). For selected values of k and t, Table N.1 of Appendix N gives the values of a quantity τ_t required to satisfy a specified P^* requirement for $t = 2, 3, 4, 5$ and $2t \leqslant k \leqslant 10$, that is, only for $t \leqslant k/2$. (The value of τ_t for a given k is symmetric about $k/2$. Hence the value of τ_t for $t > k/2$ can be found from this table using the given k and replacing t by $k - t$.) Using this value of τ_t, the corresponding common n needed for a specified δ^* is then found by computing

$$n = \left(\frac{\tau_t \sigma}{\delta^*} \right)^2. \qquad (11.2.1)$$

If the result in (11.2.1) is not an integer, we round upward to obtain a conservative result.

REMARK   It is quite possible that by using sample sizes that are "almost" common in the sense that they differ by one, we may obtain a slight reduction in the total sample size and still satisfy the basic requirement; however, we ignore such refinements here.   □

Example 11.2.1.   A company needs to purchase two machines, and $k = 5$ brands are under consideration. If two machines of the same brand are purchased there will be total reliance on one manufacturer. If that manufacturer is temporarily out of parts, or has labor difficulties, both machines may be down simultaneously. As a result, the company feels it is preferable to have one machine in each of two different brands. The goal then is to select the two best machines for purchase (rather than purchase a duplicate of the one best machine). Samples of n machines of each of the five brands are to be subjected to a life test and the operating time (of usage) to failure in kilocycles measured for each machine. The best brands of machines are defined as the ones with the largest average lifetime. Assume that the lifetimes for each brand are approximately normally distributed with common variance $\sigma^2 = 100$. In applying normal theory to life-testing data, we assume that $\mu > 3\sigma$ for each population so that the probability of a negative observation is negligible. The present question is, how large should n be so that if we base our selection rule on the sample means, we will make a correct selection with probability at least $P^* = .90$ whenever $\delta = \mu_{[4]} - \mu_{[3]} \geq \delta^* = 5.00$.

From Table $N.1$ with $P^* = .90$, $k = 5$, and $t = 2$, we find $\tau_t = 2.8505$, and from (11.2.1) we compute

$$n = \left(\frac{(2.8505)(10)}{5.00} \right)^2 = 32.49.$$

Rounding upward, we obtain the result that $n = 33$ observations must be taken from each of the five brands of machine.

REMARK.   A study of Table $N.1$ shows that for any fixed P^* and k, the value of τ_t increases as t increases for $t \leq k/2$ and therefore decreases for $t > k/2$. Further, as noted before, τ_t is symmetric about $k/2$. This property and (11.2.1) imply that for σ and δ^* fixed, the common sample size n is also symmetric about $t = k/2$, increasing for $t \leq k/2$ and decreasing for $t > k/2$. $\square$

In general, for $k \geq 3$, the problem of selecting the two best populations is at least as hard a problem as selecting the one best, and for $k \geq 4$, it is harder. More precisely, for any value of k the problem gets harder as t approaches $k/2$, and, by symmetry, the same is true if t decreases toward $k/2$ from above. This can be seen in various ways from Table $N.1$. For example, for $k = 4$ and any fixed P^*, the value of τ_t increases as t increases up to $k/2 = 2$, so that more observations per population are needed for $t = 2$ than for $t = 1$.

Now suppose that the common sample size n is fixed, and we are interested in evaluating the selection procedure for this fixed n and given values of t, k, and σ. For given t and k the entry in Table $N.1$ can be interpreted as the true value of a quantity τ for which the probability of a correct selection under the generalized least favorable configuration, that is, $P_{\mathrm{GLF}}\{\mathrm{CS}|\tau\}$, is equal to P. When the pairs $(\tau, P_{\mathrm{GLF}}\{\mathrm{CS}|\tau\})$, or equivalently (τ, P), are plotted on a graph and connected by a smooth curve, we obtain the operating characteristic curve for the selection rule R in terms of running values of τ. However, the investigator is usually more interested in the properties of the selection rule in terms of running values of $\delta = \mu_{[k-t+1]} - \mu_{[k-t]}$. The true value of τ is related to the true value of δ by

$$\tau = \frac{\delta \sqrt{n}}{\sigma}; \tag{11.2.2}$$

that is, τ is δ is expressed in the scale $\sigma/\sqrt{n}$. As a result, for fixed n and σ, the relationship in (11.2.2) can be used to convert any τ value to a δ value. In particular, we have

$$\delta\left(\tau; \sigma/\sqrt{n}\right) = \frac{\tau\sigma}{\sqrt{n}}. \tag{11.2.3}$$

Then the operating characteristic curve can be given in terms of running values of $\delta(\tau;\ \sigma/\sqrt{n})$ and the corresponding values of $P = P_{\mathrm{GLF}}\{\mathrm{CS}|\delta(\tau;\sigma/\sqrt{n})\}$.

As a numerical illustration, consider Example 11.2.1 again, but now suppose that samples of $n=33$ observations have been taken from each of the $k=5$ brands. The present aim is to calculate points on the operating characteristic curve of the selection procedure where $t=2$. From Table N.1 for $t=2$ we find the following pairs (τ, P):

τ	2.1474	2.8505	3.2805	3.6591	4.1058	5.0584
P^*	.750	.900	.950	.975	.990	.999

Since $\sigma/\sqrt{n} = \sqrt{100/33}$ here, the τ value in each of these pairs is converted to a δ value using (11.2.3) as

$$\delta\left(\tau; 10/\sqrt{33}\right) = \tau\left(10/\sqrt{33}\right) = (1.7408)\tau.$$

For example, with $P=.750$ the corresponding δ value is $\delta = (1.7408)(2.1474) = 3.738$. The points on the operating characteristic curve are then as follows:

$\delta(\tau; 10/\sqrt{33})$	3.738	4.962	5.711	6.370	7.147	8.806
P	.750	.900	.950	.975	.990	.999

These values can also be interpreted as (δ^*, P^*) pairs. For example, the probability of a correct selection is at least $P^*=.90$ whenever $\delta \geqslant \delta^* = 4.962$.

Alternatively, we can find points on the operating characteristic curve by choosing arbitrary values of δ, using (11.2.2) to compute the corresponding values of τ for the given n and σ, and interpolating in Table N.1 to find each corresponding value of P. We can avoid table interpolation by plotting the pairs (τ, P) obtained from Table N.1, connecting these points by a smooth curve, and reading off the P values correspoinding to each of the values of τ computed from (11.2.2). This approach is illustrated by Example 11.3.1 in the next section.

11.3   SAMPLE SIZES UNEQUAL

In this section we consider the problem of selecting the t best normal populations when the sample sizes are fixed and not all equal. The selection rule R for this problem was explained in Section 11.1, but the procedure explained in Section 11.2 for calculating the operating characteristic curve applies only for equal sample sizes.

An approximate solution can be found using the method explained in Section 2.4 of Chapter 2 where we had $t = 1$. Suppose that a sample of size n_j is taken from the jth population, for $j = 1, 2, \ldots, k$. Then we use the square-mean-root formula to calculate a generalized average sample size as

$$n_0 = \left(\frac{\sqrt{n_1} + \sqrt{n_2} + \cdots + \sqrt{n_k}}{k} \right)^2. \qquad (11.3.1)$$

The result in (11.3.1) need not be rounded to an integer. Using this value of n_0 in place of n, the procedure of Section 11.2 gives an approximate method of calculating the operating characteristic curve.

Example 11.3.1.   Consider the situation described in Example 11.2.1 again, but now suppose the samples have already been drawn from each of the $k = 5$ machines and the sample sizes are

$$n_1 = 20, \quad n_2 = 25, \quad n_3 = 25, \quad n_4 = 30, \quad n_5 = 30.$$

The average sample size is computed from (11.3.1) as

$$n_0 = \left(\frac{\sqrt{20} + \sqrt{25} + \sqrt{25} + \sqrt{30} + \sqrt{30}}{5} \right)^2 = (5.085)^2 = 25.86.$$

Suppose as before that the lifetimes are all normally distributed with common variance $\sigma^2 = 100$, and the goal is to choose the best $t = 2$ brands in an unordered manner, but now we want points on the operating characteristic curve for the sample sizes given earlier (or for $n_0 = 25.86$) and the selected values $\delta = 5, 6, 7, 8, 9, 10$. In this example for any δ value the corresponding τ value from (11.2.2) is

$$\tau = \frac{(5.085)\delta}{10}.$$

For the six δ values stated, the corresponding τ values are, respectively, 2.548, 3.051, 3.560, 4.068, 4.577, and 5.085. Since six interpolations would be required to determine the corresponding P values, we have plotted in Figure 11.3.1 a graph of the pairs (τ, P) given in Table N.1 for $t=2, k=5$. From this smooth curve we can readily find the P value corresponding to each of the six preceding τ values as .855, .927, .971, .990, .997, .999, and hence corresponding to each of the six δ values. The results are as follows:

δ	5	6	7	8	9	10
P	.855	.927	.971	.990	.997	.999

These points are plotted and connected by a smooth curve in Figure 11.3.2. This figure gives the operating characteristic curve for the selection procedure when $k=5, t=2, \sigma^2=100, n_0=25.86$, as a function of the true value of the distance measure δ. For example, the probability of a correct selection is equal to .90 under the generalized least favorable configuration

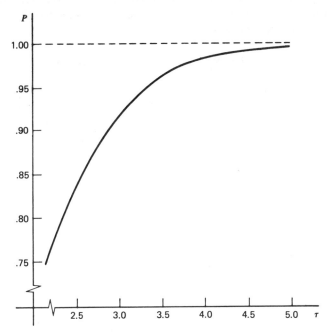

Figure 11.3.1   Graph of (τ, P) values taken from Table $N.1$ with $k=5$, $t=2$.

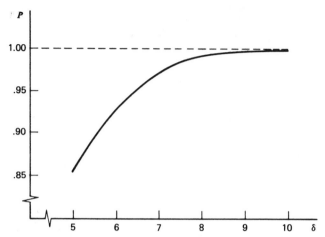

Figure 11.3.2 Graph of operating characteristic curve of (δ, P) values for a square-mean-root sample size of 25.86.

when $\delta = 5.65$, and it is .95 when $\delta = 6.5$. These points can also be interpreted as (δ^*, P^*) pairs under the least favorable configuration; thus the probability of a correct selection is at least .90 whenever $\delta \geqslant \delta^* = 5.65$ or at least .95 whenever $\delta \geqslant \delta^* = 6.5$.

11.4   NOTES, REFERENCES, AND REMARKS

The theory behind the selection of the t best populations (in the ordered as well as in the unordered cases) is considered in the earliest paper by Bechhofer (1954), and some appropriate tables appear there. Other good sources for these tabled values are the tables of Milton (1963, 1966), which are unpublished; Milton computed forward probabilities in 1963 as well as equi-percentage points in 1966 for fixed values of t, k, and P^*.

It should also be mentioned that the main sequential procedures considered in Bechhofer, Kiefer, and Sobel (1968) deal with the problem of selecting the t best of k populations for both the ordered and unordered cases. It is safe to say that more is known about this generalization of the basic problem of selecting the best population than about any other generalization. Of course, sequential procedures almost always yield more efficient procedures. However, before comparing procedures of widely different types, one should carefully check to see that both procedures are equally acceptable. Only fixed sample size and two-stage procedures are considered in this book. The reader interested in sequential procedures should consult Bechhofer, Kiefer, and Sobel (1968).

In this chapter we select exactly t populations and require that they be the t best. We could also ask for the number of observations needed such that a selected subset of exactly f populations *contains* the one best population (or t best populations). Alternatively, for a fixed number of observations we could ask for the smallest fixed-size subset of size f that will contain the t best populations with probability at least P^*. Although such problems are natural generalizations of the problem of this chapter, they are here regarded as *subset selection problems* and are covered in Chapter 13; see, for example, Section 13.4. The associated theory and tables appear in Desu and Sobel (1968) and Desu (1970); these tables overlap and supplement the tables needed for the present chapter.

PROBLEMS

11.1 Fichter (1976) states that certain military installations in the northeastern United States have been criticized for a rise in overhead (O/H) rates that is inordinate compared to installations in other parts of the country. Increased overhead rates cause a decrease in job orders and therefore in jobs, which then cause even more increase in overhead rates because the direct labor base is used to recover overhead costs. Fichter chose four installations, one in each of four regions of the United States, and reported the following data on average hourly overhead rate in dollars for the fiscal years 1968–1975.

Location of Installation

Year	Northeast	Mid-Atlantic	South	Midwest
1968	5.32	5.85	5.78	4.79
1969	5.89	6.45	6.67	5.17
1970	8.54	8.22	6.43	5.79
1971	9.10	9.49	4.76	6.25
1972	10.22	10.84	5.12	6.75
1973	11.00	11.55	5.51	7.16
1974	11.58	11.78	5.36	8.01
1975	12.97	12.20	5.96	9.63
$\bar{x}_j$	9.3275	9.5475	5.6987	6.6925
s_j	2.6855	2.4707	0.6458	1.5800

Assume that these data are normally distributed with a common known variance $\sigma^2 = 4.00$, and the goal is to select the $t = 2$ populations with the largest mean overhead rate. Make the selection and calculate the operating characteristic curve. Does it appear that the criticism of northeastern installations is justified?

11.2   Soller and Putter (1965) discuss a sire-sampling program in Israel. A fixed group of k young sires are progeny-tested each year, and the t sires whose daughters have the largest average milk production are kept for widespread use. The parameter of interest is the "transmitting ability" of each sire, denoted by μ_j for the jth sire and defined as the average performance of an infinite number of the sire's daughters. To carry out the sire-sampling program, a sample of n daughters of each sire is selected and milk production for 305 days is observed for each daughter in the sample. The production of each daughter is assumed to be normally distributed with the same variance σ^2 for each sire. This variance is known from past experience to be $\sigma^2 = 570,000 \ kg^2$. The average 305-day milk production of the daughter of the jth sire in the experiment (daughter-average) is denoted by $\bar{x}_j$ and used to estimate the parameter μ_j. The usual number k of sires tested is 4, 6, or 8 in Israel.

(a)   If it is desired to select only $t = 2$ sires for widespread use, determine for each of these values of k how many daughters n should be tested for each sire if P^* is specified to be .90 and $\delta^* = 100$ kg.

(b)   The usual value of n in Israel is 60. Determine the operating characteristic curve of the selection procedure for $n = 60$ if $t = 2$ and $k = 6$.

(c)   Do (b) if $t = 3$ and $k = 8$.

CHAPTER 12

Complete Ordering of k Populations

12.1 INTRODUCTION

All the previous chapters have dealt with statistical selection problems, since the goal in each case has been to select a group of one or more populations from some number k of given populations. In this chapter we consider statistical *ranking* problems, where we want to order or *rank* all the k populations with respect to the relative values of some parameter.

For example, in the evaluation of k competing consumer or product goods, we may wish to order or rank these goods with respect to the values of their means or variances. When we have a set of doses of increasing potency or items of supposedly increasing difficulty, we again may wish to order all the items with respect to some measure of their effectiveness. As a more specific example, Carroll and Gupta (1975) discuss a problem in which data were collected for the National Cancer Institute. The data give the weight loss (in grams) of mice inoculated with a certain type of tumor-inducing chemical (or virus). The mice were all given the same drug, but at five different dosage levels. The tumors are often internal and not visible or objectively measurable without killing the test animal; in these cases weight loss is often considered a good indication of the presence of tumors. For our purposes we can assume that the intent of the drugs is to induce weight loss, since we are not dealing with the results of the pathologist's report on the test animal. The problem then is to find the best dosage level for inducing weight loss, the second best dosage level, and so on, until the worst dosage level is reached. It should be quite clear that the problem of a complete ranking arises in many areas other than medicine. Bishop and Dudewicz (1976) consider applications in reliability and life-testing studies.

We consider this problem for the normal distribution model only, first when the desired ordering is with respect to means, and second when the desired ordering is with respect to variances.

12.2   COMPLETE ORDERING OF k NORMAL POPULATION MEANS WHEN THE COMMON VARIANCE IS KNOWN

Suppose we have the model of k normal populations with a common known variance σ^2, or

$$N\left(\mu_1,\sigma^2\right), N\left(\mu_2,\sigma^2\right), \dots, N\left(\mu_k,\sigma^2\right). \tag{12.2.1}$$

As usual, the ordered μ values are denoted by

$$\mu_{[1]} \leqslant \mu_{[2]} \leqslant \cdots \leqslant \mu_{[k]}.$$

The goal is to order the k populations with respect to their μ values.

As a first step, a "distance" measure must be defined. Because we are ranking *all* k means, we need to consider the entire set of $k-1$ distances between each successive pair of ordered μ values, or

$$\delta_1 \quad = \mu_{[k]} \quad - \mu_{[k-1]}, \tag{12.2.2}$$

$$\delta_2 \quad = \mu_{[k-1]} - \mu_{[k-2]},$$

$$\vdots$$

$$\delta_{k-1} = \mu_{[2]} \quad - \mu_{[1]}.$$

The problem is then to construct a procedure for ranking the populations such that the probability of a completely correct ranking is at least some specified value P^* whenever *each* of the distances δ_i is greater than a specified value δ_i^*, that is, whenever

$$\delta_1 \geqslant \delta_1^*, \quad \delta_2 \geqslant \delta_2^*, \quad \dots, \quad \delta_{k-1} \geqslant \delta_{k-1}^*.$$

Even if it were practicable for the investigator to specify $k-1$ different values of δ_i^* and even if the complete solution were known, the statistical solution would be quite complicated. As a result, for simplicity and/or practicality we assume a common threshold value δ^* for all the δ_i^*. In other words, we want the probability of a completely correct ranking to be at least P^* whenever each distance δ_i is greater than δ^*, or, in symbols, whenever

$$\delta_1 \geqslant \delta^*, \quad \delta_2 \geqslant \delta^*, \quad \ldots, \quad \delta_{k-1} \geqslant \delta^*. \qquad (12.2.3)$$

The procedure is to take a random sample of n observations from each of the k populations, compute the k sample means, and order them as

$$\bar{x}_{[1]} \leqslant \bar{x}_{[2]} \leqslant \cdots \leqslant \bar{x}_{[k]}.$$

The proposed rule is simply to assert that the population with the largest sample mean $\bar{x}_{[k]}$ is the one with the largest μ value, that the population with the second largest sample mean $\bar{x}_{[k-1]}$ is the one with the second largest μ value, and so on.

Now suppose that we are in the process of designing a fixed sample size experiment to rank the k populations completely with respect to their μ values. Then we specify δ^* and P^*, and the problem is to determine the minimum common sample size n to be taken from each population so that the (δ^*, P^*) requirement is satisfied. The relevant table for this problem is given here as Table P.1 in Appendix P. Table P.1 can be used to find a number τ_t that satisfies a probability requirement corresponding to a given k and P^*. Since δ^*, n, and σ enter this requirement only by virtue of the relation $\tau_t = \delta^* \sqrt{n} / \sigma$, the sample size n can be computed from

$$n = \left(\frac{\tau_t \sigma}{\delta^*} \right)^2. \qquad (12.2.4)$$

The result in (12.2.4) is always rounded upward to give a conservative result. Since the format of Table P.1 is different from our usual format, interpolation is required. The appropriate method of interpolation is the same as for Table A.2; that is, we use linear interpolation on $\log \tau$ (or $\log \tau_t$) and $\log (1 - P^*)$.

Example 12.2.1 In the application described in Carroll and Gupta (1975) there are $k = 5$ drug levels, denoted by A, B, C, D, and E. In this application normal distributions were assumed with a common unknown variance; they computed an estimate of σ^2 and treated it as if it were the actual value of σ^2. For the purposes of our example suppose that the standard deviation σ is equal to 20. Further, suppose that we wish to make a completely correct ranking at least 80% of the time whenever each difference of successive ordered means is at least $\delta^* = 10$. Entering Table P.1 with $k = 5$, we locate $P^* = .90$ and interpolate to find $\tau_t = 2.77$. Then from (12.2.4) we find

$$n = \left(\frac{\tau_I \sigma}{\delta^*} \right)^2 = \left[\frac{(2.77)(20)}{10} \right]^2 = 30.69.$$

Thus we must take $n = 31$ observations from each of the five populations. This means we must put 31 test animals on each of the five drugs, so that a total of 155 test animals are needed for the preceding (δ^*, P^*) requirement.

*12.3   A TWO-STAGE PROCEDURE FOR COMPLETE ORDERING OF k NORMAL POPULATION MEANS WHEN THE COMMON VARIANCE IS UNKNOWN

In the problem of completely ranking k normal populations with respect to their means when the common variance σ^2 is unknown, we cannot obtain a single-stage solution to satisfy the (δ^*, P^*) requirement that holds for all possible values of the unknown variance σ^2. (Here the distances are defined as in (12.2.2), and δ^* and P^* are as defined in Section 12.2.) The reason a single-stage solution does not exist is that for any reasonable procedure that requires taking n_j observations from the jth population, the probability of a correct selection depends on the value of σ^2, and as σ^2 approaches infinity, the probability approaches $1/k$. As a result, for any reasonable value of P^* that exceeds $1/k$, the P^* condition is not satisfied for σ^2 sufficiently large.

However, if we use a two-stage procedure, this difficulty is avoided. The same situation arose in Section 3.4, where we developed a two-stage procedure for selecting the population with the largest mean when the common variance is unknown. The present procedure is analogous to that of Chapter 3.

In the first stage a sample of n observations is taken from each of the k populations, and the k sample variances $s_1^2, s_2^2, \ldots, s_k^2$ are calculated. The pooled sample variance s^2 is then computed by

$$s^2 = \frac{(n-1)s_1^2 + \cdots + (n-1)s_k^2}{k(n-1)} = \frac{s_1^2 + \cdots + s_k^2}{k}. \tag{12.3.1}$$

The number of degrees of freedom for s^2 is $\nu = k(n-1)$.

In the second stage, a second sample of size $N - n$ is taken from each of the k populations. The value of N is obtained from the expression

$$N = \max\left(n, \left\{ \frac{2s^2 h^2}{\delta^{*2}} \right\}^+ \right), \tag{12.3.2}$$

where $\{a\}^+$ means the smallest integer equal to or greater than a.

In using the formula (12.3.2), we need to know δ^* and h. The value of δ^* is obtained from the usual specification on all successive pairs of ordered means. The probability of a completely correct ranking is to be at least P^* whenever the following $k-1$ inequalities between successive means hold jointly:

$$\mu_{[k]} \quad - \mu_{[k-1]} \geqslant \delta^*,$$

$$\mu_{[k-1]} - \mu_{[k-2]} \geqslant \delta^*,$$

$$\vdots$$

$$\mu_{[2]} \quad - \mu_{[1]} \quad \geqslant \delta^*.$$

The value of h is obtained from Table P.2, which covers the case $P^* = .95$, various values of n, and $k = 3, 4, 5,$ and 6.

The next step is to compute the k sample means, in each case using the *entire* sample of N observations, and order them as

$$\bar{x}_{[1]} \leqslant \bar{x}_{[2]} \leqslant \cdots \leqslant \bar{x}_{[k]}.$$

We then assert that the population corresponding to the sample mean $\bar{x}_{[k]}$ is the one with the largest μ value,..., the population corresponding to the sample with mean $\bar{x}_{[1]}$ is the one with the smallest μ value. With this procedure the probability of a correct ranking is guaranteed to be at least P^* whenever $\mu_{[i+1]} - \mu_{[i]} \geqslant \delta^*, i = 1,\ldots, k-1$, regardless of the true value of the unknown σ^2.

Example 12.3.1. Suppose we have five brands of paintbrush cleaners and have a method of measuring the effectiveness of each brand of cleaner in removing enamel paint. We wish to obtain a complete ranking of the brands with respect to average effectiveness. We specify that the probability of correctly ranking the five populations is to be at least $P^* = .95$ whenever the difference between each pair of successive means is at least $\delta^* = 1.00$.

Suppose the measurements for all brands are normally distributed with a common variance that is unknown, so that we must settle for a two-stage procedure. We begin with an initial sample of $n = 30$ observations from each population, and compute the pooled sample variance $s^2 = 4$ with $\nu = k(n-1) = 145$ degrees of freedom. To determine the size of the second sample, we first use Table P.2 to obtain $h = 2.24$ for $k = 5$, $P^* = .95$, $n = 30$.

Table 12.3.1   Approximations for the initial sample size

δ^*/σ ＼ k	3	4	5
0.3	90	100	100
0.4	50	60	60
0.5	40	40	40

Then we compute the right-hand member of (12.3.2) as

$$\left\{ \frac{2s^2h^2}{\delta^{*2}} \right\}^+ = \left\{ \frac{2(4)(2.24)^2}{1.00} \right\}^+ = \{40.14\}^+ = 41.$$

Now since max $(30, 41) = 41$, we find that $N = 41$ and that

$$N - n = 41 - 30 = 11,$$

so that an additional 11 observations must be taken from each population.

It would be useful to have a rough estimate of an optimal choice of the initial sample size n. Such an estimate may be obtained from tables or graphs of the values of the expected total sample size. Approximations for the initial sample sizes are given in Table 12.3.1 for $k = 3, 4, 5$ as a function of δ^*/σ. Thus if σ were known (or estimated) so that δ^*/σ could be computed (or estimated), this table could be used to determine (or estimate) the optimum value of n. Since σ is always unknown, we must have an estimate, based either on some data or on previous experience with the same kind of data.

The entries in Table 12.3.1 are based on the assumption that the least favorable configuration of means, found by setting each equation in (12.2.2) equal to δ^*, is the true one. These values are given to indicate an order of magnitude, and should not be taken as specific. It is better to underestimate rather than overestimate the optimal value of n, since additional observations can always be taken in the second stage.

12.4   COMPLETE ORDERING OF k NORMAL POPULATION VARIANCES

Suppose we have k normal populations and want to rank them with respect to their variances, or equivalently, their standard deviations. The ordered σ values are denoted by

$$\sigma_{[1]} \leqslant \sigma_{[2]} \leqslant \cdots \leqslant \sigma_{[k]}.$$

As before, we need to provide a "distance" measure between any two populations. Here we use the ratio of two successive pairs of standard deviations, so the $k-1$ distances are the ratios

$$\frac{\sigma_{[2]}}{\sigma_{[1]}} = \delta_{21}, \frac{\sigma_{[3]}}{\sigma_{[2]}} = \delta_{32}, \ldots, \frac{\sigma_{[k]}}{\sigma_{[k-1]}} = \delta_{k, k-1}. \tag{12.4.1}$$

Note that each of these distances is greater than or equal to one.

In most problems it is not practicable to specify a positive lower bound for the threshold value of so many distances. As a result, we simplify the problem by specifying a single constant $\delta^* > 0$ as a threshold value for each of the $k-1$ distances in (12.4.1). The preference zone is then given by

$$\frac{\sigma_{[2]}}{\sigma_{[1]}} \geqslant \delta^*, \quad \frac{\sigma_{[3]}}{\sigma_{[2]}} \geqslant \delta^*, \ldots, \quad \frac{\sigma_{[k]}}{\sigma_{[k-1]}} \geqslant \delta^*.$$

The problem is to provide a procedure for ranking all k populations with respect to variances such that the probability of a completely correct ranking is at least some specified value P^* whenever each of the distances between (i.e., ratios of) successive standard deviations is at least δ^*.

The procedure is to take a sample of common size n from each of the k populations, compute the k sample variances s_j^2 (or standard deviations s_j) and order them from smallest to largest as

$$s_{[1]}^2 \leqslant s_{[2]}^2 \leqslant \cdots \leqslant s_{[k]}^2.$$

The proposed rule is to assert that the ordering of population variances (or standard deviations) is the same as the ordering of the sample variances (or standard deviations).

Now suppose that the experiment is in the process of being designed so that we want to know the common sample size n needed to satisfy this (δ^*, P^*) requirement. Table P.3 in Appendix P gives the value of P^* as a function of the degrees of freedom ν and $c = (\delta^*)^2$. Thus we enter the table with k, P^*, and c to find ν, and the common sample size is then found from $n = \nu + 1$.

Alternatively, if we have a sample size n in mind or the common sample size is fixed at n, then we may enter Table P.3 with $\nu = n - 1$ degrees of freedom and $c = (\delta^*)^2$, and read off the minimum probability P^* of correctly ordering the k populations.

REMARK. The reader should note that we are using δ^* in (12.4.1) for comparing standard deviations and not for comparing variances. This is not always the case in the literature on ranking and selecting variances. □

Example 12.4.1. Suppose we have five brands of thermometers whose readings are normally distributed. A consumer panel wishes to rank the five brands completely with respect to their variances. This is a natural and practical goal, since, although all thermometers are calibrated, the main concern is with their variability, rather than with means. Assume we specify that the probability of a correct ranking is to be at least $P^* = .90$ whenever the four inequalities

$$\frac{\sigma_{[2]}}{\sigma_{[1]}} \geqslant \delta^*, \quad \frac{\sigma_{[3]}}{\sigma_{[2]}} \geqslant \delta^*, \ldots, \quad \frac{\sigma_{[5]}}{\sigma_{[4]}} \geqslant \delta^*$$

hold jointly, where δ^* is specified as 1.3. How many observations shall we take from each population?

We enter Table P.3 with $k = 5$, $c = (1.3)^2 = 1.69$ to find $P^* = .90$. Using $c = 1.7$, we need to interpolate between $P^* = .879$ and $P^* = .918$, which correspond to $\nu = 50$ and $\nu = 60$, respectively. With linear interpolation we obtain $\nu = 55.4$ degrees of freedom, which is rounded upward to $\nu = 56$. Thus we should take $n = \nu + 1 = 57$ observations from each population.

Suppose the experiment is performed, that is, 57 thermometers of each brand are tested, and the sample variances are computed as follows:

Brand	A	B	C	D	E
Sample variance	1.02	1.31	1.04	0.98	1.10

The complete ranking of the populations from smallest to largest with respect to variances is then D, A, C, E, B, since this is the ranking obtained from the observed sample variances.

REMARK. Clearly, if the number of populations were fewer, then the number of observations required per population would decrease, since it is easier to rank, say, three populations than five populations. ☐

12.5 SEPARATION OF POPULATION VARIANCES INTO GROUPS

In some instances we may wish to separate k normal populations into groups according to the magnitude of the true variances, which are ordered as

$$\sigma_{[1]}^2 \leqslant \cdots \leqslant \sigma_{[k]}^2.$$

The groups are to be chosen so that the variances have relatively the same magnitude within each group. Several procedures for such a separation have been proposed. We discuss two methods in this section. It is important to note that the methods differ in a basic manner from our usual approach in that there is no control of the overall error (the probability of a correct separation). Nevertheless the methods may be of use in some applications. The first method results in a separation into mutually exclusive groups, whereas the second method provides a separation into groups that are usually overlapping.

12.5.1 Separation into Nonoverlapping Groups

Using the procedure of separation of variances into nonoverlapping groups, the first step is to test the null hypothesis that all the variances are equal. Suppose that n observations are taken from each of the k populations, and the ordered sample variances are denoted by

$$s_{[1]}^2 \leqslant \cdots \leqslant s_{[k]}^2.$$

One test of homogeneity of variances is performed by computing the test statistic

$$\frac{s_{[k]}^2}{s_{[1]}^2}, \tag{12.5.1}$$

and comparing it with a critical value from Table P.5 with $v = n - 1$ degrees of freedom. The hypothesis is then rejected at level α if the value of the test statistic in (12.5.1) exceeds the critical value at level α.

If this hypothesis is rejected, then the *successive* ratios

$$\frac{s_{[2]}^2}{s_{[1]}^2}, \frac{s_{[3]}^2}{s_{[2]}^2}, \ldots, \frac{s_{[k]}^2}{s_{[k-1]}^2}$$

are computed and tested for significance by comparing them to a critical value found from Table P.4. Whenever a ratio is found to be significant, a separation into disjoint groups is made. For example, suppose that $k \geqslant 6$ and the ratios $s_{[3]}^2/s_{[2]}^2$ and $s_{[6]}^2/s_{[5]}^2$ are the only significant ratios. Let σ_j^2 denote the population variance associated with the ordered sample variance $s_{[j]}^2$. Then since $s_{[3]}^2/s_{[2]}^2$ is significant, σ_2^2 and σ_3^2 go into different groups. The case is similar for σ_5^2 and σ_6^2. Thus the population variances are separated into the following three *disjoint* groups:

$$\left(\sigma_1^2, \sigma_2^2\right), \left(\sigma_3^2, \sigma_4^2, \sigma_5^2\right), \left(\sigma_6^2, \ldots, \sigma_k^2\right).$$

12.5.2 Separation into Overlapping Groups

If we want a more elaborate ranking of the population variances, then we may compare not only successive sample variances, but *all* pairs of sample variances. Thus we start with the ratio

$$\frac{s_{[k]}^2}{s_{[1]}^2},$$

and compare it to the critical value $c(k,\nu)$ given in Table P.5. If this ratio is significant, then $s_{[k]}^2/s_{[2]}^2$ is tested against $c(k-1,\nu)$, given in the same Table P.5; if this ratio is not significant, then this procedure is terminated and the ratio $s_{[k-1]}^2/s_{[1]}^2$ is compared to $c(k-1,\nu)$, and so on. In general, for the ratio $s_{[a]}^2/s_{[b]}^2$ the critical value is $c(a-b+1,\nu)$. Whenever a ratio is found to be significant, a separation into groups is made.

For example, suppose the results of significance for an example with $k=6$ are as given in Table 12.5.1. A significant ratio is denoted by S and a nonsignificant ratio by NS. All those ratios not given are not significant.

REMARK. It is interesting to note in Table 12.5.1 that in going across any row (i.e., to the right) or down any column, if a ratio is NS then all subsequent ratios are also NS. Hence two of the entries in Table 12.5.1, those corresponding to $(5,3)$ and $(3,2)$, are superfluous; they are included only for visual completeness. □

Table 12.5.1 Results of significance for $k=6$

$\dfrac{s_{[6]}^2}{s_{[1]}^2}$	$S,$	$\dfrac{s_{[6]}^2}{s_{[2]}^2}$	$S,$	$\dfrac{s_{[6]}^2}{s_{[3]}^2}$	$NS,$
$\dfrac{s_{[5]}^2}{s_{[1]}^2}$	$S,$	$\dfrac{s_{[5]}^2}{s_{[2]}^2}$	$S,$	$\dfrac{s_{[5]}^2}{s_{[3]}^2}$	$NS,$
$\dfrac{s_{[4]}^2}{s_{[1]}^2}$	$S,$	$\dfrac{s_{[4]}^2}{s_{[2]}^2}$	$NS,$		
$\dfrac{s_{[3]}^2}{s_{[1]}^2}$	$S,$	$\dfrac{s_{[3]}^2}{s_{[2]}^2}$	$NS,$		
$\dfrac{s_{[2]}^2}{s_{[1]}^2}$	$NS.$				

Looking at the first column, we see that $s_{[1]}^2$ is significantly different from all others except $s_{[2]}^2$. Thus σ_1^2 and σ_2^2 are in a group together (again using the notation that σ_j^2 is the variance of the population that produced the sample variance $s_{[j]}^2$). From the second column we see that σ_2^2 goes in a group with σ_3^2 and σ_4^2; from the third column we note that all the other variances belong in the same group.

Alternatively, we could form the following diagram, where an underline indicates a set of nonsignificant pairs of variances:

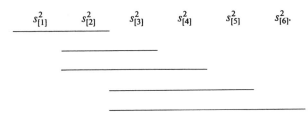

If we remove the redundancy in this diagram, namely, any lines that are proper parts of other lines, then we obtain the same groupings as indicated in the previous paragraph, or

$$s_{[1]}^2 \qquad s_{[2]}^2 \qquad s_{[3]}^2 \qquad s_{[4]}^2 \qquad s_{[5]}^2 \qquad s_{[6]}^2$$

This means that two variances are significantly different only when they are *not* underscored by a common line. In other words, the only significant differences are $(s_{[1]}^2, s_{[i]}^2)(i = 3, 4, 5, 6)$ and $(s_{[2]}^2, s_{[j]}^2)(j = 5, 6)$. The appropriate grouping of population variances under the proposed method is then

$$\left(\sigma_1^2, \sigma_2^2\right), \left(\sigma_2^2, \sigma_3^2, \sigma_4^2\right), \left(\sigma_3^2, \sigma_4^2, \sigma_5^2, \sigma_6^2\right).$$

REMARK.　The reader has to be extremely careful in making any assertion at all (let alone any joint assertion) on the basis of these disjoint or overlapping groups, since we have not controlled any error that would justify or give a statistical basis for the many statements required to describe this separation into groups.　□

12.6　NOTES, REFERENCES, AND REMARKS

The problem of correctly ordering all k populations for $k \geqslant 3$ has been considered for some time, but progress has been very slow. Although the problem is mentioned in Bechhofer (1954) and some tables for completely ordering three populations according to variances appear in Bechhofer and Sobel (1954), it was not until the recent paper of Carroll and Gupta (1975) that serious progress was made on the problem of completely ordering k populations according to means.

It is of some technical interest that for $k = 3$ the probability of a correct selection can still be set up as a single integral if we can use the normal cumulative distribution function $\Phi(x)$ in the integrand without showing it as an integral. However, for $k \geqslant 4$ this is no longer possible, and we need at least a double integral. This makes the problem of calculating tables much more difficult if the basic method of calculation is a standard form of quadrature. Thus it appears that the development of this aspect of ranking and selection will be dependent on the development of efficient computer techniques for carrying out multiple integration.

In the application treated by Carroll and Gupta (1975), the variance is not known, but their procedure will approximately satisfy the (δ^*, P^*) requirement if the pooled estimate is treated as if it were the true value. For larger sample sizes this approach is perfectly valid, but for smaller sample sizes the procedure of Section 12.3 is preferable. The latter procedure depends on a (not too well-known) table calculated fairly early by Freeman, Kuzmack, and Maurice (1967). Although this early date (1967) seems to contradict some of the preceding discussion, a critical look at this table (and the paper) shows computing difficulties. Thus the entries in Table P.2 for the column $k = 6$ are all below those for $k = 5$, even though the entries for $k = 3$, 4, and 5 show an increasing trend for every value of n. The authors of that paper indicated that the entries for $k = 6$ were close to their borderline for accurate results, so that it would be desirable to have an independent check on the entries for $k = 6$.

Section 12.4 on the complete ordering of normal population variances is due to Schafer (1974) and Table P.3 comes from Schafer and Rutemiller (1975).

Section 12.5 is on the separation of populations into groups so that those in the same group have similar variances. The separation into disjoint groups of Section 12.5.1 and into (possibly) overlapping groups of Section 12.5.2 are both due to David (1956). Neither of these precedures offers any control on the final assertion, and hence they are not compatible with most of the procedures in this book. Nevertheless we include them because the goal is still to bring about a ranking of populations. It would be difficult to assess such procedures (or make comparisons with them) until we know more about the overall probability associated with a correct grouping. In addition, it is difficult to see how one might want to use the procedure for separation into (possibly) overlapping groups.

Bayesian methods were applied to the problem of complete ranking by Govindarajulu and Harvey (1974); these procedures are not covered in this book.

PROBLEMS

12.1   *The Birmingham News* (April 7, 1976) reported that experiments have shown that L-dopa (levodopa), the primary drug used for treatment of Parkinson's disease, may also be effective in lengthening the average life span of humans. Researchers at the Memorial Sloan-Kettering Cancer Center in New York ran experiments with 2600 mice and found that certain doses of L-dopa increase the median lifetime of the animals by 50% to 70%. Apparently the drug stimulates the body's system of natural immunity to diseases of aging.

Suppose that the lifetimes are normally distributed (and thus that the mean and median are equal) with known standard deviation $\sigma = 15$ for any dosage level. If you were designing an experiment to rank $k = 6$ different dosage levels with respect to median lifetime, and specified $\delta^* = 0.15$ and $P^* = .90$, how many observations per dosage level would you take?

12.2   In the situation described in problem 5.8, suppose we have five mixed colors and want to rank them all with respect to the variances.

(a)   How many observations should be taken for each color if we specify that the value of P^* is .90 and $\delta^* = 1.5$?

(b)   Suppose the data have already been obtained using a common sample size of $n = 50$. If $\delta^* = 1.3$, what is the minimum probability of a correct ranking in the preference zone where $\delta \geqslant \delta^*$?

CHAPTER 13

Subset Selection (or Elimination) Procedures

13.1 INTRODUCTION

There are many practical situations where it may be better to select more than one population. For example, we may need to divide a number of candidates, such as applicants for a job, into two groups—one that we feel contains the one (or more) best candidates and one that we feel contains mostly inferior candidates. Subsequently, other criteria can be imposed if further screening is necessary. Thus, for example, a first screening for admission to college might eliminate those students with achievement test scores that are below a certain minimum level, or those students with high school grade point averages below a certain minimum standard.

In another context, although there may be a single variable that we want to maximize (e.g., agricultural yield), we may also wish to take into account secondary properties, such as immunity to pests, resistance to drought, and so on. This is especially important when two or more competitors have approximately the same yield. Then we may need to select several populations that have high average yield, and subsequently investigate individually the secondary characteristics (or side effects) of those populations. If only one population were selected on the basis of yield alone and it was later found to have poor secondary characteristics, then our selection rule would not be very practical.

The notion of screening becomes very important when there are a large number k of populations from which we may need to choose a small number, perhaps only one or two. Then we may need a procedure that selects nested subsets in stages, rather than all at once. Such staging is easily carried out with existing tables if it is based on new data at each stage; however this would be inefficient.

In this chapter we introduce a new goal that involves selecting more than one population. Recall that in Chapter 11 we selected a fixed number t of the best populations; here the goal is to select a subset of populations that includes the best population. With our formulation the investigator necessarily withholds judgment about asserting which one population in the selected subset is estimated to be the overall best. The statistical control extends over only this one stage, and hence these procedures could be called one-stage subset selection procedures. Of course the investigator may want to make a decision later about the relative merits of the populations within the subset selected. In such a multiple-stage scheme the probability of a correct *final* decision is not covered by our one-stage formulation and hence it can be influenced by other relevant considerations, such as economic factors, without confounding these with the statistical results; in other words, our formulation provides no statistical control over the *final* decision. The statistical control of multiple-stage procedures (that are not fully sequential) is not considered in this book.

In this one-stage subset selection approach we again have k populations with a particular specified distribution model indexed by some parameter θ, and define the best population as the one with the largest value of θ, denoted by $\theta_{[k]}$. We assume that $\theta_{[k]} > \theta_{[k-1]}$ so that there is only one best population and a correct selection is then well defined (i.e., a subset selection is correct if it contains the best population). The investigator wants a procedure R for selecting a (nonempty) subset of the k populations that is small yet large enough that, with probability at least P^*, the subset will contain the best population (or equivalently, the selection will be correct). Moreover, the investigator wants the procedure R to guarantee that P^* is the minimum probability of a correct selection, regardless of the true values of the parameters $\theta_1, \theta_2, \ldots, \theta_k$. This requirement is referred to as the P^*, or probability, requirement. Here the preassigned P^* specified by the investigator need only satisfy $(.5)^k < P^* < 1$, since the probability $(.5)^k$ can be attained by simply tossing a fair coin for each population to determine whether to include it in the subset or not. However, for practical purposes we recommend that the specified P^* be greater than $.5 + (.5/k)$; this lower bound is always greater than .5 and for k large it is close to .5.

In Sections 13.2 and 13.3 the size S of (i.e., the number of populations in) the (nonempty) selected subset is not fixed; it depends on the data. Consequently, S is a random variable and may take any of the values $1, 2, \ldots, k$. When $S = 1$ the subset consists of only one element, which we designate as best. When $S = k$ the subset consists of all populations, for which we know with certainty (and hence also with confidence level P^*)

that the best one is contained therein. Of course, by taking $S = k$ we can always satisfy the P^* requirement, but the problem is to find the subset of *smallest* (expected) size that satisfies this requirement.

Since the size S of the selected subset is a random variable, we should consider its mean value or expectation $E(S)$; we would like this mean to be small. Although we use the simple notation $E(S)$, the reader should realize that this quantity depends not only on k, P^*, and the common sample size n, but also on the true configuration of $\theta_1, \ldots, \theta_k$. To emphasize the relation to the true configuration we also use the notation $E(S|\theta_1, \ldots, \theta_k)$, which shows explicitly the dependence on the configuration. As the value of P^* diminishes from one, so that the requirement is less stringent, the expected size of the subset will become smaller.

For fixed values of n, k, and P^*, the expectation $E(S|\theta_1, \ldots, \theta_k)$ is regarded as a criterion for measuring the efficiency of any selection procedure that satisfies the P^* requirement. Thus if we have two competitive procedures that satisfy the same P^* requirement, we will use the one that yields a smaller expected subset size for any particular configuration of concern to the investigator. It usually turns out that for two competitive procedures, one is more efficient than another for certain configurations, and less efficient for other configurations. Of course, it would be ideal if one procedure were more efficient than another (or than any other) regardless of the true configuration. Such a procedure would then be called a "uniformly more efficient" (or a "uniformly most efficient") procedure.

REMARK.   The reader may note a loose analogy between this development and that of the power of a statistical test in the classical theory of hypothesis testing.   □

Suppose we have a random sample of observations from each of the k populations. In all the models considered here we compute a certain statistic T for each sample. Let T_j denote the value of T for the observations from the jth population, for $j = 1, 2, \ldots, k$. The ordered values of these statistics are denoted by

$$T_{[1]} \leqslant T_{[2]} \leqslant \cdots \leqslant T_{[k]}.$$

For procedures with a random subset size, the subset selection procedure R is to specify a region I that includes the value $T_{[k]}$, and to include in the selected subset all those populations for which the corresponding T value is included in this specified region I. For the location parameter problem, where best is defined as the largest θ value, I is usually an interval of the form $(T_{[k]} - c, \infty)$ with $c > 0$; for the scale parameter problem, where best is defined as the smallest θ value, the statistic $T_{[1]}$ is positive and the interval

I for containing the smallest parameter is usually of the form $(0, dT_{[1]})$ with $d > 1$.

As we have noted, the present formulation for the selecting-a-subset approach to statistical ranking problems is very well suited to screening problems, because those populations excluded from the selected subset are explicitly designated as being inferior to at least one of those that are included (and implicitly as being inferior to many of those included). In most screening problems we start with a large number of populations, all of which are "potentially best" at the outset. A subset selection procedure would be most useful in reducing the "potentially best" populations from a large group to a small group that can be examined for other characteristics. For example, experiments with anticancer drugs may be aimed at classifying each drug as either active or inactive and exploring further only those drugs that are competitive with the most active drug. Of course, screening problems are frequently sequential, in the sense that the number of populations is reduced in stages. The subset resulting from the first stage is reduced again in the second stage, that resulting from the second stage is reduced in the third and so on, until eventually a single population or small number of populations are selected. We are considering only one-stage reduction procedures here.

One basic difference between the usual approach to selecting a subset of random size and the approach used in all the preceding chapters in this book is that there is no indifference zone and hence no threshold value δ^* that determines the size of the indifference zone in the formulation of the problem. Further, we assume that $\theta_{[k]} > \theta_{[k-1]}$ so that the best population is uniquely defined, and for our probability calculations, we keep this inequality strict as we approach the configuration in which all θ values are equal. As a result, the so-called worst configuration is the limit when all the θ values are equal. For convenience the approach to this limit is made by always keeping one population tagged as being best. Then if we replace the probability of a correct selection by the probability of selecting the tagged population, the resulting function is continuous at the worst configuration where all the θ values are equal. Moreover, this concept of a "tagged population" is useful in probability calculations dealing with the worst configuration, since it eliminates the confusion associated with the approach to the worst configuration. Because of such basic differences it is almost impossible to make meaningful comparisons between the present approach and the indifference zone approach of the previous chapters.

Of course if the investigator must select one population and assert that it is best in accordance with a specified (δ^*, P^*) probability requirement, the indifference zone approach is appropriate for a single-stage procedure and the subset-selection procedure is not. If, on the other hand, the investigator is willing to relax his goal, in the sense that he is not concerned about selecting the one best population as long as he has high confidence that the best one is contained in some subset of reasonably small average size, then the second approach may be more appropriate.

The investigator may want to use the subset-selection approach as part of a scheme to secure the benefit of much early elimination and then test the remaining ones more carefully in order to find the best one. However, the details and theory for such combined multi-stage procedures in general have not been worked out.

REMARKS.   There are several other fundamental differences between these two formulations. The overall feature to remember is that the indifference zone approach is oriented toward designing the experiment and determining a common sample size to be used. The random size subset selection approach is geared toward making a practical elimination scheme with sample sizes that have already been decided on and data that may have already been collected. Here we simply want to screen out the overtly inferior populations, and if the selected subset contains at least two populations, we avoid making any controlled probability statements about comparisons between the populations in the selected subset.

If the selected subset should happen to contain only one population, then we are indeed in luck. On the other hand, if it should contain all the populations, then the procedure gives us no information whatsoever. Of course, such an outcome is an indication that the sample sizes used were woefully inadequate. If we were then to take more observations and combine them with the previous data, this would constitute a multiple-stage procedure and the details for this have not been worked out. On the other hand, if we do not combine such data, then there may be some serious loss of overall efficiency. Hence one should try to have sample sizes that are large enough so that this case where $S = k$ is avoided or at least has small probability of occurrence. Since $k \geqslant 2$, this is automatically accomplished if the sample size n is large enough that the probability that $S = 1$ is close to one.

In the present discussion we merely wish to note that the indifference zone is a useful and practical device either for providing a solution to an otherwise insoluble problem or for reducing the number of observations needed to attain a specified probability of a correct selection (by weakening the definition of a correct selection). The specification of the value δ^* that defines the indifference zone is such that the selection of any population in the indifference zone (instead of the best population) is not regarded as a serious error.

For some problems selecting a subset of size one may actually not be the most desirable result. For example, if there are exactly two "extreme" populations in the positive direction and one is only slightly larger than the other, we might strongly prefer to select both of them rather than one of them, especially if other considerations are pertinent before making any irretrievable decisions. Thus the fact that the size of the subset selected depends on the observations may in general be regarded as a "plus" for subset selection procedures for certain types of problems but certainly not when we have previously decided that we have to select exactly one population. □

In this chapter we consider the selection problem for a subset of random size only for the situation where the sample sizes are fixed, either because the data have already been obtained or because of extrastatistical or cost considerations. As a secondary problem, we could determine the smallest common value of n that satisfies the P^* requirement and also controls (1) the expected subset size for some specified configuration, where $\theta_{[k]} > \theta_{[k-1]}$, or (2) the maximum of the expected subset size, the maximum being taken over all configurations for which the difference $\theta_{[k]} - \theta_{[k-1]}$ is at least a certain specified nonnegative amount.

In Section 13.4 we again consider the problem of selecting a subset that contains the one best population for normal distributions, but here the size of the subset is fixed by the investigator before the experiment. Here we do have an indifference zone and specify a threshold value δ^* for the distance measure $\delta = \theta_{[k]} - \theta_{[k-1]}$. Thus the formulation of this type of subset selection approach is the same as that used in previous chapters, and we need not assume that the sample sizes are fixed in advance.

Finally, in Section 13.5 we present an application of subset selection procedures to the scoring of multiple-choice examinations.

13.2   SELECTING A SUBSET CONTAINING THE BEST POPULATION

In this section we consider the goal of selecting a subset of random size that contains the best population for two different distribution models, namely, the normal distribution with common variance and the binomial (or Bernoulli) distribution. For the former distribution the best population is defined as the one with the largest mean, and for the latter distribution the best population is defined as the one with the largest probability of success.

13.2.1 Normal Means Case

In this subsection the distribution model is that of k normal populations with common known variance σ^2, and the goal is to select a subset of random size that contains the population with the largest mean $\mu_{[k]}$. We assume that the sample sizes are fixed as $n_1, n_2, \ldots, n_k$, and the selection procedure is to be based on the sample means $\bar{x}_1, \bar{x}_2, \ldots, \bar{x}_k$. The largest of these sample means is denoted by $\bar{x}_{[k]}$.

The first step is to specify a quantity P^*, where $(.5)^k < P^* < 1$. Assuming $\mu_{[k]} > \mu_{[k-1]}$, P^* can be interpreted as the joint confidence level for the statements that the μ value for each eliminated population is smaller than the μ value of the best population. We can also say that P^* is the minimum value of the probability that the subset selected contains the population with the largest μ value. For this goal we give only one procedure, denoted by SSN, which stands for Selecting a Subset—Normal distribution case. This procedure applies only to equal sample sizes, that is, $n_1 = n_2 = \cdots = n_k = n$, say, and common known variance, but it is exact for the distribution model assumed.

Procedure SSN. For any $j = 1, 2, \ldots, k$, the jth population is placed in the selected subset if and only if the corresponding sample mean $\bar{x}_j$ is at least as large as a certain quantity, namely,

$$\bar{x}_j \geqslant \bar{x}_{[k]} - \frac{d\sigma}{\sqrt{n}}. \qquad (13.2.1)$$

The value of $d > 0$ is the entry for τ_t in Table A.1 of Appendix A for the given k and specified P^*. Note that the region I for the selection procedure SSN is the closed interval

$$I = \left[\bar{x}_{[k]} - \frac{d\sigma}{\sqrt{n}}, \bar{x}_{[k]} \right],$$

and hence $\bar{x}_{[k]}$ is necessarily within the interval. Thus the subset cannot be empty.

REMARK. Since the sample sizes are assumed to be common here, procedure SSN could also be stated in terms of the sum $\sum_{i=1}^{n} x_{ij}$ of the observations in the jth sample, rather than the mean $\bar{x}_j$. Then (13.2.1) becomes

$$\sum_{i=1}^{n} x_{ij} \geqslant \max_{j} \left(\sum_{i=1}^{n} x_{ij} \right) - d\sigma\sqrt{n}. \qquad \square$$

With this goal a correct selection is made if the subset contains the best population, that is, the one with mean $\mu_{[k]}$ (or any population with mean equal to $\mu_{[k]}$ if the best population is not unique). Then the τ_t in Table A.1 is the smallest value of d in (13.2.1) such that the probability of a correct selection under procedure SSN is at least P^*, that is, $P\{CS|SSN\} \geqslant P^*$, for all possible configurations of μ values.

Example 13.2.1. Suppose we take random samples of $n = 9$ observations from each of $k = 2$ normal populations with common variance $\sigma^2 = 1$. The goal is to select a subset containing the population with the largest mean, and $P^* = .90$ is specified as the confidence level. With only $k = 2$ populations, of course, if the subset selected contains only one population, then we can assert that it is the best population with confidence level .90, and if the subset selected contains both populations, then we cannot make any such assertion. From Table A.1 for $k = 2$ and $P^* = .90$, we obtain $\tau_t = 1.8124$. Substituting the values $d = 1.8124$, $\sigma = 1$, $n = 9$ in (13.2.1), we find that the subset consists of those populations with sample means smaller than $\bar{x}_{[2]}$ by at most $d\sigma/\sqrt{n} = 0.60$. Thus if $\bar{x}_1$ and $\bar{x}_2$ differ by 0.60 or more, then we assert with confidence level .90 that the population giving rise to the larger sample mean has the larger true mean. If the sample means differ by less than 0.60, then we do not have enough observations to make such an assertion at level .90.

To assess the efficiency of procedure SSN we need to determine $E(S|\mu_1,\ldots,\mu_k)$, the expected value of the subset size S, for certain configurations of interest. For the configuration where all μ values are equal, we obtain $E(S|\mu_1 = \cdots = \mu_k) = kP^*$, since each of the k populations has probability P^* of being selected and the expected subset size is the sum of these k probabilities. This result is the maximum value of $E(S|\mu_1,\ldots,\mu_k)$ over all configurations. For example, if $k = 5$ and $P^* = .90$, $E(S|\mu_1,\ldots,\mu_5) \leqslant E(S|\mu_1 = \cdots = \mu_5) = 4.50$ for any configuration. For the configuration where the μ values are equally spaced and each consecutive pair of μ values differs by c/σ, the expected values $E(S)$ are given in Table Q.1 of Appendix Q in terms of $c\sqrt{n}/\sigma$ for selected values of k and P^*. For example, $E(S) = 2.91$ when $c\sqrt{n}/\sigma = 1.0$, $k = 5$, and $P^* = .90$. Note that this result for any fixed c depends on σ and n only through the ratio $\sigma/\sqrt{n}$; thus $\sigma = 1$ and $n = 100$ gives the same result as $\sigma = 2$ and $n = 400$, and the same as $\sigma = 0.1$ and $n = 1$. The value $E(S) = 2.91$ may appear rather high since the maximum is 4.50; however, the common spacing between means is only $1/\sqrt{n}$, and thus 2.91 is not very large for moderate to large values of n. For $k = 5$ and $P^* = .90$, as before, if we keep the same values for σ and c so that the common spacing is the same and multiply n by a factor of 4 (so that $c\sqrt{n}/\sigma$ is multiplied by 2), then $E(S)$ reduces to 1.82; if we multiply n by another factor of 4, then $E(S)$ reduces further to 1.16.

As another numerical illustration, consider the problem in Example 13.2.1, where $k=2$, $P^*=.90$, $n=9$, and $\sigma^2=1$. The maximum value of $E(S)$ is $kP^*=1.8$ for any configuration. For the configuration with μ values equally spaced with common difference $c/\sigma=1.00$, we have $c\sqrt{n}/\sigma=3$ and the expected subset size from Table Q.1 of Appendix Q is $E(S)=1.20$. Note that if $c/\sigma=1.67$ so that $c\sqrt{n}/\sigma=5$, the expected subset size reduces to 1.01, and hence we may expect with high probability to identify the one best population using the subset-selection approach.

13.2.2 Binomial Case

In this section the model is that of k binomial (or Bernoulli) populations, and the goal is to select a subset of random size that contains the best binomial population, where *best* is defined as the one with parameter $p_{[k]}$. We assume that the sample sizes are fixed as $n_1, n_2, \ldots, n_k$, and the selection procedure is to be based on the values of $x_1, x_2, \ldots, x_k$, the corresponding numbers of successes observed in the respective samples.

The first step is to specify a quantity P^*, where $(.5)^k < P^* < 1$. Here P^* can be interpreted as the joint confidence level for the statements that the p value of each eliminated population is smaller than the p value of the best population. We give two selection procedures, denoted by SSB-I and SSB-II (Selecting a Subset-Binomial distribution case). The first procedure, SSB-I, is exact in the sense that it involves binomial theory; however, exact computations have been carried out only for equal sample sizes. This procedure gives a conservative result for all configurations, that is, $P\{CS|SSB\text{-}I\} \geqslant P^*$ for all configurations of p values. The second procedure, SSB-II, is approximate in the sense that it uses the normal distribution after employing an arcsine transformation to stabilize the variance. This procedure can be applied with unequal sample sizes. The probability of a correct selection here may be smaller than P^* for some configurations, but only slightly smaller. This may be due to the inaccuracy of the normal approximation for the n values used and/or to the use of an average common sample size as an approximation in the theory when the n_j are actually unequal.

Procedure SSB-I. For any $j=1, 2, \ldots, k$, the jth population is placed in the selected subset if and only if the corresponding number of successes x_j is at least as large as a certain quantity, namely,

$$x_j \geqslant x_{[k]} - d, \tag{13.2.2}$$

where $x_{[k]}$ is the largest number of successes observed in any of the k samples of common size n. The value of d is given in Table Q.2 of Appendix Q for selected values of k, P^*, and n. Here the interval for the procedure is

$$I = \left[x_{[k]} - d, \ x_{[k]} \right].$$

Procedure SSB-II. For any $j = 1, 2, \ldots, k$, the jth population is placed in the selected subset if and only if a quantity z_j satisfies

$$z_j \geqslant z_{[k]} - d_1, \tag{13.2.3}$$

where the z_j are computed from

$$z_j = \sqrt{n_j + .5} \left[\arcsin\sqrt{\frac{x_j}{n_j + 1}} + \arcsin\sqrt{\frac{x_j + 1}{n_j + 1}} \right], \tag{13.2.4}$$

$\arcsin y$ is found from Table Q.5 in Appendix Q, $z_{[k]}$ is the largest among $z_1, z_2, \ldots, z_k$, and the value of d_1 is given in Table Q.3 of Appendix Q for selected values of k and P^*. Here the interval for the procedure is

$$I = \left[z_{[k]} - d_1, \ z_{[k]} \right].$$

The value of d_1 in procedure SSB-II is not a function of the n_j, since the procedure is based on asymptotic results. Of course, procedure SSB-II can be used as an alternative to procedure SSB-I if the sample sizes are equal, but the results are then only approximate unless the common n is quite large.

Example 13.2.2. Suppose that $n = 50$ observations are taken independently from each of $k = 5$ binomial (or Bernoulli) populations. The investigator specifies a confidence level $P^* = .95$ for selecting a subset containing the best population. From Table Q.2 of Appendix Q, for $k = 5$, $P^* = .95$, and $n = 50$, we find $d = 11$. Thus procedure SSB-I indicates that the populations selected are those for which the number of successes x_j satisfy $x_j \geqslant x_{[5]} - 11$. Now suppose that the respective observed numbers of successes are $x_1 = 32$, $x_2 = 28$, $x_3 = 40$, $x_4 = 35$, and $x_5 = 25$. Then $x_{[5]} = 40$ and the populations selected are those for which $x_j \geqslant 29$, which includes the populations labeled 1, 3, and 4. Accordingly, the investigator asserts with confidence level $P^* = .95$ that one of these three populations is the best one, or equivalently, that the p values of populations 2 and 5, the eliminated populations, are smaller than the largest p value among populations 1, 3, and 4, the retained populations.

Even though an exact result has been obtained for this example in which the sample sizes are equal, we use the same situation to illustrate procedure SSB-II. From Table Q.3 of Appendix Q, for $k = 5$, $P^* = .95$, we find $d_1 = 3.06$. Hence the populations retained are those for which $z_j \geqslant z_{[k]} - 3.06$. The z_j are computed from (13.2.4) as

$$z_j = \sqrt{50.5} \left(\arcsin \sqrt{\frac{x_j}{51}} + \arcsin \sqrt{\frac{x_j + 1}{51}} \right),$$

and the results are

$$z_1 = 13.14, \quad z_2 = 12.00, \quad z_3 = 15.63, \quad z_4 = 14.03, \quad z_5 = 11.16.$$

Hence $z_{[5]} = 15.63$ and the populations selected for the subset are those for which $z_j \geqslant 12.57$, namely those labeled 1, 3, and 4. We do *not* recommend that the user carry out both procedures on the same data as we have done here; we are merely pointing out that if both procedures were used, the two results would usually be the same.

13.3   SELECTING A SUBSET OF POPULATIONS BETTER THAN A CONTROL

A goal related to that of selecting a subset containing the best population is that of selecting a subset of the k populations that includes all populations better than a control population or a fixed standard, say, θ_0. Although we say "better than," we are actually interested in selecting all populations as good as or better than the control; for practical purposes this distinction is inconsequential. A correct selection with this goal is defined as including all populations with θ values as large as or larger than θ_0, or equivalently, as eliminating only populations with θ values smaller than θ_0. This goal is again more flexible than the goal of selecting the one best population in that the investigator chooses a subset of populations but does not specify any ordering within the subset. In this case an empty subset indicates that the control is best or, equivalently, that none of the contenders is better than the control.

With the preceding exception (i.e., the selected subset may be empty) the format for this goal is virtually equivalent to that discussed in Sections 13.1

and 13.2, although the specific procedures are different. Many of the same tables used in the previous section can also be used for this section.

13.3.1   Normal Means Case

The distribution model here is that of k normal populations with common variance σ^2, and the goal is to select a subset of populations that includes all populations with μ values greater than or equal to a control or standard μ value, denoted by μ_0. As in Chapter 10, this reference comparison is called a standard if μ_0 is a known constant, and a control if μ_0 is unknown and is to be estimated by taking an additional random sample from the control population. In the latter case the control population is also assumed to be normally distributed with the same variance σ^2. These two cases require different procedures, which we denote by SSN-Standard and SSN-Control, respectively.

Before proceeding with the present problem, we point out that it is different but related to the problem treated in Chapter 10. There the goal was to separate those populations that are better than the control from those that are worse. There the probability of a correct selection was calculated under the least favorable configuration, which specified roughly one-half of the populations as having a common mean smaller than the control mean and the other half as having a common mean larger than the control mean. Further, we specified two constants, δ_1^* and δ_2^*, in addition to P^*, and hence that problem used an indifference zone approach. In the present formulation we have no indifference zone and need specify only the one value P^* to determine the selection procedure. Further, the sample sizes are assumed given here, and are not determined by prespecified quantities.

The procedure here controls the probability that the selected subset contains all populations with μ values larger than μ_0 regardless of the true configuration of μ values. A correct selection occurs if the subset selected *contains all* those populations better than the control, and the probability of a correct selection is at least P^* for all configurations. Hence the first step is to specify a value for P^*, where $(.5)^k < P^* < 1$. Of course, P^* can also be interpreted as the confidence level for the joint assertion that the μ value of each population eliminated is smaller than μ_0. Clearly, even if the assertion is correct, it does not follow that *each* population selected has a μ value as good as or better than μ_0.

Consider first the problem of selecting a subset containing all populations better than a known standard μ_0 when the common variance σ^2 is known. The subset selection procedure is as follows:

Procedure SSN-Standard (μ_0 Known). For any $j = 1, 2, \ldots, k$, the jth population is placed in the subset if and only if the corresponding sample mean $\bar{x}_j$ satisfies

$$\bar{x}_j \geqslant \mu_0 - \frac{d_1 \sigma}{\sqrt{n}}, \tag{13.3.1}$$

where n is the common sample size and d_1 is the solution to the equation

$$\Phi(d_1) = (P^*)^{1/k}. \tag{13.3.2}$$

In (13.3.2), $\Phi(x)$ is the standard cumulative normal distribution that is tabled in Appendix C. However, Table M.2 of Appendix M gives directly the solution to (13.3.2) for selected values of k and P^*. Note that the value of d_1 depends only on k and P^*, and not on μ_0, σ, or n.

It is worth noting that in the case where the normal populations have different but known variances $\sigma_1^2, \sigma_2^2, \ldots, \sigma_k^2$, and/or the sample sizes are not common, the procedure is very similar. We replace the inequality in (13.3.1) by

$$\bar{x}_j \geqslant \mu_0 - \frac{d_1 \sigma_j}{\sqrt{n_j}}$$

and proceed as before.

If μ_0 is unknown, a sample of n_0 observations is also taken from the control population. Let $\bar{x}_0$ denote the mean of the control sample. The procedure for selecting a subset containing all populations with μ values larger than μ_0 when all sample sizes have the common value n and σ^2 is known is called procedure SSN-Control.

Procedure SSN-Control (μ_0 Unknown). For each $j = 1, 2, \ldots, k$, population j is placed in the subset if and only if the corresponding sample mean $\bar{x}_j$ satisfies

$$\bar{x}_j \geqslant \bar{x}_0 - \frac{d\sigma}{\sqrt{n}} \tag{13.3.3}$$

where d is the entry for τ_t in Table A.1 of Appendix A that corresponds to $k + 1$, that is, the total number of populations sampled, and the specified P^*.

It is worth noting that a procedure similar to procedure SSN-Control can be used if the variances are known but not common and the sample sizes are unequal but the ratios $\sigma_j/\sqrt{n_j}$ are constant for $j=0,1,\ldots,k$. Then we replace (13.3.3) by the inequality

$$\bar{x}_j \geqslant \bar{x}_0 - \frac{d\sigma_j}{\sqrt{n_j}}$$

and proceed as before, and use the same table.

Note that with each of the procedures, SSN-Standard and SSN-Control, the control or standard is excluded if the subset selected is not empty and is included only if all k populations are excluded. In the latter case the conclusion is that none of the populations is better than the control or standard, respectively.

Example 13.3.1.   Producers of the new synthetic automobile engine oils claim that consumers never have to change their oil and that, as a result, oil filters last 24,000 miles rather than the usual 5,000 to 15,000 miles. The producers also claim that synthetic oil prevents the sludge that gums up engines and therefore saves engine wear and gas consumption (although it apparently is not expected to reduce the energy problem, because it requires more materials in its production). Suppose an experiment is to be carried out to compare five different brands of synthetic oil, on the basis of filter lifetime, against the claimed standard of 24,000 miles. Suppose further that the lifetimes are known to be normally distributed with common variance $\sigma^2 = 10,000$ and that the experiment involves making measurements on the filter lifetime for simulated runs of an automobile engine using samples of $n = 25$ from each of the five types of synthetic oil. The filters and engines used are identical for each run, so that any differences in lifetime can be attributed to oil type only. What is the procedure for selecting those types with lifetime more than 24,000 miles if we specify $P^* = .95$? From Table M.2 of Appendix M, the solution to $\Phi(d_1) = (.95)^{1/5}$ is $d_1 = 2.319$. Hence from (13.3.1), the jth type of oil is retained if and only if $\bar{x}_j \geqslant 24,000 - (2.319)\sqrt{10,000}/5$, that is, if and only if $\bar{x}_j \geqslant 23,953.62$ miles.

Example 13.3.2.   Suppose that samples of size $n = 36$ are taken from each of $k = 6$ populations and also from a control population with unknown mean. Suppose all seven populations are normally distributed with common variance $\sigma^2 = 2$, and the sample means are as follows:

$$\bar{x}_0 = 1.74, \quad \bar{x}_1 = 1.10, \quad \bar{x}_2 = 0.90, \quad \bar{x}_3 = 1.97,$$

$$\bar{x}_4 = 2.34, \quad \bar{x}_5 = 0.97, \quad \bar{x}_6 = 0.86.$$

The goal is to select a subset containing all populations better than the control such that we have confidence level .95 in this selection. Procedure SSN-Control applies, since μ_0 is unknown and we have an equal number of observations from all $k + 1$ populations. The entry in Table A.1 of Appendix A for $P^* = .95$ and $k + 1 = 7$ is $\tau_t = 3.2417$. Substituting this value for d in (13.3.3), we get $\bar{x}_j \geqslant 1.74 - (3.2417)\sqrt{2} / \sqrt{36} = 0.98$. The populations in the selected subset are those labeled 1, 3, and 4, a subset of size 3, which is then asserted to contain those populations better than the control.

13.3.2 Binomial Case

Now consider the goal of selecting a subset containing all binomial (or Bernoulli) populations whose p value is larger than (or equal to) a control or standard p value, denoted by p_0. As in the normal means case, we call the reference comparison a standard if p_0 is a known constant, and we call it a control if p_0 is unknown. In the latter case a random sample of size n_0 is taken from the control population and used to estimate p_0. These two cases require different procedures, which we denote by SSB-Standard and SSB-Control, respectively.

The investigator first specifies a value P^*, where $(.5)^k < P^* < 1$. This value P^* can be interpreted as the confidence level for the joint statement that the p value of each population eliminated is smaller than p_0, and also as the minimum probability of a correct selection for all configurations. A correct selection occurs if all the populations with p values as large or larger than p_0 are included in the selected subset. This does *not* imply that *each* population in the selected subset has a p value that is greater than or equal to p_0.

Consider first the case where p_0 is known. The subset selection procedure is as follows:

Procedure SSB-Standard (p_0 Known). For each $j = 1, 2, \ldots, k$, the jth population is placed in the selected subset if and only if the corresponding x_j satisfies

$$x_j \geqslant n_j p_0 - d_2\sqrt{n_j p_0(1 - p_0)} = M_j. \tag{13.3.4}$$

For the special case where the n_j all have the common value n, the M_j are also common (although the common value depends on n and p_0) and equal to, say, M. Table Q.4 of Appendix Q gives the exact value of M for selected values of p_0 and n, k and P^*. The M_j can also be determined approximately by substituting for d_2 in (13.3.4) the solution to the equation

$$\Phi(d_2) = (P^*)^{1/k}, \tag{13.3.5}$$

where $\Phi(x)$ is the standard cumulative normal distribution given in the table in Appendix C. However, in many situations it is easier to use Table M.2 in Appendix M, since it gives directly the solution to (13.3.5) for selected values of k and P^*. As in Chapter 10, the k in (13.3.5) is the number of populations *not* including the standard or control.

For the case of a control population where p_0 is unknown, we let x_0 denote the number of successes in the control sample so that x_0/n_0 is the sample estimate of p_0. Then we have two selection procedures, denoted by SSB-Control-I and SSB-Control-II; the former procedure is exact when $n_0, n_1, \ldots, n_k$ all have the common value n and is approximate otherwise; the latter procedure is always approximate.

Procedure SSB-Control-I (p_0 Unknown). For each $j = 1, 2, \ldots, k$, the jth population is included in the selected subset if and only if

$$\frac{x_j}{n_j} \geqslant \frac{x_0}{n_0} - \frac{d'}{2\sqrt{2}} \sqrt{\frac{1}{n_j} + \frac{1}{n_0}} \,. \tag{13.3.6}$$

If $n_j = n$ for $j = 0, 1, \ldots, k$, and we set $.5d'\sqrt{n}$ equal to d, then (13.3.6) takes the simpler form

$$x_j \geqslant x_0 - d, \tag{13.3.7}$$

where d is given in Table Q.2 of Appendix Q as a function of k, n, and P^*. If the n_j are all large and are not too different from each other, then a simple approximate solution is provided by replacing each n_j in (13.3.6) by $\bar{n}$, the square-mean-root average of the $(k+1)$ sample sizes (equation (2.4.1) of Chapter 2 with k replaced by $(k+1)$), and using for d' the value of τ_t in Table A.1 of Appendix A for the given k and P^*. Then (13.3.6) takes the simpler form

$$x_j \geqslant x_0 - \frac{d'\sqrt{\bar{n}}}{2} \,. \tag{13.3.8}$$

The approximation that leads to (13.3.8) is based on the asymptotic normality of the binomial distribution, and the value of d' here does not depend on $\bar{n}$.

The second procedure gives another approximation for large, not necessarily equal, values of n_j. This approximation also uses the asymptotic normality of the binomial distribution, but it should be an improvement over procedure SSB-Control-I for unequal n_j because it also uses the variance-stabilizing transformation.

Procedure SSB-Control-II (p_0 Unknown). For each $j = 1, 2, \ldots, k$, the jth population is included in the selected subset if and only if

$$z_j \geqslant z_0 - d_1, \tag{13.3.9}$$

where z_j is calculated from (13.2.4) and d_1 is given in Table Q.3 of Appendix Q for selected values of k and P^*. Note that the value of d_1 does not depend on the n values.

With each of the procedures SSB-Standard, SSB-Control-I, and SSB-Control-II, the control or standard is excluded if the subset selected is not empty and is included only if all k populations are excluded. In the latter case the conclusion is that none of the populations is better than the control or standard.

Example 13.3.3. Paulson (1952b) considered the problem where the investigator is interested in determining the effect of treatments on a certain disease where effect is measured by the probability of survival. There are three experimental treatments and a standard treatment for which the probability of survival is known to be about .80. The purpose of the experiment is to select those experimental treatments that are better than the known standard $p_0 = .80$. Suppose that 150 animals are all inoculated with the specific disease under consideration and are randomly divided into three groups of $n = 50$ each and the three experimental treatments are applied to these groups. Each group is given a different experimental treatment. What is the rule for inclusion in the subset if $P^* = .95$? We use procedure SSB-Standard, and since the sample sizes are equal here, the solution is exact. From Table Q.4 of Appendix Q, with $k = 3$, $P^* = .95$, $p_0 = .80$, and $n = 50$, we find $M = 34$. Hence the rule is to retain treatment j if and only if $x_j \geqslant 34$. Alternately, using the approximate method based on (13.3.4) and (13.3.5), we find from Table M.2 of Appendix M for $k = 3$, $P^* = .95$, that $d_2 = B = 2.122$. Then from (13.3.4) the approximate rule is to retain treatment j if and only if $x_j \geqslant 50(.8) - 2.122\sqrt{50(.8)(.2)} = 34.00$. In this case the result of the approximate procedure is identical to that of the exact procedure.

Example 13.3.4. In Example 13.3.3, even though we know the order of magnitude of the p value of the standard treatment, we may wish to consider p_0 as unknown and take observations from a control sample also. Suppose now there are five experimental treatments. Then we would divide 300 animals randomly into six groups of 50 each and give each group a different treatment. We use procedure SSB-Control-I. The value of d for $k = 5$, $P^* = .95$, $n = 50$ from Table Q.2 of Appendix Q is $d = 11$. From (13.3.7) treatment j is retained if and only if $x_j \geqslant x_0 - 11$, where x_0 is the observed number of survivors in the control sample.

13.4   A FIXED SUBSET SIZE APPROACH FOR THE NORMAL DISTRIBUTION

For all the subset selection procedures described in the previous sections of this chapter, the size of the subset was a random variable depending on the outcome of the experiment. Thus in some instances the size of the subset is large, whereas in others it may be small.

For some problems we may wish to have a subset of fixed size so that the number of populations in the subset is fixed, rather than a varying number. This is primarily a question of what the experimenter wants. If he wants a subset of fixed size, there is no point in trying to convince him of the desirability of obtaining any slight increase in efficiency by using a procedure with a random subset size. We would rather give him the best fixed subset size procedure available. In this section we consider the formulation for selecting a subset of fixed size. The model is k normal populations and the goal is to select a subset of fixed size s that contains the (one) best population. We consider this problem only where best is defined as the one with the largest mean.

13.4.1   Selection Procedures Based on Means

The model is k normal populations with unknown means and a common *known* variance σ^2, and ordered means

$$\mu_{[1]} \leqslant \mu_{[2]} \leqslant \cdots \leqslant \mu_{[k]}.$$

The goal is to choose a subset of fixed size f containing the population with the largest mean $\mu_{[k]}$, such that the probability of a correct selection is at least P^* whenever the distance measure is at least δ^*, that is,

$$\mu_{[k]} - \mu_{[k-1]} \geqslant \delta^*.$$

Suppose that the experiment is in the process of being designed, so that the problem is to determine the common sample size n required per population such that the specified (δ^*, P^*) requirement is satisfied. Then for a given k and f, Table Q.6 of Appendix Q gives a value of λ for selected values of P^*. Then the required sample size n is given by

$$n = \left(\frac{\sigma\lambda}{\delta^*}\right)^2. \tag{13.4.1}$$

If the right-hand side of (13.4.1) is not an integer, then we round upward to the next larger integer to find the common value of n.

Example 13.4.1. Suppose we have $k = 6$ normal populations with a common variance $\sigma^2 = 100$. We wish to select a subset of size $f = 3$ containing the best population such that the probability of correct selection is at least $P^* = .95$ whenever $\mu_{[6]} - \mu_{[5]} \geqslant \delta^* = 2.0$. What is the common sample size that should be taken from each population?

We enter Table Q.6 of Appendix Q with $P^* = .95$, $k = 6$, $f = 3$, to obtain the value $\lambda = 1.8659$. Then from (13.4.1), we find that

$$n = \left[\frac{(10)(1.8659)}{2.0} \right]^2 = 87.04.$$

Thus we need to take 88 observations from each of the six populations.

If we make our requirements less stringent, by decreasing P^*, increasing f, or both, we would expect to be able to use a smaller common sample size. For example, if $P^* = .90$ instead of .95, with the same f and δ^*, then $\lambda = 1.4536$, and hence

$$n = \left[\frac{(10)(1.4536)}{2.0} \right]^2 = 52.82.$$

Then we require only $n = 53$ observations per population.

If we keep $P^* = .90$ and now change f from 3 to 4, then $\lambda = 0.9709$, so that

$$n = \left[\frac{(10)(.9709)}{2.0} \right]^2 = 23.57.$$

Now only $n = 24$ observations per population are needed.

Similarly, if δ^* is increased, the required common sample size is decreased since δ^* appears in the denominator of the expression for n in (13.4.1).

To complete the solution to the problem we take n observations from each population, compute the means, and order them as

$$\bar{x}_{[1]} \leqslant \bar{x}_{[2]} \leqslant \cdots \leqslant \bar{x}_{[k]}.$$

We then choose the f largest means and designate these populations as the subset containing the largest mean $\mu_{[k]}$.

As a variant on the preceding procedure, suppose that the sample size was determined from extrastatistical considerations and we are given n observations from each of k populations. The selection rule for a fixed f is the same as before. But what can we say now about the selection procedure?

For the given k and f, Table Q.6 can be used to obtain the (λ, P^*) pairs that are satisfied. For each of these values of λ, we can find the value of δ^*

that corresponds to the given n by solving (13.4.1) for δ^* as

$$\delta^* = \frac{\sigma\lambda}{\sqrt{n}}. \qquad (13.4.2)$$

Then these (δ^*, P^*) pairs can be plotted on a graph to obtain the operating characteristic curve of the selection procedure for the fixed n and f, and known σ.

As another variant on this procedure, suppose that only n is fixed or given. We may still use Table Q.6 to obtain a triplet (f, λ, P^*). If we specify any two members of this triplet, the third can be determined. It may be most natural to consider this problem where δ^* (and therefore λ) and P^* are fixed, and the problem is to find the value of f, the fixed size of the subset to be selected. For example, suppose that $k = 8$, $\sigma^2 = 100$, and $n = 50$ observations are taken from each population. Now let us specify $\delta^* = 2.00$ and $P^* = .95$. The explicit expression for λ from (13.4.1) is

$$\lambda = \frac{\delta^*\sqrt{n}}{\sigma}. \qquad (13.4.3)$$

Substituting the appropriate values in (13.4.3) gives

$$\lambda = \frac{2.00\sqrt{50}}{10} = 1.414.$$

Now from the $P^* = .95$ column of Table Q.6, we find that this λ value falls between $f = 5$ and $f = 6$. We round upward to obtain $f = 6$ as the solution. This value of f may be too large, since choosing a subset of size six from eight populations eliminates only two populations. Then we may decide to relax P^* to, say, .90, in which case $f = 5$. If this value is still too large, then we can make δ^* larger, say, $\delta^* = 3.0$, in which case

$$\lambda = \frac{3.0\sqrt{50}}{10} = 2.121,$$

and then $f = 3$. Of course, it must be remembered that we are paying a price for being able to select a smaller subset, since P^* is smaller or δ^* is larger, or both.

*13.5 APPLICATION OF SUBSET SELECTION TO THE SCORING OF TESTS

In the scoring of tests that provide only multiple-choice answers, a variety of scoring techniques have been used in order to take into account and/or to discourage guessing. In this section we present a new method of scoring that uses some ideas of ranking and selection.

In particular, consider a test consisting of n questions. Each question has k possible answers, of which only one answer is correct. The student (or subject) is instructed to select a subset of answers containing the unique correct answer. If this subset (or answer) contains only one element or choice, the test format is the traditional one. However, if the subset size is not restricted to one, what is to prevent the student from including all k possible answers in his subset, so that the subset always contains the correct answer? We must devise a scoring technique that decreases with the size of the selected subset and is weighted in such a manner that any simple strategy of pure guessing (for example, (1) selecting two answers at random, or (2) tossing a fair coin for each of the k choices to decide whether it should be included or not) has an expected score of zero.

To be more specific, suppose that $k=5$ so that each question has five possible answers, only one of which is correct. The student is instructed to select a subset of the five possible answers that he thinks contains the one correct answer. The method of scoring given in Table 13.5.1 is one set of weights that provides an expected score of zero on each question, regardless of how the guessing is carried out.

To illustrate the main idea, suppose a student guesses the answer using a strategy of choosing two answers at random. Then the probability that the two answers chosen contain the one correct answer is $\frac{2}{5}$, and the probability that these two answers do not contain the correct answer is $\frac{3}{5}$. Consequently, the expected score S per question is

$$E(S)=3\left(\frac{2}{5}\right)-2\left(\frac{3}{5}\right)=0. \tag{13.5.1}$$

If he tosses a fair coin for each of the five choices, the expectation is $+2$ if the correct answer is included and -2 if it is not; hence the overall result is again zero. The same expectation results for any strategy of guessing.

Table 13.5.1 Scoring method for $k=5$ possible answers

Size of subset	Score if subset includes the correct answer	Score if subset does not include the correct answer
0	—	0
1	4	−1
2	3	−2
3	2	−3
4	1	−4
5	0	—

Note that the scoring method of Table 13.5.1 can be obtained by grading each of the five possible answers separately as follows. If the one correct answer is included in the subset, four points are scored; if the correct answer is not included, zero points are scored. From the appropriate number, four or zero, one point is subtracted for each incorrect answer that is also in the subset to give the net score. Note also that this scoring method has large negative scores as well as an expected score of zero for "pure" guessing. These properties have a tendency to keep the student honest.

REMARK 1. This scoring method differs significantly from the system suggested by deFinetti (1972), in which the student is required to designate his own a priori probability distribution of correctness on all the answers. The present scoring appears to measure the ability of the student properly to screen out wrong answers and to include the correct answer without explicitly bringing in his prior (or posterior) probabilities of an answer being correct.   □

REMARK 2. The property that the expected score is zero for pure guessing determines the negative scores for any fixed values (or scoring scheme) given for the positive scores. For example, if the positive score for a subset of size two in Table 13.5.1 were, say, a, then the negative score must be $-2a/3$ so that $a(2/5) - (2a/3)(3/5)$ is equal to zero. In the present discussion no criterion or theory for an optimal scoring method is considered.   □

Now suppose that the student is not necessarily guessing but has probability p of screening out (i.e., not including) a wrong answer and probability p' of including (i.e., not screening out) the correct answer on any question. With $k = 5$ possible answers and the scoring method of Table 13.5.1, the expected score S per question is

$$E(S) = 4pp' - 4(1-p)(1-p') = 4(p + p' - 1) \qquad (13.5.2)$$

out of a maximum of four points. Thus the relative expected score in percent is $100(p + p' - 1)\%$.

We now sketch the derivation of (13.5.2) since it helps explain the ideas. The variance of S is also derived. We have two cases: (1) the right answer is kept, which occurs with probability p', and (2) the right answer is not kept, which occurs with probability $(1-p')$. For case (1) the scoring method gives one point for each wrong answer not included in the subset and there are always exactly four wrong answers; thus the conditional expected gain is $4p$. For case 2, the scoring method takes away one point for each wrong answer included in the subset, and thus the conditional expected gain is $-4(1-p)$ (i.e., a loss of $4(1-p)$). The sum $4pp' - 4(1-p) \times (1-p')$ then yields (13.5.2).

To illustrate the use of (13.5.2), suppose that a student has probability .9 of screening out a wrong answer, and .8 of including a correct answer. Then the expected score is $4(.9+.8-1)=2.8$ per question out of a maximum of 4 points, or 70%.

If $p=p'=1$, then the expected score is 100%; if $p=p'=.5$, then the expected score is zero. The lowest possible score of -100% occurs with $p=p'=0$. Because scores are usually positive, we can make a simple transformation of our scores to achieve positive values. For example, for any score X that ranges between -1 and $+1$, if we make the transformation $Y=(X+1)/2$, or if X is a percentage, $Y=[(X+100)/2]\%$, then the expected score for Y is $(p+p')/2$, which lies between 0 and 1. $\square$

We now derive the mean and variance of S, the score per question, when there are, say, five multiple-choice answers and the scoring method of Table 13.5.1 is used. Let X_0, X_1, X_2, X_3, X_4 denote the five Bernoulli random variables associated with the answers to the five possibilities per question; here X_0 denotes the $0-1$ random variable associated with the single correct answer (taking the value 1 if it is included and 0 otherwise), and the other four X_i are $0-1$ random variables associated with the four remaining (wrong) answers (taking the value 1 if screened out and zero otherwise). Thus each X is equal to 1 for a correct action on any single choice. Assume that these five trials (corresponding to the five answers in each item) are all independent. On any single trial, X_0 has probability p' of success, since p' is the probability of including a correct answer. Each of X_1, X_2, X_3, X_4 has probability p of success, since p is the probability of screening out a wrong answer. The score per question, S, can be written as

$$S = 4X_0 - (1-X_1) - (1-X_2) - (1-X_3) - (1-X_4)$$

$$= 4(X_0 - 1) + X_1 + X_2 + X_3 + X_4. \tag{13.5.3}$$

This follows from the discussion after (13.5.1) or can be verified by looking at each possible occurrence in Table 13.5.1. Since

$$E(X_0) = p', \quad E(X_1) = E(X_2) = E(X_3) = E(X_4) = p,$$

$$V(X_0) = p'(1-p'), \quad V(X_1) = V(X_2) = V(X_3) = V(X_4) = p(1-p),$$

and since S is the linear combination in (13.5.3), we obtain

$$E(S) = 4(p'-1) + 4p = 4(p+p'-1),$$

$$V(S) = 16p'(1-p') + 4p(1-p).$$

Note that S ranges between -4 and 4. If we want to transform S so that its range is from 0 to 1, we define $Y=(S+4)/8$ so that Y ranges between 0 and 1. Then

$$E(Y)=\frac{4+E(S)}{8}=\frac{(p+p')}{2}, \quad V(Y)=\frac{p'(1-p')}{4}+\frac{p(1-p)}{16}. \quad (13.5.4)$$

Now consider a test of n multiple-choice questions, where each question has exactly five choices. Let $\overline{Y}$ be the total average score for all questions, where the scoring method of Table 13.5.1 is employed for each question and Y is the score defined as previously to range between 0 and 1. We may now ask how many questions should be on the test such that the following requirement is satisfied–for $p_0=(p+p')/2$, the probability of getting a total average score $\overline{Y}$ that lies between $p_0-\varepsilon^*$ and $p_0+\varepsilon^*$ is at least $P^*=.95$, say, for some specified ε^*, $0<\varepsilon^*<1$. In other words, we seek the number n of questions that is sufficiently large that the average score $\overline{Y}$ can be regarded as a true reflection of the student's p_0 value.

To answer the question for any given p_0, we use the fact that $\overline{Y}$ has the binomial distribution and for large n the binomial distribution can be well approximated by the normal distribution. In symbols, we want to determine the value of n such that

$$P\left\{p_0-\varepsilon^*<\overline{Y}<p_0+\varepsilon^*\right\}\geqslant P^*. \quad (13.5.5)$$

In order to use the normal distribution the middle term in this inequality must be transformed to the standardized variable $Z=(\overline{Y}-E(\overline{Y}))/\sqrt{V(\overline{Y})}$ Now $E(\overline{Y})=E(Y)$ and $V(\overline{Y})=V(Y)/n$, where Y is the total score per question, and $E(Y)$ and $V(Y)$ are given in (13.5.4). Thus we know that

$$E(\overline{Y})=\frac{p+p'}{2}=p_0, \quad V(\overline{Y})=\frac{1}{n}\left[\frac{p(1-p)}{16}+\frac{p'(1-p')}{4}\right]. \quad (13.5.6)$$

Since $E(\overline{Y})=p_0$, the standardized form of (13.5.5) is

$$P\left\{\frac{-\varepsilon^*}{\sqrt{V(\overline{Y})}}<\frac{\overline{Y}-E(\overline{Y})}{\sqrt{V(\overline{Y})}}<\frac{\varepsilon^*}{\sqrt{V(\overline{Y})}}\right\}\geqslant P^*, \quad (13.5.7)$$

where $V(\overline{Y})$ is calculated from (13.5.6). In order to solve for n we set the right-hand end point of the interval in (13.5.7) equal to the point on the standard normal curve that has a right-tail probability of $(1-P^*)/2$, for example, equal to 1.96 if $P^*=.95$.

To illustrate the calculation of n consider an example where $p=.7$ and $p'=.9$ so that $p_0=.8$, and we specify $\varepsilon^*=.05$ and $P^*=.95$. Then from

(13.5.6) we have

$$V(\overline{Y}) = \frac{1}{n}\left[\frac{(.7)(.3)}{4} + \frac{(.9)(.1)}{16}\right] = \frac{.035625}{n}, \qquad \sqrt{V(\overline{Y})} = \frac{.189}{\sqrt{n}},$$

and the right-hand end point of (13.5.7) is

$$\frac{\varepsilon^*}{\sqrt{V(\overline{Y})}} = \frac{.05}{.189/\sqrt{n}} = .265\sqrt{n}.$$

Setting $.265\sqrt{n} = 1.96$, we obtain $n = 54.7$, which is rounded upward to $n = 55$. Thus the multiple-choice test should consist of at least 55 questions in order to satisfy the given (ε^*, P^*) requirement.

REMARK 3.   Since each score per question S is the sum of five binomial variables, and there are 55 questions, computing $\overline{Y}$ is equivalent to adding 275 independent binomial variables. For such a large number of variables, the normal approximation to the binomial distribution gives a very good approximation. If ε^* were increased to $.1$, say, then (13.5.7) becomes

$$P\{-.529\sqrt{n} < Z < .529\sqrt{n}\} \geqslant .95,$$

in which case $n = 13.7$. Then we need a test with at least 14 items. Even here we are adding $5(14) = 70$ binomial variates and the normal approximation yields accurate results.   □

REMARK 4.   In the previous development we started with a p_0 value and required a probability of at least P^* that the $\overline{Y}$ score is between $p_0 - \varepsilon^*$ and $p_0 + \varepsilon^*$. The result depended on the given p_0. We may, however, determine the sample size regardless of the p_0. Obviously, the n value will be conservative (i.e., larger than what would result for any given p_0 value). This general result can be found by making $V(\overline{Y})$ as large as possible. The maximum value of $V(\overline{Y})$ in (13.5.6) is achieved for $p = p' = 1/2$, in which case $V(\overline{Y}) = 5/(64n)$. With $P^* = .95$, we set $\varepsilon^*/\sqrt{V(\overline{Y})} = 1.96$, so that

$$n = \frac{(1.96)^2 5}{64\varepsilon^{*2}} = \frac{.300}{\varepsilon^{*2}}.$$

When $\varepsilon^* = .05$ we obtain $n = 120.0$, which is to be compared with the value $n = 54.7$ that resulted when we assumed the given value $p_0 = .8$.   □

It is of interest to compare the preceding results with the usual method of scoring in which only one choice can be selected as an answer for each question. In treating the case $k = 5$ (as we have done earlier), the score S_1

per question would be 4 if correct and -1 if incorrect. We use p_0 to denote the student's probability of answering any one question correctly. The quantity $Y = (S_1 + 1)/5$ transforms the score to the interval $(0, 1)$, and we are interested in the average $\overline{Y}$ of the Y values. In this case

$$E(S_1) = 5p_0 - 1 \quad \text{and} \quad E(Y) = p_0.$$

Note that we do not use p and p' in this formulation. For the variance we obtain

$$V(S_1) = 25p_0(1 - p_0) \quad \text{and} \quad V(Y) = p_0(1 - p_0).$$

Then we want the smallest integer n such that

$$P\left\{ \frac{-\varepsilon^* \sqrt{n}}{\sqrt{p_0(1 - p_0)}} < \frac{(\overline{Y} - p_0)\sqrt{n}}{\sqrt{p_0(1 - p_0)}} < \frac{\varepsilon^* \sqrt{n}}{\sqrt{p_0(1 - p_0)}} \right\} \geqslant P^*.$$

Taking the same entries as before, namely, $p_0 = .8$, $\varepsilon^* = .05$, and $P^* = .95$, we obtain $n = 247.24$ (as opposed to the value 54.7 obtained for subset scoring). More generally for any ε^* and P^* and $p = .7$, $p' = .9$ (so that $p_0 = .8$), the ratio of the sample sizes is 4.52 to 1, with the subset scoring method requiring the smaller number of questions. Even if we allow p_0 to be near .5, as in Remark 4, this ratio is still 3.2 to 1 with the subset scoring on the lower side. Indeed if p and p' both approach p_0 the same ratio $\frac{16}{5} = 3.2$ is obtained for any p_0. The explanation for these results lies in the fact that under the subset-scoring scheme we are observing five random variables for each question (i.e., we make use of the result on each choice as a random variable) as opposed to the usual scheme where each question corresponds to one random variable only.

REMARK 5.   The ideas of this section could easily be extended to different types of judging contests, such as beauty contests, athletic contests, and so on, and we could even allow the judges to come up with ordered subsets. The problem would get even more interesting if there was more than one correct answer (i.e., ties for first place); details of such problems have yet to be worked out. □

13.6 NOTES, REFERENCES, AND REMARKS

The area of subset selection procedures (which is equivalent to the idea of elimination) is well developed. Here the theory is different; no indifference zone is usually brought to bear and the orientation is toward working with data already collected, rather than toward determining a sample size for designing the experiment. Since the methods have different goals, different input, and so on, it is very difficult to make any meaningful comparisons, and any comparisons that are made should be questioned. One usually does not criticize a fixed sample size procedure because a more efficient sequential procedure is available even when the two procedures test the same hypothesis; if the two procedures test *different* hypotheses, there is no point in making a comparison.

The original papers for subset selection are Gupta (1956, 1965). Note that the first illustrations in Section 13.2.1 all utilize Table A.1 of Appendix A. The values of $E(S)$ given in Table Q.1 are obtained from Gupta (1965).

It is interesting to note that for some problems the table needed for a subset selection approach is exactly the same (usually rearranged with table entries and captions interchanged) as the table needed for the corresponding approach of selecting the one best population. This occurs for the normal means problem and also for the normal variances problem, but is not a general phenomenon.

The corresponding subset selection problem for binomial populations is treated in Gupta and Sobel (1960), and several of our tables, such as Q.3 and Q.4, associated with the binomial populations come from Gupta, Huyett, and Sobel (1957). The other problems treated here for selecting a subset of populations better than a control generally do not require new tables.

The theory and tables for the problem in Section 13.4 of selecting a fixed size subset of normal populations with a common known variance are in Desu and Sobel (1968). This paper is related to an earlier paper of Desu (Mahamunulu, (1967)) for selecting the s best populations, since the two problems are inverse to each other. More extensive goals are also considered there (e.g., selecting s populations and asserting that these s contain the t best populations, or in the dual problem that these s are contained among the t best populations). In another paper Sobel (1969) considers the problem of selecting s populations and asserting that they contain at least one of the t best populations.

The original problem of Gupta (1956) has been abstracted and generalized; the principal references here are Deverman (1969), Deverman and Gupta (1969), Gupta and Panchapakesan (1972). A procedure that controls the probability of eliminating those populations that are distinctly inferior is treated by Desu (1970), and a similar idea is used for eliminating a "strictly non-t-best" population in Carroll, Gupta, and Huang (1975). Two-stage procedures for the subset selection problem in the case of normal distributions with unknown (not necessarily equal) variances are given in Dudewicz and Dalal (1975). A different idea that has been used in subset selection procedures for binomial populations is to condition on the number of successes obtained in a fixed number of trials (see Gupta and Nagel (1971), Gupta, Huang, and Huang (1976)).

The nonparametric analogue of the problem of selecting a fixed size subset containing the population with the largest α-quantile is treated by Desu and Sobel (1971). Unfortunately the tables are not very extensive and we have included only the case of the median in this book.

Goel (1972) uses inverse sampling for selecting a subset of Poisson populations containing the best one.

The use of subset selection procedures to select (or eliminate) regression variables is described by McCabe and Arvesen (1974).

Section 13.5 deals with a new application of subset selection to the scoring of tests. It was somewhat of a surprise to the authors that this model is three or four times more efficient in scoring tests than the usual method of scoring the result of each multiple-choice question as being either correct or wrong.

PROBLEMS

13.1 A market research organization plans to investigate the effectiveness of various types of advertising of a new household product. Six cities that are relatively homogeneous in population are selected for test areas. In one city (city 0) the product is not advertised at all. The types of advertising campaign used in each of the other five cities are as follows:

City	Type of advertising
1	Free sample placed in mailbox
2	Mail advertising with cents-off coupon
3	Promotional display in supermarkets
4	Outdoor billboard advertising
5	Newspaper advertising with cents-off coupon

Two months after the advertising campaign, random samples of $n = 400$ persons are selected from each of the six cities and interviewed to determine whether they have purchased the product. The numbers of persons claiming they purchased the product are shown here.

City j	x_j
0	128
1	195
2	189
3	170
4	125
5	132

Determine for $P^* = .95$ which types of advertising campaign are more effective than the control campaign with no advertising.

13.2   A manufacturer of electric irons is considering which of six different suppliers of thermostats to use. Although the manufacturer would like complete accuracy between the setting on the iron and the actual temperature, the consumers may object more to an actual temperature higher than the setting than they will to one that is lower, since too high a temperature may ruin the fabric when the iron is applied. As a result, the problem is to determine which suppliers should not be considered because their thermostats provide an actual temperature that is greater than the setting on the iron. Samples of 25 thermostats are taken from each supplier and the irons are tested at the 500°F setting. The actual mean temperatures, measured by a thermocouple, are as follows:

Supplier j	$\bar{x}_j$
1	514.4
2	512.3
3	509.1
4	498.6
5	499.2
6	502.3

Assuming that the actual temperatures for each supplier are normally distributed with common variance $\sigma^2 = 150$, what subset of suppliers should be eliminated from consideration since their actual temperature is larger than the standard of 500°F?

13.3   A medical research worker plans an experiment to see which of three new types of muscle relaxant are more effective than the standard treatment. It is known that 60% of patients suffering from a particular neurological disorder obtain beneficial results from the standard treatment. A sample of 75 patients suffering from this neurological disorder is divided randomly into three groups of size 25 each for treatment with the three different muscle relaxants. The respective numbers of patients claiming beneficial results after a three-month treatment period are 10, 15, and 10. Which treatments are better than the standard for $P^* = .90$? What if $P^* = .95$?

13.4   Gatewood and Perloff (1973) reported an experiment to evaluate and compare three different methods of presenting information about food prices to consumers in a supermarket. The criteria for evaluating these methods were the speed and accuracy with which the consumer could process the information given to choose the lowest-priced package size within each of nine different product groups. The three methods of presentation were as follows.

I. The current supermarket method (given only total price and net weight).

II. The current supermarket method, but each subject was also given a computational device (a small wheel that divides total price by net weight) to aid in calculating the unit price.

III. The current supermarket method but with unit price also displayed.

Seventy-five subjects were randomly divided into three groups of equal size, one group to be used for each presentation method. In a simulated supermarket setting the members of each group were instructed to choose the most economical package for each of the nine product groups. The means and standard deviations for number of correct choices, and also for number of minutes required to complete their choices, are shown here.

	Number of correct choices		
Method j	I	II	III
$\bar{x}_j$	5.72	5.96	8.04
s_j	1.31	1.57	0.45

	Number of minutes required		
Method j	I	II	III
$\bar{x}_j$	23.93	31.72	3.60
s_j	10.00	9.57	1.11

Assume that the number of correct choices for each method follows the normal distribution with common known variance $\sigma^2 = 1.50$.

(a) Find the procedure for selecting a subset of the three methods that contains the most accurate one if $P^* = .90$, and make the selection.

(b) Determine the expected subset size for the configuration with μ values equally spaced with common difference $c/\sigma = 1.00$.

(c) Regard method I as a control population. Find the procedure for selecting a subset of the other two methods that contains those methods that are better than method I if $P^* = .90$, and make the selection.

13.5   The effectiveness of direct mail advertising is influenced by many factors, including the color of the material. An early study by Birren (1945) indicated that the best colors were shades of the three primary colors, red, yellow, and blue. A later study by Dunlap (1950) compared cherry red, yellow, and blue with white as a control. The experiment consisted of sending a card to all 572 members of the Kansas State Alumni Association to notify them that their membership was about to expire and should be renewed. All cards used black print, and the number of cards sent of each color was approximately the same. Only one mailing was used. The actual data are given here.

Color	White	Yellow	Blue	Red
Number sent	147	144	141	140
Number returned	60	73	65	54

(*a*)   What is the procedure for selecting a subset containing all colors that evoke a better response than the control color of white if we specify $P^* = .90$? Which colors are selected?

(*b*)   If we do not regard white as a control, what is the procedure for selecting a subset containing the color that evokes the best response if $P^* = .90$? What colors are selected?

(*c*)   Since the yellow color evoked the best response here, consider a similar experiment where five shades of yellow are to be compared with the color white as a control and the goal is to select a subset containing all shades better than the control color in terms of promoting a response. For the following data, what is the selection procedure for $P^* = .90$? What shades are selected?

Color	Number sent	Number returned
White	25	18
Shade 1 (palest)	25	17
Shade 2	25	18
Shade 3	25	16
Shade 4	25	15
Shade 5 (darkest)	25	10

(*d*)   Now consider another similar experiment where only the five shades of yellow are to be compared (no white cards are sent), and the goal is to select a subset containing the shade that evokes the best response. For the following data what is the selection procedure for $P^* = .90$? What shades are selected?

Shade	Number sent	Number returned
1	30	18
2	30	22
3	30	20
4	30	14
5	30	12

13.6 The standard typewriter keyboard, designed by a Milwaukee printer named C. L. Sholes in 1873, reflects an effort to divide the most frequently used letters about equally among the four quadrants of the keyboard, because at that time adjacent keys often jammed. These mechanical problems of typewriters are now essentially eliminated. Nevertheless the standard keyboard remains unchanged. A very large number of keyboard arrangements, including a sequential alphabetical arrangement, have been considered over the years as alternatives. Some experiments have shown that typists can double their speed on a simplified keyboard that places the more common letters (such as a, s, e) in more strategic locations. Of course, it would be difficult for an experienced typist to become accustomed to a new keyboard. Consider an experiment to determine which of eight different keyboard arrangements are better than the standard arrangement for persons who have no typing skill and no knowledge of the standard keyboard and hence are forced to use the "hunt and peck" system. A group of 225 such persons will be divided randomly into nine groups of equal size, and each person will be evaluated on typing performance (speed and accuracy) using the keyboard assigned. Assume that the evaluation scores are normally distributed with common variance $\sigma^2 = 20$. What is the selection procedure if we specify $P^* = .90$?

13.7 In the situation of problem 13.6 consider an experiment where the goal is to select a fixed-size subset of the eight alternative keyboards that contains the keyboard that produces the best evaluation score on typing performance. Suppose that the scores are normally distributed with $\sigma^2 = 20$.

(a) If the subset size is $f = 2$, how many observations are needed on each keyboard if we specify $\delta^* = 1.5$, $P^* = .90$?

(b) What if the subset size is $f = 3$?

CHAPTER 14

Selecting the Best Gamma Populations

14.1 INTRODUCTION

The ranking and selection problems in this chapter apply to populations that follow the gamma distribution. The gamma distribution model is used widely in many fields of applications, particularly in engineering and the social and physical sciences. Since the gamma distribution is seldom discussed in elementary textbooks, we provide some background material about the distribution itself before proceeding to a discussion about the ranking and selection procedures for this model.

The gamma distribution is a family of continuous distributions with density function given by

$$f(x) = \frac{1}{\theta^r \Gamma(r)} x^{r-1} e^{-x/\theta} \qquad \text{for} \quad x \geqslant 0, \qquad (14.1.1)$$

where $\Gamma(r)$ denotes the so-called gamma function. If r is an integer, $\Gamma(r)$ can be computed as $\Gamma(r) = (r-1)!$. For any value of r we have the relation $\Gamma(r+1) = r\Gamma(r)$, so that $\Gamma(r)$ needs to be tabulated only for values of r between 0 and 1. One value of particular interest is $\Gamma(1/2) = \sqrt{\pi} = \sqrt{3.1416\ldots} = 1.772\ldots$, which then gives the answer for all half-integers. Tables of the gamma function $\Gamma(r)$ are widely available.

The gamma distribution is indexed by two parameters, denoted here by θ and r. Both θ and r must be positive numbers. The parameter θ is called the *scale parameter*; the quantity $2r$ is called the *degrees of freedom*. We do not assume that r or $2r$ is an integer. The mean and variance of a random variable X that follows the gamma distribution with parameters r and θ are

$$\mu_x = r\theta, \qquad \sigma_x^2 = r\theta^2.$$

Note that r is unitless and that θ has the same units as x.

The gamma distribution with $\theta = 1$ is sometimes called the *standardized gamma distribution*. Figure 14.1.1 provides graphs of the standardized gamma density function for selected values of r. Figure 14.1.2 shows how the graph of the gamma density function for $r = 2$ changes with different values of θ.

One special member of the gamma distribution family has already been introduced in this book. This is the chi-square distribution with ν degrees

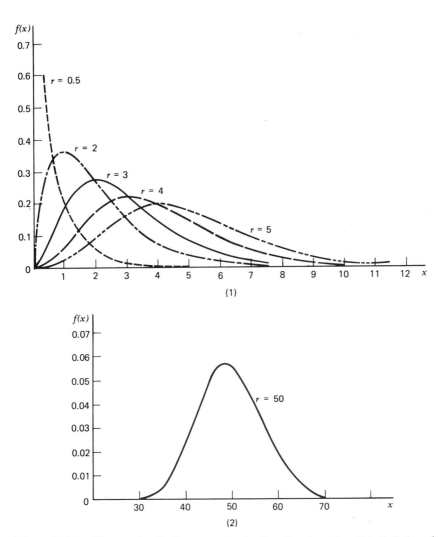

Figure 14.1.1   The standardized gamma density function for (1) $r = 0.5$, 2, 3, 4, and 5; (2) $r = 50$.

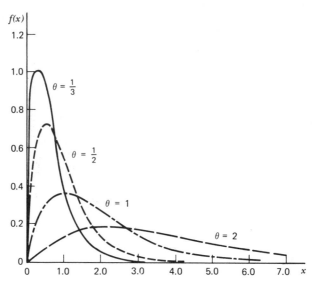

Figure 14.1.2   Graphs of the gamma density function for $r=2$ and $\theta = \frac{1}{3}, \frac{1}{2}$, 1, and 2.

of freedom. In other words, if we let $r = \nu/2$ and $\theta = 2$ in the general form of the distribution given in (14.1.1), the result is the chi-square distribution.

Another special case of the gamma distribution results when $r = 1$. This distribution is so important that it has a special name, the *exponential distribution*. Since r is a constant the exponential distribution is indexed by the single parameter $\theta > 0$; the mean is θ and the variance is θ^2. Figure 14.1.3 gives graphs of the exponential density function for selected values of θ. When the exponential distribution is used as the law governing time to failure (or disrepair), it is also customary to use the symbol $\lambda = 1/\theta$, since λ represents the rate at which failures occur in the so-called Poisson process.

Before continuing our discussion we present some specific examples of applications where the gamma distribution (with particular values of θ and r) is the appropriate distribution model.

Example 14.1.1 (Poisson Process).   The most common theoretical application of the gamma distribution arises in the situation where an (unlimited) sequence of events occurs in time in accordance with the Poisson process at the average rate of $\lambda = 1/\theta$ events per unit time. If the variable X represents the waiting time to the rth occurrence of the event, then X follows the gamma distribution with parameters θ and r. In particular, the density of the waiting time to the first occurrence of the event follows the exponential distribution with parameter θ.

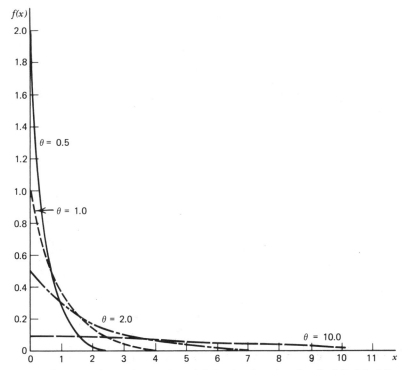

Figure 14.1.3   Graphs of the exponential density function for $\theta = 0.5$, 1.0, 2.0, and 10.0.

The application in the preceding paragraph is justified on a theoretical basis using the Poisson process for the occurrence of the events. There are many other examples where applications of the gamma distribution are justified on an empirical basis, that is, the gamma distribution simply provides a good fit to a set of data. In these examples we continue to refer to the parameter $2r$ as degrees of freedom without requiring that r or $2r$ be an integer.

Example 14.1.2 (Time between failures).   Proschan (1963) provides records of the duration of time (in units of hours) between successive failures of the air conditioning system of each member of a fleet of 13 Boeing 720 jet airplanes. A frequency distribution of these data shows that the exponential distribution (i.e., the gamma distribution with $r = 1$) with $\lambda = 1/\theta = 0.0107$ (or $\theta = 93.14$) provides a good fit. Figure 14.1.4 shows the agreement between this fitted distribution and the histogram representing the frequency distribution of the data. Note that since θ is the mean duration of time between successive failures in this application (sometimes denoted

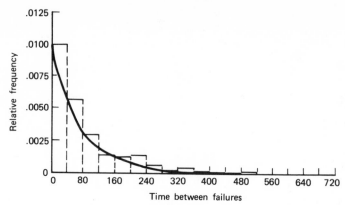

Figure 14.1.4 Histogram showing relative frequency distribution for data representing durations of time between successive failures of the air conditioning systems of each member of a fleet of 13 Boeing 720 jet airplanes.

by MTBF), a large value of θ would indicate that successive failures occur infrequently.

Example 14.1.3 (Interarrival Times). Both psychologists and biophysicists have been interested in the time between successive electrical impulses received by measuring devices implanted in the spinal cords of various mammals (mice, cats, monkeys, and the like). Figure 14.1.5 shows a frequency distribution of interresponse times in the spontaneous activity of a single spinal interneurone. The frequency distribution is based on 391 intervals recorded by a micropipette inserted in the spinal cord of a cat.

Example 14.1.4 (Psychology). Psychologists have studied the length of time between successive presses of a bar by a rat placed in a Skinner box, when a certain reinforcement schedule for operant conditioning has been applied to modify the rat's behavior. Mueller (1950) found that an exponential distribution with parameter $\theta = 20$ provides a good fit to the data.

Example 14.1.5 (Lifetimes). Wilk, Gnanadesikan, and Huyett (1962) fit a gamma distribution to failure time (in weeks) observed for 31 transistors in an accelerated life test (actually 34 transistors were under observation, but three survived the entire period of observation). Here a fit to the data yielded $\theta = 9.01$ and $r = 1.74$. Note that in this experiment θ is proportional to the time to failure (when r is considered as fixed). Thus a large value of θ would indicate a large time to failure or a long expected lifetime.

In all the ranking and selection problems under consideration here, we have k gamma populations, each with the *same* known value of r, but with

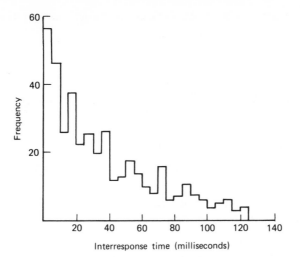

Figure 14.1.5 Frequency distribution of interresponse times in the spontaneous activity of a single spinal interneurone. The distribution consists of 391 intervals recorded by a micropipette inserted in the spinal cord of a cat. Reproduced from W. J. McGill (1963), Stochastic latency mechanism, in R. D. Luce, R. R. Bush, and E. Galanter, Eds., *Handbook of Mathematical Psychology*, John Wiley, New York, with permission.

possibly different values of the scale parameter θ. For example, in the situation of Example 14.1.2 we might have different types of air conditioning systems; in the situation of Example 14.1.4 we might have different reinforcement schedules. We let θ_j denote the scale parameter for the jth population. Then the ordered θ values are denoted by

$$\theta_{[1]} \leqslant \theta_{[2]} \leqslant \cdots \leqslant \theta_{[k]}.$$

The following three different goals are covered in this chapter: (1) selecting the gamma population with the largest θ value, (2) selecting the gamma population with the smallest θ value, and (3) selecting a subset containing the gamma population with the largest θ value.

14.2   SELECTING THE GAMMA POPULATION WITH THE LARGEST SCALE PARAMETER

In this section the goal is to identify the gamma population with the largest value of θ, namely, $\theta_{[k]}$. Note that because r is the same for each population, choosing the population with the largest θ value is equivalent to selecting the population with the largest mean.

The distance measure that we use for this problem is the ratio of the two largest θ values, that is,

$$\delta = \frac{\theta_{[k]}}{\theta_{[k-1]}}. \tag{14.2.1}$$

Note that since $\theta_{[k]} \geqslant \theta_{[k-1]}$, δ is always greater than or equal to 1. The generalized least favorable configuration is

$$\theta_{[1]} = \theta_{[2]} = \cdots = \theta_{[k-1]} \quad \text{and} \quad \theta_{[k]} = \delta\theta_{[k-1]}, \tag{14.2.2}$$

and the least favorable configuration is the same as (14.2.2) with δ replaced by δ^*.

The selection procedure is to compute the mean of each set of sample data and order them from smallest to largest as

$$\bar{x}_{[1]} \leqslant \bar{x}_{[2]} \leqslant \cdots \leqslant \bar{x}_{[k]}.$$

For r known, the sample estimate of θ_j is $\bar{x}_j/r$, the arithmetic mean of the jth sample divided by r. But since r is common, the ordering of the $\bar{x}_j/r$ is the same as the ordering of the $\bar{x}_j$. As a result, the selection rule is to assert that the population that generated the largest sample mean $\bar{x}_{[k]}$ is the one with the largest population mean, or equivalently, is the one with the largest scale parameter, or is the one with θ value $\theta_{[k]}$.

Suppose that the experiment is in the process of being designed and the problem is to determine the smallest common sample size n such that if a random sample of n observations is drawn from each of the gamma distributions with common known parameter r ($2r$ degrees of freedom), the best population, that is, the one with the largest scale parameter $\theta_{[k]}$, can be selected with probability at least P^* whenever $\delta \geqslant \delta^*$. Here P^*, where $1/k < P^* < 1$, and δ^*, where $\delta^* > 1$, are preassigned numbers. Then Table R.1 in Appendix R gives the value of $M = nr$ for the appropriate value of k, and the value of n is found by dividing this result by the known value of r. In other words, we need $n = M/r$ observations per gamma population where M is the entry in Table R.1. (When $r = 1$ we have the exponential distribution and $n = M$.)

REMARK.   The determination of $M = nr$ is based on the following well-known fact about the gamma distribution. If U and V are independent random variables, each having the gamma distribution with the same scale parameter θ, and degrees of freedom a and b, respectively, then the random variable $(U + V)$ also has the gamma distribution with the same scale parameter θ and degrees of freedom equal to $(a + b)$. Thus if we have a random sample of size n from a gamma distribution with scale parameter

θ and $2r$ degrees of freedom, the sum of these random variables also has a gamma distribution with scale parameter θ and $2rn$ degrees of freedom.

□

Example 14.2.1.   Consider a modification of Example 14.1.5 where $k=4$ brands of transistors are to be subjected to a life test. In each test the time to failure is measured. A longer time to failure means that the transistor has better performance. Suppose that we know from previous experience that the distribution of times to failure is gamma with degrees of freedom $2r=3.5$ (or parameter $r=1.75$) for this type of data, but the θ values may differ. How large a sample of each type of transistor should be put into the test if we specify $P^*=.95$ and $\delta^*=2.00$? From Table R.1 for $k=4$, we find $M=18$. Thus $n=18/r=18/1.75=10.3$. We round upward to get a requirement $n=11$ gamma observations from each brand of transistor.

In the same situation suppose that we have more than 11 transistors of each of the four brands—say, N transistors of each brand—and want to subject them *all* to a life test. Then for each brand of transistor we can wait until $f=11$ failures occur and estimate the mean time to failure for the jth brand by

$$y_j = \frac{x_{(1)j}+x_{(2)j}+\cdots+x_{(f)j}+(N-f)x_{(f)j}}{f},$$

where $x_{(1)j} \leqslant x_{(2)j} \leqslant \cdots \leqslant x_{(f)j}$ are the order statistics for the times of the first f failures, all measured from time zero. Since f is a positive integer, the preceding statistic y_j makes sense; if we were sampling for a fixed time period, it would not make sense, partly because we could end up with zero failures, in which case y_j would be undefined. Then we order the y values for the four brands as

$$y_{[1]} \leqslant y_{[2]} \leqslant y_{[3]} \leqslant y_{[4]},$$

and assert that the brand that produced $y_{[4]}$ is the one with the longest lifetime.

The reason for stopping after $f=11$ failures of each type, instead of waiting until all N fail, is of course to save time and money. Thus one may be interested in computing the expected time it takes to obtain $f=11$ failures from each of the four brands, but this is a separate problem that we do not discuss here.

Example 14.2.2.   As explained in Example 14.1.1, the waiting time to the rth event in a series of events occurring in a Poisson process with a rate of $\lambda=1/\theta$ follows the gamma distribution with parameters θ and r.

Suppose we have $k = 5$ toll collection stations at different points on a turnpike (not necessarily in series) and automobiles arrive at them in accordance with a Poisson process but with rate $\lambda_j = 1/\theta_j$ for the jth station. The problem is to determine the sample size needed at each toll station to select the toll station that will have the longest average wait to collect the $r = 100$th toll. This means selecting the population with the largest θ (or smallest λ) value where $r = 100$. Suppose we specify $P^* = .90$ and $\delta^* = 1.10$. Then Table R.1 indicates that $M = 745$, so that $n = \frac{745}{100} = 7.45$, and we need eight gamma observations at each of the $k = 5$ toll stations. Recall that each gamma observation indicates the collection of 100 tolls. Thus if the observations are collected in batches of 100 tolls, the first toll station that completes eight batches of 100 collections is selected. Alternatively, we can say that we need 745 exponential observations at each of the $k = 5$ stations. Then if the observations are not collected in batches of size 100, we can select the first toll station that collects its 745th toll.

14.3 SELECTING THE GAMMA POPULATION WITH THE SMALLEST SCALE PARAMETER

The problem of selecting the gamma population with the smallest scale parameter θ is related to the problem of selecting the normal population with the smallest variance (discussed in Chapter 5). This follows because if we have a random sample of size n from a population that is normal with mean μ and variance σ^2, then the sample variance s^2 (computed with denominator $n - 1$, say), multiplied by $(n - 1)/\sigma^2$, has a chi-square distribution with $\nu = n - 1$ degrees of freedom.

As remarked earlier, the chi-square distribution with ν degrees of freedom is exactly the gamma distribution with $\theta = 2$ and $r = \nu/2$. This fact, along with the one given in the previous paragraph, permits us to make the statement that $(n - 1)s^2/\sigma^2 = \nu s^2/\sigma^2$ has the gamma distribution with $\theta = 2$ and $\nu = n - 1$ degrees of freedom. Further, if we have a random variable X that has the gamma distribution with parameters θ and r, then the transformed variable $Y = X/\theta$ has the standardized gamma distribution with parameter r. Thus $(\nu s^2/\sigma^2)/2 = \nu s^2/2\sigma^2$ has the standardized gamma distribution with $\nu = n - 1$ (or $2r$) degrees of freedom.

As a result, the theory and tables of Chapter 5 can be used to select the gamma distribution with the smallest scale parameter θ. The distance function introduced there is

$$\Delta = \frac{\theta_{[1]}}{\theta_{[2]}} = \frac{1}{\delta}.$$

In order to determine the sample size we specify the threshold value δ^*

where $\delta^* > 1$, or Δ^* where $0 < \Delta^* < 1$, and a value of P^* where $1/k < P^* < 1$, and find the degrees of freedom ν as the corresponding entry in the Table G.1 of Appendix G.

Now the degrees of freedom for the gamma distribution is $2nr$ where n is the number of gamma observations (not to be confused with the n in the discussion of the first two paragraphs of this section); thus we set $\nu = 2nr$ and find the common gamma sample size n from $n = \nu/2r$. This means we need $\nu/2r$ gamma observations from each population. If we take exponential observations, then we need $\nu/2 = nr$ observations from each population.

The selection procedure is to compute the k sample means and order them as

$$\bar{x}_{[1]} \leqslant \bar{x}_{[2]} \leqslant \cdots \leqslant \bar{x}_{[k]}.$$

Then we assert that the population that gave rise to $\bar{x}_{[1]}$ has the smallest θ value $\theta_{[1]}$.

It is important to emphasize here that for the problem of selecting the smallest θ value, we cannot use Table R.1, which applies to selection of the largest θ value. These two problems are *not* equivalent in small sample or exact theory for the gamma distribution model. In large sample theory (i.e., when n is quite large) the tables *are* interchangeable.

Example 14.3.1.   A psychologist is studying the length of time it takes to complete $k = 4$ different complex tasks. Each task is made up in such a way that it requires the solution of $r = 5$ simpler tasks, and the length of time required to complete each simple task that is part of the jth complex task has an exponential distribution with scale parameter θ_j. As a result, the length of time it takes to complete the jth complex task follows the gamma distribution with scale parameter θ_j and $2r = 10$ degrees of freedom. The mean length of time to completion of the jth complex task is then $r\theta_j$. Thus if the goal is to select the complex task with the smallest mean length of time, this is equivalent to selecting the population with the smallest θ value.

Suppose we specify $P^* = .99$ and $\delta^* = 1.50$ so that $\Delta^* = 1/\delta^* = .67$. From Table G.1 with $k = 4$ we find $\nu = 50$ (using the method of interpolation described in Appendix G). Then since $\nu = 2nr$ and $r = 5$ here, we find the common gamma sample size required is $n = 50/10 = 5$ for each of the $k = 4$ complex tasks.

To carry out the experiment, for each complex task the length of time required to completion is observed for each of the five replications and the sample mean is computed. These four means are ordered, and the task that produces the smallest mean is asserted to be the one for which learning time is fastest. Let θ_s denote the true average learning time of the task selected by this procedure. Then with confidence level $P^* = .99$, we can state that

$$\Delta^* = .67 \leqslant \frac{\theta_{[1]}}{\theta_s},$$

or equivalently, that

$$\theta_s \leqslant 1.5\theta_{[1]}.$$

Consequently, θ_s is at most $(\delta^* - 1)100\% = 50\%$ greater than the smallest θ value $\theta_{[1]}$.

14.4   SELECTING A SUBSET CONTAINING THE GAMMA POPULATION WITH THE LARGEST SCALE PARAMETER

Subset selection procedures were covered in Chapter 13 for the normal distribution and the binomial distribution models. For the gamma distribution model with known degrees of freedom $2r$, we may also be interested in selecting a subset containing the population with the largest θ value. As before, this subset is of random size, the size depending on the outcome of the experiment.

We take a random sample of size n from each of the gamma populations, compute the sample means, and order them as

$$\bar{x}_{[1]} \leqslant \bar{x}_{[2]} \leqslant \cdots \leqslant \bar{x}_{[k]}.$$

At this stage we must specify a quantity P^*, where $(.5)^k < P^* < 1$. Here P^* can be interpreted as the minimum probability that the subset selected contains the population with the largest θ value, or alternatively, as the joint confidence level for the statements that the θ value for each eliminated population is smaller than the θ value of the best population.

The subset selection procedure is to consider each j $(j = 1, 2, \ldots, k)$ separately, and to place population j in the selected subset if and only if

$$\bar{x}_j \geqslant b\bar{x}_{[k]},$$

where b is given in Table R.2 as a function of P^*, k, and $\nu = 2nr$.

Example 14.4.1. Suppose we have samples of $n = 10$ observations from each of $k = 4$ gamma populations, each with $2r = 4$ degrees of freedom and we wish to select a subset containing the gamma population with the largest θ value. Then we have a total of $\nu = 2nr = 2(10)(2) = 40$ degrees of freedom per population. For $P^* = .90$, Table R.2 gives the value $b = .579$. Thus the selection rule is to place in the subset all populations for which the corresponding mean is greater than or equal to $.579\bar{x}_{[k]}$. For example, if the means are

$$\bar{x}_1 = 2.7, \quad \bar{x}_2 = 1.3, \quad \bar{x}_3 = 3.9, \quad \bar{x}_4 = 2.1,$$

then $\bar{x}_{[k]} = 3.9$ and $(.579)(3.9) = 2.26$. Thus the rule here is to retain only those populations that give rise to a sample mean value of at least 2.26, and the selected subset consists of the populations labeled 1 and 3, for which the sample means are 2.7 and 3.9. We then assert that one of these two is the best population, that is, the one with the largest θ value, although we do not know which one.

The size S of the subset selected is a random variable that may take on any of the values $1, 2, \ldots, k$. The larger the size of the subset, the less information we have. Thus it is important that we have a procedure for which the subset size is small on the average. The expected size $E(S)$ is not easily computed for the gamma distribution problem. It depends not only on n, k, r, and P^*, but also on the true configuration of $\theta_1, \theta_2, \ldots, \theta_k$. For the configuration where all θ values are equal, we obtain $E(S) = kP^*$, and this is the largest value of $E(S)$ for all configurations with $\theta_{[i]}/\theta_{[k]} \leqslant 1/\delta^*$ for $i = 1, 2, \ldots, k-1$. This may not be very satisfactory in that since P^* is usually close to 1, kP^* is usually close to k. However, as $\theta_{[i]}$ decreases from $\theta_{[k]}/\delta^*$ (whenever and for any i for which this is possible), the value of $E(S)$ also decreases.

As a secondary problem, we may want to determine the smallest value of n that satisfies a specified (δ^*, P^*) requirement and also keeps $E(S) < 1 + \varepsilon^*$ for a specified $\varepsilon^* > 0$. We do not consider this problem here.

14.5   NOTES, REFERENCES, AND REMARKS

As mentioned earlier, this chapter is related to Chapter 5 in that the problem of selecting the gamma population with the *smallest* scale parameter is equivalent to that of selecting the normal population with the smallest variance that is treated in Chapter 5, and the procedure for selecting the gamma population with the *largest* scale parameter could be used for selecting the normal population with the largest variance if the sample sizes are large. However, some more important applications of the gamma distribution deal with waiting time, reliability, and life testing. The tables given here as Table R.1 for this latter problem are equivalent to those used in Gupta (1963), but it was necessary to compute new tables for this book in order to cover areas of more practical application.

The theory for selecting a subset containing the gamma population with the largest scale parameter in Section 14.4 is from Gupta (1963), and this is the origin of Table R.2 for $\nu \leqslant 50$; the remainder of Table R.2 is original and was calculated using an asymptotic approximation. These entries could be in error by as much as one or two units in the last digit shown. Since these entries are identical with those in Table 2.3 of Guttman and Milton (1969), it was possible to use their table to obtain more exact values for the special cases $k = 2$, 3, and 4. These constants are referred to as δ_3^* values in a forthcoming survey of ranking and selection by Gupta (1976).

CHAPTER *15

Selection Procedures for Multivariate Normal Distributions

15.1 INTRODUCTION

A multivariate distribution is the joint probability distribution of p variables or components where $p \geqslant 2$. A single observation from a multivariate population is a measurement on each of these components, that is, a pair of measurements if $p = 2$, a triplet of measurements if $p = 3$, and, in general, a p-tuple of measurements. Therefore a random sample of size n from a multivariate distribution consists of a number n of p-tuples of measurements, one p-tuple for each of the n sample observations. We sometimes refer to these as *vector observations*. For example, the multinomial distribution with p categories can be considered as a multivariate distribution with p components, and each observation is then recorded as a p-tuple (or vector) in which the one component that corresponds to the category observed is equal to one and the other components are all equal to zero. The term *component* emphasizes the fact that two measurements are in the same p-tuple.

The most important multivariate distribution is the *multivariate normal distribution*. This distribution is an extension of the univariate normal distribution introduced in Chapter 2. The multivariate normal distribution has as parameters not only the means and variances of each of the p variables (as in the univariate case), but also covariances (or correlations) between pairs of these components; clearly the latter parameters are all zero for variables in different p-tuples. In this chapter we consider several selection problems for the multivariate normal distribution model.

In order to familiarize the reader with the concept of a multivariate distribution, we first give some examples of situations where two or more components would usually be measured simultaneously on each subject.

Example 15.1.1 (Graduate Record Examination Scores). A student who takes the Graduate Record Examination (GRE) is given two scores, called Verbal and Quantitative. Successful graduate study in a field such as economics requires both verbal and quantitative skills. As a result, a committee considering applicants for graduate study in economics might wish to take into account both scores for each applicant, rather than just the total score. Hence if the committee is considering 50 applicants for graduate study, they would have 50 pairs of scores, each pair consisting of a verbal score and a quantitative score so that $n=50$ and $p=2$. (In this application and others the committee could avoid pairs of scores by using the minimum of these two scores for each applicant, but this may not yield the best method for ranking individuals.)

Example 15.1.2   (Memory).   Psychologists consider the term *memory* as including several distinct processes. Three of these processes are known as recall, recognition, and motor memory (an example of the latter is remembering how to ride a bicycle after not having done so for 10 years). A group of 20 persons are chosen as subjects for an experiment involving memory. Each subject "learns" a fixed set of nonsense words, a fixed set of visual forms, and a fixed set of motor tasks. After a fixed period of time, each subject is tested and scored for recall on the nonsense words, recognition on the visual forms, and motor memory on the motor tasks. Thus we have a triplet of scores for each of the 20 subjects so that $n=20$ and $p=3$. (Here again a single score, such as the minimum, or the sum, may not yield the best type of classification and a multivariate analysis would be required.)

Before proceeding to a discussion of the ordering and selection procedures for a multivariate normal distribution with p variables, we introduce some of the notation to be used in this chapter.

Let $X_1, X_2, \ldots, X_p$ denote p variables (or components) that have a multivariate normal distribution. The respective means are denoted by $\mu_1, \mu_2, \ldots, \mu_p$, the respective variances by $\sigma_1^2, \sigma_2^2, \ldots, \sigma_p^2$, and the covariance between the two components X_c and X_d is σ_{cd} for all $c = 1, 2, \ldots, p$ and $d = 1, 2, \ldots, p$ (thus $\sigma_{cc} = \sigma_c^2$). The means are usually given in the form of a row vector as $\mu = (\mu_1, \mu_2, \ldots, \mu_p)$. The corresponding column vector is denoted by μ'. The variances and covariances for any p-tuple are usually presented in the form of a p by p (symmetric and positive definite) matrix called the variance-covariance matrix and denoted by Σ; in symbols,

$$\Sigma = \begin{bmatrix} \sigma_{11} & \sigma_{12} & \cdots & \sigma_{1p} \\ \sigma_{21} & \sigma_{22} & \cdots & \sigma_{2p} \\ \cdot & \cdot & \cdot & \cdot \\ \cdot & \cdot & \cdot & \cdot \\ \cdot & \cdot & \cdot & \cdot \\ \sigma_{p1} & \sigma_{p2} & \cdots & \sigma_{pp} \end{bmatrix}.$$

Since Σ is positive-definite, it is nonsingular and in fact the determinant of Σ is positive. The inverse of Σ, denoted by Σ^{-1}, then exists; we denote the elements of Σ^{-1} by σ^{cd} for $c = 1, 2, \ldots, p$ and $d = 1, 2, \ldots, p$, so that we have

$$\Sigma^{-1} = \begin{bmatrix} \sigma^{11} & \sigma^{12} & \cdots & \sigma^{1p} \\ \sigma^{21} & \sigma^{22} & \cdots & \sigma^{2p} \\ \cdot & \cdot & \cdot & \cdot \\ \cdot & \cdot & \cdot & \cdot \\ \cdot & \cdot & \cdot & \cdot \\ \sigma^{p1} & \sigma^{p2} & \cdots & \sigma^{pp} \end{bmatrix}.$$

Of course, Σ^{-1} has the property that $\Sigma\Sigma^{-1} = \Sigma^{-1}\Sigma = I$ where I is the p by p identity matrix. Readers not acquainted with these concepts related to the inverse of a matrix are referred to any basic book on matrices.

The correlation between any two components X_c and X_d of the same p-tuple is defined as

$$\rho_{cd} = \frac{\sigma_{cd}}{\sqrt{\sigma_c^2 \sigma_d^2}} = \frac{\sigma_{cd}}{\sigma_c \sigma_d}$$

for all $c = 1, 2, \ldots, p$ and $d = 1, 2, \ldots, p$. In particular, $\rho_{cc} = 1$.

15.2 SELECTING THE BEST COMPONENT FROM p CORRELATED COMPONENTS

Our aim is to discuss some ranking and selection problems concerning k multivariate normal distributions, each having p variables where $p \geqslant 2$. However, before doing this we consider the simpler model of a *single* p-variate normal population where the goal is to select the variable or component with (1) the largest mean, and (2) the smallest variance. If the correlation (or equivalently, the covariances) between each pair of components are equal to zero (that is, $\rho_{cd} = 0$ for all $c \neq d$), then the variables are independent and the model reduces to that of p *independent* normal populations. This problem was discussed in Chapters 2 and 3 (with p replaced by k). In order to emphasize this distinction we call the present case the *model of p correlated normal populations*, or we may speak of the problem of *selecting the best component of a single multivariate normal population*.

15.2.1 Selecting the Component with the Largest Mean

In this section the goal is to select the component with the largest mean. We assume that the variances of the p variables have the common known value σ^2, that is

$$\sigma_1^2 = \sigma_2^2 = \cdots = \sigma_p^2 = \sigma^2,$$

and also that the correlations are all common and known so that $\rho_{cd} = \rho_I$ for all $c \neq d$. (This common correlation is labeled ρ_I because it is usually referred to as the intraclass correlation.) As mentioned earlier, when $\rho_I = 0$ the p components are independent.

As an example, suppose we have an examination consisting of p approximately parallel parts such that the variances of the scores on the p parts are equal, and the correlation between the scores of the same person on any two different parts is the same. The value of the correlation between different people is, of course, zero; we frequently also assume that the variance for different people has the common value σ^2. However, the mean scores on the various parts may differ, and indeed, we wish to choose the part that has the largest true mean if a (large) homogeneous population of people are given this examination.

Thus instead of having p independent normal populations, we now have a single p-variate normal population. Let the ordered means be denoted by

$$\mu_{[1]} \leqslant \mu_{[2]} \leqslant \cdots \leqslant \mu_{[p]},$$

the common variance by σ^2, and a common correlation between any two components by ρ_I. We suppose that both σ^2 and ρ_I are known (or can be estimated fairly accurately from previous experience or other data). The goal is to define a selection procedure for choosing the variable that has the largest mean in such a way that the probability of a correct selection is at least P^* regardless of the value of ρ_I provided only that

$$\delta = \mu_{[p]} - \mu_{[p-1]} \geqslant \delta^*,$$

where $\delta^* > 0$ and P^* (with $1/p < P^* < 1$) are both preassigned.

The procedure is to take a random sample of n *vector* observations from the p-variate normal population, compute the sample mean $\bar{x}_j$ for the jth variable for $j = 1, 2, \ldots, p$, order these means, and denote them by

$$\bar{x}_{[1]} \leqslant \bar{x}_{[2]} \leqslant \cdots \leqslant \bar{x}_{[p]}.$$

Then the selection rule is to assert that the variable or component that gives rise to the largest sample mean is the one with the largest μ value.

The sample size n is obtained from the expression

$$n = \left(\frac{\tau_t \sigma}{\delta^*} \right)^2 (1 - \rho_I), \tag{15.2.1}$$

where σ and ρ_I are known, δ^* is given, and τ_t is obtained from Table A.1 of Appendix A as the entry that corresponds to the given values of $k = p$ and P^*. If the value of n from (15.2.1) is not an integer, we round upward to obtain a conservative result. Note that if $\rho_I = 0$, the result of (15.2.1) is the same as in the independent case discussed in Chapter 2.

REMARK.   A comparison of the value of the sample size n from (15.2.1) with ρ_I and the n from Chapter 2 shows that the gain is in the factor $(1 - \rho_I)$. Thus, if the correlation between the parts is high, we require fewer observations. Since the independent case yields the largest value of n, we can ignore the correlational effect, in which case we would have a conservative procedure; this would be wasteful of observations.   □

Example 15.2.1.   Kaneko (1975) studies frames of reinforced concrete in a variety of shapes, as, for example, those shown in Figure 15.2.1. A single load is placed at the fulcrum of the frame (this is the center in both case a and case b) and the resulting force (or stress) imposed on each of the different extremities is measured. We wish to choose the extremity with the largest stress so that it can be treated first and reinforced if necessary. If the frame has p sides, how many observations of the stress on each side should be taken when $\sigma = 75, \rho_I = .5$, and we specify $P^* = .95, \delta^* = 25$?

For Figure 15.2.1a, we have $p = 3$, and Table A.1 gives $\tau_t = 2.7101$. Substituting the appropriate values in (15.2.1), we obtain

$$n = \left[\frac{(2.7101)(75)}{25} \right]^2 (.5) = 33.05,$$

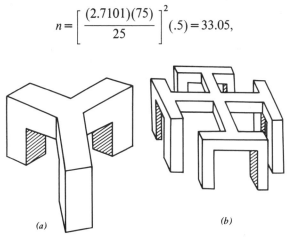

(a) (b)

Figure 15.2.1   Symmetric frames of reinforced concrete for (a) $p = 3$ sides, and (b) $p = 4$ sides. Reproduced from I. Kaneko (1975), A reduction theorem for the linear complementarity problem with a x-matrix, *Technical Report* 75-24, Department of Industrial Engineering, University of Wisconsin, Madison, Wisconsin, with permission.

so that a sample of 34 observations is required, where each observation is a 3-tuple or triplet.

If $p = 4$, as in Figure 15.2.1b, then $\tau_t = 2.9162$ and

$$n = \left[\frac{(2.9162)(75)}{25} \right]^2 (.5) = 38.27,$$

so that we take a sample of 39 observations, each of which is a 4-tuple.

REMARK. The assumption of a common covariance may be more appropriate for a than b in Figure 15.2.1, since in b opposite sides could have a different covariance than adjacent sides, whereas in a there is complete symmetry between all three sides. $\square$

15.2.2 Selecting the Component with the Smallest Variance

As a parallel to the problem of selecting that single component (out of p correlated components) with the largest mean, we may wish to select that single component (out of p correlated components) with the smallest variance. Recall that in Chapter 5 we treated the problem of selecting that single population (out of k *independent* populations) with the smallest variance. We now treat the analogous problem for a multivariate population with correlated components.

As an example, suppose we have three alternative machines for testing blood, say, and wish to select the one with the greatest precision, that is, the smallest standard deviation. If we take blood samples and use a *different* sample for testing by each machine, then the procedure of Chapter 5 is applicable because the measurements are independent. On the other hand, if for each blood sample we use some of the *same* blood sample for each of the testing machines, then the results are correlated.

The exact statistical analysis for carrying out this procedure is quite complicated. However, it has been shown that a useful conservative approach is obtained by employing the techniques of Chapter 5 and simply ignoring the correlation between different components, that is, by treating them as if they were independent components.

15.3 PROPERTIES OF THE MULTIVARIATE NORMAL DISTRIBUTION

In this section we discuss some well-known distribution properties of the multivariate normal distribution and introduce some additional notation required for the ranking and selection of k multivariate normal distributions.

We make use of the result that when $X_1, X_2, \ldots, X_p$ follow the p-variate normal distribution, the linear combination (or weighted score)

$$L = b_1 X_1 + b_2 X_2 + \cdots + b_p X_p, \tag{15.3.1}$$

where $b_1, b_2, \ldots, b_p$ are any constants, has a (univariate) normal distribution with mean θ_L (or simply θ) given by

$$\theta = b_1 \mu_1 + b_2 \mu_2 + \cdots + b_p \mu_p, \tag{15.3.2}$$

and variance $\text{var}(L)$ given by

$$\text{var}(L) = \sum_{c=1}^{p} \sum_{d=1}^{p} b_c b_d \sigma_{cd}. \tag{15.3.3}$$

Now suppose we have k different multivariate normal populations. In our notation, the jth population has mean vector

$$\mu^{(j)} = \left(\mu_1^{(j)}, \mu_2^{(j)}, \ldots, \mu_p^{(j)} \right),$$

variance-covariance matrix Σ_j with elements $\sigma_{cd}^{(j)}$, and the inverse Σ_j^{-1} with elements σ_j^{cd}, for each of $j = 1, 2, \ldots, k$. In other words, $\mu_c^{(j)}$ is the mean of the cth component for the jth population and $\sigma_{cd}^{(j)}$ is the covariance between the cth and dth components for the jth population. Then we take a random sample of n observations from each of these k multivariate normal populations, each observation being a p-tuple of measurements. We let $x_{ci}^{(j)}$ denote the ith measurement on the cth component of the jth population. Then the data consist of kpn measurements, that is, n observations from k populations with p measurements in each observation, which might be recorded as in Table 15.3.1.

The last line of Table 15.3.1 gives the corresponding sample means for the cth component, which are computed as

$$\bar{x}_c^{(j)} = \frac{\displaystyle\sum_{i=1}^{n} x_{ci}^{(j)}}{n}, \tag{15.3.4}$$

for $c = 1, 2, \ldots, p$ and each of $j = 1, 2, \ldots, k$. The covariances for the jth population are computed as

$$s_{cd}^{(j)} = \frac{\displaystyle\sum_{i=1}^{n} \left(x_{ci}^{(j)} - \bar{x}_c^{(j)} \right)\left(x_{di}^{(j)} - \bar{x}_d^{(j)} \right)}{n-1} \tag{15.3.5}$$

Table 15.3.1   Presentation of sample data consisting of n observations from each of k populations, each having p components

Population j	1				2			$\cdots$	k		
Component c of the p variables	1	2	$\cdots$	p	1	$\cdots$	p	$\cdots$	1	$\cdots$	p
$i=1$	$x^{(1)}_{11}$	$x^{(1)}_{21}$	$\cdots$	$x^{(1)}_{p1}$	$x^{(2)}_{11}$	$\cdots$	$x^{(2)}_{p1}$	$\cdots$	$x^{(k)}_{11}$	$\cdots$	$x^{(k)}_{p1}$
$i=2$	$x^{(1)}_{12}$	$x^{(1)}_{22}$	$\cdots$	$x^{(1)}_{p2}$	$x^{(2)}_{12}$	$\cdots$	$x^{(2)}_{p2}$	$\cdots$	$x^{(k)}_{12}$	$\cdots$	$x^{(k)}_{p2}$
Observation i $\vdots$	$\vdots$	$\vdots$		$\vdots$	$\vdots$		$\vdots$		$\vdots$		$\vdots$
$i=n$	$x^{(1)}_{1n}$	$x^{(1)}_{2n}$	$\cdots$	$x^{(1)}_{pn}$	$x^{(2)}_{1n}$	$\cdots$	$x^{(2)}_{pn}$	$\cdots$	$x^{(k)}_{1n}$	$\cdots$	$x^{(k)}_{pn}$
Sample mean of component c	$\bar{x}^{(1)}_{1}$	$\bar{x}^{(1)}_{2}$	$\cdots$	$\bar{x}^{(1)}_{p}$	$\bar{x}^{(2)}_{1}$	$\cdots$	$\bar{x}^{(2)}_{p}$	$\cdots$	$\bar{x}^{(k)}_{1}$	$\cdots$	$\bar{x}^{(k)}_{p}$

for $c = 1, 2, \ldots, p$, for $d = 1, 2, \ldots, p$, and for $j = 1, 2, \ldots, k$. The corresponding sample variance-covariance matrix for the jth population is denoted by S_j and its inverse by S_j^{-1}. Since each S_j is positive-definite (with probability one), we need not be concerned with the question of whether S_j^{-1} exists.

15.4   SELECTING THE BEST POPULATION WITH RESPECT TO MEANS

In this section we consider the problem of defining what is meant by selecting the best multivariate normal distribution under the assumption that our main interest is in the μ values. The question of what is meant by *best* is much more elusive for multivariate than for univariate distributions; in order to clarify this, we first consider a specific example.

Suppose we have three candidates who are auditioning for a part in a dance group. During the audition each candidate is given scores on several attributes, such as execution of steps, expressiveness, movement, and so on. For simplicity, we limit ourselves to scores x_1 and x_2 on only $p=2$ attributes for each different candidate, with corresponding true mean scores, μ_1 and μ_2. Suppose there are $k=3$ candidates. We can assume that a sample of size 1 is taken from each of three bivariate distributions. The notation is shown in Table 15.4.1.

We assume that larger μ values are better for each of the attributes or components (or that a simple transformation makes this the case), and the problem is to select the overall best candidate. How can we define best?

Table 15.4.1   Notation for multivariate data with $p = 2$ and $k = 3$

Candidate j (or population j)	Observations $c = 1, \quad c = 2$	Means $c = 1, \quad c = 2$
1	$x_1^{(1)}, \quad x_2^{(1)}$	$\mu_1^{(1)}, \quad \mu_2^{(1)}$
2	$x_1^{(2)}, \quad x_2^{(2)}$	$\mu_1^{(2)}, \quad \mu_2^{(2)}$
3	$x_1^{(3)}, \quad x_2^{(3)}$	$\mu_1^{(3)}, \quad \mu_2^{(3)}$

One possibility is to say that one candidate is better than another if the former is superior to the other on *all* attributes. For our example this means that candidate i is better than candidate j if $\mu_1^{(i)} > \mu_1^{(j)}$ and $\mu_2^{(i)} > \mu_2^{(j)}$. To illustrate this, suppose the μ values are as follows:

$$\left(\mu_1^{(1)}, \mu_2^{(1)} \right) = (17, 42),$$

$$\left(\mu_1^{(2)}, \mu_2^{(2)} \right) = (20, 44),$$

$$\left(\mu_1^{(3)}, \mu_2^{(3)} \right) = (16, 40).$$

Then candidate 2 is better than both candidates 1 and 3 since the μ values for candidate 2 are larger than the corresponding μ values for each of the other two candidates. However, with this definition of *best*, there may well be no candidate who is best. For example, suppose the μ values are as follows:

$$\left(\mu_1^{(1)}, \mu_2^{(1)} \right) = (17, 42),$$

$$\left(\mu_1^{(2)}, \mu_2^{(2)} \right) = (20, 40),$$

$$\left(\mu_1^{(3)}, \mu_2^{(3)} \right) = (23, 38).$$

Then candidate 3 is better than either of the other candidates on attribute 1, but is worse than either of the other candidates on attribute 2. Thus no candidate is better than any other in all respects, although each candidate is better than some other in at least one respect.

As a result, the definition of *best* must consider all attributes simultaneously. One possibility is to use a weighted average of the μ values for each candidate as the criterion for best. That is, we assume that there exist constants b_1 and b_2 (that are known or determinable) such that the criterion of primary interest for ranking the candidates can be written as

$$\theta_1 = b_1 \mu_1^{(1)} + b_2 \mu_2^{(1)},$$

$$\theta_2 = b_1 \mu_1^{(2)} + b_2 \mu_2^{(2)},$$

$$\theta_3 = b_1 \mu_1^{(3)} + b_2 \mu_2^{(3)}.$$

With this *linear combination criterion*, our goal is to select the candidate with the largest value of θ. This may be a natural approach to the problem, since the candidates are to be compared on several (in the present case two) attributes simultaneously, and a dancer who is judged outstanding in execution of steps but inferior in expressiveness may not be as good overall as one who is judged better than average on both attributes.

Of course, the statement of the goal in terms of θ values assumes that the fixed constant weights b_1 and b_2 are given, or that we know how to find them. Since larger values of μ_1 and μ_2 are considered better, it follows that b_1 and b_2 can be presumed to be positive; we can also assume without loss of generality that b_1 and b_2 sum to one. As a result, the ratio, say b_2/b_1, determines the values of b_1 and b_2. (Note that in the degenerate case where $b_1 = 1, b_2 = 0$, or $b_1 = 0, b_2 = 1$, this present criterion in terms of θ values reduces to the univariate normal means problem that was introduced in Chapter 2.)

A second possibility is to use a combination of the μ values for each candidate that is quadratic in form and also utilizes known variances and covariances in the criterion for best. In particular, we might define the θ value for population j ($j = 1, 2, 3$) by

$$\theta_1 = \frac{\sigma_{22}\left(\mu_1^{(1)}\right)^2 - 2\sigma_{12}\mu_1^{(1)}\mu_2^{(1)} + \sigma_{11}\left(\mu_2^{(1)}\right)^2}{\sigma_{11}\sigma_{22} - \sigma_{12}^2},$$

$$\theta_2 = \frac{\sigma_{22}\left(\mu_1^{(2)}\right)^2 - 2\sigma_{12}\mu_1^{(2)}\mu_2^{(2)} + \sigma_{11}\left(\mu_2^{(2)}\right)^2}{\sigma_{11}\sigma_{22} - \sigma_{12}^2},$$

$$\theta_3 = \frac{\sigma_{22}\left(\mu_1^{(3)}\right)^2 - 2\sigma_{12}\mu_1^{(3)}\mu_2^{(3)} + \sigma_{11}\left(\mu_2^{(3)}\right)^2}{\sigma_{11}\sigma_{22} - \sigma_{12}^2}.$$

Here, of course, we are assuming that the σ_{cd} are all known and do not depend on j. For candidate j, θ_j measures a certain generalized 'distance' between the μ values of the two attributes considered. This generalized 'distance' was considered in 1930 by the Indian statistician P. C. Mahalanobis (who was instrumental in establishing the Indian Statistical Institute and in fostering the development and application of statistics in India). Such a measure is now called the Mahalanobis 'distance'. With this *Mahalanobis 'distance' criterion* our goal is to select the candidate with the largest value of θ.

These two criteria for best are each considered in more detail in the following subsections, and the corresponding selection procedures are described fully for an arbitrary number p of variables (attributes or components) and k of populations (candidates).

Before proceding to this discussion, we give some examples of actual experiments in which the observations follow a multivariate normal distribution and one of these goals might be of interest. These experiments deal with meteorological data and are discussed in greater detail in Crutcher and Falls (1976).

In one experiment we have data on wind components in different directions:

$$x_1 = \text{east-west},$$

$$x_2 = \text{north-south},$$

$$x_3 = \text{vertical}.$$

The measurements are in meters per second and are taken at a number of different sites. The number of components is $p = 3$ and the number of populations k is the number of sites where the data are collected.

A second application deals with hurricane motions. Apparently sets of hurricane movements for the North Atlantic, Carribbean, and Gulf regions can be described by the bivariate normal distribution. The $p = 2$ variables are the components along two sets of (orthogonal) axes imposed over the usual latitude-longitude coordinates and k is the number of regions under consideration.

A third example again deals with wind velocity measured as follows:

$$x_1 = \text{longitudinal},$$

$$x_2 = \text{lateral},$$

$$x_3 = \text{vertical}.$$

The measurements are in meters per second in gusts a few meters above the ground at k different locations. For each location we have a three-variate distribution that is assumed to be a multivariate normal distribution with unknown mean vector

$$\mu = (\mu_1, \mu_2, \mu_3),$$

and a known matrix of variances and covariances

$$\Sigma = \begin{bmatrix} \sigma_{11} & \sigma_{12} & \sigma_{13} \\ \sigma_{21} & \sigma_{22} & \sigma_{23} \\ \sigma_{31} & \sigma_{32} & \sigma_{33} \end{bmatrix}$$

that is common for all locations. The best location could be defined as the one with the largest linear combination of means

$$L = b_1 \mu_1 + b_2 \mu_2 + b_3 \mu_3$$

for some specified constant set of positive weights b_1, b_2, b_3, or as the one with the largest Mahalanobis 'distance'

$$\theta = \mu \Sigma^{-1} \mu' = \sum_{c=1}^{3} \sum_{d=1}^{3} \mu_c \mu_d \sigma^{cd}.$$

Here the purpose of selecting the largest linear combination or Mahalanobis 'distance' is to identify that location where the gust effect is greatest. Alternatively, we might be interested in identifying the site where the gust effect is smallest. Indeed, this might be particularly useful in comparing various locations for airplane landing fields.

15.4.1 Selecting the Best Population Based on a Linear Combination

In this subsection we consider the general problem of k normal populations, each with p variables, when the goal is to select the population with the largest θ value and θ_j is defined as a given linear combination of $\mu^{(j)}$ values, with positive coefficients not depending on j. (If any of the coefficients were zero, we would remove that component from all the observations.)

The linear combination for the jth population with p variables is

$$\theta_j = b_1 \mu_1^{(j)} + b_2 \mu_2^{(j)} + \cdots + b_p \mu_p^{(j)}, \tag{15.4.1}$$

where the b values are all positive and known and sum to one. The ordered values of θ_j are denoted by

$$\theta_{[1]} \leqslant \theta_{[2]} \leqslant \cdots \leqslant \theta_{[k]}.$$

The distance measure used is

$$\delta = \theta_{[k]} - \theta_{[k-1]}.$$

Clearly, $\delta \geq 0$ and for $\delta > 0$ a correct selection is well defined; here the preference zone is defined by $\delta \geq \delta^*$ and the indifference zone is defined by $\delta < \delta^*$, for some $\delta^* > 0$.

A random sample of n observations is taken from each of the k multivariate normal distributions, where now, of course, each observation is a p-tuple of measurements. For the sample from the jth population, the sample mean for the cth component is $\bar{x}_c^{(j)}$ computed from (15.3.4). Note that in the present discussion, all the sample means are based on the same number n of measurements. The case of unequal n_j is considered later as a variation of the present discussion. The analogous sample statistic to estimate θ_j is an average score for the sample from the jth population, which is the same linear combination of the $\bar{x}_c^{(j)}$ as (15.4.1) is of the μ_j, namely,

$$\bar{L}_j = b_1 \bar{x}_1^{(j)} + b_2 \bar{x}_2^{(j)} + \cdots + b_p \bar{x}_p^{(j)}, \tag{15.4.2}$$

for $j = 1, 2, \ldots, k$. These average scores are then ordered from smallest to largest as

$$\bar{L}_{[1]} \leq \bar{L}_{[2]} \leq \cdots \leq \bar{L}_{[k]},$$

and the selection rule is to assert that the best population is the one that gives rise to the largest $\bar{L}$ value, $\bar{L}_{[k]}$. (Here equalities occur with zero probability and we need not be concerned with that possibility.) We can also define a linear combination L_{ji} of the ith observation for the sample from the jth population by (15.4.2) if we replace $\bar{x}_c^{(j)}$ by $x_{ci}^{(j)}$ for each $c = 1, 2, \ldots, p$; we use the notation L_j to refer to the typical value (or random variable) corresponding to L_{ji}.

Since the $\bar{x}_c^{(j)}$ for $c = 1, 2, \ldots, p$ follow the multivariate normal distribution, the discussion in Section 15.3 indicates that for each j, $\bar{L}_j$ has a univariate normal distribution with mean θ_j and variance

$$\mathrm{var}(\bar{L}_j) = \frac{1}{n} \sum_{c=1}^{p} \sum_{d=1}^{p} b_c b_d \sigma_{cd}^{(j)} = \frac{1}{n} \mathrm{var}(L_j). \tag{15.4.3}$$

As a result, the procedures and tables appropriate for a univariate normal means selection problem for a specified (δ^*, P^*) pair can be used here also. In particular, if the variance-covariance matrices Σ_j for $j = 1, 2, \ldots, k$ are all known, the tables in Appendix A apply exactly. If the variances of the L_j's are approximately the same for each $j = 1, 2, \ldots, k$, we denote the common value by var(L); then the number of observations required per population to satisfy the (δ^*, P^*) requirement is given by

$$n = \frac{\tau_t^2 \, \text{var}(L)}{\delta^{*2}},$$ (15.4.4)

where τ_t is found from Table A.1 of Appendix A with k replaced by p and P^* unchanged.

REMARK. The condition that the var(L_j) are common holds if the covariance matrices of the j populations are common, that is, $\Sigma_1 = \Sigma_2 = \ldots = \Sigma_k$. This is frequently the case for many applications. □

Unknown Covariance Matrices. If Σ_j is unknown, it is estimated by the sample variance-covariance matrix S_j defined by

$$S_j = \begin{bmatrix} s_{11}^{(j)} & s_{12}^{(j)} & \cdots & s_{1p}^{(j)} \\ s_{21}^{(j)} & s_{22}^{(j)} & \cdots & s_{2p}^{(j)} \\ \vdots & \vdots & \vdots & \vdots \\ s_{p1}^{(j)} & s_{p2}^{(j)} & \cdots & s_{pp}^{(j)} \end{bmatrix},$$ (15.4.5)

where each element is calculated from (15.3.5). The $\sigma_{cd}^{(j)}$ in (15.4.3) are replaced by $s_{cd}^{(j)}$ to calculate the estimates of var(L_j) and the same tables of Appendix A that were used can again be used to give approximate results. If the Σ_j are unknown but assumed to have a common value Σ, then this common value is estimated by S, computed from

$$S = \frac{S_1 + S_2 + \cdots + S_k}{k},$$ (15.4.6)

and the $\sigma_{cd}^{(j)}$ in (15.4.3) are replaced by the corresponding entry in S for each $j = 1, 2, \ldots, k$.

Unequal Sample Sizes. If the sample sizes are unequal, say n_j for the jth population, then (15.4.3) is replaced by

$$\text{var}\left(\overline{L}_j\right) = \frac{1}{n_j} \sum_{c=1}^{p} \sum_{d=1}^{p} b_c b_d \sigma_{cd}^{(j)} = \frac{1}{n_j} \text{var}(L_j).$$ (15.4.7)

If Σ_j is unknown, it is estimated by S_j as above but the n in (15.3.5) is replaced by n_j. If the Σ_j are unknown but assumed common, the common value Σ is estimated by

$$S = \frac{(n_1-1)\,S_1 + (n_2-1)\,S_2 + \cdots + (n_k-1)\,S_k}{(n_1-1) + (n_2-1) + \cdots + (n_k-1)} \tag{15.4.8}$$

Example 15.4.1. Rulon, Tiedeman, Tatsuoka, and Langmuir (1967) have made a detailed study of personnel classification in the airline industry. The personnel officer of World Airlines has developed an Activity Preference Inventory that contains 30 pairs of indoor-outdoor activities, 35 pairs of solitary-convivial activities, and 25 items of liberal-conservative activities. Since the preference of, say, 22 outdoor activities out of 30 pairs implies the preference for indoor in the remaining 8 activities, we need only one label for each pair. We use the captions *Outdoor Score, Social Score,* and *Political Score* for the component variables.

These tests were administered to three groups of employees of World Airlines, namely, to passenger agents, mechanics, and operations control agents. Suppose that we wish to choose that group with the largest linear combination $\theta = (\mu_1 + \mu_2 + \mu_3)/3$ (so that $b_1 = b_2 = b_3 = \frac{1}{3}$) in (15.4.1), where μ_1 is the true mean Outdoor Score, μ_2 is the true mean Social Score, and μ_3 is the true mean Political Score. Assume that the test scores for all groups follow the trivariate normal distribution with a common unknown variance-covariance matrix.

A sample of 85 passenger agents, 93 mechanics, and 66 operations control agents were chosen. The results of the basic calculations are shown in Table 15.4.2.

The pooled covariance matrix is calculated using (15.4.8) as

$$S = \begin{bmatrix} 16.4640 & 1.4587 & 0.3179 \\ 1.4587 & 18.2834 & 0.9770 \\ 0.3179 & 0.9770 & 11.1339 \end{bmatrix}$$

The estimated variances of the average L scores $\bar{L}_1, \bar{L}_2, \bar{L}_3$ are then computed from (15.4.7) with $\sigma_{cd}^{(j)}$ replaced by $s_{cd}^{(j)}$ as

$$\text{var}(\bar{L}_1) = \frac{51.3885}{9(85)} = 0.67175,$$

$$\text{var}(\bar{L}_2) = \frac{51.3885}{9(93)} = 0.61396,$$

$$\text{var}(\bar{L}_3) = \frac{51.3885}{9(66)} = 0.86513.$$

Table 15.4.2   Sample data calculations for example 15.4.1

Population		Components			
		Outdoor score	Social score	Political score	
j	n_j	$\bar{x}_1$	$\bar{x}_2$	$\bar{x}_3$	Sample covariance matrices
Passenger agents	85	12.5882	24.2235	9.0235	$S_1 = \begin{bmatrix} 20.2451 & 4.6169 & -2.4070 \\ 4.6169 & 18.7947 & 2.5066 \\ -2.4070 & 2.5066 & 9.8804 \end{bmatrix}$
Mechanics	93	18.5376	21.1398	10.1398	$S_2 = \begin{bmatrix} 12.7078 & -1.5759 & 2.5436 \\ -1.5759 & 20.7085 & 0.1759 \\ 2.5436 & 0.1759 & 10.5128 \end{bmatrix}$
Operations control agents	66	15.5758	15.4545	13.2424	$S_3 = \begin{bmatrix} 16.8942 & 1.6727 & 0.6890 \\ 1.6727 & 14.1902 & 0.1342 \\ 0.6890 & 0.1342 & 13.6327 \end{bmatrix}$
Total	244				

The values of the $\bar{L}$ scores from (15.4.2) with $b_1 = b_2 = b_3 = \frac{1}{3}$ are

$$\bar{L}_1 = 15.2784,$$

$$\bar{L}_2 = 16.6057,$$

$$\bar{L}_3 = 14.7576.$$

Since $\bar{L}_2$ is the largest $\bar{L}$ score, we assert that the group with the best θ score is the mechanics.

Example 15.4.2. Consider again the situation of Example 15.4.1, but suppose that we are in the process of designing an experiment to administer the World Airlines test to passenger agents, mechanics, and operations control agents who are employees of another airline. We decide to use a common sample size n for the three groups, and the question is how large should n be if we specify $\delta^* = 0.50, P^* = .95$. From Table A.1, with k replaced by $p = 3$, we obtain $\tau_f = 2.7101$. Based on the World Airlines data, we assume that a good estimate of the common variance of L_1, L_2, L_3 is $51.3885/9 = 5.71$ From (15.4.4), we find n as

$$n = \frac{(2.7101)^2(5.71)}{(0.50)^2} = 167.7,$$

and we need 168 observations from each of the $p = 3$ groups.

15.4.2   Selecting the Best Population Based on a Mahalanobis 'Distance'

For an arbitrary number p of variables, the Mahalanobis 'distance' θ from the origin of coordinates for a single population is the quadratic form

$$\theta = \sum_{c=1}^{p} \sum_{d=1}^{p} \mu_c \mu_d \sigma^{cd}$$

$$= (\mu_1, \mu_2, \ldots, \mu_p) \begin{bmatrix} \sigma^{11} & \cdots & \sigma^{1p} \\ \vdots & & \vdots \\ \sigma^{p1} & \cdots & \sigma^{pp} \end{bmatrix} \begin{bmatrix} \mu_1 \\ \vdots \\ \mu_p \end{bmatrix}.$$

Using the matrix notation introduced in Section 15.1, this quadratic form can be written as

$$\theta = \mu \Sigma^{-1} \mu', \tag{15.4.9}$$

where μ is the mean vector, Σ is the p by p positive-definite variance-covariance matrix, and Σ^{-1} is its inverse. This quadratic form has dimension two and is measured in square units, not in the unit of the μ values.

If the matrix Σ were the identity matrix, that is, $\sigma_{cc} = 1$ for all $c = 1, 2, \ldots, p$ and $\sigma_{cd} = 0$ for all $c \neq d$ so that all the variances equal one and all the covariances equal zero, then (15.4.9) becomes

$$\theta = \mu_1^2 + \mu_2^2 + \cdots + \mu_p^2,$$

which represents the square of the ordinary Euclidean distance from the origin to the point $(\mu_1, \mu_2, \ldots, \mu_p)$. For an arbitrary positive-definite Σ matrix, θ is a generalized (square) distance, being a sum of squares and cross products of μ values weighted by the elements of Σ^{-1}. The value of θ is always nonnegative because Σ and Σ^{-1} are positive-definite matrices. (Single quotes are inserted around the word *distance* as a reminder that it is really the square of a distance.)

When Σ is arbitrary but known, the Mahalanobis 'distance' in (15.4.9) is a function of $\mu_1, \mu_2, \ldots, \mu_p$. When $p = 2$, the 'distance' may be depicted in two dimensions using a contour diagram that shows the curve of different combinations of μ_1 and μ_2 that have the same value of θ. These contours are equivalent to the height contours of the bivariate normal density for various pairs of means μ_1, μ_2. (A contour curve of any bivariate distribution $f(x_1, x_2)$ consists of all values of the variables (x_1, x_2) that have constant density.) Thus for the bivariate normal density shown as three-dimensional

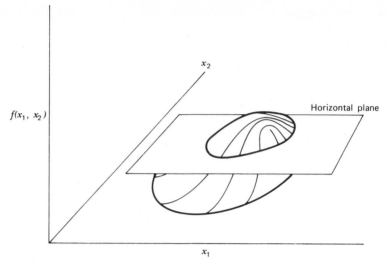

Figure 15.4.1   Contour ellipse for bivariate normal surface or density.

in Figure 15.4.1, the contours are the cross sections obtained by cutting the density by a horizontal plane at a fixed distance (say, d_0) above the (x_1, x_2) plane.

For $p = 2$ the shape of these contours is determined only by the variances σ_1^2 and σ_2^2 and the correlation coefficient ρ_{12} (or, equivalently, the covariance σ_{12}) and does not depend on d_0. The contours are circles with center at (μ_1, μ_2) when $\sigma_1^2 = \sigma_2^2$ and $\rho_{12} = 0$ (i.e., when Σ is proportional to the identity matrix); in general, they are ellipses with center at (μ_1, μ_2) and with major and minor axes whose direction (and ratio of lengths) does not depend on d_0. The slope of the major axis has the same sign as the correlation coefficient ρ_{12}. Figure 15.4.2 shows a contour diagram for the bivariate normal distribution depicted in Figure 15.4.1.

For $p = 2$ if contours of the bivariate normal density are graphed in two dimensions with axes μ_1 and μ_2, we obtain contours of the Mahalanobis 'distance' θ. Figure 15.4.3 gives the contours of θ for three different bivariate normal distributions. For population 3, $\rho = 0$ and the variances are equal so that the contours are circles. For populations 1 and 2 the major axis of the ellipses has positive slope, that is, $\rho_{12} > 0$. As mentioned earlier, the Mahalanobis 'distance' is the square of the distance from the origin to the point (μ_1, μ_2) when $\sigma_1^2 = \sigma_2^2 = 1$ and $\rho_{12} = 0$. When the contours are ellipses, we wish to measure distance from the center of the ellipse in a manner that, in some sense, equalizes points on the same contour in a

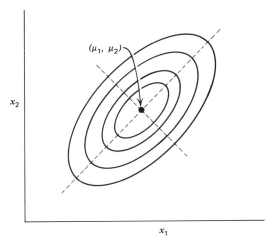

Figure 15.4.2 Contour diagram for bivariate normal surface (or density) in Fig. 15.4.1.

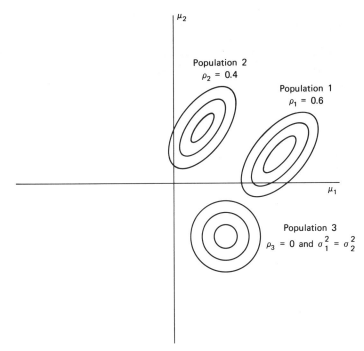

Figure 15.4.3 Three bivariate normal distributions, referred to a common origin.

given family of ellipses; the Mahalanobis 'distance' measure is designed to do this.

Now suppose that we have k multivariate normal populations, each with p variables. The Mahalanobis 'distance' measure from the origin for the jth population is

$$\theta_j = \mu^{(j)} \Sigma^{-1} \mu^{(j)'}; \qquad (15.4.10)$$

this is the analogue of (15.4.9) for the jth population. In this formulation we define the best population as the one with the largest θ value. Then with the ordered θ values

$$\theta_{[1]} \leqslant \theta_{[2]} \leqslant \cdots \leqslant \theta_{[k]},$$

the goal is to identify the population with θ value $\theta_{[k]}$.

REMARK.   The geometric interpretation of the Mahalanobis 'distance' is that we have a family of ellipses

$$\sum_{c=1}^{p} \sum_{d=1}^{p} \mu_c \mu_d \sigma^{cd} = \varepsilon$$

for varying ε values. These ellipses have the same contours and are, so to speak, parallel to each other with common center at the origin. Two populations (or points) on the same ellipse are equally desirable. The present goal is to select the population that is *furthest* from the origin and hence has the largest ε value. In some problems the goal is to select the population that is closest to the origin.   □

The ranking and selecting measure of distance that we use here to find the largest θ value consists of two parts:

$$\delta_1 = \theta_{[k]} - \theta_{[k-1]} \qquad \text{and} \qquad \delta_2 = \frac{\theta_{[k]}}{\theta_{[k-1]}}.$$

Since the θ values in (15.4.10) are nonnegative numbers, it is always true that $\delta_1 \geqslant 0$ and $\delta_2 \geqslant 1$. Thus the preference zone here is described by the intersection of the regions

$$\delta_1 \geqslant \delta_1^* \qquad \text{and} \qquad \delta_2 \geqslant \delta_2^*,$$

where $\delta_1^* > 0$ and $\delta_2^* > 1$.

In order to obtain a better understanding of the preference zone, we consider the special case where $k = 2$. The preference zone is the shaded region depicted in Figure 15.4.5. The boundaries of the region are the lines

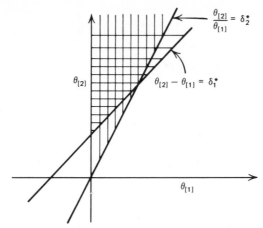

Figure 15.4.4 Derivation of the preference zone. The horizontally hatched portion indicates the region $\theta_{[2]} - \theta_{[1]} \geqslant \delta_1^*$; the vertically hatched portion indicates the region $\theta_{[2]}/\theta_{[1]} \geqslant \delta_2^*$.

where equality holds, that is,

$$\theta_{[2]} - \theta_{[1]} = \delta_1^* \qquad \text{and} \qquad \frac{\theta_{[2]}}{\theta_{[1]}} = \delta_2^*.$$

In Figure 15.4.4 the horizontally hatched region consists of all points $(\theta_{[1]}, \theta_{[2]})$ for which $\theta_{[2]} - \theta_{[1]} \geqslant \delta_1^*$, and the vertically hatched region consists of all points $(\theta_{[1]}, \theta_{[2]})$ for which $\theta_{[2]}/\theta_{[1]} \geqslant \delta_2^*$. The preference zone in Figure 15.4.5 consists of all those points common to both regions, or equivalently, the intersection òf the two hatched regions in Figure 15.4.4.

In carrying out the procedure, we must distinguish between the two cases where (1) the Σ_j are known, and (2) the Σ_j are unknown. In many situations we may not actually know the Σ_j matrices but we have good estimates, so that for practical purposes we may wish to assume that the Σ_j are known. When the Σ_j are unknown, they must be estimated; if the Σ_j are assumed common, the estimation procedure must take this information into account. The two cases require different tables to carry out the procedure.

Case of Known Covariance Matrices. We first deal with the problem of determining a common sample size n to take from each of the k populations when the population covariance matrices are known. To determine n we specify values of $P^*, \delta_1^* > 0$, and $\delta_2^* > 1$; the number p of variables is also fixed. The value P^* is interpreted as the minimum probability of a correct selection whenever we have *both* $\delta_1 \geqslant \delta_1^*$ and $\delta_2 \geqslant \delta_2^*$.

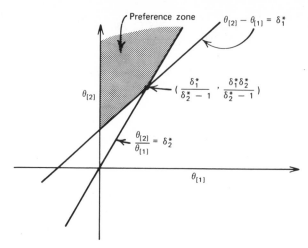

Figure 15.4.5 Preference zone defined by the intersection of $\theta_{[2]} - \theta_{[1]} \geqslant \delta_1^*$ *and* $\theta_{[2]}/\theta_{[1]}\delta_2^*$.

For the goal of selecting the population with the *largest* Mahalanobis 'distance' from the origin when the covariance matrices Σ_j are known, we enter Table S.1 with the values of P^* and δ_2^* for the appropriate number of populations k and number of variables p to find the value of $n\delta_1^*$; this value is divided by δ_1^* to find n. For example, if $k=4$ and $p=5$ and we specify $P^* = .90, \delta_1^* = 0.50, \delta_2^* = 1.50$, Table S.1 gives the value $n\delta_1^* = 60.29$. When this quantity is divided by the given value $\delta_1^* = 0.50$, we obtain $n = 60.29/0.50 = 120.6$, so that we should take 121 observations from each of the $k=4$ populations, each observation being a vector with $p=5$ components.

For selecting the population with the *smallest* Mahalanobis 'distance' from the origin when the covariance matrices Σ_j are known, we use Table S.2 in a manner similar to that of Table S.1. Note that Table S.2 is not as extensive as Table S.1. For example, suppose that $k=4$ and $p=5$ and we specify $P^* = .90, \delta_1^* = 0.50, \delta_2^* = 1.50$, then Table S.2 gives the value $n\delta_1^* = 60.31$, from which we obtain $n = 60.31/0.50 = 120.6$. We should take 121 observations from each of the four populations, each observation being a vector with $p=5$ components.

Different tables are needed for these two problems because the theory is not the same. Nevertheless we have noticed that corresponding entries in Tables S.1 and S.2 differ by at most 0.07% in all cases covered. For the special case $k=2$, the results are exactly the same. As a result, if Table S.2 does not include the particular values of k and/or δ_2^* needed for a problem of selecting the population with the smallest Mahalanobis 'distance' from the origin for Σ_j known, we recommend for practical purposes that the entry be found from Table S.1.

Once n is determined and n observations are taken on the p characteristics for each of the k individuals, the procedure is completed by computing the sample mean vectors $\bar{x}^{(j)}$ for each of the k populations and computing a quantity u_j (not to be confused with μ_j) for the jth population from the formula

$$u_j = \left(\bar{x}_1^{(j)}, \ldots, \bar{x}_p^{(j)} \right) \Sigma_j^{-1} \begin{bmatrix} \bar{x}_1^{(j)} \\ \vdots \\ \bar{x}_p^{(j)} \end{bmatrix} = \sum_{c=1}^{p} \sum_{d=1}^{p} \bar{x}_c^{(j)} x_d^{(j)} \sigma_j^{cd}, \quad (15.4.11)$$

for each $j = 1, 2, \ldots, k$. Then the u values are ordered as

$$u_{[1]} \leqslant u_{[2]} \leqslant \cdots \leqslant u_{[k]},$$

and the selection rule is to assert that the population corresponding to the largest (smallest) value of u is the one with the largest (smallest) θ value, where θ_j is the Mahalanobis 'distance' from the origin to the jth population defined by (15.4.10) for the jth population.

Example 15.4.3.   Each year the book *American Universities and Colleges* (known as The College Handbook) lists over 800 colleges in the United States and Canada, and provides a profile of the students in attendance at the colleges. Among the measurements provided for each college are the scores on the Scholastic Aptitude Test (SAT), both verbal (SAT-V) and mathematical (SAT-M). Colleges frequently use these tests in making a variety of comparisons and decisions.

Both the SAT-V and the SAT-M tests are defined to have a true mean of 500 and a standard deviation $\sigma = 100$. The correlation p between the SAT-V and the SAT-M is approximately equal to .5. This means that the covariance matrix Σ is known and has diagonal entries $\sigma^2 = 100$ and off-diagonal entries $\rho\sigma^2 = .5(100)^2$, or

$$\Sigma = \begin{bmatrix} (100)^2 & .5(100)^2 \\ .5(100)^2 & (100)^2 \end{bmatrix} = (10,000) \begin{bmatrix} 1 & .5 \\ .5 & 1 \end{bmatrix}.$$

(Of course, if $\rho = .5$ is an approximation, then Σ is also an approximation.)

Suppose we wish to compare the incoming applicants to five different departments. How many applicants shall we choose from each department in order to make a correct complete ranking of the five departments with respect to the aptitude of incoming students? We wish to have probability at least $P^* = .90$ whenever a certain appropriate distance condition (to be defined) is satisfied.

As stated, this problem is not well defined since we do not have a specific ranking and selection measure for ordering populations. One way to handle this problem is to define "best" as meaning the population (in this case the college department) that is furthest away from the origin (in the sense of Mahalanobis 'distance'); the second best population is the one second furthest from the origin; and so on. Thus we order the populations in accordance with the ordering of the squared distances from the origin

$$\theta_j = \left(\mu_1^{(j)}, \mu_2^{(j)} \right) \Sigma^{-1} \begin{bmatrix} \mu_1^{(j)} \\ \mu_2^{(j)} \end{bmatrix},$$

where $\mu_1^{(j)}$ is the true mean SAT-V for the jth department, $\mu_2^{(j)}$ is the true mean SAT-M for the jth department, and Σ is the known common variance-covariance matrix.

We take a sample of n observations from the jth department and compute the sample means $\bar{x}_1^{(j)}$ and $\bar{x}_2^{(j)}$ of the SAT-V and SAT-M scores of the candidates for the jth department, for each $j = 1, 2, \ldots, 5$. Then we compute

$$u_j = \left(\bar{x}_1^{(j)}, \bar{x}_2^{(j)} \right) \Sigma^{-1} \begin{bmatrix} \bar{x}_1^{(j)} \\ \bar{x}_2^{(j)} \end{bmatrix}$$

for each $j = 1, 2, \ldots, 5$.

If we set $P^* = .90$ and $\delta_2^* = 2.00$, then linear interpolation for $p = 2$ in Table S.1 gives the value $n\delta_1^* = 39.73$. If $\delta_1^* = 0.80$ then $n = 49.7$, so that a sample of 50 observations is required for each of the five populations, and each observation consists of a pair (SAT-V and SAT-M).

Now suppose that a sample of 50 observations is taken from each of the five departments, and the sample mean scores are as follows:

Department	Sample means SAT-V	Sample means SAT-M
1	560	480
2	620	570
3	680	610
4	670	630
5	630	550

To carry out the procedure the u values must be computed. The first step is to compute Σ^{-1} as

$$\Sigma^{-1} = \frac{1}{10,000}\begin{bmatrix} 1.33 & -0.67 \\ -0.67 & 1.33 \end{bmatrix}.$$

Then the u values are

$$u_1 = \frac{1}{10,000}(560, 480)\begin{bmatrix} 1.33 & -0.67 \\ -0.67 & 1.33 \end{bmatrix}\begin{bmatrix} 560 \\ 480 \end{bmatrix}$$

$$= \frac{1}{10,000}\left[(560)^2(1.33) + (480)^2(1.33) - 2(560)(480)(0.67)\right] = 36.3.$$

$$u_2 = \frac{1}{10,000}(620, 570)\begin{bmatrix} 1.33 & -0.67 \\ -0.67 & 1.33 \end{bmatrix}\begin{bmatrix} 620 \\ 570 \end{bmatrix} = 47.0,$$

$$u_3 = 55.4,$$

$$u_4 = 55.9,$$

$$u_5 = 46.6.$$

We note that $u_4 = 55.9$ is the largest of the u values, so the population of applicants to the department labeled 4 is asserted to be the one with the largest Mahalanobis 'distance' from the origin of SAT scores. The probability that this assertion is correct is at least .90 whenever $\delta_1 = \theta_{[5]} - \theta_{[4]} \geq 0.80$ and $\delta_2 = \theta_{[5]}/\theta_{[4]} \geq 2.00$.

Example 15.4.4. Consider a shooting contest in which each of k contestants shoots n arrows (or bullets) at the center of a target. The object is to choose the best contestant or the one with the best shooting ability. In this model each shot yields an x and a y coordinate (which are not necessarily independent), and what counts is how close the composite score is to the center of the target. If the x and y coordinates were independent, we would use ordinary Euclidean distance (or the square of this distance) as a measure of departure from the center or origin. But since x and y are correlated, we use the Mahalanobis 'distance'. Thus we wish to order

$$\theta_j = \left(\mu_x^{(j)}, \mu_y^{(j)}\right)\Sigma_j^{-1}\begin{bmatrix} \mu_x^{(j)} \\ \mu_y^{(j)} \end{bmatrix}$$

for $j = 1, 2, \ldots, k$ and select the *smallest* θ value, where $(\mu_x^{(j)}, \mu_y^{(j)})$ is the mean vector for the jth contestant and Σ_j is the covariance matrix for the jth contestant.

Suppose we have 10 contestants and wish to select the best shot, that is, the one who comes closest to the center of the target. We want the probability of a correct selection to be at least $P^* = .90$ whenever both $\delta_1 \geqslant \delta_1^* = 2.0$ and $\delta_2 \geqslant \delta_2^* = 3.0$. Further, suppose that previous data provide good estimates of the covariance matrices. How many observations shall we take on each contestant?

From Table S.2 with $k = 10$ and $\delta_2^* = 3.0$, we obtain a value of $n\delta_1^*$ between 33.21 for $p = 1$ and 35.44 for $p = 5$, which by linear interpolation for $p = 2$ yields $n\delta_1^* = 33.77$. Consequently, we obtain $n = 33.77/2.0 = 16.9$, in which case each of the 10 contestants should shoot 17 times.

Case of Unknown Covariance Matrices. When the population covariance matrices $\Sigma_1, \Sigma_2, \ldots, \Sigma_k$ are unknown, they must be estimated by the sample covariance matrices $S_1, S_2, \ldots, S_k$ respectively, as given by (15.4.5). If the Σ_j are assumed common for $j = 1, 2, \ldots, k$, the sample estimate of the common Σ is given in (15.4.8) or, if the n_j are also common, by (15.4.6). The procedures are similar to the case of known covariance matrices except that here we compute

$$v_j = \left(\bar{x}_1^{(j)}, \ldots, \bar{x}_p^{(j)} \right) S_j^{-1} \begin{pmatrix} \bar{x}_1^{(j)} \\ \vdots \\ \bar{x}_p^{(j)} \end{pmatrix} = \sum_{c=1}^{p} \sum_{d=1}^{p} \bar{x}_c^{(j)} \bar{x}_d^{(j)} s_j^{cd}, \qquad (15.4.12)$$

where s_j^{cd} is the (c, d) element of the inverse of the matrix S_j; that is, $S_j^{-1} = (s_j^{cd})$.

The value of n required for selecting the population with the *largest* Mahalanobis 'distance' from the origin when the Σ_j are unknown can be found from Table S.3. Unfortunately, these tables are difficult to compute, so that the entries are somewhat sparse. Tables for selecting the population with the *smallest* Mahalanobis 'distance' from the origin when the Σ_j are unknown apparently have not yet been calculated.

To find n from Table S.3, we use the specified values of P^* and δ_2^* for the appropriate k and p to determine a single column of the table, which we refer to as the *column determined*. Each entry in the column labeled $n\delta_1^*$ is divided by the specified value δ_2^* and compared with the corresponding entry in the column determined. The entry in the column determined that agrees with the corresponding n found by dividing $n\delta_1^*$ by δ_1^* is the required value of n.

For example, suppose that $k = 2$, $p = 4$, $P^* = .90$, $\delta_1^* = 2.00$, and $\delta_2^* = 1.50$. These values determine that n will be an entry in the first column of the first table of Table S.3. The entries in the column labeled $n\delta_1^*$ are each divided by 2.00 until an agreement is obtained. When 120 is divided by 2 we obtain 60, and the corresponding entry in the column determined is 61; thus the required sample size is very slightly smaller than 61, so that $n = 61$ observations are required for each of the $k = 2$ populations.

Example 15.4.5. Harris and Hárris (1973) provide an extensive study of (1) the relationship between concept attainment in several different school subjects, and (2) the relationship between concept learning in various school subjects and basic cognitive abilities. This study has a wealth of interesting data. In one section (Tables III.15, III.16, A.9, A.11, A.12) data are given on a variety of concepts for boys and girls in 1970 and 1971. For simplicity of computation we consider only four concepts in the area of physical science:

Concept 1 = conduction,
Concept 2 = expansion,
Concept 3 = friction,
Concept 4 = melting.

As a first step, suppose we wish to choose the population (from among boys and girls in 1970 and 1971) that has the largest Mahalanobis 'distance' from the origin. How many observations should we take so that the probability of a correct selection is at least $P^* = .95$ whenever $\delta_1 \geqslant \delta_1^* = 2.0$ and $\delta_2 \geqslant \delta_2^* = 2.0$?

Suppose first that the population covariance matrices are known. From Table S.1 with $k = 4$ (Boys 1970, Boys 1971, Girls 1970, Girls 1971) and $p = 4$ (concepts 1, 2, 3, 4), we obtain (by linear interpolation) the value $n\delta_1^* = 36.02$. Hence $n = 36.02/2.0 = 18.01$, and we need to take 19 observations from each of the four populations. Each observation consists of measurements on four variables.

Now suppose that these measurements are taken and the reults for the sample means are as follows:

Sample Means

Concept	Boys		Girls	
	1970	1971	1970	1971
1	6.62	7.36	6.04	6.83
2	7.51	7.69	7.80	7.74
3	7.69	7.77	7.49	7.40
4	7.75	8.37	8.34	8.69

and that the true variance-covariance matrices Σ_B and Σ_G for boys and girls, respectively, are

$$\Sigma_B = \begin{bmatrix} 7.50 & 4.50 & 4.00 & 4.00 \\ 4.50 & 7.50 & 4.00 & 4.00 \\ 4.00 & 4.00 & 5.50 & 3.50 \\ 4.00 & 4.00 & 3.50 & 5.50 \end{bmatrix} \qquad \Sigma_G = \begin{bmatrix} 7.00 & 3.75 & 3.50 & 3.50 \\ 3.75 & 7.00 & 3.50 & 3.50 \\ 3.50 & 3.50 & 5.00 & 2.75 \\ 3.50 & 3.50 & 2.75 & 5.00 \end{bmatrix}.$$

Then we need to compute the inverses of these matrices as

$$\Sigma_B^{-1} = \begin{bmatrix} 0.2689 & -0.0644 & -0.0909 & -0.0909 \\ -0.0644 & 0.2689 & -0.0909 & -0.0909 \\ -0.0909 & -0.0909 & 0.3864 & -0.1136 \\ -0.0909 & -0.0909 & -0.1136 & 0.3864 \end{bmatrix},$$

$$\Sigma_G^{-1} = \begin{bmatrix} 0.2668 & -0.0409 & -0.1020 & -0.1020 \\ -0.0409 & 0.2668 & -0.1020 & -0.1020 \\ -0.1020 & -0.1020 & 0.3789 & -0.0656 \\ -0.1020 & -0.1020 & -0.0656 & 0.3789 \end{bmatrix},$$

and from (15.4.11) the u values are computed as

$$u_1 = (6.62, 7.51, 7.69, 7.75)\Sigma_B^{-1} \begin{bmatrix} 6.62 \\ 7.51 \\ 7.69 \\ 7.75 \end{bmatrix} = 13.39,$$

$$u_2 = (7.36, 7.69, 7.77, 8.37)\Sigma_B^{-1} \begin{bmatrix} 7.36 \\ 7.69 \\ 7.77 \\ 8.37 \end{bmatrix} = 14.63,$$

$$u_3 = (6.04, 7.80, 7.49, 8.34)\Sigma_G^{-1} \begin{bmatrix} 6.04 \\ 7.80 \\ 7.49 \\ 8.34 \end{bmatrix} = 16.83,$$

$$u_4 = (6.83, 7.74, 7.40, 8.69)\Sigma_G^{-1} \begin{bmatrix} 6.83 \\ 7.74 \\ 7.40 \\ 8.69 \end{bmatrix} = 17.20.$$

Since u_4 is the largest and this u value corresponds to Girls 1971, we assert that Girls 1971 is the population with the largest Mahalanobis 'distance' from the origin. The probability that this assertion is a correct selection is at least .95 whenever $\delta_2 = \theta_{[4]}/\theta_{[3]} \geqslant 2.00$ and $\delta_1 = \theta_{[4]} - \theta_{[3]} \geqslant 2.00$, where θ_1, θ_2, θ_3, θ_4 are the Mahalanobis 'distances' from the origin for Boys 1970, Boys 1971, Girls 1970, and Girls 1971, respectively.

Now suppose that Σ_B and Σ_G are unknown; then we use Table S.3. Entering Table S.3 with $k=4$, $p=4$, and $\delta_2^* = 2.00$, we obtain the entries bracketing $\delta_1^* = 2.00$ as $\delta_1^* = 1.780$, $n = 50.555$, and $\delta_1^* = 2.106$, $n = 47.473$. Linear interpolation gives $n = 48.5$ for $\delta_1^* = 2.00$, so we should take 49 observations from each population.

The selection procedure is now similar to the case where the Σ_j are known, except that we compute v_j from (15.4.12) using the sample covariance matrices instead of the u_j that use the population covariance matrices. Suppose the results of the data are as follows:

Sample means

Concept	Boys		Girls	
	1970	1971	1970	1971
1	6.70	7.42	6.01	6.87
2	7.50	7.71	7.85	7.70
3	7.65	7.75	7.55	7.45
4	7.73	8.35	8.30	8.61

Sample covariance matrices:

Boys   1970   Boys   1971

$$S_1 = \begin{bmatrix} 7.45 & 4.86 & 3.72 & 3.41 \\ 4.86 & 7.51 & 4.06 & 4.34 \\ 3.72 & 4.06 & 5.52 & 3.44 \\ 3.41 & 4.34 & 3.44 & 5.76 \end{bmatrix}, \quad S_2 = \begin{bmatrix} 7.84 & 4.56 & 4.30 & 4.18 \\ 4.56 & 7.13 & 3.97 & 3.73 \\ 4.30 & 3.97 & 5.95 & 3.70 \\ 4.18 & 3.73 & 3.70 & 5.43 \end{bmatrix},$$

Girls   1970   Girls   1971

$$S_3 = \begin{bmatrix} 6.76 & 3.89 & 2.92 & 3.20 \\ 3.89 & 7.67 & 3.89 & 3.71 \\ 2.92 & 3.89 & 4.67 & 2.71 \\ 3.20 & 3.71 & 2.71 & 4.67 \end{bmatrix}, \quad S_4 = \begin{bmatrix} 7.67 & 3.76 & 3.30 & 3.85 \\ 3.76 & 6.81 & 2.82 & 3.06 \\ 3.30 & 2.82 & 5.06 & 2.93 \\ 3.85 & 3.06 & 2.93 & 4.88 \end{bmatrix}.$$

The inverse matrices are

Boys   1970

$$S_1^{-1} = \begin{bmatrix} 0.2547 & -0.1111 & -0.0767 & -0.0213 \\ -0.1111 & 0.3248 & -0.0836 & -0.1290 \\ -0.0767 & -0.0836 & 0.3612 & -0.1073 \\ -0.0213 & -0.1290 & -0.1073 & 0.3475 \end{bmatrix},$$

Boys   1971

$$S_2^{-1} = \begin{bmatrix} 0.2694 & -0.0735 & -0.0834 & -0.1000 \\ -0.0735 & 0.2718 & -0.0821 & -0.0742 \\ -0.0834 & -0.0821 & 0.3612 & -0.1255 \\ -0.1000 & -0.0742 & -0.1255 & 0.3976 \end{bmatrix},$$

Girls   1970

$$S_3^{-1} = \begin{bmatrix} 0.2468 & -0.0495 & -0.0570 & -0.0968 \\ -0.0495 & 0.2767 & -0.1382 & -0.1057 \\ -0.0570 & -0.1382 & 0.4199 & -0.0949 \\ -0.0968 & -0.1057 & -0.0949 & 0.4195 \end{bmatrix},$$

Girls   1971

$$S_4^{-1} = \begin{bmatrix} 0.2441 & -0.0558 & -0.0565 & -0.1237 \\ -0.0558 & 0.2302 & -0.0518 & -0.6920 \\ -0.0565 & -0.0518 & 0.3353 & -0.1243 \\ -0.1237 & -0.6920 & -0.1243 & 0.4205 \end{bmatrix}.$$

Finally, the sample Mahalanobis 'distances' from the origin are

$$v_1 = (6.70, 7.50, 7.65, 7.73)\,S_1^{-1} \begin{bmatrix} 6.70 \\ 7.50 \\ 7.65 \\ 7.73 \end{bmatrix} = 13.13,$$

$$v_2 = (7.42, 7.71, 7.75, 8.35)\,S_2^{-1} \begin{bmatrix} 7.42 \\ 7.71 \\ 7.75 \\ 8.35 \end{bmatrix} = 14.40,$$

$$v_3 = (6.01, 7.85, 7.55, 8.30)\,S_3^{-1} \begin{bmatrix} 6.01 \\ 7.85 \\ 7.55 \\ 8.30 \end{bmatrix} = 17.25,$$

$$v_4 = (6.87, 7.70, 7.45, 8.61) S_4^{-1} \begin{bmatrix} 6.87 \\ 7.70 \\ 7.45 \\ 8.61 \end{bmatrix} = 17.57.$$

The largest of these v values is v_4, which corresponds to Girls 1971. Thus we assert that Girls 1971 is the population with the largest Mahalanobis 'distance' from the origin. With this procedure, the probability of a correct selection is at least .95 whenever $\delta_2 = \theta_{[4]}/\theta_{[3]} \geqslant 2.00$ and $\delta_1 = \theta_{[4]} - \theta_{[3]} \geqslant 2.00$.

Example 15.4.6. Consider again the situation of Example 15.4.1 but suppose that the goal is to select the group with the largest Mahalanobis 'distance' from the origin. The inverse of the estimated common covariance matrix given there is

$$S^{-1} = \frac{1}{10^3} \begin{bmatrix} 61.9521 & -4.8711 & -1.3415 \\ -4.8711 & 56.0192 & -4.7767 \\ -1.3415 & -4.7767 & 91.3914 \end{bmatrix}.$$

The sample statistics from (15.4.12) are

$$v_1 = 44.77,$$

$$v_2 = 49.35,$$

$$v_3 = 39.58,$$

and thus mechanics is again the group selected as best.

15.4.3 The Folded Normal Distribution

This section is prompted by an interesting application that deals with the folded normal distribution. The model is really not a multivariate one, but it is allied to the problems using Mahalanobis 'distance' and requires the same set of tables used for the other multivariate models in this chapter. For this reason, this material is included here.

The application is given by Leone, Nelson, and Nottingham (1961) and deals with measuring the straightness of a lead wire. In automatic manufacturing equipment the straightness of a lead wire can be important, even if only to prevent the jamming of the feeding machines. The straightness (or lack of it) is measured by a nonnegative quantity called *camber*. One end of the lead wire is clamped in a well-balanced rotating chuck and as the chuck rotates the maximum and minimum positions of the free end (perpendicular to the direction of the wire) are measured with an optical measuring device. The diameter D of the uniformly thick lead wire is the straight-line crosssectional distance through the wire. The camber C is defined by

$$C = |\text{maximum} - \text{minimum}| - D, \qquad (15.4.13)$$

which is necessarily nonnegative; it can be zero only if the lead wire is perfectly straight (and the chuck is perfectly balanced). If we try to fit a normal distribution based on large values of C, we find that the histogram is too high for small values of C–as if to compensate for the absence of negative values. An illustration of this is given in Leone, Nelson, and Nottingham (1961), where a figure shows that the model of a normal distribution folded at zero fits a histogram of 497 sample C values. Here $\mu_0 = 0$ and μ_j is the population mean of the jth normal distribution "before any folding," and we assume a common variance σ^2 before folding. As a result of folding, there is a new population mean $\mu_j(f)$ that is larger than μ_j and a new nonnegative population standard deviation $\sigma(f)$ that is smaller than σ and these can be computed; tables to do this are provided in Leone, Nelson, and Nottingham (1961).

Now we consider a ranking and selection procedure in the context of this application. Suppose that we have k different methods of producing lead wires, and we wish to select the one that is producing the straightest wires, that is, the one whose population camber mean value is closest to zero, or equivalently, the one for which the absolute value $|\mu|$ is smallest since all the μ values are positive here. This is equivalent to selecting the one with the smallest μ^2 value among $\mu_1^2, \mu_2^2, \ldots, \mu_k^2$, or using the usual ordering notation, selecting $\mu_{[1]}^2$ from

$$\mu_{[1]}^2 \leqslant \mu_{[2]}^2 \leqslant \cdots \leqslant \mu_{[k]}^2.$$

The first step is to determine a common sample size n to take from each of the k methods of production. For a specified triplet $(\delta_1^*, \delta_2^*, P^*)$, the value of n is determined from Table S.2 exactly as in Section 15.4.2 but always with $p = 1$. Here P^* is interpreted as the minimum probability required for a correct selection of the production method that makes the straightest lead wires when

$$\delta_1 = \mu_{[2]}^2 - \mu_{[1]}^2 \geqslant \delta_1^* \quad \text{and} \quad \delta_2 = \frac{\mu_{[2]}^2}{\mu_{[1]}^2} \geqslant \delta_2^* \quad \text{for} \quad \delta_1^* > 0 \quad \text{and} \quad \delta_2^* > 1.$$

For each lead wire in each sample, we observe the maximum and minimum positions of the free end as explained earlier, take their absolute difference, and subtract the appropriate value of D in order to obtain a C value as defined by (15.4.13). Let C_{ij} denote the C value for the ith observation in the jth sample, for $i = 1, 2, \ldots, n$ and $j = 1, 2, \ldots, k$. The C values for the jth sample are then squared and summed to obtain a quantity

$$M_j = \sum_{i=1}^{n} C_{ij}^2 \qquad (15.4.14)$$

for each $j = 1, 2, \ldots, k$. These M values are ordered as

$$M_{[1]} \leqslant M_{[2]} \leqslant \cdots \leqslant M_{[k]},$$

and the production method that gives rise to the smallest M value, $M_{[1]}$, is asserted to be the one that is producing the straightest lead wires.

As an example, suppose that we specify $P^* = .90$, $\delta_1^* = 0.5$, and $\delta_2^* = 1.50$ and there are $k = 4$ methods of production. Then we enter Table S.2 with $k = 4$ and $p = 1$ to obtain $n\delta_1^* = 59.49$ and $n = 59.49/0.5 = 118.98$, so that we need to take $n = 119$ observations on each of the $k = 4$ methods. The 119 observations are squared and summed as in (15.4.14) to obtain four M values. The method that gives rise to $M_{[1]}$ is asserted to be the best one.

15.5 SELECTING THE POPULATION WITH THE LARGEST CORRELATION COEFFICIENT

In this section we consider two separate models for the problem of selecting the population with the largest correlation. In the first model we have k bivariate normal populations with one variate X common to all of them and possibly unequal correlation coefficients, and in the second we have k multivariate normal populations each with its own common correlation coefficient between any two components (and, of course, these may not be equal). These two models deal with different types of problems and are treated separately.

15.5.1 Selecting the Treatment that Has the Largest Correlation with the Overall Effect

Suppose we have k bivariate variables

$$(X, Y_1), (X, Y_2), \ldots, (X, Y_k),$$

where $Y_1, \ldots, Y_k$ denote measurements on k different treatments, say, and X is a measure of the overall effect of all these treatments. If these k bivariate variables are normally distributed with correlation coefficients

$\rho_1, \rho_2, \ldots, \rho_k$, respectively, and ordered values

$$\rho_{[1]} \leqslant \rho_{[2]} \leqslant \cdots \leqslant \rho_{[k]},$$

then we may wish to select the treatment (or population) with the largest ρ value, $\rho_{[k]}$. This means that we wish to select the treatment that has the highest correlation with the effect; here the treatments are our populations.

We first take a sample of size n from each of the k bivariate populations. Let $(x_i^{(j)}, y_i^{(j)})$ denote the ith observation, for $i = 1, 2, \ldots, n$, from the jth population, $j = 1, 2, \ldots, k$, and let $\bar{x}^{(j)}$ and $\bar{y}^{(j)}$ denote the respective sample means for the jth population. We then compute the correlation coefficient r_j for the sample from the jth population using

$$r_j = \frac{\sum\limits_{i=1}^{n} \left(x_i^{(j)} - \bar{x}^{(j)} \right) \left(y_i^{(j)} - \bar{y}^{(j)} \right)}{\sqrt{\left[\sum\limits_{i=1}^{n} \left(x_i^{(j)} - \bar{x}^{(j)} \right)^2 \right] \left[\sum\limits_{i=1}^{n} \left(y_i^{(j)} - \bar{y}^{(j)} \right)^2 \right]}}, \qquad (15.5.1)$$

$$= \frac{s_{xy}^{(j)}}{\sqrt{s_{xx}^{(j)} s_{yy}^{(j)}}} = \frac{s_{xy}^{(j)}}{s_x^{(j)} s_y^{(j)}}, \qquad (15.5.2)$$

where $s_{xy}^{(j)}$, $s_x^{(j)}$, and $s_y^{(j)}$ in (15.5.2) are, respectively, the covariance of x and y, standard deviation of the x values, and standard deviation of the y values, and all are obtained for the observations from the jth population. These k sample correlation coefficients are ordered as

$$r_{[1]} \leqslant r_{[2]} \leqslant \cdots \leqslant r_{[k]},$$

and we assert that the population that gives rise to the largest r value, $r_{[k]}$, is the one with the largest ρ value, $\rho_{[k]}$.

Suppose that we are in the process of designing an experiment to select the bivariate population with the largest ρ value, and the problem is to determine the sample size n needed to satisfy a certain probability requirement. This requirement is that the probability of a correct selection should be at least P^* whenever the distance measure

$$\delta = \rho_{[k]} - \rho_{[k-1]}$$

satisfies $\delta \geqslant \delta^*$ for some specified $\delta^* > 0$.

The required value of n can be determined approximately from Table A.1 of Appendix A. We enter the table with the values of k and P^* to obtain a quantity τ_t. Then n is found by substituting τ_t and the specified δ^*

value into the expression

$$n = \frac{\tau_t^2}{\delta^{*2}}. \tag{15.5.3}$$

Example 15.5.1 Suppose that a personnel manager of a construction company wishes to determine which one of four tests given to trainees is the best predictor of ability to set rivets. Whichever test is the most useful predictor will be used to evaluate prospective employees. Since the riveter has to catch the hot rivet, set it, and drive it into the metal plate with a heavy riveting tool, he must be good in (1) finger dexterity, (2) manual dexterity, (3) visual depth perception, and (4) strength of the arm and wrist. As a result, the four tests are designed to measure these four abilities. In each case larger scores indicate better ability. The manager plans to take a random sample of n riveting personnel after a certain period of training and observe the number of rivets (Y) set by each subject. The scores X_1, X_2, X_3, and X_4 on the four skill tests are available for each subject from the files.

The personnel manager specifies that he wants the probability of a correct selection of the most useful predictor test to be at least $P^* = .90$ when the true difference of correlation coefficients satisfies $\delta = \rho_{[4]} - \rho_{[3]} \geqslant 0.30$. We enter Table A.1 with $k = 4$ and $P^* = .90$ to obtain $\tau_t = 2.4516$. Then n is obtained from (15.5.3) as

$$n = \frac{(2.4516)^2}{(0.30)^2} = 66.78,$$

and thus the sample should consist of 67 trainees. To carry out the selection procedure, we compute the four correlation coefficients between the number of rivets set and the scores on each of the four skill tests, order them, and determine $r_{[4]}$. The test that gives rise to the largest sample correlation is asserted to be the best predictor of rivet-setting ability.

Some Refinements. Determination of the exact least favorable configuration is difficult in this problem, even for the asymptotic result. However, we do have some pertinent information, and we now show how this information can be used to pinpoint the asymptotic answer more exactly. Consider the two configurations C_0 and C_1 defined by

$$C_0: \quad \rho_{[1]} = \rho_{[2]} = \cdots = \rho_{[k-1]} = \frac{-\delta^*}{2}, \qquad \rho_{[k]} = \frac{\delta^*}{2}, \tag{15.5.4}$$

$$C_1: \quad \rho_{[1]} = \rho_{[2]} = \cdots = \rho_{[k-1]} = 0, \qquad \rho_{[k]} = \delta^*, \tag{15.5.5}$$

and the configuration C_α that is between C_0 and C_1 defined by

$$C_\alpha: \quad \rho_{[1]} = \rho_{[2]} = \cdots = \rho_{[k-1]} = \frac{-\delta^*}{2}(1-\alpha), \quad \rho_{[k]} = \frac{\delta^*}{2}(1+\alpha), \quad (15.5.6)$$

for $0 \leqslant \alpha \leqslant 1$. When $\alpha = 0$ we obtain C_0 and when $\alpha = 1$ we obtain C_1. We claim that for any number k of populations, the least favorable configuration occurs for a configuration C_α. In fact, we conjecture that the least favorable configuration is C_0 for $k = 2$ and that the least favorable configuration C_α approaches C_1 as k increases. We therefore recommend obtaining the n value for each of the preceding three configurations (C_0, C_1 and C_α with $\alpha = 1/2$) and using them to fit a simple parabola

$$n = a + b\rho + c\rho^2. \quad (15.5.7)$$

In particular, we use each of the three pairs of values for (ρ, n), say $(-\delta^*/2, n_0)$, $(0, n_1)$, and $(-\delta^*/4, n_{.5})$, and solve the resulting three equations for the three unknowns, a, b, and c. The n value required, denoted by n_{max}, is then found from the formula

$$n_{max} = a - \frac{b^2}{4c}, \quad (15.5.8)$$

since this gives the maximum value of n on the parabola in (15.5.7).

The value n_{max} is a better estimate of the number of observations required to satisfy the (δ^*, P^*) requirement than that found from (15.5.3). Tables A.1 and A.3 are used for this parabola method, and we note that no new tables are required. The computations are illustrated by examples after we give the relevant formulas.

For $\rho = -\delta^*/2$ as in the configuration C_0 in (15.5.4), we use Table A.1 for the appropriate k and specified P^* to find a value τ_t; then n_0 is calculated from the formula

$$n_0 = \left(\frac{\tau_t}{\delta^*}\right)^2 \frac{\left[1 - (\delta^*/2)^2\right]^2}{1 + (\delta^*/2)^2}. \quad (15.5.9)$$

For $\rho = 0$ as in configuration C_1 in (15.5.5), we first use the specified δ^* to calculate a quantity ρ_t from the formula

$$\rho_t = \frac{\left[1 - (\delta^*)^2\right]^2}{1 + \left[1 - (\delta^*)^2\right]^2}. \quad (15.5.10)$$

With this value of ρ_t for the caption ρ, the specified P^* and appropriate k, we use linear interpolation in Table A.3 to obtain an h value. Then n_1 is calculated from the formula

$$n_1 = \left(\frac{h}{\delta^*}\right)^2 \frac{1}{1-\rho_t} = \left(\frac{h}{\delta^*}\right)^2 \left\{1 + \left[1-(\delta^*)^2\right]^2\right\}. \qquad (15.5.11)$$

For $\rho = -\delta^*/4$ as in configuration $C_{.5}$ in (15.5.6), the first step is to calculate the intermediate positive quantity

$$Q = \left[1 + \left(\frac{\delta^*}{4}\right)^2\right]\left[\frac{1-(3\delta^*/4)^2}{1-(\delta^*/4)^2}\right]^2 + \left(\frac{\delta^*}{4}\right)^2$$

$$-2\left(\frac{\delta^*}{4}\right)\left(\frac{3\delta^*}{4}\right)\left[\frac{1-(3\delta^*/4)^2}{1-(\delta^*/4)^2}\right]\sqrt{\frac{1+(\delta^*/4)^2}{1+(3\delta^*/4)^2}} \ . \qquad (15.5.12)$$

Then ρ_t is found from

$$\rho_t = \frac{Q}{1+Q}. \qquad (15.5.13)$$

With this value of ρ_t for the caption ρ, the specified P^* and appropriate k, we again use linear interpolation in Table A.3 to obtain an h value. Then $n_{.5}$ is given by

$$n_{.5} = \left(\frac{h}{\delta^*}\right)^2 \frac{\left[1-(\delta^*/4)^2\right]^2}{(1-\rho_t)\left[1+(\delta^*/4)^2\right]}. \qquad (15.5.14)$$

To illustrate the computations, suppose that $k=5$, $P^*=.90$, and $\delta^*=0.2$. From Table A.1 we obtain $\tau_t = 2.5997$ and n_0 (for configuration C_0 where $\rho = -\delta^*/2 = -.10$) is found from (15.5.9) as

$$n_0 = \left(\frac{2.5997}{0.2}\right)^2 \frac{\left[1-(0.1)^2\right]^2}{1+(0.1)^2} = 163.96.$$

For $\rho = 0$ (C_1) we calculate ρ_t from (15.5.10) as

$$\rho_t = \frac{\left[1-(0.2)^2\right]^2}{1+\left[1-(0.2)^2\right]^2} = .4796,$$

and the h value from Table A.3 is $h = 1.8453$. Then from (15.5.11) we obtain n_1 as

$$n_1 = \left(\frac{1.8453}{0.2}\right)^2 \frac{1}{(1 - .4796)} = 163.58.$$

For $\rho = -\delta^*/4 = -.05$ (C_5) we use (15.5.12) to obtain $Q = .95065$ and use (15.5.13) to find $\rho_t = .48735$. With this ρ_t, $k = 5$ and $P^* = .90$, linear interpolation in Table A.3 gives $h = 1.8426$. Finally, these values are substituted in (15.5.14) to obtain

$$n_{.5} = \left(\frac{1.8426}{0.2}\right)^2 \frac{\left[1 - (0.05)^2\right]^2}{(.51265)\left[1 + (0.05)^2\right]} = 164.33.$$

We now substitute the three pairs of (n, ρ) values (163.96, .10), (163.58, 0), and (164.33, .05) in (15.5.7) to obtain the three equations

$$163.96 = a - .10b + .01c$$

$$163.58 = a$$

$$164.33 = a - .05b + .0025c.$$

The solution is $a = 163.58, b = -26.2, c = -224.0$, which from (15.5.8) gives the solution for n as

$$n_{\max} = 163.58 - \frac{(26.2)^2}{4(-224)} = 164.35.$$

Thus we need 165 observations from each of the $k = 5$ populations. (The result given by (15.5.14) happens to be the correct answer, but this need not always be the case.)

In this example with $k = 5$, $P^* = .90$, and $\delta^* = 0.20$, the simplified result based on (15.5.3) and Table A.1 is $n = (2.5997/0.20)^2 = 169$. Although this result is easy to obtain and also conservative, we have not proved that it will be conservative in all cases. The more complicated method of fitting a parabola gives a surer means of obtaining a good, conservative result.

For larger values of k, the difference between the two results should be larger. For $k = 25$, $\delta^* = .5$, and $P^* = .90$, for example, the simplified solution from (15.5.3) is $n = 46$. The parabola solution yields $n_0 = 38.05$, $n_1 = 38.85$, and $n_{.5} = 39.49$ for $\rho = -.25$, 0, and $-.125$, respectively, which give the result $n_{\max} = 39.53$; thus only 40 observations are required from each of the $k = 25$ populations. Here again, the simplified result is conservative, but it involves an increase of 15% in the number of observations.

REMARK.   The reader is reminded that the preceding discussion applies to the asymptotic case; small sample results based on exact distribution theory are not available.   □

Suppose that we have some prior information about the ρ values. For example, suppose we are given that at least one ρ value is greater than $\rho_0 > 0$ so that the least favorable configuration is

$$\rho_{[1]} = \rho_{[2]} = \cdots = \rho_{[k-1]} = \rho_0, \qquad \rho_{[k]} = \rho_0 + \delta^*. \tag{15.5.15}$$

This value ρ_0 is used to compute the quantity

$$Q = (1 + \rho_0^2)\left[\frac{1 - (\rho_0 + \delta^*)^2}{1 - \rho_0^2}\right]^2 + \rho_0^2$$

$$- 2|\rho_0(\rho_0 + \delta^*)|\left[\frac{1 - (\rho_0 + \delta^*)^2}{1 - \rho_0^2}\right]\sqrt{\frac{1 + \rho_0^2}{1 + (\rho_0 + \delta^*)^2}}, \tag{15.5.16}$$

after which we compute

$$\rho_t = \frac{Q}{1 + Q}, \tag{15.5.17}$$

Finally, this value of ρ_t is used for the caption ρ in Table A.3, with the given k and specified P^*, to find an h value. Then the required value of n is

$$n = \left(\frac{h}{\delta^*}\right)^2 \frac{(1 - \rho_0^2)^2}{(1 + \rho_0^2)(1 - \rho_t)}. \tag{15.5.18}$$

REMARK.   Here the least favorable configuration is at $\rho = \rho_0$ and we do not use the parabola method to obtain the required value of n.   □

Consider again the preceding example where $k = 5, \delta^* = .2, P^* = .90$, but now assume we know that at least one ρ value is greater than $\rho_0 = .5$. From (15.5.16) we obtain $Q = 0.3920$, so that $\rho_t = .2816$ and $h = 1.8999$. Then (15.5.18) gives $n = 56.53$ so that only 57 observations are required from each population to meet the same (δ^*, P^*) requirement. This is a reduction of about 65% from the previous result of 165. This reduction is brought about by using a priori information about the ρ values. Thus any bona fide prior information about the ρ values (of the type described earlier) can and should be used by the method described.

It should be pointed out that for the present problem the *total* number of observations taken is $n(k+1)$, since for each individual in the sample we observe the vector $(X, Y_1, Y_2, \ldots, Y_k)$.

Up to this point in this section, we have assumed that the variables $Y_1, Y_2, \ldots, Y_k$ are independent. In practical applications these variables are usually correlated; the preceding theory can be extended to cover this case also. We assume either that all pairs of Y variables have common known correlation ρ_y or that none of the pairwise correlations exceeds the given value ρ_y, where ρ_y is not necessarily zero. The derivation requires only the assumption that $\rho_y + \rho^2 \geqslant 0$, which of course holds automatically for $\rho_y = 0$ as in the model studied earlier in this section. The entire procedure reduces to the independent case when $\rho_y = 0$.

The least favorable configuration here is

$$\rho_{[1]} = \rho_{[2]} = \cdots = \rho_{[k-1]} = \rho, \qquad \rho_{[k]} = \rho + \delta^*,$$

and the distance measure is $\delta = \rho_{[k]} - \rho_{[k-1]}$. For any given ρ, known ρ_y, and specified δ^*, the first step is to calculate

$$A^2 = \left[1 - (\rho + \delta^*)^2 \right]^2 + (1 - \rho^2)^2 \left(\frac{\rho_y + \rho^2}{1 + \rho^2} \right)$$

$$- \frac{2|\rho_y + \rho(\rho + \delta^*)|(1 - \rho^2)\left[1 - (\rho + \delta^*)^2 \right]}{\sqrt{(1 + \rho^2)\left[1 + (\rho + \delta^*)^2 \right]}}, \qquad (15.5.19)$$

and use A^2 to calculate

$$Q = \frac{A^2(1 + \rho^2)}{(1 - \rho_y)(1 - \rho^2)^2}, \qquad (15.5.20)$$

which is always positive. This Q value is used in (15.5.17) to determine a value of ρ_t. With this ρ_t value, we enter Table A.3 for the appropriate k and given P^*, and use linear interpolation to obtain an h value. Then the required value of n is

$$n = \left(\frac{h}{\delta^*} \right)^2 \frac{(1 - \rho^2)^2(1 - \rho_y)}{(1 + \rho^2)(1 - \rho_t)}. \qquad (15.5.21)$$

As before the total number of observations taken is $n(k+1)$. If we have no a priori information about ρ, we can again use the parabola method to obtain n_{max}.

REMARK 1. The model that allows the variables $Y_1, Y_2, \ldots, Y_k$ to be dependent has been worked out only recently. Further extensions of this model are possible, but they are not included in this book. □

REMARK 2. This idea of fitting a parabola for those cases where the theory gets complicated can be used in other problems also, although such methods are not included elsewhere in this book. □

15.5.2 Selecting the Population with the Largest Intraclass Correlation

Consider a situation in which we have k examinations, each examination consisting of $p \geqslant 2$ subtests. We assume that the subtests are homogeneous and that the correlations between any two subtests are approximately equal. If the correlations are all equal (within each examination), then the p-dimensional correlation matrix for the jth examination is

$$
\begin{bmatrix}
1 & \rho_j & \rho_j & \cdots & \rho_j \\
  & 1 & \rho_j & \cdots & \rho_j \\
  &   & 1 & \cdots & \rho_j \\
  &   &   & \ddots & \vdots \\
  &   &   &   & 1
\end{bmatrix},
$$

and the common value ρ_j is called the *intraclass correlation* for the jth examination. Although we show only part of the correlation matrix, the rest can easily be filled in since the matrix is symmetric. We order these ρ values for different examinations as

$$
\rho_{[1]} \leqslant \rho_{[2]} \leqslant \cdots \leqslant \rho_{[k]}.
$$

The problem is to select the examination (population) with the largest common intraclass correlation coefficient $\rho_{[k]}$.

In a more general model, the correlations between the $p \geqslant 2$ tests (within each examination) are not all equal, so that the correlation matrix for the jth population (candidate) is

$$
\begin{bmatrix}
1 & \rho_{12}^{(j)} & \cdots & \rho_{1p}^{(j)} \\
  & 1 & \cdots & \rho_{2p}^{(j)} \\
  &   & \ddots & \vdots \\
  &   &   & 1
\end{bmatrix}.
$$

Now the problem is to select the population with the largest average correlation; the average correlation for the jth population is defined by

$$\bar{\rho}_j = \frac{1}{p(p-1)} \sum_{\substack{c=1 \\ c \neq d}}^{p} \sum_{d=1}^{p} \rho_{cd}^{(j)}$$

so that $\bar{\rho}_j$ takes the place of the intraclass correlation ρ_j in the preceding paragraph. There is a $\bar{\rho}$ value for each of the k populations. We denote the ordered values by

$$\bar{\rho}_{[1]} \leqslant \bar{\rho}_{[2]} \leqslant \cdots \leqslant \bar{\rho}_{[k]},$$

and we wish to select the population corresponding to $\bar{\rho}_{[k]}$.

A sample of size n is taken from each population. For the sample from the jth population, the correlation coefficient $r_{cd}^{(j)}$ between subsets c and d is computed from

$$r_{cd}^{(j)} = \frac{s_{cd}^{(j)}}{\sqrt{s_{cc}^{(j)} s_{dd}^{(j)}}} = \frac{s_{cd}^{(j)}}{s_c^{(j)} s_d^{(j)}}, \qquad (15.5.22)$$

for each $c = 1, 2, \ldots, p, d = 1, 2, \ldots, p, c \neq d$ (there are $p(p-1)/2$ correlation coefficients). The formulas for $s_{cd}^{(j)}$, $s_{cc}^{(j)}$, and $s_{dd}^{(j)}$ in (15.5.22) are given in (15.3.5). Then the sample average for all pairs of components (c, d) in the jth population is computed from

$$\bar{r}_j = \frac{1}{p(p-1)} \sum_{\substack{c=1 \\ c \neq d}}^{p} \sum_{d=1}^{p} r_{cd}^{(j)}. \qquad (15.5.23)$$

These computations are repeated for each $j = 1, 2, \ldots, k$, and the $\bar{r}$ values are ordered as

$$\bar{r}_{[1]} \leqslant \bar{r}_{[2]} \leqslant \cdots \leqslant \bar{r}_{[k]}.$$

Then the population that gives rise to the largest $\bar{r}$ value, $\bar{r}_{[k]}$, is asserted to be the one with the largest $\bar{\rho}$ value, $\bar{\rho}_{[k]}$.

Suppose we are in the process of designing an experiment to select the population with the largest average correlation between components, and the problem is to determine the common sample size n needed to satisfy the requirement that the probability of a correct selection is at least P^* whenever the distance measure

$$\delta = \bar{\rho}_{[k]} - \bar{\rho}_{[k-1]}$$

satisfies $\delta \geqslant \delta^*$ for some specified $\delta^* > 0$.

The value of n can be determined using Table A.1 of Appendix A in the following manner. (This solution is based on a large-sample approximation.) As a first step, we enter Table A.1 with the number of populations k and the specified P^* value to find a value τ_t. For $p \geqslant 4$, τ_t is related to n by the formula

$$\tau_t = \sqrt{n}\ \delta^* \sqrt{\frac{(p+3)(p-3)}{p(p-1)}}\ , \qquad (15.5.24)$$

and for $p = 2$ the relationship is

$$\tau_t = \delta^* \sqrt{n}\ . \qquad (15.5.25)$$

The required value of n is then found by solving (15.5.24) for n if $p \geqslant 4$, or solving (15.5.25) for n if $p = 2$.

REMARK.   In the special case where $p = 2$, both of the problems considered in this subsection reduce to the problem considered in Section 15.5.1 of selecting the bivariate population with the largest ρ value. This follows because when $p = 2$, we have

$$\rho_{12}^{(j)} = \rho_{21}^{(j)} = \rho_j \qquad \text{and} \qquad \bar{\rho} = \tfrac{1}{2}(2\rho_j) = \rho_j$$

for each $j = 1, 2, \ldots, k$. The reader should note that the solution for n found from the expression in (15.5.25) is equivalent to that given in (15.5.3).   □

Example 15.5.2.   Harris and Harris (1973, Table III.5) list the following $k = 3$ areas (biological science, earth science, physical science), each with $p = 10$ science concepts:

Areas	Concepts
Biological science	Bird, cell, fish, human heart, invertebrate, eye lens, lungs, mammal, muscle, pore
Earth science	Cloud, earth core, fossil, glacier, meteor, moon, planet, sedimentary rock, volcano, wind
Physical science	Conductor, evaporation, expansion, friction, liquid, melting, molecule, solid, sound, thermometer

These concepts are all highly correlated within areas. Suppose we wish to select the population (area) with the largest average correlation between concepts. If we specify $P^* = .95$ and $\delta^* = 0.25$, then Table A.1 for $k = 3$ gives the value $\tau_t = 2.7101$. The value of n is obtained from (15.5.24) by solving for n in the equation

$$2.7101 = \sqrt{n} \ (0.25)\sqrt{\frac{(13)(7)}{10(9)}} \ .$$

The result is

$$n = \frac{(2.7101)^2(10)(9)}{(0.25)^2(13)(7)} = 116.4,$$

so that we need to take 117 vector observations (each of which is a 10-tuple) from each of the $k=3$ populations.

15.6   SELECTING THE POPULATION WITH THE LARGEST MULTIPLE CORRELATION COEFFICIENT

Suppose we have some two variables, say X and Y, where X is a predictor (or independent) variable and Y is a variable to be predicted (or dependent variable). The simple correlation between X and Y provides a measure of the strength of the linear relationship between X and Y. The square of the simple correlation coefficient, sometimes called the coefficient of determination, is the proportion of the total variation in Y that is explained by its linear relationship with X. This measure ranges between zero and one, with increasing values representing increasing degrees of association. The multiple correlation coefficient is a straightforward extension of a simple correlation coefficient to the case where there are p predictor (independent) variables, $X_1, X_2, \ldots, X_p$, and a single predicted (dependent) variable Y for each population. Its square can be interpreted as the maximum proportion of the total variation in Y that is explained by the p predictor variables.

Suppose that we have k populations and that the same predictor variables, $X_1, X_2, \ldots, X_p$, are being used for each population. We denote the predicted variable for the jth population by $Y^{(j)}$, for $j = 1, 2, \ldots, k$. (Note that the predictor variables do *not* vary with j.) Then the goal might be to determine for which population the predictability of Y is best, based on these predictor variables. The predictors here are regarded as fixed values (although they may be set by the investigator), and they are not counted as observations in the following discussion; only the Y values are counted as observations.

Before proceeding with the general discussion, we give an example from the literature where this goal is of interest.

Example 15.6.1   (Consumer Expenditure).   An analysis of this type is discussed in Rizvi and Solomon (1973) for data based on a study by Sudman and Ferber (1971). The $k = 3$ population variables to be predicted all represent consumer household expenditures. These variables were defined as

$$Y^{(1)} = \text{expenditures on dairy products,}$$

$$Y^{(2)} = \text{expenditures on bakery products,}$$

$$Y^{(3)} = \text{expenditures on auto supplies.}$$

For each of these populations, the predictor variables were

$$X_1 = \text{age of head of household,}$$

$$X_2 = \text{family income,}$$

$$X_3 = \text{education of head of household.}$$

The purpose of the study was to determine which one of the three values of $Y^{(j)}$ is best explained by these three predictor variables.

In classical multiple correlation analysis the $p + 1$ variables $Y, X_1, X_2, \ldots, X_p$ would be assumed to follow a multivariate normal distribution with $p + 1$ variables, and there is a multiple correlation coefficient between $Y^{(j)}$ and $(X_1, X_2, \ldots, X_p)$ for each j $(j = 1, 2, \ldots, k)$. This coefficient is usually denoted by $\rho_{0(12\ldots p)}^{(j)}$, where the superscript (j) indicates that the coefficient applies to the jth population, and in the subscript, 0 represents the dependent variable $Y^{(j)}$, and $(12\ldots p)$ (usually written without separating commas) represents the set of predictor variables. For simplicity of notation, we write $\theta_j = \rho_{0(12\ldots p)}^{(j)}$. These θ values can be ordered from the smallest to largest as

$$\theta_{[1]} \leqslant \theta_{[2]} \leqslant \cdots \leqslant \theta_{[k]},$$

and the goal is to select the population with the largest θ value $\theta_{[k]}$. Further, we wish to make this choice such that the probability of a correct selection is at least P^* whenever the θ values are in some well-defined preference zone.

Proceeding as before, we need to specify a preference zone. The distance measure is the pair of quantities

$$\delta_1 = \theta_{[k]} - \theta_{[k-1]}, \qquad \delta_2 = \frac{\theta_{[k]}}{\theta_{[k-1]}},$$

and the preference zone is defined as $\delta_1 \geqslant \delta_1^*$ and $\delta_2 \geqslant \delta_2^*$, where $0 < \delta_1^* < 1$ and $\delta_2^* > 1$ are specified. The least favorable configuration for this problem is given by

$$\theta_{[1]} = \cdots = \theta_{[k-1]} = \frac{\delta_1^*}{\delta_2^* - 1}, \qquad \theta_{[k]} = \frac{\delta_1^* \delta_2^*}{\delta_2^* - 1}. \tag{15.6.1}$$

(This is the point where the lines cross in Figures 15.4.4 and 15.4.5.)

To carry out the selection we compute all the sample multiple correlation coefficients, $r_{0(12\ldots p)}^{(1)}, \ldots, r_{0(12\ldots p)}^{(k)}$, each based on n observations, order them, and assert that the population that gives rise to the largest sample multiple correlation coefficient is the one with the largest θ value (i.e., the one that has the largest population multiple correlation coefficient). The formula for calculating the sample multiple correlation coefficient for population j is

$$r_{0(12\ldots p)}^{(j)} = \sqrt{1 - \frac{s_{Y(12\ldots p)}^{(j)}}{s_Y^{(j)}}}, \tag{15.6.2}$$

where

$$s_{Y(12\ldots p)}^{(j)} = \frac{\sum_{i=1}^{n} \left(y_i^{(j)} - \hat{y}_i^{(j)} \right)^2}{n - p - 1}, \tag{15.6.3}$$

and

$$s_Y^{(j)} = \frac{\sum_{i=1}^{n} \left(y_i^{(j)} - \bar{y}^{(j)} \right)^2}{n - 1}; \tag{15.6.4}$$

in (15.6.4) $\bar{y}^{(j)}$ is the sample mean of the n observations on $y^{(j)}$, and in (15.6.3) $\hat{y}_i^{(j)}$ is the predicted y value from the regression equation for population j fitted to the observations in the ith sample.

The sample multiple correlation coefficient for the jth population can also be calculated from the formula

$$r_{0(12\ldots p)}^{(j)} = 1 - \frac{\det R^{(j)}}{\det R_{11}^{(j)}}, \tag{15.6.5}$$

where $R^{(j)}$ is the $(p+1)$-dimensional matrix of simple correlations of $Y^{(j)}, X_1, X_2, \ldots, X_p$, and $R_{11}^{(j)}$ is the p-dimensional matrix that is found by deleting the first row and first column of $R^{(j)}$ (the matrix of simple correlations of $X_1, X_2, \ldots, X_p$), and $\det A$ denotes the determinant of the matrix A.

Suppose that we wish to determine the common sample size n required per population such that the probability of correctly selecting the population with the largest multiple correlation coefficient is at least P^* whenever the true values of the multiple correlations satisfy both conditions $\delta_1 \geqslant \delta_1^*$ and $\delta_2 \geqslant \delta_2^*$. The procedure is to enter Table S.1 of Appendix S with the appropriate values of k (the number of populations), p (the number of predictor variables), and the constants $P^*, \delta_1^*, \delta_2^*$.

As a numerical illustration, consider Example 15.6.1, where $k = 3$ (for y_1, y_2, y_3) and $p = 3$ (for x_1, x_2, x_3), and suppose that we specify $P^* = .90$. To choose δ_1^* and δ_2^*, note the least favorable configuration given by (15.6.1); here δ_1^* and δ_2^* depend on both $\theta_{[k-1]}$ and $\theta_{[k]}$, which are multiple correlations. We might feel that a pair of values like $\theta_{[k-1]} = .3$ and $\theta_{[k]} = .6$ should be in the preference zone. If this is the case, then we might be conservative and choose $\delta_1^* = 0.25$ and $\delta_2^* = 1.50$ so that there is some leeway when $\delta_1 \geqslant \delta_1^*$ and $\delta_2 \geqslant \delta_2^*$. From Table S.1, we use linear interpolation on the entries for $p = 1$ and $p = 5$ at $P^* = .90$ to obtain the value of $n\delta_1^* = 49.64$. Thus $n = 49.64/0.25 = 198.6$, and a sample of 199 vector observations is required from each of the $k = 3$ populations, or a total of 597 vector observations, each of which is a 3-tuple. Once these data are obtained, the sample multiple correlation coefficients are calculated from (15.6.2) or (15.6.5) and the population that produces the largest sample correlation coefficient is asserted to be the one with the largest θ value.

It is interesting to note that the conservative approach in the last paragraph has cost quite a bit. If we had used $\theta_{[k-1]} = .3$ and $\theta_{[k]} = .6$ without any "leeway," then we would have $\delta_1^* = 0.30$ and $\delta_2^* = 2.00$. These values yield a common sample size $29.66/0.30 = 98.9$ or 99 observations from each of the $k = 3$ populations; almost exactly one-half of the previous result.

If the data have already been collected with some fixed common sample size n, a probability statement can still be made. For example, suppose we have $k = 3$, $p = 3$, and a common sample size of $n = 100$ from each of the three populations. We can then make a statement such as the following: "The probability of a correct selection is at least .90 when the true values of the multiple correlations are such as to satisfy both $\theta_{[3]} - \theta_{[2]} \geqslant \delta_1^*$ and $\theta_{[3]}/\theta_{[2]} \geqslant \delta_2^*$." The method for determining the pairs (δ_1^*, δ_2^*) is described below.

For each value of δ_2^* we obtain from Table S.1 a value of $n\delta_1^*$, from which δ_1^* can be determined. To illustrate, Table 15.6.1 provides some values of (δ_1^*, δ_2^*) for this example with $P^* = .90$, $p = 3, k = 3$, for both $n = 100$ and $n = 50$.

Table 15.6.1   Excerpt of Table S.1 for $P^* = .90, k = 3, p = 3,$ *and values of (δ_1^*, δ_2^*) for $n = 100$ and $n = 50$*

δ_2^*	$n\delta_1^*$	$n = 100$ δ_1^*	$n = 50$ δ_1^*
1.50	49.64	0.4964	0.9928
1.75	36.33	0.3633	0.7266
2.00	29.66	0.2966	0.5932
2.50	22.96	0.2296	0.4592
3.00	19.58	0.1958	0.3916

These (δ_1^*, δ_2^*) pairs are points on the operating characteristic curve of the selection procedure for $n = 100$, $P^* = .90$, and for $n = 50$, $P^* = .90$.

The preceding discussion assumed that the sample size is the same for all k populations. If the sample sizes are not equal, say n_j observations are obtained from the jth population for $j = 1, 2, \ldots, k$, we can use the square-mean-root method to compute

$$n_0 = \left(\frac{\sqrt{n_1} + \cdots + \sqrt{n_k}}{k} \right)^2. \qquad (15.6.6)$$

Then this value n_0 can be used as if it were the common sample size to calculate the approximate operating characteristic curve.

To illustrate the procedure, suppose that $k = 3$, $p = 3$, and $n_1 = 40$, $n_2 = 45$, $n_3 = 62$, and we arbitrarily take $\delta_2^* = 2.50$. Then using (15.6.6) we compute

$$n_0 = \left(\frac{\sqrt{40} + \sqrt{45} + \sqrt{62}}{3} \right)^2 = 48.57$$

and 48.57 is used for n in all further computations. For example, we can say that the probability of a correct selection is at least .90 whenever $\delta_2 \geqslant 2.50$ and $\delta_1 \geqslant \delta_1^*$. The value of δ_1^* is obtained by entering Table S.1 (or our excerpt in Table 15.6.1) with $k = 3$, $\delta_2^* = 2.50$, and interpolating for $p = 3$ to obtain $n_0 \delta_1^* = 22.96$, from which $\delta_1^* = 22.96/48.57 = 0.47$. Then $(\delta_1^*, \delta_2^*) = (0.47, 2.50)$ is a point on the operating characteristic curve for $n_0 = 48.57$, $P^* = .90$.

REMARK.   Since the value n_0 is used only for further computation, it can be left in decimal form; it need not be rounded to an integer.   □

15.7   NOTES, REFERENCES, AND REMARKS

The main problem with applications of ranking and selection procedures to multivariate analysis is that the definition of best and the measure of distance must be integrated to form a combination that relates to the goal in a meaningful way. This problem is greatly simplified if we start with a specific linear combination (or more general function) of the components of the crucial unknown vector of θ values, $(\theta_1, \theta_2, \ldots, \theta_p)$, that we wish to maximize (or minimize). In general, goals involving partial rankings have not been successfully developed for a multivariate situation. The associated calculations are usually difficult and the impetus to develop meaningful formulations has been slow. Perhaps this is due to the fact that there is a paucity of examples in which investigators agree that the Mahalanobis (square) distance or some other well-defined distance measure is precisely what they want to maximize. Multivariate analysis techniques are effective for testing a null hypothesis (e.g., homogeneity), but there are generally too many ways of generating an alternative hypothesis. As presently developed, the methods of ranking and selection are concentrated on particular alternatives of interest. It would also be desirable to know more about the robustness of multivariate ranking and selection methods.

Alam and Rizvi (1966) used both the ratio and the difference of u values in the multivariate problem, and Rizvi and Solomon (1973) applied the resulting procedure to the problem of selecting the largest multiple correlation coefficient. We note that they use q for what we call p. The idea of using both a ratio and a difference was first introduced by Sobel (1968) in a paper dealing with ranking Poisson populations. In a more recent paper Alam, Rizvi, and Solomon (1976) introduced another measure of distance but the necessary tables have not been developed.

Frischtak (1973) investigated several problems dealing with ranking means (and separately ranking variances) in a single k-variate normal distribution. Some of these problems were also considered (with a Wald sequential rule) in the book by Bechhofer, Kiefer, and Sobel (1968).

The problem of ranking multivariate populations according to their generalized variance is difficult to interpret, and, in general, either the required probability of a correct selection is known only approximately or asymptotically or the required tables are not available. Some papers in this area are Eaton (1967), Frischtak (1973), and for selecting a subset, Gnanadesikan and Gupta (1970).

A procedure that averages all the pairwise correlations off the diagonal of a correlation matrix is due to Govindarajulu and Gore (1971); we use this procedure for the problem of selecting the population with the largest intraclass correlation in Section 15.5.2. However, we should note that their formula, given here as (15.5.24), is presented in an unclear exposition, with no restriction on the size of p and no solution for $p = 2$ or $p = 3$.

Procedures for selecting multivariate normal populations better than a control are considered by Krishnaiah and Rizvi (1966).

This whole field is as yet undeveloped, and the reader is encouraged to regard this chapter as an introduction to a wide area that will see considerable development in the future as more meaningful models are formulated.

PROBLEMS

15.1   Cooley and Lohnes (1962, p. 119) refer to a Scientific Careers Study to determine personality dimensions associated with different post-college career decisions of college majors. Sophomore and senior males were selected (in 1958) from science and engineering majors at six colleges in eastern Massachusetts. They were administered the Study of Values test measuring six attributes that were hypothesized as related to membership in the following three groups :

1. Research group. Those who enter graduate work to conduct funda-mental research as part of their future work.
2. Applied science group. Those who continue in science and engineer-ing, but do not plan a research career.
3. Nonscience group. Those who have done science work and enter fields having more direct involvement with people.

The following attribute scores are for the Study of Values variables:

$$X_1 = \text{theoretical}, \quad X_4 = \text{social},$$
$$X_2 = \text{economic}, \quad X_5 = \text{political},$$
$$X_3 = \text{aesthetic}, \quad X_6 = \text{religious}.$$

Sample means

Groups	$\bar{x}_1$	$\bar{x}_2$	$\bar{x}_3$	$\bar{x}_4$	$\bar{x}_5$	$\bar{x}_6$	Sample Sizes
Research	56.73	33.24	41.55	34.58	37.24	36.82	33
Applied	51.56	41.52	38.60	29.84	41.76	36.96	25
Nonscience	48.95	38.21	34.95	36.58	41.53	39.38	38

Variance-covariance matrix for each group

Research group

	1 Theoretical	2 Economic	3 Aesthetic	4 Social	5 Political	6 Religious
1	13.89	−2.31	10.03	−0.74	1.00	−22.24
2	−2.31	51.19	−11.79	−33.43	0.35	−4.61
3	10.03	−11.79	83.26	−14.64	−6.61	−58.59
4	−0.74	−33.43	−14.64	68.63	−6.08	−13.95
5	1.00	0.35	−6.61	−6.08	27.06	−15.67
6	−22.24	−4.61	−58.59	−13.95	−15.67	114.72

Applied group

	1	2	3	4	5	6
1	46.17	−1.80	8.11	−23.53	−11.78	−17.73
2	−1.80	51.01	−29.70	−13.12	13.30	19.85
3	8.11	−29.70	102.17	−23.23	−8.72	−49.81
4	−23.53	−13.12	−23.23	50.14	−6.12	−17.20
5	−11.78	13.30	−8.72	−6.12	54.61	−40.63
6	−17.73	19.85	−49.81	−17.20	−40.63	111.12

Nonscience group

	1	2	3	4	5	6
1	48.59	−6.02	12.56	−8.51	0.33	−46.49
2	−6.02	60.39	−32.12	−30.99	20.45	−11.71
3	12.56	−32.12	77.94	−2.78	−21.43	−33.98
4	−8.51	−30.99	−2.78	49.82	−19.23	11.39
5	0.33	20.45	−21.43	−19.23	35.99	−16.29
6	−46.49	−11.71	−33.98	11.39	−16.29	97.07

Assume that these variance-covariance matrices are good enough estimates to make the assumption that the Σ_j are known but unequal.

(a) Select the population with the largest Mahalanobis 'distance' from the origin.

(b) Select the population with the largest linear combination of means where $b_1 = b_2 = \cdots = b_6 = \frac{1}{6}$.

(c) Suppose you were designing such an experiment for the goal stated in (a). How large a sample would you take from each population if it was specified that the probability of a correct selection is to be at least $P^* = .95$ whenever $\delta_1 = \theta_{[6]} - \theta_{[5]} \geqslant 5.00$ and $\delta_2 = \theta_{[6]}/\theta_{[5]} \geqslant 1.25$?

15.2   Sir Ronald Fisher (1936) presents data consisting of four measurements on samples of flowers from the three species of iris: virginica, versicolor, and setosa. The four measurements are sepal width, sepal length, petal length, and petal width. Thus we have $k = 3$ populations, each with $p = 4$ variables to be measured. We assume that for each species these variables follow the multivariate normal distribution. Suppose that the goal is to determine the best species of iris, where *best* is defined as the largest value of the Mahalanobis 'distance' from the origin.

(a) If the Σ_j are assumed known, how large a sample should we take from each population if it is specified that the probability of a correct selection is to be at least $P^* = .90$ whenever $\delta_1 = \theta_{[3]} - \theta_{[2]} \geqslant 4.00$ and $\delta_2 = \theta_{[3]} / \theta_{[1]} \geqslant 3.00$?

(b) If the Σ_j are assumed unknown, Table S.3 must be used, but this table does not cover the case where $k = 3$, $p = 4$. As a result, suppose we consider only the first two of these species of iris so that $k = 2, p = 4$. Then how large should n be to satisfy the requirements analogous to those stated in (a)?

15.3   Tatsuoka (1971, p. 201) gives a numerical example of an experiment designed in part to compare the effectiveness of concentrated versus distributed practice of shorthand, represented by the following three types of instruction:

(1) Two hours of instruction per day for six weeks,
(2) Three hours of instruction per day for four weeks,
(3) Four hours of instruction per day for three weeks.

Note that all three methods have the same total hours of instruction. The subjects were female seniors in a vocational high school. When the instruction is completed, all subjects are to be given a standardized shorthand test that consists of two subtests that evaluate the subjects on $X_1 =$ speed, $X_2 =$ accuracy. Suppose that the goal is to determine which one of the three methods of instruction has the largest average correlation $\bar{\rho}$.

(a) How many subjects should be used so that the probability of a correct selection is at least $P^* = .90$ when $\delta = \rho_{[3]} - \rho_{[2]} \geqslant 0.20$?

(b) The experiment as reported was carried out with 30 subjects, 10 in each instruction group. The subtest scores follow, where larger scores mean greater proficiency.

Type of instruction

1		2		3	
Subtest Scores		Subtest Scores		Subtest Scores	
x_1	x_2	x_1	x_2	x_1	x_2
36	26	46	17	26	14
34	22	34	21	31	14
28	21	31	17	30	16
34	23	31	18	34	16
34	21	36	23	30	13
29	19	26	19	27	13
48	25	35	16	21	12
28	20	33	19	31	15
34	21	23	15	37	14
38	20	30	14	29	14

Compute the sample correlation between subtest scores for each type of instruction and make a selection of the instruction group that has the largest average correlation.

15.4   Tintner (1946) gives an example with three price variables

$$Y^{(1)} = \text{farm prices,}$$

$$Y^{(2)} = \text{food prices,}$$

$$Y^{(3)} = \text{other prices,}$$

and four production variables

$$X_1 = \text{durable goods,}$$

$$X_2 = \text{nondurable goods,}$$

$$X_3 = \text{minerals,}$$

$$X_4 = \text{agricultural products.}$$

The observations for the production variables are the annual price indexes for the period 1919–1939 (all with base year 1926) and the corresponding observations for the price variables are the annual production indexes published by the Federal Reserve Board for X_1, X_2, X_3, and by the Department of Agriculture for X_4 (all with base 1935–1939). Suppose that we are interested in determining which one of the price indexes is best explained by the three production indexes.

(a) How many observations should be taken (number of different years represented by the study) if we specify that the probability of a correct selection is to be at least $P^* = .95$ when $\delta_1^* = 2.0$, $\delta_2^* = 2.0$?

(b) The study was carried out with $n = 21$ observation and the given matrix of correlations between all variables is shown below. Construct the three correlation matrices R_j, one for each of the price indexes to be predicted, compute the multiple correlation coefficients from (15.6.5), and make a selection.

	x_1	x_2	x_3	x_4	$y^{(1)}$	$y^{(2)}$	$y^{(3)}$
x_1	1.000	0.496	0.873	0.481	-0.436	-0.427	-0.203
x_2	0.496	1.000	0.768	0.710	0.426	0.430	0.584
x_3	0.873	0.768	1.000	0.712	-0.038	-0.044	0.139
x_4	0.481	0.710	0.712	1.000	0.261	0.267	0.378
$y^{(1)}$	-0.436	0.426	-0.038	0.261	1.000	0.987	0.905
$y^{(2)}$	-0.427	0.430	-0.044	0.267	0.987	1.000	0.914
$y^{(3)}$	-0.203	0.584	0.139	0.378	0.905	0.914	1.000

15.5   Rao (1948) studies three human populations in India; these populations are the Brahmin caste, the Artisan caste, and the Korwa caste. For each caste, the following four anthropomorphic measurements are of interest:

$$X_1 = \text{stature},$$

$$X_2 = \text{sitting height},$$

$$X_3 = \text{nasal depth},$$

$$X_4 = \text{vertical nasal length}.$$

For each caste we have a four-variate distribution that we assume to be a multivariate normal distribution with $p = 4$ variables, and we have $k = 3$ populations (castes). Suppose that the covariance matrices are common and known. If the goal is to determine the population with the largest Mahalanobis 'distance' from the origin of anthropomorphic measurements, how many observations should be taken from each population so that we can assert with probability .95 that our selection is correct whenever $\delta_1 = \theta_{[3]} - \theta_{[2]} \geq 2.00$ and $\delta_2 = \theta_{[3]}/\theta_{[2]} \geq 2.00$?

Tables for Normal Means Selection Problems

Tables A.1 and A.2 relate the values of P (or P^*), the probability of a correct selection, and the value of the slippage τ (or τ_t), for the problem of selecting the one population with the best mean out of k normal populations with known variance. Each entry in Table A.1 is the value of τ (or τ_t) required to attain the specified probability P (or P^*) of a correct selection for a slippage τ (or τ_t), for selected k. For best defined as the population with the *largest* mean, τ satisfies $\tau = \delta \sqrt{n} / \sigma$ for P calculated under the generalized least favorable configuration, where $\delta = \mu_{[k]} - \mu_{[k-1]}$. Under the least favorable configuration with P^* specified and n fixed, the determination of δ^* is equivalent to determining $\tau_c = \delta^* \sqrt{n} / \sigma$; since this is given in Table A.1 we also use the notation τ_t for it.

Tables A.3–A.9 apply to other selection problems involving means of normal distributions.

SOURCES FOR APPENDIX A

Tables A.1 and A.2 are adapted from R. C. Milton (1963), Tables of the equally correlated multivariate normal probability integral, *Technical Report* No. 27, University of Minnesota, Minneapolis, Minnesota, with permission.

Table A.4 is adapted from P. R. Krishnaiah and J. V. Armitage (1966), Tables for the multivariate *t*-distribution, *Sankhyā*, B, **28**, Parts 1 and 2 (part of Aerospace Research Laboratories Report 65–199, Wright Patterson Air Force Base, Dayton, Ohio), with permission.

Table A.5 is adapted from H. J. Chen (1974), Percentage points of the multivariate *t* distribution and their application, *Technical Report* 74–14, Department of Mathematics, Memphis State University, Memphis, Tennessee, with permission.

Table A.6 is adapted from E. J. Dudewicz and Y. L. Tong (1971), Optimum confidence intervals for the largest location parameter, in S. S.

Gupta and J. Yackel, Eds., *Statistical Decision Theory and Related Topics*, Academic Press, New York, pp. 363–376, with permission of the publisher and the author.

Tables A.7 and A.8 are adapted from M. H. Rizvi (1971), Some selection problems involving folded normal distribution, *Technometrics*, **13**, 355–369, with permission.

Table A.9 is adapted from E. J. Dudewicz and S. R. Dalal (1975), Allocation of observations in ranking and selection with unequal variances, *Sankhyā*, B, **37**, 28–78, with permission.

TABLE INTERPOLATION FOR APPENDIX A

Interpolation in Tables A.1 and A.2

Tables A.1 and A.2 relate the two quantities P (or P^*) and τ (which here may be regarded as either τ_l, τ_c, or τ) for selected values of k. If either one of the values τ and P^* is not listed specifically in these tables, the value of the other can be found by interpolation. Although ordinary linear interpolation could be used, more accurate results can be obtained by using linear interpolation for the transformed quantities $\log \tau$ and $\log (1 - P^*)$ (both logarithms are natural logarithms, i.e., to the base $e = 2.71828\ldots$). Then the antilogarithm of the interpolated value gives the desired quantity, τ or $1 - P^*$. The specific interpolation instructions for a particular k are as follows:

1. To find a value of τ for a specified P that is not listed specifically in Table A.1, use linear interpolation on $\log \tau$ and $\log (1 - P^*)$ to yield $\log (1 - P^*)$ and then take the antilog. This gives $(1 - P^*)$, from which P^* is readily obtained.

2. To find a value of P^* for a specified τ that is not listed specifically in Table A.2, use linear interpolation on $\log \tau$ and $\log (1 - P^*)$ to yield $\log \tau$ and then take the antilog.

The following examples 1 and 2 illustrate these interpolation procedures respectively for $k = 3$.

Example 1. The problem is to find τ for $k = 3$, $P^* = .920$. The case $P^* = .920$ is not included as a caption in Table A.1, but there are table entries for τ on both sides of $P^* = .920$, namely, $P^* = .900$ and $P^* = .950$. In the following table we abstract these entries on the left. The corresponding logarithms (to the base e) are given in the table on the right in preparation for interpolation.

Table excerpt			Ready for interpolation	
τ	P^*		$\log \tau$	$\log (1 - P^*)$
2.2302	.900		0.802	-2.303
$\boxed{\tau}$	.920		$\boxed{\log \tau}$	-2.526
2.7101	.950		0.997	-2.996

Linear interpolation in the table on the right is carried out as follows:

$$\frac{\log \tau - 0.997}{-2.526 - (-2.996)} = \frac{0.802 - 0.997}{-2.303 - (-2.996)},$$

from which we obtain the solution $\log \tau = 0.865$. The antilog is $\tau = 2.375$, so that $\tau = 2.38$ after rounding upward.

Example 2. The problem is to find P^* for $k = 3$, $\tau = 2.30$. Table A.2 does not include the case $\tau = 2.30$ as a caption, but there are table entries for P^* on both sides of $\tau = 2.30$, namely, $\tau = 2.20$ and $\tau = 2.40$. In the table on the left we abstract these entries. The corresponding logarithms (to the base e) are given on the right in preparation for interpolation.

Table excerpt			Ready for interpolation	
τ	P^*		$\log \tau$	$\log (1 - P^*)$
2.20	.896		0.788	-2.263
2.30	$\boxed{P^*}$		0.833	$\boxed{\log (1 - P^*)}$
2.40	.921		0.875	-2.538

Linear interpolation in the table on the right is carried out as follows:

$$\frac{\log (1 - P^*) - (-2.538)}{0.833 - 0.875} = \frac{-2.263 - (-2.538)}{0.788 - 0.875},$$

and the solution is $\log (1 - P^*) = -2.405$. The antilog is $1 - P^* = .0903$, so that $P^* = .910$ after rounding P^*.

Interpolation in Table A.3

Although the first differences of entries in Table A.3 are not constant, simple linear interpolation (without any transformation) will provide answers for ρ values not in this table that are sufficiently accurate for most practical purposes. Hence no special instructions are needed for interpolation here.

REMARK. The following points should be noted: (Here $\Phi(x)$ denotes the standard normal cumulative distribution function.)

1. For $k=2$ the results do not depend on ρ because of the identity

$$\int_{-\infty}^{\infty} \Phi\left[\frac{x\sqrt{\rho}+h}{\sqrt{1-\rho}}\right] d\Phi(x) = \Phi(h),$$

for all ρ.

2. The entries for $\rho=0$ are solutions of $\Phi(x)=(P^*)^{1/m}$ and hence are the same as the entries in Table M.2 for $m=k-1$.

3. The entries for $\rho=1.0$ are solutions of $\Phi(x)=P^*$ and hence do not depend on k.

Interpolation in Table A.4

For both $P^*=.95$ and $P^*=.99$, linear interpolation with $1/\nu$ provides an easy and good approximation to entries in Table A.4 for large ν values and ν values that are not listed specifically in the table.

Thus, for example, if we want the entries for $P^*=.95$ and $k=10$ that correspond to $\nu=90$ and $\nu=360$, the calculations are as follows:

1. For $\nu=360$

ν	$1/\nu$	t value
120	.0083	2.45
360	.0028	$\boxed{t}$
∞	0	2.42

$$\frac{t-2.42}{2.45-2.42} = \frac{.0028-0}{.0083-0},$$

$$t=2.43.$$

2. For $\nu = 90$

ν	$1/\nu$	t value
60	.0167	2.48
90	.0111	$\boxed{t}$
120	.0083	2.45

$$\frac{t - 2.45}{2.48 - 2.45} = \frac{.0111 - .0083}{.0617 - .0083},$$

$$t = 2.45.$$

It should be noted that the entries for $\nu = \infty$ are in agreement with the corresponding entries for $\rho = \frac{1}{2}$, $P^* = .95$, $.99$ in Table A.3. Hence for $\nu = \infty$, accuracy can be obtained to an extra decimal by using the corresponding entry in Table A.3. In the preceding illustration 1, the result would be 2.427.

Interpolation in Table A.5

Table A.5 is only a slight variation of Table A.4 and the same method of interpolation is useful here, namely, linear interpolation on $1/\nu$.

The values for $\nu = \infty$ were added to this table by solving the following equation for x:

$$\Phi^k(x) + \Phi^k(-x) = P^*.$$

(Unfortunately the case $P^* = .75$ was not included here.) These values for $\nu = \infty$ are quite useful for large values of ν and they also simplify the interpolation problem.

REMARK. The values in Table A.5 differ only slightly from the one-sided percentiles of the multivariate t tables (Table A.4) for the case $\rho = 0$. In fact, for $k \geqslant 3$, $P^* = .95$, $.99$ and $\nu \geqslant 5$, there is agreement to two decimal places, and for $k = 2$, the error is at most one in the second decimal for $\nu \geqslant 13$. Thus if the special tables given here as Table A.5 were not available, then our Table A.4 of one-sided equi-percentiles for the multivariate t (based on a common denominator) with a common $\rho = 0$ could be used for any reasonable values of ν. Unfortunately these latter tables are available (at the time of this writing) only for $P^* = .95$ and $.99$.

Interpolation in Table A.6

In Table A.6 both d and P are increasing with c in a manner close to linear. As a result, ordinary linear interpolation will suffice for most practical purposes.

Table A.1 Values of τ (or τ_t) for fixed P (or P^*)

k	.750	.900	.950	.975	.990	.999
			P (or P^*)			
2	0.9539	1.8124	2.3262	2.7718	3.2900	4.3702
3	1.4338	2.2302	2.7101	3.1284	3.6173	4.6450
4	1.6822	2.4516	2.9162	3.2220	3.7970	4.7987
5	1.8463	2.5997	3.0552	3.4532	3.9196	4.9048
6	1.9674	2.7100	3.1591	3.5517	4.0121	4.9855
7	2.0626	2.7972	3.2417	3.6303	4.0860	5.0504
8	2.1407	2.8691	3.3099	3.6953	4.1475	5.1046
9	2.2067	2.9301	3.3679	3.7507	4.1999	5.1511
10	2.2637	2.9829	3.4182	3.7989	4.2456	5.1916
15	2.4678	3.1734	3.6004	3.9738	4.4121	5.3407
20	2.6009	3.2986	3.7207	4.0899	4.5230	5.4409
25	2.6987	3.3911	3.8099	4.1761	4.6057	5.5161

Table A.2 Values of P (or P^*) for fixed τ (or τ_t)

k	0.0	0.2	0.4	0.6	0.8	1.0	1.2	1.4	1.6	1.8	2.0	2.2	2.4
						τ							
2	.500	.556	.611	.664	.714	.760	.802	.839	.871	.898	.921	.940	.955
3	.333	.391	.452	.513	.574	.634	.690	.742	.789	.830	.866	.896	.921
4	.250	.304	.363	.425	.488	.552	.614	.674	.729	.779	.823	.861	.893
5	.200	.250	.305	.365	.429	.494	.559	.622	.682	.738	.788	.832	.869
6	.167	.212	.264	.322	.384	.449	.516	.581	.645	.704	.758	.807	.848.
7	.143	.185	.234	.289	.350	.414	.481	.548	.613	.676	.733	.785	.830
8	.125	.164	.210	.263	.322	.385	.452	.520	.587	.651	.711	.766	.814
9	.111	.148	.191	.242	.299	.361	.427	.495	.563	.629	.691	.748	.799
10	.100	.134	.176	.224	.280	.341	.406	.474	.543	.610	.674	.732	.785
15	.067	.093	.126	.167	.215	.271	.332	.398	.467	.537	.606	.671	.731
20	.050	.072	.100	.135	.178	.228	.286	.349	.417	.488	.558	.626	.691
25	.040	.058	.083	.114	.153	.200	.254	.315	.381	.451	.522	.592	.659
50	.020	.042[a]	.064[a]	.086[a]	.108[a]	.130	.172	.223	.282	.347	.416	.488	.560

[a]These four entries were inserted by linear interpolation.

Table A.2 (*Continued*)

k	2.6	2.8	3.0	3.2	3.4	3.6	3.8	4.0	4.2	4.4	4.6	4.8	5.0
2	.967	.976	.983	.988	.992	.995	.996	.998	.999	.999	.999	.9997	.9998
3	.941	.957	.969	.978	.985	.990	.993	.996	.997	.998	.999	.9993	.9996
4	.919	.940	.956	.969	.978	.985	.990	.993	.996	.997	.998	.999	.999
5	.900	.925	.945	.961	.972	.981	.987	.992	.995	.997	.998	.999	.999
6	.883	.912	.935	.953	.967	.977	.985	.990	.993	.996	.997	.998	.999
7	.869	.900	.926	.946	.962	.974	.982	.988	.992	.995	.997	.998	.999
8	.855	.890	.918	.940	.957	.970	.980	.986	.991	.994	.996	.998	.999
9	.843	.880	.910	.934	.953	.967	.977	.985	.990	.994	.996	.998	.999
10	.831	.870	.902	.928	.948	.964	.975	.983	.989	.993	.996	.997	.998
15	.785	.832	.871	.904	.930	.950	.965	.976	.984	.990	.993	.996	.998
20	.750	.802	.847	.884	.915	.938	.957	.970	.980	.987	.992	.995	.997
25	.721	.777	.826	.867	.901	.928	.949	.965	.976	.984	.990	.994	.996
50	.630	.696	.756	.809	.853	.891	.920	.943	.961	.974	.983	.989	.993

Table A.3 *Percentiles (h values) of the standard normal ranking integral for selected values of the correlation coefficient ρ*

k = 2

ρ	.750	.900	.950	.975	.990
			P*		
0 − 1	0.674	1.282	1.645	1.960	2.326

k = 3

ρ	.750	.900	.950	.975	.990
			P*		
0	1.108	1.632	1.955	2.239	2.575
.1	1.094	1.626	1.951	2.237	2.574
.2	1.078	1.617	1.946	2.234	2.572
.3	1.060	1.607	1.938	2.229	2.569
.4	1.039	1.594	1.929	2.222	2.565
.5	1.014	1.577	1.916	2.212	2.558
.6	0.985	1.556	1.900	2.199	2.548
.7	0.949	1.529	1.877	2.180	2.532
.8	0.904	1.493	1.846	2.152	2.509
.9	0.842	1.440	1.798	2.108	2.469
1.0	0.674	1.282	1.645	1.960	2.326

k = 4

ρ	.750	.900	P* .950	.975	.990
0	1.332	1.818	2.121	2.391	2.712
.1	1.312	1.809	2.116	2.388	2.710
.2	1.288	1.796	2.108	2.383	2.708
.3	1.260	1.780	2.097	2.375	2.703
.4	1.228	1.759	2.082	2.364	2.696
.5	1.189	1.734	2.062	2.349	2.685
.6	1.144	1.701	2.036	2.328	2.668
.7	1.090	1.659	2.001	2.298	2.644
.8	1.022	1.603	1.952	2.254	2.607
.9	0.927	1.521	1.877	2.185	2.544
1.0	0.674	1.282	1.645	1.960	2.326

k = 5

ρ	.750	.900	P* .950	.975	.990
0	1.480	1.943	2.234	2.494	2.806
.1	1.456	1.932	2.228	2.491	2.804
.2	1.427	1.917	2.218	2.485	2.801
.3	1.393	1.896	2.204	2.475	2.795
.4	1.353	1.871	2.185	2.461	2.786
.5	1.306	1.838	2.160	2.442	2.772
.6	1.250	1.797	2.127	2.415	2.750
.7	1.183	1.745	2.083	2.377	2.719
.8	1.099	1.676	2.022	2.322	2.672
.9	0.982	1.574	1.929	2.236	2.594
1.0	0.674	1.282	1.645	1.960	2.326

k = 6

ρ	.750	.900	P* .950	.975	.990
0	1.590	2.036	2.319	2.572	2.877
.1	1.563	2.024	2.312	2.568	2.875
.2	1.530	2.007	2.301	2.561	2.871
.3	1.491	1.983	2.285	2.550	2.864
.4	1.445	1.954	2.263	2.534	2.854

Table A.3 (*Continued*)

k = 6

ρ	P*				
	.750	.900	.950	.975	.990
.5	1.391	1.916	2.234	2.511	2.837
.6	1.328	1.869	2.195	2.480	2.812
.7	1.251	1.809	2.144	2.436	2.776
.8	1.155	1.730	2.074	2.373	2.721
.9	1.023	1.614	1.967	2.274	2.631
1.0	0.674	1.282	1.645	1.960	2.326

k = 7

ρ	P*				
	.750	.900	.950	.975	.990
0	1.677	2.111	2.386	2.634	2.934
.1	1.647	2.097	2.378	2.630	2.932
.2	1.611	2.078	2.366	2.623	2.928
.3	1.568	2.052	2.349	2.611	2.920
.4	1.518	2.019	2.325	2.593	2.908
.5	1.458	1.978	2.292	2.567	2.889
.6	1.389	1.926	2.249	2.532	2.862
.7	1.305	1.859	2.193	2.483	2.821
.8	1.200	1.772	2.115	2.413	2.760
.9	1.055	1.645	1.998	2.304	2.660
1.0	0.674	1.282	1.645	1.960	2.326

k = 8

ρ	P*				
	.750	.900	.950	.975	.990
0	1.748	2.172	2.442	2.686	2.981
.1	1.717	2.157	2.434	2.682	2.979
.2	1.678	2.137	2.421	2.674	2.975
.3	1.632	2.109	2.402	2.660	2.967
.4	1.577	2.074	2.376	2.641	2.953
.5	1.514	2.029	2.340	2.613	2.933
.6	1.439	1.972	2.294	2.575	2.902
.7	1.349	1.901	2.233	2.521	2.858
.8	1.236	1.807	2.149	2.446	2.792
.9	1.081	1.670	2.022	2.328	2.684
1.0	0.674	1.282	1.645	1.960	2.326

Table A.3 (Continued)

k = 9

ρ	.750	.900	P* .950	.975	.990
0	1.808	2.224	2.490	2.731	3.022
.1	1.775	2.209	2.481	2.726	3.020
.2	1.735	2.187	2.467	2.717	3.015
.3	1.686	2.158	2.447	2.703	3.006
.4	1.628	2.120	2.419	2.682	2.992
.5	1.560	2.072	2.381	2.652	2.970
.6	1.481	2.012	2.332	2.611	2.937
.7	1.386	1.936	2.267	2.554	2.890
.8	1.266	1.836	2.177	2.474	2.819
.9	1.103	1.691	2.043	2.349	2.705
1.0	0.674	1.282	1.645	1.960	2.326

k = 10

ρ	.750	.900	P* .950	.975	.990
0	1.860	2.269	2.531	2.769	3.057
.1	1.826	2.253	2.522	2.764	3.055
.2	1.783	2.230	2.508	2.755	3.050
.3	1.732	2.200	2.486	2.740	3.041
.4	1.671	2.160	2.457	2.718	3.026
.5	1.601	2.109	2.417	2.686	3.002
.6	1.517	2.046	2.365	2.643	2.967
.7	1.417	1.966	2.296	2.583	2.917
.8	1.293	1.861	2.202	2.498	2.843
.9	1.122	1.710	2.062	2.367	2.722
1.0	0.674	1.282	1.645	1.960	2.326

Table A.4 One-sided multivariate t distribution with common correlation $\rho = \frac{1}{2}$ (equi-coordinate percentage points)

$P^* = .95$

ν	k 2	3	4	5	6	7	8	9	10
5	2.01	2.44	2.68	2.85	2.98	3.08	3.16	3.24	3.30
6	1.94	2.34	2.56	2.71	2.83	2.92	3.00	3.06	3.12
7	1.89	2.27	2.48	2.62	2.73	2.81	2.89	2.95	3.00
8	1.86	2.22	2.42	2.55	2.66	2.74	2.81	2.87	2.92
9	1.83	2.18	2.37	2.50	2.60	2.68	2.75	2.81	2.86
10	1.81	2.15	2.34	2.47	2.56	2.64	2.70	2.76	2.81
12	1.78	2.11	2.29	2.41	2.50	2.58	2.64	2.69	2.73
14	1.76	2.08	2.25	2.37	2.46	2.53	2.59	2.64	2.69
16	1.75	2.06	2.23	2.34	2.43	2.50	2.56	2.61	2.65
18	1.73	2.04	2.21	2.32	2.41	2.48	2.53	2.58	2.62
20	1.72	2.03	2.19	2.30	2.39	2.46	2.51	2.56	2.60
25	1.71	2.00	2.16	2.27	2.36	2.42	2.48	2.52	2.56
30	1.70	1.99	2.15	2.25	2.33	2.40	2.45	2.50	2.54
60	1.67	1.95	2.10	2.21	2.28	2.35	2.39	2.44	2.48
120	1.66	1.93	2.08	2.18	2.26	2.32	2.37	2.41	2.45
∞	1.64	1.92	2.06	2.16	2.23	2.29	2.34	2.38	2.42

$P^* = .99$

ν	k 2	3	4	5	6	7	8	9	10
5	3.36	3.90	4.21	4.43	4.60	4.73	4.85	4.94	5.03
6	3.14	3.61	3.88	4.06	4.21	4.32	4.42	4.51	4.58
7	3.00	3.42	3.66	3.83	3.96	4.06	4.15	4.22	4.29
8	2.90	3.29	3.51	3.66	3.78	3.88	3.96	4.03	4.09
9	2.82	3.19	3.40	3.54	3.66	3.75	3.82	3.89	3.94
10	2.76	3.11	3.31	3.45	3.56	3.64	3.72	3.78	3.83
12	2.68	3.01	3.19	3.32	3.42	3.50	3.56	3.62	3.67
14	2.62	2.93	3.11	3.23	3.32	3.40	3.46	3.51	3.56
16	2.58	2.88	3.05	3.17	3.26	3.33	3.39	3.44	3.48
18	2.55	2.84	3.01	3.12	3.20	3.27	3.33	3.38	3.42
20	2.53	2.81	2.97	3.08	3.16	3.23	3.29	3.34	3.38
25	2.48	2.76	2.91	3.01	3.10	3.16	3.21	3.26	3.30
30	2.46	2.72	2.87	2.97	3.05	3.11	3.16	3.20	3.24
60	2.39	2.64	2.78	2.87	2.94	3.00	3.04	3.08	3.12
120	2.36	2.60	2.73	2.82	2.89	2.94	2.99	3.03	3.06
∞	2.33	2.56	2.68	2.77	2.84	2.89	2.93	2.97	3.00

Table A.5 Values of d for a two-sided confidence interval for the largest mean of k normal populations with common unknown variance

	$k=2$				$k=3$				$k=4$				$k=5$			
P^* v	.900	.950	.975	.990	.900	.950	.975	.990	.900	.950	.975	.990	.900	.950	.975	.990
2	2.92	4.30	6.21	9.92	3.34	4.89	7.03	11.22	3.71	5.41	7.77	12.39	4.02	5.85	8.38	13.35
3	2.35	3.18	4.18	5.84	2.66	3.57	4.65	6.48	2.93	3.89	5.06	7.02	3.14	4.15	5.38	7.46
4	2.13	2.78	3.50	4.60	2.40	3.09	3.86	5.05	2.62	3.34	4.15	5.41	2.80	3.54	4.39	5.70
5	2.02	2.57	3.16	4.03	2.26	2.84	3.47	4.39	2.46	3.06	3.71	4.67	2.62	3.23	3.90	4.90
6	1.94	2.45	2.97	3.71	2.18	2.70	3.24	4.02	2.36	2.89	3.45	4.25	2.50	3.05	3.62	4.44
7	1.89	2.36	2.84	3.50	2.12	2.60	3.09	3.78	2.29	2.78	3.28	3.98	2.43	2.93	3.44	4.15
8	1.86	2.31	2.75	3.36	2.08	2.53	2.99	3.61	2.24	2.70	3.17	3.80	2.37	2.84	3.31	3.95
9	1.83	2.26	2.68	3.25	2.04	2.48	2.91	3.49	2.21	2.64	3.08	3.66	2.33	2.77	3.21	3.80
10	1.81	2.23	2.63	3.17	2.02	2.44	2.85	3.39	2.18	2.60	3.01	3.56	2.30	2.72	3.14	3.69
12	1.78	2.18	2.56	3.05	1.98	2.38	2.76	3.26	2.14	2.53	2.91	3.41	2.25	2.65	3.03	3.53
14	1.76	2.14	2.51	2.98	1.96	2.34	2.70	3.17	2.11	2.48	2.85	3.31	2.22	2.60	2.96	3.42
16	1.75	2.12	2.47	2.92	1.94	2.31	2.66	3.11	2.08	2.45	2.80	3.24	2.20	2.56	2.90	3.35
18	1.73	2.10	2.44	2.88	1.93	2.29	2.63	3.06	2.07	2.43	2.76	3.19	2.18	2.53	2.86	3.29
20	1.72	2.09	2.42	2.85	1.91	2.27	2.60	3.02	2.06	2.41	2.73	3.15	2.16	2.51	2.83	3.24
25	1.71	2.06	2.38	2.79	1.89	2.24	2.56	2.95	2.03	2.37	2.68	3.07	2.14	2.47	2.78	3.16
30	1.70	2.04	2.36	2.75	1.88	2.22	2.53	2.91	2.02	2.35	2.65	3.03	2.12	2.44	2.74	3.11
40	1.68	2.02	2.33	2.70	1.87	2.19	2.49	2.86	2.00	2.32	2.61	2.97	2.10	2.41	2.70	3.05
50	1.68	2.01	2.31	2.68	1.86	2.18	2.47	2.83	1.99	2.30	2.59	2.93	2.09	2.39	2.67	3.01
60	1.67	2.00	2.30	2.66	1.85	2.17	2.46	2.81	1.98	2.29	2.57	2.91	2.08	2.38	2.65	2.99
∞	1.64	1.96	2.24	2.58	1.82	2.12	2.39	2.71	1.94	2.23	2.49	2.81	2.04	2.32	2.57	2.88

	k=6				k=7				k=8				k=9			
P^* / v	.900	.950	.975	.990	.900	.950	.975	.990	.900	.950	.975	.990	.900	.950	.975	.990
2	4.28	6.21	8.89	14.15	4.50	6.52	9.32	14.84	4.69	6.78	9.69	15.43	4.85	7.01	10.02	15.94
3	3.32	4.37	5.66	7.83	3.47	4.56	5.89	8.14	3.60	4.72	6.09	8.41	3.71	4.86	6.26	8.64
4	2.94	3.71	4.58	5.94	3.06	3.85	4.75	6.15	3.17	3.97	4.89	6.33	3.26	4.08	5.02	6.48
5	2.74	3.38	4.06	5.08	2.85	3.50	4.20	5.24	2.94	3.60	4.31	5.38	3.03	3.69	4.41	5.50
6	2.62	3.17	3.76	4.59	2.72	3.28	3.87	4.73	2.81	3.37	3.97	4.84	2.88	3.46	4.06	4.94
7	2.54	3.04	3.56	4.28	2.63	3.14	3.66	4.40	2.71	3.22	3.75	4.50	2.78	3.30	3.83	4.59
8	2.48	2.95	3.42	4.07	2.57	3.04	3.52	4.17	2.64	3.12	3.60	4.27	2.71	3.18	3.67	4.34
9	2.43	2.88	3.32	3.91	2.52	2.96	3.41	4.01	2.59	3.04	3.49	4.09	2.66	3.10	3.55	4.17
10	2.40	2.82	3.24	3.80	2.48	2.90	3.32	3.89	2.55	2.98	3.40	3.96	2.61	3.04	3.46	4.03
12	2.35	2.74	3.12	3.63	2.43	2.82	3.20	3.71	2.49	2.89	3.27	3.78	2.55	2.95	3.33	3.84
14	2.31	2.68	3.05	3.51	2.39	2.76	3.12	3.59	2.45	2.82	3.18	3.65	2.51	2.88	3.24	3.71
16	2.28	2.64	2.99	3.43	2.36	2.72	3.06	3.50	2.42	2.78	3.12	3.56	2.48	2.83	3.18	3.62
18	2.26	2.61	2.95	3.37	2.34	2.68	3.02	3.44	2.40	2.74	3.07	3.50	2.45	2.80	3.13	3.55
20	2.25	2.59	2.91	3.32	2.32	2.66	2.98	3.39	2.38	2.72	3.04	3.44	2.43	2.77	3.09	3.49
25	2.22	2.55	2.85	3.24	2.29	2.61	2.92	3.30	2.35	2.67	2.97	3.35	2.40	2.72	3.02	3.40
30	2.20	2.52	2.81	3.18	2.27	2.58	2.88	3.24	2.33	2.64	2.93	3.29	2.38	2.69	2.98	3.34
40	2.18	2.48	2.77	3.12	2.24	2.55	2.83	3.17	2.30	2.60	2.88	3.22	2.35	2.65	2.92	3.26
50	2.16	2.46	2.74	3.08	2.23	2.53	2.80	3.13	2.28	2.58	2.85	3.18	2.33	2.62	2.89	3.22
60	2.16	2.45	2.72	3.05	2.22	2.51	2.78	3.11	2.27	2.56	2.83	3.15	2.32	2.61	2.87	3.19
∞	2.11	2.39	2.63	2.93	2.17	2.44	2.69	2.98	2.22	2.49	2.73	3.02	2.27	2.53	2.77	3.06

Table A.5   (*Continued*)

$$k = 10$$

v \ P^*	.900	.950	.975	.990
2	5.00	7.22	10.31	16.40
3	3.81	4.98	6.42	8.86
4	3.34	4.18	5.13	6.62
5	3.10	3.77	4.51	5.61
6	2.95	3.53	4.14	5.03
7	2.84	3.37	3.91	4.67
8	2.77	3.25	3.74	4.42
9	2.71	3.16	3.62	4.23
10	2.67	3.10	3.52	4.09
12	2.60	3.00	3.38	3.90
14	2.56	2.93	3.29	3.76
16	2.53	2.88	3.22	3.67
18	2.50	2.85	3.17	3.60
20	2.48	2.82	3.13	3.54
25	2.45	2.76	3.06	3.44
30	2.42	2.73	3.02	3.38
40	2.39	2.69	2.96	3.30
50	2.38	2.66	2.93	3.26
60	2.36	2.65	2.91	3.23
∞	2.31	2.57	2.80	3.09

Table A.6 *Values of d for a fixed-length confidence interval for the largest mean of k normal populations with common known variance*

c	k = 3		k = 4		k = 5	
	d	P*	d	P*	d	P*
1.0	0.309	.377	0.177	.365	0.077	.353
1.1	0.360	.411	0.228	.399	0.130	.386
1.2	0.411	.444	0.280	.432	0.182	.418
1.3	0.461	.477	0.332	.464	0.236	.450
1.4	0.512	.508	0.385	.495	0.289	.480
1.5	0.564	.539	0.437	.525	0.343	.511
1.6	0.615	.569	0.490	.554	0.397	.540
1.7	0.666	.597	0.542	.583	0.451	.568
1.8	0.717	.624	0.595	.610	0.505	.596
1.9	0.769	.650	0.648	.636	0.559	.622
2.0	0.820	.675	0.701	.661	0.614	.647
2.1	0.872	.699	0.754	.685	0.669	.672
2.2	0.923	.721	0.808	.708	0.723	.695
2.3	0.975	.743	0.861	.730	0.778	.718
2.4	1.027	.763	0.915	.751	0.833	.739
2.5	1.079	.782	0.968	.771	0.888	.759
2.6	1.130	.800	1.022	.789	0.944	.778
2.7	1.182	.817	1.076	.807	0.989	.797
2.8	1.234	.833	1.129	.823	1.054	.814
2.9	1.286	.848	1.183	.839	1.109	.830
3.0	1.338	.861	1.237	.853	1.164	.845
3.1	1.390	.874	1.291	.866	1.219	.859
3.2	1.442	.886	1.344	.879	1.275	.872
3.3	1.494	.897	1.398	.890	1.330	.884
3.4	1.546	.907	1.452	.901	1.385	.895
3.5	1.598	.916	1.506	.911	1.440	.905
3.6	1.651	.925	1.560	.920	1.495	.915
3.7	1.703	.933	1.613	.928	1.550	.924
3.8	1.755	.940	1.667	.936	1.604	.932
3.9	1.807	.946	1.720	.943	1.659	.939
4.0	1.859	.952	1.774	.949	1.714	.946

	$k = 6$		$k = 8$		$k = 10$	
c	d	P^*	d	P^*	d	P^*
1.0	− 0.003	.341	− 0.126	.320	− 0.219	.302
1.1	0.050	.373	− 0.071	.351	− 0.162	.332
1.2	0.104	.405	− 0.015	.382	− 0.105	.362
1.3	0.159	.436	0.041	.412	− 0.047	.392
1.4	0.213	.467	0.098	.442	0.011	.422
1.5	0.268	.497	0.154	.472	0.070	.451
1.6	0.323	.526	0.212	.501	0.129	.481
1.7	0.378	.554	0.269	.530	0.188	.509
1.8	0.434	.582	0.327	.558	0.247	.537
1.9	0.490	.609	0.384	.585	0.306	.565
2.0	0.545	.634	0.442	.611	0.366	.592
2.1	0.601	.659	0.500	.637	0.426	.618
2.2	0.658	.683	0.558	.661	0.485	.643
2.3	0.714	.706	0.617	.685	0.545	.667
2.4	0.770	.728	0.675	.708	0.605	.691
2.5	0.826	.749	0.733	.730	0.664	.714
2.6	0.883	.768	0.791	.750	0.724	.735
2.7	0.939	.787	0.850	.770	0.783	.756
2.8	0.995	.805	0.908	.789	0.843	.775
2.9	1.052	.821	0.966	.806	0.902	.794
3.0	1.108	.837	1.024	.823	0.961	.811
3.1	1.164	.851	1.082	.839	1.020	.827
3.2	1.221	.865	1.139	.853	1.079	.843
3.3	1.277	.878	1.197	.867	1.138	.857
3.4	1.333	.889	1.255	.879	1.196	.870
3.5	1.389	.900	1.312	.891	1.255	.883
3.6	1.445	.910	1.369	.902	1.313	.894
3.7	1.501	.919	1.426	.912	1.371	.905
3.8	1.556	.928	1.484	.921	1.429	.915
3.9	1.612	.935	1.540	.929	1.487	.924
4.0	1.667	.942	1.597	.937	1.545	.932

Table A.7 Value[a] of $\delta^*\sqrt{n}$ needed to determine the common sample size n so that P* is the minimum probability of correctly selecting the population with the smallest value of $|\mu|$

				P*		
k	.750	.900	.950	.975	.990	.999
2	1.487	2.289	2.756	3.163	3.640	4.653
3	1.979	2.689	3.120	3.500	3.950	4.916
4	2.226	2.901	3.316	3.684	4.122	5.064
5	2.388	3.043	3.448	3.809	4.239	5.166
6	2.507	3.149	3.548	3.903	4.327	5.244
7	2.601	3.233	3.627	3.979	4.398	5.307
8	2.678	3.302	3.692	4.041	4.457	5.359
9	2.742	3.361	3.748	4.094	4.508	5.404
10	2.798	3.412	3.796	4.141	4.552	5.443

[a]Note that for $k=2$ the entries in Tables A.7 and A.8 are the same.

Table A.8 Value[a] of $\delta^*\sqrt{n}$ needed to determine the common sample size n so that P* is the minimum probability of correctly selecting the population with the largest value of $|\mu|$

				P*		
k	.750	.900	.950	.975	.990	.999
2	1.487	2.289	2.756	3.163	3.640	4.653
3	1.890	2.633	3.080	3.472	3.932	4.910
4	2.094	2.819	3.258	3.643	4.095	5.054
5	2.230	2.947	3.380	3.760	4.207	5.154
6	2.332	3.042	3.472	3.849	4.291	5.230
7	2.414	3.119	3.546	3.920	4.359	5.291
8	2.481	3.182	3.607	3.979	4.416	5.343
9	2.538	3.237	3.660	4.030	4.465	5.386
10	2.588	3.284	3.705	4.074	4.507	5.425

[a]Note that for $k=2$ the entries in Tables A.7 and A.8 are the same.

Table A.9 *Value of* h *needed in the two-stage procedure for selecting the population with the largest mean of* k *normal populations with variances unknown and not necessarily equal*

k	n	$P^* = .750$	$P^* = .900$	$P^* = .950$	$P^* = .975$	$P^* = .990$
2	5	1.14	2.29	3.11	3.94	5.14
	10	1.03	2.00	2.61	3.18	3.89
	15	1.00	1.93	2.50	3.02	3.64
	20	0.99	1.90	2.45	2.95	3.54
	25	0.98	1.88	2.42	2.91	3.48
	30	0.98	1.87	2.41	2.88	3.45
	∞	0.96	1.82	2.33	2.78	3.29
3	5	1.74	2.90	3.75	4.63	5.91
	10	1.56	2.48	3.08	3.64	4.34
	15	1.51	2.39	2.94	3.43	4.04
	20	1.49	2.34	2.87	3.35	3.92
	25	1.48	2.32	2.84	3.30	3.85
	30	1.47	2.30	2.81	3.27	3.81
	∞	1.44	2.24	2.72	3.13	3.62
4	5	2.08	3.26	4.14	5.05	6.40
	10	1.84	2.75	3.34	3.90	4.60
	15	1.78	2.63	3.17	3.67	4.27
	20	1.75	2.58	3.10	3.57	4.13
	25	1.74	2.55	3.06	3.51	4.05
	30	1.73	2.54	3.03	3.48	4.01
	∞	1.69	2.46	2.92	3.23	3.80
5	5	2.32	3.53	4.42	5.37	6.78
	10	2.03	2.94	3.53	4.08	4.79
	15	1.96	2.81	3.34	3.83	4.43
	20	1.93	2.75	3.26	3.72	4.28
	25	1.91	2.72	3.21	3.66	4.20
	30	1.90	2.69	3.18	3.62	4.14
	∞	1.85	2.60	3.06	3.46	3.92
6	5	2.52	3.74	4.66	5.64	7.09
	10	2.18	3.08	3.67	4.22	4.93
	15	2.10	2.93	3.46	3.95	4.55
	20	2.06	2.87	3.38	3.83	4.39
	25	2.04	2.84	3.33	3.77	4.30
	30	2.03	2.81	3.30	3.73	4.25
	∞	1.97	2.71	3.16	3.56	4.02
7	5	2.67	3.92	4.85	5.86	7.35
	10	2.30	3.19	3.79	4.34	5.05
	15	2.21	3.04	3.57	4.05	4.64
	20	2.17	2.97	3.47	3.93	4.48
	25	2.14	2.93	3.42	3.86	4.39
	30	2.13	2.91	3.39	3.82	4.33
	∞	2.07	2.80	3.25	3.64	4.09

k	n	$P^* = .750$	$P^* = .900$	$P^* = .950$	$P^* = .975$	$P^* = .990$
8	5	2.81	4.07	5.03	6.05	7.58
	10	2.40	3.29	3.88	4.44	5.15
	15	2.30	3.12	3.65	4.13	4.73
	20	2.25	3.05	3.55	4.00	4.55
	25	2.23	3.01	3.50	3.93	4.46
	30	2.21	2.98	3.46	3.89	4.40
	∞	2.15	2.87	3.31	3.70	4.15
9	5	2.93	4.21	5.18	6.22	7.79
	10	2.48	3.37	3.96	4.52	5.23
	15	2.37	3.20	3.72	4.20	4.80
	20	2.33	3.12	3.62	4.07	4.62
	25	2.30	3.08	3.56	4.00	4.53
	30	2.28	3.05	3.53	3.93	4.46
	∞	2.21	2.94	3.37	3.76	4.20
10	5	3.04	4.33	5.32	6.38	7.98
	10	2.55	3.45	4.04	4.60	5.31
	15	2.44	3.26	3.79	4.26	4.86
	20	2.39	3.18	3.68	4.13	4.68
	25	2.36	3.14	3.62	4.05	4.58
	30	2.35	3.11	3.58	4.01	4.51
	∞	2.27	2.99	3.42	3.80	4.25
15	5	3.46	4.82	5.87	7.01	8.75
	10	2.83	3.72	4.32	4.88	5.60
	15	2.69	3.50	4.02	4.50	5.09
	20	2.63	3.41	3.90	4.34	4.88
	25	2.59	3.35	3.83	4.26	4.78
	30	2.57	3.32	3.79	4.20	4.71
	∞	2.47	3.18	3.61	3.98	4.42
20	5	3.77	5.18	6.28	7.49	9.34
	10	3.02	3.91	4.51	5.07	5.80
	15	2.85	3.66	4.18	4.66	5.25
	20	2.78	3.56	4.04	4.49	5.03
	25	2.74	3.50	3.97	4.40	4.91
	30	2.72	3.46	3.92	4.34	4.84
	∞	2.61	3.30	3.73	4.09	4.53
25	5	4.02	5.47	6.62	7.88	9.82
	10	3.16	4.06	4.66	5.23	5.96
	15	2.98	3.78	4.30	4.78	5.37
	20	2.90	3.67	4.16	4.60	5.14
	25	2.85	3.61	4.08	4.50	5.01
	30	2.83	3.57	4.03	4.44	4.94
	∞	2.70	3.40	3.81	4.18	4.61

APPENDIX B

Figures for Normal Means Selection Problems

The following figures provide graphs of $P\{CS|\tau\}$ for given values of k and τ (or τ_t). Thus each figure in Appendix B is a continuous version of the entries for the corresponding value of k in Table A.2 of Appendix A. For example, for $k=5$ the τ values in Table A.2 range from 0.0 to 5.0 in steps of 0.2, whereas the corresponding graph is a continuous curve through these points.

Figure B.1   *Graph of $P_{LF}\{CS|\tau\}$ or $CL\{\tau|R\}$ for $k=2, 4, 6, 8$ 10, 20, 50.*

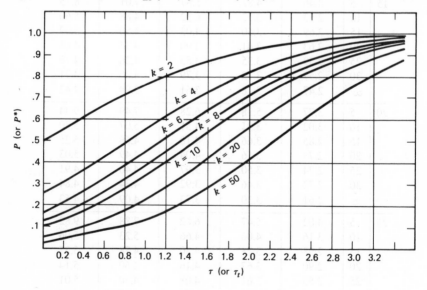

Figure B.2 *Graph of* $P_{LF}\{CS|\tau\}$ *or* $CL\{\tau|R\}$ *for* $k = 3, 5, 7, 9, 15, 25$.

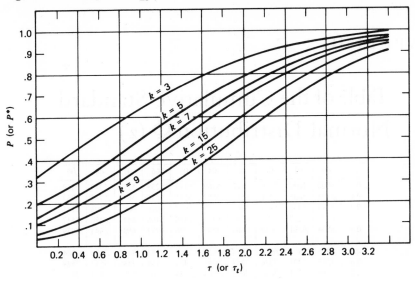

Table of the Cumulative Standard Normal Distribution $\Phi(z)$

z	.00	.01	.02	.03	.04	.05	.06	.07	.08	.09
0.0	.5000	.5040	.5080	.5120	.5160	.5199	.5239	.5279	.5319	.5359
0.1	.5398	.5438	.5478	.5517	.5557	.5596	.5636	.5675	.5714	.5753
0.2	.5793	.5832	.5871	.5910	.5948	.5987	.6026	.6064	.6103	.6141
0.3	.6179	.6217	.6255	.6293	.6331	.6368	.6406	.6443	.6480	.6517
0.4	.6554	.6591	.6628	.6664	.6700	.6736	.6772	.6808	.6844	.6879
0.5	.6915	.6950	.6985	.7019	.7054	.7088	.7123	.7157	.7190	.7224
0.6	.7257	.7291	.7324	.7357	.7389	.7422	.7454	.7486	.7517	.7549
0.7	.7580	.7611	.7642	.7673	.7704	.7734	.7764	.7794	.7823	.7852
0.8	.7881	.7910	.7939	.7967	.7995	.8023	.8051	.8078	.8106	.8133
0.9	.8159	.8186	.8212	.8238	.8264	.8289	.8315	.8340	.8365	.8389
1.0	.8413	.8438	.8461	.8485	.8508	.8531	.8554	.8577	.8599	.8621
1.1	.8643	.8665	.8686	.8708	.8729	.8749	.8770	.8790	.8810	.8830
1.2	.8849	.8869	.8888	.8907	.8925	.8944	.8962	.8980	.8997	.9015
1.3	.9032	.9049	.9066	.9082	.9099	.9115	.9131	.9147	.9162	.9177
1.4	.9192	.9207	.9222	.9236	.9251	.9265	.9279	.9292	.9306	.9319
1.5	.9332	.9345	.9357	.9370	.9382	.9394	.9406	.9418	.9429	.9441
1.6	.9452	.9463	.9474	.9484	.9495	.9505	.9515	.9525	.9535	.9545
1.7	.9554	.9564	.9573	.9582	.9591	.9599	.9608	.9616	.9625	.9633
1.8	.9641	.9649	.9656	.9664	.9671	.9678	.9686	.9693	.9699	.9706
1.9	.9713	.9719	.9726	.9732	.9738	.9744	.9750	.9756	.9761	.9767
2.0	.9772	.9778	.9783	.9788	.9793	.9798	.9803	.9808	.9812	.9817
2.1	.9821	.9826	.9830	.9834	.9838	.9842	.9846	.9850	.9854	.9857
2.2	.9861	.9864	.9868	.9871	.9875	.9878	.9881	.9884	.9887	.9890
2.3	.9893	.9896	.9898	.9901	.9904	.9906	.9909	.9911	.9913	.9916
2.4	.9918	.9920	.9922	.9925	.9927	.9929	.9931	.9932	.9934	.9936
2.5	.9938	.9940	.9941	.9943	.9945	.9946	.9948	.9949	.9951	.9952
2.6	.9953	.9955	.9956	.9957	.9959	.9960	.9961	.9962	.9963	.9964
2.7	.9965	.9966	.9967	.9968	.9969	.9970	.9971	.9972	.9973	.9974
2.8	.9974	.9975	.9976	.9977	.9977	.9978	.9979	.9979	.9980	.9981
2.9	.9981	.9982	.9982	.9983	.9984	.9984	.9985	.9985	.9986	.9986
3.0	.9987	.9987	.9987	.9988	.9988	.9989	.9989	.9989	.9990	.9990
3.1	.9990	.9991	.9991	.9991	.9992	.9992	.9992	.9992	.9993	.9993
3.2	.9993	.9993	.9994	.9994	.9994	.9994	.9994	.9995	.9995	.9995
3.3	.9995	.9995	.9995	.9996	.9996	.9996	.9996	.9996	.9996	.9997
3.4	.9997	.9997	.9997	.9997	.9997	.9997	.9997	.9997	.9997	.9998

APPENDIX D

Table of Critical Values for the Chi-Square Distribution

$1 - P_1^*$ is the probability that a chi-square variable with ν degrees of freedom is less than the corresponding table entry.

ν	.010	.025	$1 - P_1^*$ .050	.100	.250
1	0.000	0.001	0.004	0.016	0.102
2	0.020	0.051	0.103	0.211	0.575
3	0.115	0.216	0.352	0.584	1.213
4	0.297	0.484	0.711	1.064	1.923
5	0.554	0.831	1.145	1.610	2.675
6	0.872	1.237	1.635	2.204	3.455
7	1.239	1.690	2.167	2.833	4.255
8	1.646	2.180	2.733	3.490	5.071
9	2.088	2.700	3.325	4.168	5.899
10	2.588	3.247	3.940	4.865	6.737
12	3.571	4.404	5.226	6.304	8.438
14	4.660	5.629	6.571	7.790	10.165
16	5.812	6.908	7.962	9.312	11.912
18	7.015	8.231	9.390	10.865	13.675
20	8.260	9.591	10.851	12.443	15.452
22	9.542	10.982	12.338	14.042	17.240
24	10.856	12.401	13.848	15.659	19.037
26	12.198	13.844	15.379	17.292	20.843
28	13.565	15.308	16.928	18.939	22.657
30	14.954	16.791	18.493	20.599	24.478
32	16.362	18.291	20.072	22.271	26.304
34	17.789	19.806	21.664	23.952	28.136
36	19.233	21.336	23.269	25.643	29.973
38	20.691	22.878	24.884	27.343	31.815
40	22.164	24.433	26.509	29.051	33.660

417

ν	$1 - P_1^*$				
	.010	.025	.050	.100	.250
45	25.901	28.366	30.612	33.350	38.291
50	29.707	32.357	34.764	37.689	42.942
55	33.570	36.398	38.958	42.060	47.610
60	37.485	40.482	43.188	46.459	52.294
65	41.444	44.603	47.450	50.883	56.990
70	45.442	48.758	51.739	55.329	61.698
75	49.475	52.942	56.054	59.795	66.417
80	53.540	57.153	60.391	64.278	71.145
85	57.634	61.389	64.749	68.777	75.881
90	61.754	65.647	69.126	73.291	80.625
95	65.898	69.925	73.520	77.818	85.376
100	70.065	74.222	77.929	82.358	90.133
110	78.458	82.867	86.792	91.471	99.666
120	86.923	91.573	95.705	100.624	109.220
130	95.451	100.331	104.662	109.811	118.792
140	104.034	109.137	113.659	119.029	128.380
150	112.668	117.985	122.692	128.275	137.983
200	156.432	162.728	168.279	174.835	186.172
250	200.939	208.098	214.392	221.806	234.577
300	245.972	253.912	260.878	269.068	283.135
400	337.155	346.482	354.651	364.207	380.577
500	429.388	439.936	449.147	459.926	478.323
600	522.365	534.019	544.180	556.056	576.286
700	615.907	628.577	639.613	652.497	674.413
800	709.897	723.513	735.362	749.185	772.669
900	804.252	818.756	831.370	846.075	871.032
1000	898.912	914.257	927.594	943.133	969.484

Tables for Binomial Selection Problems

The entries in Table E.1, for $k = 2$, 3, 4, and 10 binomial (or Bernoulli) populations, respectively, give the minimum integer value of n, the common sample size, required to attain the specified (δ^*, P^*) pair in selecting the one best population. Best is defined as the population with the largest p value. The true difference $\delta = p_{[k]} - p_{[k-1]}$ satisfies the condition $\delta \geqslant \delta^*$ in the preference zone and $\delta = \delta^*$ under the least favorable configuration.

SOURCE FOR APPENDIX E

Table E.1 is adapted from M. Sobel and M. Huyett (1957), Selecting the best one of several binomial populations, *Bell System Technical Journal*, **36**, 537–576, with permission.

TABLE INTERPOLATION FOR APPENDIX E

Table E.1 relates the three quantities δ^*, P^*, and n, for a selected value of k. If any two of these quantities are given but at least one is not listed specifically in the table for the particular k, the other quantities can be found by interpolation. Although ordinary linear interpolation could be used, more accurate results can be obtained by using linear interpolation with the transformed quantities $\log \delta^*$, $\log (1 - P^*)$, and $\log n$ (all logarithms are natural logarithms, that is, to the base $e = 2.71828...$). Then the antilogarithm of the interpolated value gives the desired quantity δ^*, $1 - P^*$, or n.

The specific interpolation instructions for a particular k are as follows:

1. To find a value of n for specified δ^* and P^* use linear interpolation on $\log \delta^*$ and $\log (1 - P^*)$ to yield $\log n$ and then take the antilog.
2. To find a value of δ^* for specified n and P^* use linear interpolation on $\log n$ and $\log (1 - P^*)$ to yield $\log \delta^*$ and then take the antilog.
3. To find a value of P^* for specified n and δ^* use linear interpolation on $\log n$ and $\log \delta^*$ to yield $\log (1 - P^*)$ and then take the antilog. This gives $(1 - P^*)$, from which P^* is readily obtained.

If any one of the specified values P^*, δ^*, or n is a stub or a caption in the appropriate table, the interpolation procedure is quite simple. This procedure is illustrated by Examples 1a, 2a, and 3a for $k = 3$. Otherwise a kind of double interpolation is required. Examples 1b, 2b, and 3b, which follow, illustrate this interpolation procedure for $k = 3$.

Example 1a. The problem is to find n for $k = 3$, $\delta^* = .08$, $P^* = .90$. The case $\delta^* = .08$ is not included as a caption in these tables, but there are table entries for n on both sides of $\delta^* = .08$ when $P^* = .90$, namely, $\delta^* = .05$ and $\delta^* = .10$. In the following table on the left we abstract these entries for $k = 3$. The corresponding logarithms (to the base e) are given in the table on the right in preparation for interpolation.

Table excerpt		Ready for interpolation	
δ^*	n	$\log \delta^*$	$\log n$
.05	498	-2.996	6.211
.08	$\boxed{n}$	-2.526	$\boxed{\log n}$
.10	125	-2.303	4.828

Linear interpolation in the table on the right is carried out as follows:

$$\frac{\log n - 4.828}{-2.526 - (-2.303)} = \frac{6.211 - 4.828}{-2.996 - (-2.303)},$$

$$\log n = 5.273.$$

The antilog is $n = 196$, after rounding upward to obtain a conservative result.

Example 2a. The problem is to find δ^* for $k = 3$, $P^* = .90$, $n = 50$. An abstract of the table for $k = 3$ follows with the entries for δ^* on both sides of $n = 50$ when $P^* = .90$, and the corresponding logarithms of n and δ^*.

Table excerpt		Ready for interpolation	
n	δ^*	$\log n$	$\log \delta^*$
55	.15	4.007	-1.897
50	$\boxed{\delta^*}$	3.912	$\boxed{\log \delta^*}$
31	.20	3.434	-1.609

Linear interpolation of the table on the right yields the following:

$$\frac{\log \delta^* - (-1.609)}{3.912 - 3.434} = \frac{-1.897 - (-1.609)}{4.007 - 3.434},$$

$$\log \delta^* = -1.849.$$

The antilog is $\delta^* = .16$, after rounding upward. (*Remark*: Note that this use of δ^* (instead of δ_t) is contrary to our general plan since δ^* is not a specified quantity here.)

Example 3a. The problem is to find P^* for $k = 3$, $\delta^* = .15$, $n = 50$. The following is an abstract of the table for $k = 3$ with the entries for n on both sides of $n = 50$ (when $\delta^* = .15$) and the corresponding P^*, and the corresponding logarithms of n and $(1 - P^*)$.

Table excerpt		Ready for interpolation	
n	P^*	$\log n$	$\log (1 - P^*)$
41	.85	3.714	-1.897
50	$\boxed{P^*}$	3.912	$\boxed{\log (1 - P^*)}$
55	.90	4.007	-2.303

Linear interpolation in the table on the right yields the following:

$$\frac{\log (1 - P^*) - (-2.303)}{3.912 - 4.007} = \frac{-1.897 - (-2.303)}{3.714 - 4.007},$$

$$\log (1 - P^*) = -2.171.$$

The antilog is $1 - P^* = .114$, so that $P^* = .88$ after rounding downward on P^*. For some purposes we may want to round upward and the result is then .89.

Example 1b. The problem is to find n for $k = 3$, $\delta^* = .12$, $P^* = .82$. The table for $k = 3$ lists neither $\delta^* = .12$ nor $P^* = .82$, but there are entries for n

on both sides of each value. We abstract these entries for $\delta^* = .10$ and $\delta^* = .15$, $P^* = .80$ and $P^* = .85$, in the table on the left. The corresponding logarithms needed are given on the right.

Table excerpt   Ready for interpolation

δ^*	P^* .80	.82	.85	$\log \delta^*$	$\log(1 - P^*)$ -1.609	-1.715	-1.897
.10	69		91	-2.303	4.234	b_1	4.511
.12		n		-2.120	a_1	$\log n$	a_2
.15	31		41	-1.897	3.434	b_2	3.714

In the table on the right either we can find a_1 and a_2 using vertical linear interpolation or we can find b_1 and b_2 using horizontal linear interpolation. Knowledge of either a_1 and a_2, or b_1 and b_2, enables us to find $\log n$ by another linear interpolation. (By either method the result for n is always the same.)

We find a_1 and a_2 by solving the following two equations:

$$\frac{a_1 - 3.434}{-2.120 - (-1.897)} = \frac{4.234 - 3.434}{-2.303 - (-1.897)},$$

$$\frac{a_2 - 3.714}{-2.120 - (-1.897)} = \frac{4.511 - 3.714}{-2.303 - (-1.897)}.$$

The results are

$$a_1 = 3.873, \qquad a_2 = 4.152.$$

We then insert these values in the proper place in the preceding table on the right and use horizontal linear interpolation to find $\log n$ as follows:

$$\frac{\log n - 3.873}{-1.715 - (-1.609)} = \frac{4.152 - 3.873}{-1.897 - (-1.609)},$$

$$\log n = 3.976.$$

The antilog is $n = 53.3$, which is rounded upward to the next larger integer as $n = 54$ for a conservative result.

REMARK. The reader may verify that horizontal interpolation for b_1 and b_2 yields $b_1 = 4.336$ and $b_2 = 3.537$. Then vertical interpolation with these numbers yields $\log n = 3.976$, exactly as just demonstrated. □

Example 2b. The problem is to find δ^* for $k=3$, $P^*=.87$, $n=50$. The table abstract for $k=3$ and the corresponding appropriate logarithms follow.

Table excerpt				Ready for interpolation			
		P^*				$\log(1-P^*)$	
δ^*	.85	.87	.90	$\log \delta^*$	-1.897	-2.040	-2.303
.10	91		125	-2.303	4.511	b_1	4.828
δ^*		50		$\log \delta^*$	a_1	3.912	a_2
.15	41		55	-1.897	3.714	b_2	4.007

The interpolation proceeds as in Example 1b.

$$\frac{b_1-4.511}{-2.040-(-1.897)} = \frac{4.828-4.511}{-2.303-(-1.897)},$$

$$\frac{b_2-3.714}{-2.040-(-1.897)} = \frac{4.007-3.714}{-2.303-(-1.897)},$$

$$b_1=4.623, \qquad b_2=3.817.$$

From these values we obtain δ^* as follows:

$$\frac{\log \delta^*-(-1.897)}{-2.303-(-1.897)} = \frac{3.912-3.817}{4.623-3.817} \qquad \text{or} \qquad \log \delta^* = -1.9449.$$

The antilog is $\delta^*=.14$, after rounding downward.

Example 3b. The problem is to find P^* for $k=3$, $\delta^*=.12$, $n=60$. The abstract from the table for $k=3$ and the corresponding appropriate logarithms follow.

Table excerpt				Ready for interpolation			
		P^*				$\log(1-P^*)$	
δ^*	.80	P^*	.85	$\log \delta^*$	-1.609	$\log(1-P^*)$	-1.897
.10	69		91	-2.303	4.234	b_1	4.511
.12		60		-2.120	a_1	4.094	a_2
.15	31		41	-1.897	3.434	b_2	3.714

From Example 1b, we know that $a_1 = 3.873$ and $a_2 = 4.152$. Hence we need only interpolate for $\log(1 - P^*)$ as follows:

$$\frac{\log(1 - P^*) - (-1.609)}{4.094 - 3.873} = \frac{-1.897 - (-1.609)}{4.152 - 3.873},$$

$$\log(1 - P^*) = -1.837.$$

The antilog is $1 - P^* = .159$, so that $P^* = .84$ after rounding downward on P^*.

Table E.1 Smallest integer sample size n needed to satisfy the (δ^*, P^*) requirement in selecting the binomial population with the largest probability

$k = 2$

δ^* \ P^*	.50	.60	.75	.80	.85	.90	.95	.99
.05	0	14	92	142	215	329	541	1082
.10	0	4	23	36	54	83	135	270
.15	0	2	11	16	24	37	60	120
.20	0	1	6	9	14	21	34	67
.25	0	1	4	6	9	13	22	43
.30	0	1	3	4	6	9	15	29
.35	0	1	2	3	5	7	11	21
.40	0	1	2	3	4	5	9	16
.45	0	1	2	2	3	4	7	13
.50	0	1	1	2	3	4	5	10

$k = 3$

δ^* \ P^*	.50	.60	.75	.80	.85	.90	.95	.99
.05	31	79	206	273	364	498	735	1308
.10	8	20	52	69	91	125	184	327
.15	4	9	23	31	41	55	82	145
.20	3	5	13	17	23	31	46	81
.25	2	4	9	11	15	20	29	52
.30	2	3	6	8	10	14	20	35
.35	2	2	5	6	8	10	15	26
.40	1	2	4	5	6	8	11	20
.45	1	2	3	4	5	6	9	15
.50	1	2	3	3	4	5	7	12 $k = 4$

$k = 4$

δ^* \ P^*	.50	.60	.75	.80	.85	.90	.95	.99
.05	71	134	283	359	458	601	850	1442
.10	18	34	71	90	114	150	212	360
.15	8	15	32	40	51	67	94	160
.20	5	9	18	23	29	38	53	89
.25	3	6	12	14	18	24	34	57
.30	3	4	8	10	13	17	23	39
.35	2	3	6	7	9	12	17	28
.40	2	3	5	6	7	9	13	21
.45	2	2	4	5	6	7	10	17
.50	2	2	3	4	5	6	8	13

$k = 10$

δ^* \ P^*	.50	.60	.75	.80	.85	.90	.95	.99
.05	218	314	513	606	725	890	1169	1803
.10	55	79	128	151	181	222	291	449
.15	25	35	57	67	80	98	129	198
.20	14	20	32	38	45	55	72	111
.25	9	13	20	27	29	35	46	70
.30	7	9	14	17	20	24	32	48
.35	5	7	11	13	15	18	23	35
.40	4	5	8	10	11	13	17	26
.45	3	4	6	8	9	11	14	20
.50	3	4	5	6	7	9	11	16

Figures for Binomial Selection Problems

The sets of graphs in Figure F.1 and Figure F.2 relate the values of P^*, the probability of a correct selection calculated under the least favorable configuration, the lower bound δ^* of the true difference $\delta = p_{[k]} - p_{[k-1]}$, and the minimum integer value of n, the common sample size, for k binomial (or Bernoulli) populations, where the problem is to select the one population with the largest p value. Figure F.1 is for interpolation to values of k that are not in our tables for selected fixed values of δ^*. Figure F.2 is for interpolation to values of n and/or δ^* that are not in our tables for selected fixed values of k. In almost all cases extrapolation to values not on the figure is also possible.

SOURCE FOR APPENDIX F

Figures F.1 and F.2 are adapted from M. Sobel and M. Huyett (1957), Selecting the best one of several binomial populations, *Bell System Technical Journal*, **36**, 537–576, with permission.

Figure F.1. *Figures for interpolation to values of k for selected (δ*, P*) pair in the binomial selection problem.*

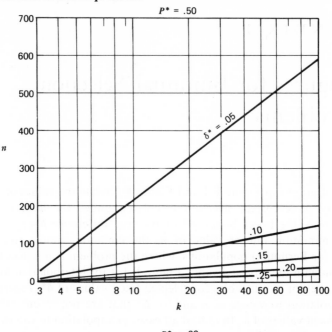

P* = .50

δ* = .05

.10

.15

.20

.25

n

k

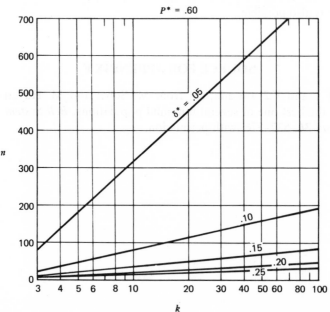

P* = .60

δ* = .05

.10

.15

.20

.25

n

k

Figure F.1. (Continued)

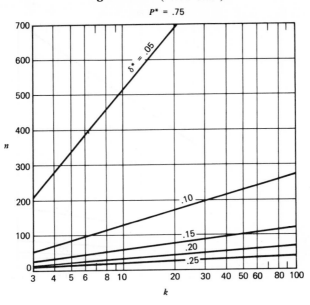

$P^* = .75$

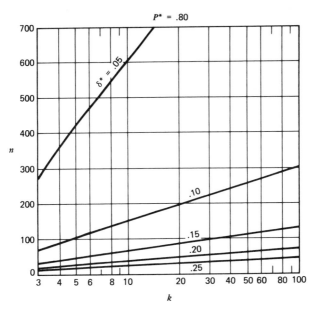

$P^* = .80$

429

Figure F.1. (Continued)

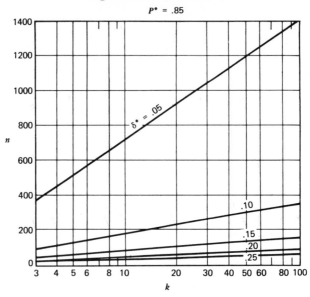

$P^* = .85$

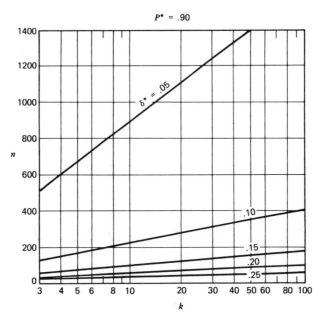

$P^* = .90$

430

Figure F.1. (Continued)

$P^* = .95$

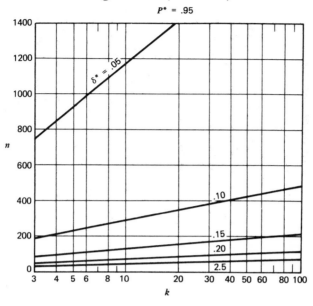

$P^* = .99$

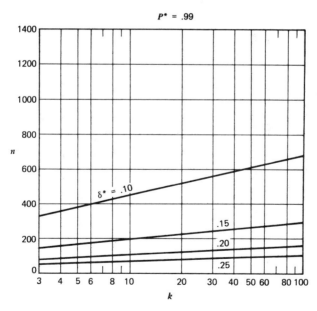

Figure F.2. *Figures for interpolation to values of n and/or* δ* *for selected values of* k *in the binomial selection problem.*

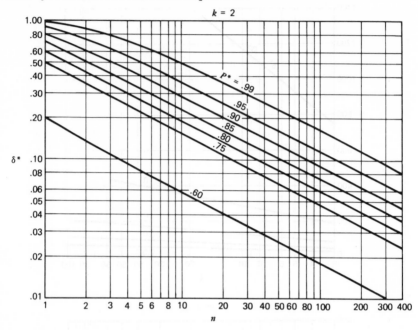

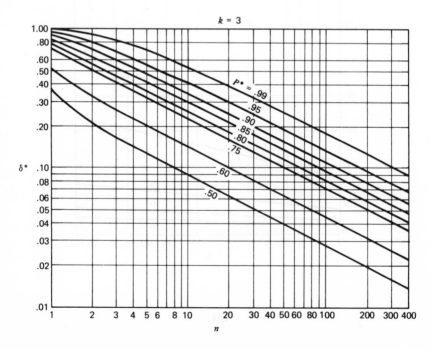

Figure F.2. (Continued)

k = 4

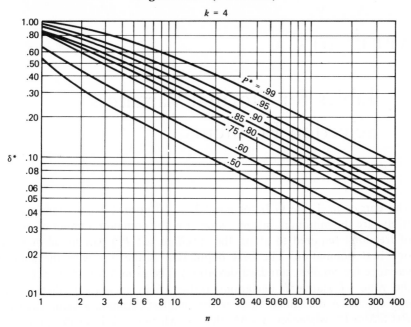

k = 5

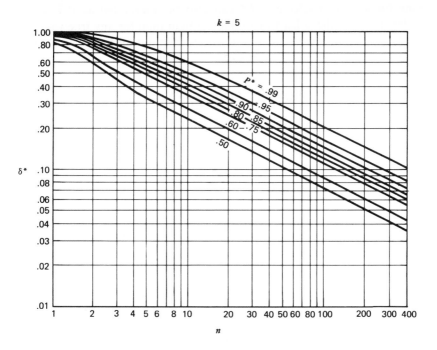

Tables for Normal Variances Selection Problems

The entries in Table G.1, for $k = 2, 3, \ldots, 10$ normal populations, respectively, are the minimum integer value of ν, the common number of degrees of freedom, required to attain the specified (Δ^*, P^*) pair in selecting the one best population. The best population is the one with the smallest variance (or smallest standard deviation). The ratio Δ (or Δ^*) is defined as the ratio of standard deviations (rather than variances), that is, $\Delta = \sigma_{[1]}/\sigma_{[2]}$. Then the tabled value is the smallest common ν value such that $P(CS|\Delta) \geqslant P^*$ whenever $\Delta \leqslant \Delta^*$, that is, whenever Δ is in the preference zone for a correct selection. P^* can also be interpreted as P calculated under the generalized least favorable configuration for any true value Δ. If the k population means are all known, $n = \nu$ and if all unknown, $n = \nu + 1$, where n is the common sample size.

SOURCE FOR APPENDIX G

Table G.1 is adapted from S. S. Gupta and M. Sobel (1962), On the smallest of several correlated F statistics, *Biometrika*, **49**, 509–523, with permission of the Biometrika Trustees and the authors.

TABLE INTERPOLATION FOR APPENDIX G

Table G.1 relates the three quantities Δ^*, P^*, and ν, for a selected value of k. If any two of these quantities are given but at least one is not listed specifically in the table for the particular k, the other quantities can be found by interpolation. Ordinary linear interpolation is not recommended generally because it does not provide good accuracy. Rather, we use linear interpolation on the quantities $-\log \Delta^*$, $\log(1 - P^*)$ and $1/\sqrt{\nu}$ (all logarithms are natural logarithms, that is, to the base $e = 2.71828\ldots$).

The specific interpolation instructions for a particular k are as follows:

1. To find a value of ν for specified Δ^* and P^* use linear interpolation on $-\log \Delta^*$ and $\log (1-P^*)$ to yield $1/\sqrt{\nu}$, take the reciprocal and then square it.

2. To find a value of Δ^* for specified ν and P^* use linear interpolation on $1/\sqrt{\nu}$ and $\log (1-P^*)$ to yield $-\log \Delta^*$, change the sign and then take the antilog.

3. To find a value of P^* for a specified ν and Δ^*, use linear interpolation on $1/\sqrt{\nu}$ and $-\log \Delta^*$ to yield $\log (1-P^*)$ and then take the antilog. This gives $(1-P^*)$, from which P^* is readily obtained.

If any one of the specified values P^*, Δ^*, or ν is a caption in the appropriate table, the interpolation procedure is quite simple. This procedure is illustrated by Examples 1a, 2a, and 3a for $k=4$. Otherwise a simple kind of double interpolation is required. Examples 1b, 2b, and 3b illustrate this interpolation procedure for $k=4$.

Example 1a. The problem is to find ν for $k=4$, $\Delta^*=0.50$, $P^*=.990$. The case $\Delta^*=0.50$ is not included as a caption in these tables, but there are table entries for ν on both sides of $\Delta^*=0.50$ when $P^*=.990$, namely, $\Delta^*=0.40$ and $\Delta^*=0.60$. In the following table on the left, we abstract these entries for $k=4$. The corresponding logarithms (to the base e) of Δ^* and square root of the reciprocal of ν are given in the table on the right in preparation for interpolation.

Table excerpt		Ready for interpolation	
Δ^*	ν	$-\log \Delta^*$	$1/\sqrt{\nu}$
0.40	11	.9163	.3015
0.50	ν	.6931	$1/\sqrt{\nu}$
0.60	30	.5108	.1826

Linear interpolation in the table on the right is carried out as follows:

$$\frac{(1/\sqrt{\nu})-.1826}{.6931-.5108} = \frac{.3015-.1826}{.9163-.5108},$$

$$1/\sqrt{\nu} = .2361, \qquad \sqrt{\nu} = 4.235.$$

The square of 4.235 is 17.9, which we round upward to get $\nu=18$.

Example 2a. The problem is to find Δ^* for $k=4$, $P^*=.750$, $\nu=40$. The following is an abstract of the table for $k=4$ with the entries for Δ^* on both sides of $\nu=40$ when $P^*=.750$, and the corresponding values, $-\log \Delta^*$ and $1/\sqrt{\nu}$.

Table excerpt		Ready for interpolation	
Δ^*	ν	$-\log \Delta^*$	$1/\sqrt{\nu}$
0.80	31	.2231	.1796
$\boxed{\Delta^*}$	40	$\boxed{-\log \Delta^*}$	.1581
0.85	56	.1625	.1336

Linear interpolation in the table on the right yields the following:

$$\frac{-\log \Delta^* - .1625}{.1581 - .1336} = \frac{.2231 - .1625}{.1796 - .1336},$$

$$-\log \Delta^* = .1948.$$

The antilog of $-.1948$ is .823, so that $\Delta^* = 0.82$ after rounding downward.

Example 3a. The problem is to find P^* for $k=4$, $\Delta^*=0.75$, $\nu=40$. The following is an abstract of the table for $k=4$ with the entries for ν on both sides of $\nu=40$ (when $\Delta^*=0.75$) and the corresponding P^*, and the corresponding values of $1/\sqrt{\nu}$ and $\log(1-P^*)$.

Table excerpt		Ready for interpolation	
ν	P^*	$1/\sqrt{\nu}$	$\log(1-P^*)$
39	.900	.1601	-2.303
40	$\boxed{P^*}$	.1581	$\boxed{\log(1-P^*)}$
54	.950	.1361	-2.996

Linear interpolation in the table on the right yields the following:

$$\frac{\log(1-P^*) - (-2.996)}{.1581 - .1361} = \frac{-2.303 - (-2.996)}{.1601 - .1361},$$

$$\log(1-P^*) = -2.361.$$

The antilog is $1-P^*=.0943$ so that $P^*=.906$ after rounding upward on P^*.

Example 1*b*. The problem is to find ν for $k=4$, $\Delta^*=0.72$, $P^*=.920$. The table for $k=4$ lists neither $\Delta^*=0.72$ nor $P^*=.920$, but there are entries for ν on both sides of each value. The following abstracts these entries for $\Delta^*=0.70$ and $\Delta^*=0.75$, $P^*=.900$ and $P^*=.950$, on the left. The corresponding quantities needed for interpolation are given on the right.

Table excerpt

Δ^*	.900	P^* .920	.950
0.70	26		36
0.72		ν	
0.75	39		54

Ready for interpolation

$-\log \Delta^*$	-2.303	$\log(1-P^*)$ -2.526	-2.996
.3567	.1961	b_1	.1667
.3285	a_1	$1/\sqrt{\nu}$	a_2
.2877	.1601	b_2	.1361

In the table on the right, we can either find a_1 and a_2 using vertical linear interpolation or find b_1 and b_2 using horizontal linear interpolation. Knowledge of either a_1 and a_2 or b_1 and b_2 enables us to find $1/\sqrt{\nu}$ by another linear interpolation. (By either method the result for n is always the same.)

We find a_1 and a_2 by solving the following two equations:

$$\frac{a_1-.1601}{.3285-.2877}=\frac{.1961-.1601}{.3567-.2877},$$

$$\frac{a_2-.1361}{.3285-.2877}=\frac{.1667-.1361}{.3567-.2877}.$$

The results are

$$a_1=.1814, \qquad a_2=.1542.$$

We then insert these values in the proper place in the preceding table on the right and use horizontal linear interpolation to find $1/\sqrt{\nu}$ as follows:

$$\frac{1/\sqrt{\nu}-.1814}{-2.526-(-2.303)}=\frac{.1542-.1814}{-2.996-(-2.303)},$$

$$1/\sqrt{\nu}=.1726, \qquad \sqrt{\nu}=5.794.$$

The square of 5.794 is 33.6, which we round upward to get $\nu=34$.

Example 2b. The problem is to find Δ^* for $k=4$, $P^*=.920$, $\nu=50$. The table abstract for $k=4$ and the corresponding quantities needed for interpolation follow.

Table excerpt

Δ^*	P^*		
	.900	.920	.950
0.75	39		54
Δ^*		50	
0.80	63		88

Ready for interpolation

$-\log \Delta^*$	$\log(1-P^*)$		
	-2.303	-2.526	-2.996
.2877	.1601	b_1	.1361
$-\log \Delta^*$	a_1	.1414	a_2
.2231	.1260	b_2	.1066

The interpolation proceeds as in Example 1b:

$$\frac{b_1-.1601}{-2.526-(-2.303)} = \frac{.1361-.1601}{-2.996-(-2.303)},$$

$$\frac{b_2-.1260}{-2.526-(-2.303)} = \frac{.1066-.1260}{-2.996-(-2.303)},$$

$$b_1=.1524, \qquad b_2=.1198.$$

From these values we obtain Δ^* as follows:

$$\frac{-\log \Delta^*-.2231}{.1414-.1198} = \frac{.2877-.2230}{.1524-.1198},$$

$$-\log \Delta^*=.2660.$$

The antilog of -2.660 is 0.766, so that $\Delta^*=0.77$ after rounding upward.

Example 3b. The problem is to find P^* for $k=4$, $\Delta^*=0.72$, $\nu=35$. The abstract from the table for $k=4$ and the corresponding quantities needed for interpolation follow.

Table excerpt

Δ^*	P^*		
	.900	P^*	.950
0.70	26		36
0.72		35	
0.75	39		54

Ready for interpolation

$-\log \Delta^*$	$\log(1-P^*)$		
	-2.303	$\log(1-P^*)$	-2.996
.3567	.1961	b_1	.1667
.3285	a_1	.1690	a_2
.2877	.1601	b_2	.1361

From Example 1b we know that $a_1 = .1814$ and $a_2 = .1542$. Hence we need only interpolate for $\log (1 - P^*)$ as follows:

$$\frac{\log (1 - P^*) - (-2.303)}{.1690 - .1814} = \frac{-2.996 - (-2.303)}{.1542 - .1814},$$

$$\log (1 - P^*) = -2.619.$$

The antilog is $1 - P^* = .0729$, so that $P^* = .927$ after rounding downward on P^*.

This interpolation procedure appears to be uniformly accurate throughout the tables and gives good (slightly conservative) results even for Δ^* close to 1. For example, the correct value in Example 1a is 17, and the interpolated value is 18. (On the other hand, using ordinary linear interpolation with the values of Δ^* and ν, we would obtain the result 21, which is a considerably larger error.) Another excellent reason for using this procedure is that we can set $1/\sqrt{\nu} = 0$ at $-\log \Delta^* = 0$ and regard this additional point as a value in the tables. Then we can obtain ν values for $0.95 < \Delta^* < 1.00$ by interpolation rather than by extrapolation. Thus for $k = 4$, $P^* = .990$, from the values for $\Delta^* = 0.90$ and $\Delta^* = 1.00$, the interpolated value is $\nu = 1154$ and the correct result is $\nu = 1145$; this is an error of less than 1%, again in the conservative direction.

Table G.1 *Smallest number of degrees of freedom v per population needed to select the population with the smallest variance such that* $P\{CS|\Delta\} \geqslant P^*$

$k = 2$

Δ (or Δ^*)	.750	.900	.950	.975	.990
			P(or P^*)		
0.25	1	2	3	4	4
0.40	1	3	5	6	8
0.50	2	5	7	10	13
0.60	3	8	12	16	23
0.70	5	14	23	31	44
0.75	7	21	34	47	66
0.80	10	34	55	78	109
0.85	18	63	103	146	205
0.90	42	148	244	347	488
0.95	173	625	1029	1461	2057

$k = 3$

Δ (or Δ^*)	.750	.900	.950	.975	.990
			P(or P^*)		
0.25	1	3	3	4	5
0.40	2	4	6	8	10
0.50	4	7	9	12	16
0.60	5	11	16	21	27
0.70	10	21	31	40	53
0.75	14	32	47	61	81
0.80	23	52	76	100	133
0.85	41	96	141	187	250
0.90	94	226	333	443	591
0.95	392	947	1398	1862	2489

$k = 4$

Δ (or Δ^*)	.750	.900	.950	.975	.990
			P(or P^*)		
0.25	2	3	4	5	6
0.40	3	5	7	9	11
0.50	5	8	11	14	17
0.60	7	14	18	24	30
0.70	14	26	36	46	59
0.75	19	39	54	69	90
0.80	31	63	88	113	147
0.85	56	116	164	212	276
0.90	130	273	386	500	652
0.95	540	1145	1619	2100	2742

$k = 5$

Δ (or Δ^*)	.750	.900	.950	.975	.990
			P (or P^*)		
0.25	2	3	4	5	6
0.40	4	6	8	9	11
0.50	5	9	12	15	18
0.60	9	15	20	26	32
0.70	16	29	39	50	63
0.75	23	44	60	75	96
0.80	37	71	97	123	157
0.85	68	131	180	229	294
0.90	157	308	424	540	695
0.95	651	1288	1777	2269	2923

$k = 6$

Δ (or Δ^*)	.750	.900	.950	.975	.990
			P (or P^*)		
0.25	2	4	4	5	6
0.40	4	6	8	10	12
0.50	6	10	13	16	19
0.60	10	17	22	28	34
0.70	18	32	42	53	67
0.75	26	48	64	80	101
0.80	42	77	104	130	165
0.85	77	143	193	243	308
0.90	178	334	453	572	729
0.95	739	1399	1900	2401	3063

$k = 7$

Δ (or Δ^*)	.750	.900	.950	.975	.990
			P (or P^*)		
0.25	3	4	5	5	7
0.40	4	7	8	10	12
0.50	6	10	13	16	20
0.60	11	18	23	29	35
0.70	20	34	45	56	70
0.75	29	51	68	84	105
0.80	47	83	110	136	172
0.85	85	152	203	254	320
0.90	195	357	477	598	756
0.95	813	1491	2001	2509	3177

Table G.1 (Continued)

k = 8

Δ (or Δ*)	P (or P*)				
	.750	.900	.950	.975	.990
0.25	3	4	5	6	7
0.40	5	7	9	11	13
0.50	7	11	14	17	21
0.60	11	19	24	31	36
0.70	21	36	46	58	72
0.75	31	54	71	87	108
0.80	50	87	114	142	177
0.85	91	160	212	263	330
0.90	211	375	498	619	779
0.95	875	1569	2086	2600	3273

k = 9

Δ (or Δ*)	P (or P*)				
	.750	.900	.950	.975	.990
0.25	3	4	5	6	7
0.40	5	7	9	11	13
0.50	7	11	14	17	21
0.60	12	19	25	32	37
0.70	22	37	48	60	74
0.75	33	56	73	90	111
0.80	54	91	119	146	182
0.85	97	167	219	271	339
0.90	224	391	516	638	799
0.95	930	1636	2160	2678	3357

k = 10

Δ (or Δ*)	P (or P*)				
	.750	.900	.950	.975	.990
0.25	3	4	5	6	7
0.40	5	8	9	11	13
0.50	8	12	15	18	22
0.60	13	20	25	33	38
0.70	24	39	50	62	76
0.75	35	58	75	92	114
0.80	56	94	122	150	186
0.85	102	173	226	278	346
0.90	236	406	531	655	817
0.95	979	1696	2225	2748	3430

Tables for Multinomial Selection Problems

The entries in Table H.1 for $k = 2, 3, \ldots, 10$ categories in a multinomial distribution give the minimum integer value of n, the total sample size, required to satisfy the (δ^*, P^*) requirement for selecting the one best category. When *best* is defined as the category with the largest probability, the distance δ is defined by $\delta = p_{[k]}/p_{[k-1]}$ and δ satisfies $\delta \geqslant \delta^*$. In the least favorable configuration $\delta = \delta^*$. P^* can also be interpreted as the probability of a correct selection under the least favorable configuration. Entries less than or equal to 100 have been calculated exactly, whereas entries greater than 100 are conservative and are in excess by at most 3; thus the error is at most 3% for all entries.

The entries in Table H.2 for $k = 2, 3, \ldots, 10, 15, 20, 25$ categories in a multinomial distribution give the minimum integer value of n, the total sample size, required to satisfy the corresponding (δ^*, P^*) requirement when *best* is defined as the category with the smallest probability, the distance δ is defined by $\delta = p_{[2]} - p_{[1]}$, and δ satisfies $\delta \geqslant \delta^*$.

TABLE INTERPOLATION FOR APPENDIX H

For a fixed P^* we may wish to find the n corresponding to a δ^* that is not listed specifically in Table H.1. We recommend using linear interpolation on $\log \delta^*$ and $1/\sqrt{n}$. This generally gives an error of at most 2% in the value of n and this error could be in either direction, that is, either too large or too small. This same method can be used to find the δ^* for a given value of n.

If we have a fixed δ^* then we use $\log(1 - P^*)$ with $1/\sqrt{n}$ and again use linear interpolation. This same routine can be used for given P^* to find n and for given n to find P^*.

Either one of these methods can be used to obtain an operating characteristic (OC) curve for fixed values of k and n. Suppose $k = 2$ and $n = 40$. For each of the five P^* values in the table for $k = 2$, we can take $n = 40$ and use the method in the preceding paragraph to find δ^*. The resulting five pairs (δ^*, P^*) form the basic five points on a graph of P^* versus δ^*, which we can connect to form a curve that approximates the OC curve for $k = 2$ and $n = 40$.

*Table H.1   Smallest integer sample size n needed to satisfy the (δ^*, P^*)
requirement for selecting the category with the largest probability*

$k = 2$

δ^*	P^*				
	.750	.900	.950	.975	.990
1.2	55	199	327	463	653
1.4	17	59	97	137	193
1.6	9	31	49	71	101
1.8	5	19	33	45	65
2.0	5	15	23	33	47
2.2	3	11	19	25	37
2.4	3	9	15	21	29
2.6	3	7	13	17	25
2.8	3	7	11	15	21
3.0	1	7	9	13	19

$k = 3$

δ^*	P^*				
	.750	.900	.950	.975	.990
1.2	181	437	645	859	1148
1.4	52	126	186	247	330
1.6	26	64	94	125	167
1.8	17	40	59	79	106
2.0	12	29	42	56	75
2.2	9	22	32	43	58
2.4	7	18	26	35	46
2.6	6	15	22	29	39
2.8	6	13	19	25	33
3.0	5	11	17	22	29

$k = 4$

δ^*	P^*				
	.750	.900	.950	.975	.990
1.2	326	692	979	1271	1660
1.4	92	196	278	361	471
1.6	46	98	139	180	235
1.8	29	61	87	112	147
2.0	20	43	61	79	104
2.2	15	33	46	60	79
2.4	12	26	37	48	63
2.6	10	22	31	40	53
2.8	9	19	26	34	45
3.0	8	16	23	30	39

$$k = 5$$

δ^*			P^*		
	.750	.900	.950	.975	.990
1.2	486	964	1331	1701	2191
1.4	137	271	374	478	616
1.6	68	134	185	237	305
1.8	41	83	115	147	189
2.0	29	58	81	103	133
2.2	22	44	61	78	100
2.4	17	35	48	62	80
2.6	14	29	40	51	66
2.8	12	24	34	43	56
3.0	11	21	29	38	48

$$k = 6$$

δ^*			P^*		
	.750	.900	.950	.975	.990
1.2	658	1249	1697	2145	2737
1.4	184	349	475	600	766
1.6	90	172	233	295	377
1.8	56	106	144	182	233
2.0	38	74	101	127	163
2.2	29	56	76	96	122
2.4	23	44	60	76	97
2.6	19	36	49	63	80
2.8	16	30	41	53	68
3.0	14	26	36	45	58

$$k = 7$$

δ^*			P^*		
	.750	.900	.950	.975	.990
1.2	840	1545	2075	2603	3297
1.4	234	430	578	725	918
1.6	114	211	283	355	450
1.8	70	130	174	219	277
2.0	48	90	121	152	193
2.2	36	68	91	114	145
2.4	28	53	72	90	114
2.6	23	43	59	74	94
2.8	20	36	49	62	79
3.0	17	31	42	54	68

$k = 8$

δ^*			P^*		
	.750	.900	.950	.975	.990
1.2	1030	1851	2464	3071	3869
1.4	286	514	684	853	1074
1.6	140	251	334	417	525
1.8	86	154	205	256	322
2.0	59	107	142	177	224
2.2	43	80	106	133	167
2.4	34	63	84	105	132
2.6	28	51	69	86	108
2.8	23	43	58	72	91
3.0	20	37	49	62	78

$k = 9$

δ^*			P^*		
	.750	.900	.950	.975	.990
1.2	1228	2166	2862	3550	4451
1.4	340	600	793	983	1234
1.6	166	292	386	479	601
1.8	101	179	237	294	368
2.0	70	124	164	203	255
2.2	51	92	122	152	191
2.4	40	73	96	120	150
2.6	32	59	79	98	123
2.8	27	49	66	82	103
3.0	23	42	57	70	88

$k = 10$

δ^*			P^*		
	.750	.900	.950	.975	.990
1.2	1433	2489	3269	4038	5043
1.4	396	688	903	1116	1394
1.6	192	335	440	543	679
1.8	118	204	269	332	415
2.0	81	141	186	229	287
2.2	60	105	138	171	214
2.4	46	83	109	135	168
2.6	37	67	89	110	137
2.8	31	57	74	92	115
3.0	27	47	64	79	99

Table H.2 *Smallest integer sample size n needed to satisfy the (δ^*, P^*) requirement for selecting the category with the smallest probability* [a][b]

$k=2$

δ^*	P*				
	.750	.900	.950	.975	.990
.10	45	163	269	381^a	536^a
.15	21	73	119	169	239
.20	11	41	67	95	133
.25	7	25	43	61	85

$k=3$

δ^*	P*				
	.750	.900	.950	.975	.990
.10	66	158	232	308	411
.15	29	68	100	132	176
.20	16	37	54	71	95
.25	10	23	33	43	57

$k=4$

δ^*	P*				
	.750	.900	.950	.975	.990
.05	272	578^a	813^a	1051^a	1370^a
.10	65	135	190	246	321
.15	28	56	79	101	132
.20	15	29	40	52	67

$k=5$

δ^*	P*				
	.750	.900	.950	.975	.990
.05	256	507^a	696^a	885^a	1137^a
.10	60	115	157	200	256
.15	24	45	62	78	99
.20	12	22	29	36	46

k = 6

δ*	.750	.900	P* .950	.975	.990
.025	1011[a]	1888[a]	2552[a]	3215[a]	4092[a]
0.50	237	443	608[a]	763[a]	968[a]
.075	100	185	250	315	400
.100	53	97	131	164	208

k = 7

δ*	.750	.900	P* .950	.975	.990
.025	947[a]	1709[a]	2280[a]	2847[a]	3594[a]
.050	218	394	526	666[a]	838[a]
.075	91	162	215	268	337
.100	47	83	109	135	170

k = 8

δ*	.750	.900	P* .950	.975	.990
.025	887[a]	1559[a]	2058[a]	2552[a]	3201[a]
.050	200	352	465	589[a]	735[a]
.075	82	141	186	230	287
.100	42	70	91	112	139

k = 9

δ*	.750	.900	P* .950	.975	.990
.025	833[a]	1431[a]	1873[a]	2310[a]	2882[a]
.050	184	317	415	525[a]	651[a]
.075	74	124	161	198	246
.100	36	58	75	91	112

Table H.2 (Continued)

k = 10

δ*	.750	.900	.950	.975	.990
			P*		
.025	784[a]	1322[a]	1718[a]	2108[a]	2617[a]
.050	169	286	372	457	581[a]
.075	66	108	140	170	210
.100	31	48	60	72	87

k = 15

δ*	.750	.900	.950	.975	.990
			P*		
.01	3947	6417	8206	9954	12225
.02	956	1526	1937	2338	2860
.03	410	641	807	969	1180
.04	221	338	421	503	610

k = 20

δ*	.750	.900	.950	.975	.990
			P*		
.01	3251	5110	6443	7738	9413
.02	776	1188	1482	1768	2113
.03	327	484	597	707	849
.04	171	345	298	349	417

k = 25

δ*	.750	.900	.950	.975	.990
			P*		
.01	1186	1819	2269	2705	3267
.02	651	963	1184	1399	1676
.03	267	378	457	535	635
.04	133	179	213	246	291

[a]Indicates all entries in this table that are obtained by a normal approximation.
[b]All entries for k = 15, 20, 25 are obtained by a normal approximation.

Curtailment Tables for the Multinomial Selection Problem

Table I.1 for $k = 2, 3, 4, 5,$ and 10 gives $E(N|R')$, the expected number of observations needed to terminate (using curtailment) a multinomial selection problem based on k categories and at most n observations for $n \leq 20$, as a function of δ under the generalized least favorable configuration (or δ^* under the least favorable configuration), where $\delta = p_{[k]}/p_{[k-1]}$. From the table entries for $n \leq 20$ the value of S_n, the percentage saved, can be computed for any δ (or δ^*) from

$$S_n = \left[\frac{n - E(N|R')}{n} \right] 100 = 100 \left[1 - \frac{E(N|R')}{n} \right].$$

Table I.2 gives the limit S_∞ (as $n \to \infty$) of the percentage saved as a function of δ (or δ^*).

TABLE INTERPOLATION FOR APPENDIX I

For any $n > 20$ we can interpolate in Table I.2 between the values of S_{20} and S_∞ to estimate a value for S_n. If we use linear interpolation on $1/n$ and S_n, the result is generally an overestimate. If we use linear interpolation on $1/n$ and $(S_\infty - S_n)^2$, the result is generally an underestimate. These two values can be presented together as an interval estimate of S_n, or their simple average can be presented as a point estimate of S_n. A point estimate of $E(N|R')$ can then be found from

$$E(N|R') = n \left(1 - \frac{S_n}{100} \right).$$

These procedures are now illustrated for $n = 40$, $k = 3$, and $\delta = 2$. The table gives $S_\infty = 20.00$ and $E(N|R') = 17.46$ for $n = 20$, and thus $S_{20} = 100$ $[1 - 17.46/20] = 12.70$. Now we must interpolate for S_{40} in the following table:

n	S_n
20	12.70
40	S_{40}
∞	20.00

The two interpolation procedures required follow.

Ready for interpolation		Ready for interpolation	
$1/n$	S_n	$1/n$	$(S_\infty - S_n)^2$
.050	12.70	.050	$(7.3)^2 = 53.29$
.025	S_{40}	.025	$(20 - S_{40})^2$
.000	20.00	.000	.00

$$\frac{S_{40} - 20.00}{.025 - .000} = \frac{12.70 - 20.00}{.050 - .000}$$

$$S_{40} = 16.35$$

$$\frac{(20 - S_{40})^2 - .000}{.025 - .000} = \frac{53.29 - .00}{.050 - .000}$$

$$(20 - S_{40})^2 = 26.645$$

$$20 - S_{40} = \sqrt{26.645} = 5.16,$$

$$S_{40} = 14.84$$

Thus we either say that S_{40} is between 14.84 and 16.35, or estimate it as $(14.84 + 16.35)/2 = 15.6$. The estimate of $E(N|R')$ for $n = 40$ is then $40(1 - 15.6/100) = 33.76$.

Note that we use S_{20} as the lower value for interpolation because it is the largest value available. The values of $E(N|R')$ are more erratic for smaller values of n. Of course, $n = 18$ or 19 could also be used.

REMARK. A more sophisticated four-point Lagrangian interpolation (see any book on numerical analysis or on interpolation) based on $1/n$ and S_n gives the result 15.95 in the preceding numerical illustration, but this method requires more work and we omit an explanation of it here.

Table I.1 Expected sample size needed to terminate when curtailment is used

k = 2

n \ δ	1.0	1.2	1.4	1.6	1.8	2.0	2.2	2.4	2.6	2.8	3.0
1	1.00	1.00	1.00	1.00	1.00	1.00	1.00	1.00	1.00	1.00	1.00
2	2.00	2.00	2.00	2.00	2.00	2.00	2.00	2.00	2.00	2.00	2.00
3	2.50	2.50	2.49	2.47	2.46	2.44	2.43	2.42	2.40	2.39	2.38
4	3.75	3.74	3.73	3.71	3.69	3.67	3.64	3.62	3.60	3.58	3.56
5	4.13	4.11	4.08	4.05	4.01	3.96	3.92	3.88	3.84	3.81	3.77
6	5.50	5.48	5.44	5.39	5.34	5.28	5.23	5.18	5.12	5.08	5.03
7	5.81	5.79	5.73	5.66	5.58	5.50	5.43	5.35	5.29	5.22	5.16
8	7.26	7.24	7.17	7.08	6.98	6.88	6.78	6.69	6.61	6.53	6.45
9	7.54	7.50	7.41	7.29	7.17	7.05	6.93	6.82	6.72	6.63	6.54
10	9.05	9.00	8.89	8.75	8.61	8.46	8.32	8.19	8.06	7.95	7.85
11	9.29	9.24	9.10	8.94	8.77	8.60	8.43	8.28	8.15	8.02	7.91
12	10.84	10.78	10.62	10.43	10.23	10.03	9.84	9.67	9.50	9.36	9.22
13	11.07	10.99	10.81	10.59	10.36	10.14	9.93	9.74	9.56	9.41	9.26
14	12.65	12.56	12.36	12.11	11.84	11.59	11.35	11.13	10.93	10.75	10.59
15	12.86	12.76	12.53	12.25	11.96	11.68	11.42	11.19	10.98	10.79	10.62
16	14.47	14.35	14.10	13.78	13.45	13.14	12.85	12.59	12.35	12.13	11.94
17	14.66	14.54	14.25	13.91	13.55	13.22	12.91	12.63	12.38	12,16	11.96
18	16.29	16.15	15.84	15.45	15.06	14.69	14.34	14.03	13.76	13.51	13.29
19	16.48	16.32	15.98	15.57	15.15	14.75	14.39	14.07	13.78	13.53	13.31
20	18.12	17.96	17.58	17.12	16.66	16.23	15.83	15.48	15.16	14.88	14.64

k = 3

n \ δ	1.0	1.2	1.4	1.6	1.8	2.0	2.2	2.4	2.6	2.8	3.0
1	1.00	1.00	1.00	1.00	1.00	1.00	1.00	1.00	1.00	1.00	1.00
2	2.00	2.00	2.00	2.00	2.00	2.00	2.00	2.00	2.00	2.00	2.00
3	2.67	2.66	2.66	2.65	2.64	2.63	2.61	2.60	2.59	2.57	2.56
4	3.89	3.89	3.88	3.87	3.86	3.84	3.83	3.81	3.80	3.78	3.77
5	4.56	4.55	4.53	4.50	4.47	4.43	4.39	4.35	4.31	4.28	4.24
6	5.58	5.57	5.56	5.54	5.51	5.48	5.44	5.41	5.37	5.34	5.31
7	6.53	6.51	6.48	6.42	6.36	6.29	6.22	6.15	6.08	6.01	5.94
8	7.39	7.37	7.34	7.29	7.23	7.17	7.10	7.03	6.97	6.90	6.84
9	8.34	8.32	8.27	8.20	8.12	8.02	7.93	7.83	7.73	7.63	7.53
10	9.31	9.29	9.22	9.14	9.04	8.93	8.82	8.70	8.60	8.49	8.39

Table I.1 (Continued)

k = 3

n＼δ	1.0	1.2	1.4	1.6	1.8	2.0	2.2	2.4	2.6	2.8	3.0
11	10.19	10.16	10.09	9.99	9.88	9.75	9.62	9.48	9.35	9.22	9.10
12	11.15	11.12	11.04	10.92	10.78	10.63	10.48	10.33	10.18	10.04	9.91
13	12.11	12.07	11.98	11.84	11.67	11.50	11.32	11.14	10.97	10.80	10.65
14	13.01	12.97	12.86	12.71	12.53	12.34	12.14	11.95	11.76	11.59	11.42
15	13.97	13.92	13.80	13.63	13.42	13.21	12.98	12.76	12.55	12.35	12.16
16	14.93	14.88	14.74	14.54	14.31	14.07	13.82	13.58	13.35	13.13	12.93
17	15.84	15.79	15.64	15.42	15.17	14.91	14.64	14.38	14.12	13.88	13.67
18	16.80	16.74	16.57	16.34	16.06	15.77	15.47	15.18	14.91	14.66	14.42
19	17.77	17.70	17.51	17.25	16.94	16.62	16.30	15.99	15.69	15.42	15.16
20	18.69	18.61	18.41	18.13	17.80	17.46	17.11	16.78	16.47	16.18	15.91

k = 4

n＼δ	1.0	1.2	1.4	1.6	1.8	2.0	2.2	2.4	2.6	2.8	3.0
1	1.00	1.00	1.00	1.00	1.00	1.00	1.00	1.00	1.00	1.00	1.00
2	2.00	2.00	2.00	2.00	2.00	2.00	2.00	2.00	2.00	2.00	2.00
3	2.75	2.75	2.74	2.74	2.73	2.72	2.71	2.70	2.69	2.68	2.67
4	3.94	3.94	3.93	3.92	3.92	3.91	3.90	3.89	3.88	3.87	3.86
5	4.73	4.74	4.72	4.70	4.68	4.65	4.62	4.59	4.56	4.53	4.50
6	5.69	5.68	5.67	5.66	5.64	5.61	5.59	5.56	5.53	5.50	5.48
7	6.65	6.65	6.63	6.60	6.57	6.52	6.48	6.43	6.38	6.33	6.28
8	7.63	7.62	7.60	7.56	7.52	7.47	7.41	7.35	7.29	7.23	7.18
9	8.53	8.52	8.50	8.45	8.40	8.33	8.26	8.19	8.11	8.04	7.96
10	9.50	9.49	9.46	9.40	9.34	9.26	9.18	9.09	9.01	8.92	8.83
11	10.46	10.44	10.40	10.34	10.26	10.17	10.06	9.96	9.85	9.74	9.64
12	11.43	11.41	11.36	11.29	11.19	11.08	10.97	10.84	10.72	10.60	10.48
13	12.36	12.34	12.28	12.20	12.09	12.97	11.83	11.69	11.55	11.41	11.28
14	13.33	13.30	13.24	13.14	13.02	12.88	12.73	12.57	12.41	12.26	12.10
15	14.29	14.26	14.19	14.08	13.94	13.78	13.61	13.43	13.25	13.07	12.90
16	15.25	15.23	15.14	15.02	14.86	14.68	14.49	14.30	14.10	13.91	13.72
17	16.20	16.17	16.08	15.94	15.77	15.57	15.36	15.14	14.93	14.72	14.51
18	17.17	17.13	17.03	16.88	16.69	16.47	16.24	16.01	15.77	15.54	15.32
19	18.13	18.09	17.98	17.81	17.60	17.36	17.11	16.85	16.60	16.35	16.11
20	19.10	19.06	18.94	18.75	18.52	18.26	17.99	17.71	17.43	17.17	16.91

Table I.1 (Continued)

$$k = 5$$

δ n	1.0	1.2	1.4	1.6	1.8	2.0	2.2	2.4	2.6	2.8	3.0
1	1.00	1.00	1.00	1.00	1.00	1.00	1.00	1.00	1.00	1.00	1.00
2	2.00	2.00	2.00	2.00	2.00	2.00	2.00	2.00	2.00	2.00	2.00
3	2.80	2.80	2.80	2.79	2.78	2.78	2.77	2.76	2.75	2.74	2.73
4	3.96	3.96	3.96	3.95	3.95	3.94	3.94	3.93	3.92	3.92	3.91
5	4.82	4.82	4.81	4.80	4.79	4.77	4.75	4.73	4.70	4.68	4.65
6	5.77	5.76	5.76	5.74	5.73	5.71	5.69	5.67	5.64	5.62	5.60
7	6.72	6.72	6.71	6.69	6.66	6.63	6.61	6.57	6.53	6.49	6.45
8	7.72	7.72	7.70	7.68	7.65	7.61	7.57	7.53	7.48	7.43	7.38
9	8.68	8.67	8.65	8.62	8.58	8.53	8.48	8.42	8.36	8.30	8.23
10	9.62	9.61	9.59	9.56	9.51	9.45	9.39	9.33	9.26	9.18	9.11
11	10.58	10.57	10.55	10.51	10.45	10.38	10.31	10.23	10.15	10.06	9.97
12	11.56	11.55	11.52	11.47	11.40	11.32	11.24	11.14	11.04	10.94	10.84
13	12.53	12.52	12.48	12.42	12.35	12.26	12.15	12.04	11.93	11.81	11.69
14	13.49	13.48	13.44	13.37	13.28	13.18	13.06	12.94	12.81	12.68	12.55
15	14.45	14.44	14.39	14.32	14.22	14.10	13.97	13.83	13.69	13.54	13.39
16	15.42	15.41	15.35	15.27	15.16	15.03	14.88	14.73	14.57	14.40	14.24
17	16.40	16.38	16.32	16.23	16.10	15.96	15.79	15.62	15.44	15.26	15.08
18	17.37	17.35	17.28	17.18	17.04	16.88	16.70	16.51	16.31	16.12	15.92
19	18.34	18.31	18.24	18.13	17.98	17.80	17.61	17.40	17.18	16.97	16.75
20	19.30	19.28	19.20	19.08	18.92	18.72	18.51	18.28	18.05	17.82	17.59

Table I.1 (*Continued*)

$$k = 10$$

n \ δ	1.0	1.2	1.4	1.6	1.8	2.0	2.2	2.4	2.6	2.8	3.0
1	1.00	1.00	1.00	1.00	1.00	1.00	1.00	1.00	1.00	1.00	1.00
2	2.00	2.00	2.00	2.00	2.00	2.00	2.00	2.00	2.00	2.00	2.00
3	2.90	2.90	2.90	2.90	2.90	2.89	2.89	2.89	2.88	2.88	2.88
4	3.99	3.99	3.99	3.99	3.99	3.99	3.99	3.98	3.98	3.98	3.98
5	4.95	4.95	4.95	4.95	4.95	4.94	4.94	4.93	4.92	4.92	4.91
6	5.93	5.92	5.92	5.92	5.91	5.91	5.90	5.89	5.88	5.87	5.86
7	6.88	6.88	6.88	6.87	6.87	6.86	6.85	6.84	6.82	6.81	6.79
8	7.86	7.85	7.85	7.85	7.84	7.83	7.82	7.80	7.78	7.77	7.75
9	8.85	8.85	8.84	8.83	8.83	8.81	8.80	8.78	8.76	8.74	8.71
10	9.85	9.85	9.84	9.83	9.82	9.80	9.78	9.76	9.74	9.71	9.68
11	10.84	10.84	10.83	10.82	10.81	10.79	10.76	10.74	10.70	10.67	10.63
12	11.82	11.82	11.81	11.80	11.78	11.76	11.73	11.70	11.66	11.62	11.58
13	12.80	12.80	12.79	12.77	12.75	12.73	12.69	12.66	12.61	12.57	12.51
14	13.78	13.77	13.76	13.75	13.72	13.69	13.66	13.61	13.56	13.51	13.45
15	14.76	14.75	14.74	14.72	14.70	14.66	14.62	14.57	14.52	14.46	14.39
16	15.74	15.74	15.73	15.71	15.68	15.64	15.59	15.54	15.47	15.40	15.33
17	16.73	16.73	16.72	16.69	16.66	16.62	16.56	16.50	16.43	16.35	16.27
18	17.72	17.72	17.70	17.68	17.64	17.59	17.53	17.46	17.38	17.30	17.20
19	18.71	18.71	18.69	18.66	18.62	18.57	18.50	18.42	18.33	18.24	18.13
20	19.70	19.69	19.67	19.64	19.60	19.54	19.46	19.38	19.28	19.18	19.06

Table I.2 Limit of percent saving as a result of curtailment
as $n \to \infty$ (selecting the cell with the
largest probability from a multinomial distribution)

k \ δ	1.0	1.2	1.4	1.6	1.8	2.0	2.2	2.4	2.6	2.8	3.0
2	0	8.3	14.3	18.8	22.2	25.0	27.3	29.2	30.8	32.1	33.3
3	0	5.9	10.5	14.3	17.4	20.0	22.2	24.1	25.8	27.3	28.6
4	0	4.6	8.3	11.5	14.5	16.7	18.8	20.6	22.2	23.7	25.0
5	0	3.7	6.9	9.7	12.1	14.3	16.2	18.0	19.5	20.9	22.2
10	0	1.9	3.7	5.4	6.9	8.3	9.7	10.9	12.1	13.2	14.3

Tables of the Incomplete Beta Function

Table J.1 gives values of the incomplete beta function $I_x(w+1,w)$ and Table J.2 gives values of the incomplete beta function $I_x(w,w)$ for selected values of x and w.

TABLE INTERPOLATION FOR APPENDIX J

To interpolate in Tables J.1 and J.2, we calculate the normal approximation to each of two entries in the table and then (by subtraction) obtain the errors at each of these two points. Then we do a simple linear interpolation on these errors to approximate the error at any intermediate point in the table. (This can be between two successive entries in a row *or* in a column.) We then add this correction to the normal approximation at the appropriate point in the table to obtain the desired result. The mean and variance needed for the normal approximation case of Table J.1 are

$$E(X) = \frac{w+1}{2w+1}, \qquad \sigma^2(X) = \frac{w}{2(2w+1)^2},$$

and in the case of Table J.2 are

$$E(X) = \tfrac{1}{2}, \qquad \sigma^2(X) = \frac{1}{4(2w+1)}.$$

For example, suppose we want to calculate the entry in Table J.1 for $w = 18$ and $x = .65$ (assuming we did not have this entry) on the basis of the two nearest entries in the same column. Letting z denote the standardized

x values, we would calculate

$$z_1 = \frac{(.65 - 18/35)35}{\sqrt{17/2}} = 1.629,$$

$$z_2 = \frac{(.65 - 20/39)39}{\sqrt{19/2}} = 1.736,$$

and use the table in Appendix C to obtain the cumulative values $\Phi(z)$ as $\Phi(z_1) = .9483$ and $\Phi(z_2) = .9587$. Since the exact values are given in Table J.1 we know that the errors at these two points are $+.0003$ and $+.0006$, so that we can expect an error of $+.00045$ at $w = 18$. The normal approximation at $w = 18$, $x = .65$ gives

$$z = \frac{(.65 - 19/37)37}{3} = 1.683 \quad \text{and} \quad \Phi(z) = .9538.$$

Hence our interpolated answer is

$$I_{.65}(19, 18) = .9538 + .0005 = .9543,$$

which agrees exactly with the entry in our table.

Interpolation in Table J.2 can be done similarly with an accuracy of the same order of magnitude.

Table J.1   Tables of the incomplete beta function $I_x(w + 1, w)$

w \ x	.100	.200	.300	.400	.500	.600	.700	.800	.900
5	.0001	.0064	.0473	.1662	.3770	.6331	.8497	.9672	.9984
6	.0001	.0039	.0386	.1582	.3872	.6652	.8822	.9806	.9995
7	.0000	.0024	.0315	.1501	.3953	.6925	.9067	.9884	.9998
8	.0000	.0015	.0257	.1423	.4018	.7161	.9256	.9930	.9999
9	.0000	.0009	.0210	.1347	.4073	.7368	.9404	.9957	1.0000

w \ x	.300	.350	.400	.450	.500	.550	.600	.650	.700
10	.0171	.0532	.1275	.2493	.4119	.5914	.7553	.8782	.9520
11	.0140	.0474	.1207	.2457	.4159	.6037	.7720	.8930	.9613
12	.0115	.0423	.1143	.2420	.4194	.6151	.7870	.9058	.9686
13	.0094	.0377	.1082	.2383	.4225	.6257	.8007	.9168	.9745
14	.0077	.0337	.1025	.2346	.4253	.6356	.8132	.9264	.9792

w \ x	.300	.350	.400	.450	.500	.550	.600	.650	.700
15	.0064	.0301	.0971	.2309	.4278	.6448	.8246	.9348	.9831
16	.0052	.0269	.0920	.2272	.4300	.6536	.8352	.9422	.9862
17	.0043	.0241	.0872	.2236	.4321	.6619	.8450	.9486	.9887
18	.0036	.0215	.0826	.2200	.4340	.6698	.8540	.9543	.9907
19	.0029	.0193	.0784	.2165	.4357	.6773	.8624	.9593	.9924
20	.0024	.0173	.0744	.2130	.4373	.6844	.8702	.9637	.9937
22	.0017	.0139	.0670	.2063	.4402	.6978	.8843	.9711	.9958
24	.0011	.0112	.0604	.1998	.4427	.7102	.8966	.9769	.9971
26	.0008	.0090	.0545	.1935	.4449	.7217	.9074	.9815	.9980
28	.0005	.0073	.0492	.1875	.4469	.7324	.9169	.9851	.9987

w \ x	.400	.425	.450	.475	.500	.525	.550	.575	.600
30	.0445	.0964	.1817	.3022	.4487	.6026	.7424	.8518	.9254
35	.0347	.0828	.1682	.2948	.4525	.6182	.7649	.8743	.9426
40	.0271	.0714	.1559	.2875	.4555	.6320	.7846	.8928	.9555
45	.0213	.0617	.1448	.2805	.4581	.6445	.8019	.9082	.9653
50	.0168	.0534	.1346	.2738	.4602	.6560	.8173	.9211	.9729
55	.0132	.0464	.1252	.2672	.4620	.6666	.8311	.9319	.9787
60	.0105	.0403	.1167	.2609	.4637	.6765	.8436	.9412	.9833
65	.0083	.0351	.1088	.2549	.4651	.6858	.8549	.9490	.9868
70	.0066	.0306	.1015	.2490	.4663	.6945	.8652	.9558	.9896
75	.0052	.0267	.0948	.2434	.4675	.7027	.8746	.9615	.9917
80	.0042	.0233	.0886	.2380	.4685	.7105	.8832	.9665	.9934
85	.0033	.0203	.0829	.2327	.4694	.7179	.8911	.9708	.9948
90	.0026	.0178	.0776	.2276	.4703	.7250	.8984	.9745	.9959
95	.0021	.0156	.0727	.2227	.4711	.7317	.9051	.9778	.9967
100	.0017	.0137	.0681	.2180	.4718	.7382	.9113	.9806	.9974

Table J.2 Tables of the incomplete beta function $I_x(w, w)$

w \ x	.100	.200	.300	.400	.500	.600	.700	.800	.900
5	.0009	.0196	.0988	.2666	.5000	.7334	.9012	.9804	.9991
6	.0003	.0117	.0782	.2465	.5000	.7535	.9218	.9883	.9997
7	.0001	.0070	.0624	.2288	.5000	.7712	.9376	.9930	.9999
8	.0000	.0042	.0500	.2131	.5000	.7869	.9500	.9958	1.0000
9	.0000	.0026	.0403	.1989	.5000	.8011	.9597	.9974	1.0000

Table J.2 (*Continued*)

w \ x	.300	.350	.400	.450	.500	.550	.600	.650	.700
10	.0326	.0875	.1861	.3290	.5000	.6710	.8139	.9125	.9674
11	.0264	.0772	.1744	.3210	.5000	.6790	.8256	.9228	.9736
12	.0214	.0682	.1636	.3135	.5000	.6865	.8364	.9318	.9786
13	.0175	.0604	.1538	.3063	.5000	.6937	.8462	.9396	.9825
14	.0143	.0536	.1447	.2995	.5000	.7005	.8553	.9464	.9857
15	.0117	.0476	.1362	.2930	.5000	.7070	.8638	.9524	.9883
16	.0095	.0424	.1284	.2868	.5000	.7132	.8716	.9576	.9905
17	.0078	.0377	.1211	.2809	.5000	.7191	.8789	.9623	.9922
18	.0064	.0336	.1143	.2751	.5000	.7249	.8857	.9664	.9936
19	.0053	.0300	.1080	.2696	.5000	.7304	.8920	.9700	.9947
20	.0043	.0268	.1021	.2643	.5000	.7357	.8979	.9732	.9957
22	.0029	.0214	.0913	.2542	.5000	.7458	.9087	.9786	.9971
24	.0020	.0171	.0819	.2448	.5000	.7552	.9181	.9829	.9980
26	.0014	.0138	.0735	.2359	.5000	.7641	.9265	.9862	.9986
28	.0009	.0111	.0661	.2276	.5000	.7724	.9339	.9889	.9991

w \ x	.400	.425	.450	.475	.500	.525	.550	.575	.600
30	.0596	.1223	.2197	.3498	.5000	.6502	.7803	.8777	.9404
35	.0461	.1043	.2016	.3383	.5000	.6617	.7984	.8957	.9540
40	.0358	.0893	.1857	.3278	.5000	.6722	.8143	.9107	.9642
45	.0280	.0768	.1715	.3180	.5000	.6820	.8285	.9232	.9720
50	.0219	.0662	.1587	.3089	.5000	.6911	.8413	.9338	.9781
55	.0172	.0572	.1471	.3003	.5000	.6997	.8529	.9428	.9828
60	.0136	.0496	.1366	.2922	.5000	.7078	.8634	.9504	.9864
65	.0107	.0430	.1270	.2846	.5000	.7154	.8730	.9570	.9893
70	.0085	.0374	.1182	.2773	.5000	.7227	.8818	.9626	.9915
75	.0067	.0326	.1101	.2704	.5000	.7296	.8899	.9654	.9933
80	.0054	.0284	.1027	.2637	.5000	.7363	.8973	.9716	.9946
85	.0043	.0248	.0959	.2574	.5000	.7426	.9041	.9752	.9957
90	.0034	.0216	.0896	.2513	.5000	.7487	.9104	.9784	.9966
95	.0027	.0189	.0838	.2455	.5000	.7545	.9162	.9811	.9973
100	.0022	.0166	.0784	.2399	.5000	.7601	.9216	.9834	.9978

Tables for Nonparametric Selection Problems

The entries in Table K.1 are the minimum number n of observations required per population for selecting from k populations the one with the largest median.

SOURCE FOR APPENDIX K

Table K.1 is adapted from M. Sobel (1967), Nonparametric procedures for selecting the t populations with the largest α-quantiles, *The Annals of Mathematical Statistics*, **38**, 1804–1816, with permission.

TABLE INTERPOLATION FOR APPENDIX K

To find a value of n for a fixed P^* and a d^* value that is not listed specifically in Table K.1 we act as if n is proportional to $(1/d^*)^2$ and hence use linear interpolation on n and $(1/d^*)^2$. This generally gives remarkably good results; namely, the error in n is well within 1%.

To find an n value for a fixed d^* and a value of P^* that is not listed specifically in Table K.1 we use linear interpolation on $\log(1-P^*)$ and $(1/d^*)^2$.

To obtain an approximate operating characteristic (OC) curve of the selection procedure for k populations and a fixed n we obtain five pairs (d^*, P^*) by taking the five P^* values given in the table and interpolating to find the d^* value corresponding to our given n. This interpolation uses n and $(1/d^*)^2$. For example, suppose that $k=4$, $n=100$ and we want to find the d^* value corresponding to $P^*=.90$. The entries needed from Table K.1

are given in the table on the left, and the table on the right prepares for interpolation.

Table excerpt		Ready for interpolation	
$d*$	n	$(1/d*)^2$	n
.10	169	100	169
$\boxed{d*}$	100	$\boxed{(1/d*)^2}$	100
.15	75	44.44	75

Linear interpolation in the table on the right is carried out as follows:

$$\frac{(1/d*)^2 - 44.44}{100 - 44.44} = \frac{100 - 75}{169 - 75}$$

from which we obtain $(1/d*)^2 = 59.22$ and $d* = .13$.

Table K.1 Sample size needed to satisfy the (d^*, P^*) requirement in selecting the population with the largest median

$k = 2$

d^*	.750	.900	P^* .950	.975	.990
.05	131	361	563	781	1087
.10	33	89	139	195	271
.15	15	39	61	85	119
.20	9	21	35	47	67

$k = 3$

d^*	.750	.900	P^* .950	.975	.990
.05	217	503	737	979	1307
.10	53	125	183	243	325
.15	23	55	81	107	143
.20	13	31	45	59	79

$k = 4$

d^*	.750	.900	P^* .950	.975	.990
.05	285	601	851	1103	1439
.10	71	149	211	275	359
.15	31	67	93	121	159
.20	17	37	51	67	89

$k = 5$

d^*	.750	.900	P^* .950	.975	.990
.05	341	675	933	1191	1535
.10	85	169	233	297	383
.15	37	75	103	131	169
.20	21	41	57	73	93

Table K.1 (Continued)

k = 6

d*	.750	.900	P* .950	.975	.990
.05	387	733	997	1259	1607
.10	97	183	249	313	401
.15	43	81	109	139	177
.20	23	45	61	77	99

k = 7

d*	.750	.900	P* .950	.975	.990
.05	425	781	1049	1317	1667
.10	105	195	261	327	415
.15	47	85	115	145	183
.20	25	47	65	81	103

k = 8

d*	.750	.900	P* .950	.975	.990
.05	457	823	1095	1363	1717
.10	113	205	273	339	427
.15	51	91	121	149	189
.20	27	51	67	83	105

k = 9

d*	.750	.900	P* .950	.975	.990
.05	487	857	1133	1405	1761
.10	121	213	281	349	439
.15	53	95	125	155	193
.20	29	53	69	87	107

k = 10

d*	.750	.900	P* .950	.975	.990
.05	511	889	1167	1441	1801
.10	127	221	291	359	449
.15	55	97	129	159	197
.20	31	55	71	89	111

Tables for Paired–Comparison Selection Problems

The entries in Table L.1 for $k = 2, 3, \ldots, 10$, and 12 to 20 in increments of 2, respectively, give the minimum integer value of r, the number of replications, required to attain the specified $(\pi^*, P^*, .5)$ triple in selecting the best object out of k in an experiment with a balanced paired-comparison design. Here P^* is a lower bound for the probability of a correct selection and is calculated under the least favorable slippage configuration, that is, $\pi_{ij} \geqslant \pi^*$ for $i = [k]$ and $j \neq [k]$, $\pi_{ij} = .5$ for $i \neq [k]$, $j \neq [k]$, and $i \neq j$.

The entries in Table L.2 for $k_0 = 3$ and $k_0 = 4, 5, \ldots$, are the values of $\tilde{k} \leqslant k_0$ for which r is a maximum under the least favorable slippage configuration. To obtain the conjectured least favorable solution for the round robin design problem we take $\tilde{k}$ players as having positive merit and $k_0 - \tilde{k}$ players as having zero merit. Once $\tilde{k}$ is found from Table L.2 the corresponding value of r is found from Table L.1 using $\tilde{k}$.

SOURCE FOR APPENDIX L

Table L.1 is adapted from H. A. David (1963), *The Method of Paired Comparisons*, Charles Griffin, London, England, with permission of the publisher and author.

TABLE INTERPOLATION FOR APPENDIX

Table L.1 relates the three quantities, π^*, P^*, and r, for a selected value of k. If any two of these quantities are given but at least one is not listed specifically in the table for the particular k, the other quantity can be found by interpolation. The interpolation of primary interest is to find the value of π^* that corresponds to a specified P^* that is a caption in these tables for a value of r that is not an entry in these tables. Although ordinary linear interpolation could be used, more accurate results can be obtained by using linear interpolation for the transformed quantities log π^* and $r^{-.461}$ (natural logarithms). The exponent $-.461$ for r is recommended because it gives both conservative and uniformly close results for all values of k and all P^*. In particular, it gives accuracy in the interval $.50 < \pi^*$ $\leqslant .55$, where r is very large and errors tend to be magnified. For $k \leqslant 20$, the maximum error is about 10%.

This interpolation procedure is illustrated below for $k = 4$, $P^* = .90$, and $r = 7$. The appropriate entries on both sides of $r = 7$ are $r = 6$, $\pi^* = .75$ and $r = 9$, $\pi^* = .70$. In the following table on the left we abstract these entries. The corresponding values of $r^{-.461}$ and log π^* (to the base e) are given in the table on the right in preparation for interpolation.

Table excerpt		Ready for interpolation	
r	π^*	$r^{-.461}$	log π^*
9	.70	.3632	$-.3567$
7	$\boxed{\pi^*}$	.4078	$\boxed{\log \pi^*}$
6	.75	.4378	$-.2877$

Linear interpolation in the table on the right is carried out as follows:

$$\frac{\log \pi^* - (-.2877)}{.4078 - .4378} = \frac{-.3567 - (-.2877)}{.3632 - .4378},$$

$$\log \pi^* = -.3154.$$

The antilog is $\pi^* = .73$, after rounding upward.

The same procedure can be used to interpolate for P^* with a given π^* except that we now take log $(1 - P^*)$.

Table L.1 Smallest number of replications needed to satisfy the $(\pi^*, P^*, .5)$ requirement

		k = 2						k = 3		
			P^*						P^*	
π^*	.75	.90	.95	.99		π^*	.75	.90	.95	.99
.55	45	163	269	537		.55	69	165	243	433
.60	11	41	67	133		.60	17	41	60	106
.65	5	17	29	57		.65	8	18	26	45
.70	3	9	15	31		.70	4	10	15	24
.75	1	7	9	19		.75	3	6	9	15
.80	1	5	7	13		.80	2	4	6	10
.85	1	3	5	9		.85	1	3	4	7
.90	1	1	3	5		.90	1	2	3	5
.95	1	1	1	3		.95	1	1	2	3

		k = 4						k = 5		
			P^*						P^*	
π^*	.75	.90	.95	.99		π^*	.75	.90	.95	.99
.55	71	150	212	358		.55	68	135	186	306
.60	18	37	52	88		.90	17	33	46	75
.65	8	16	23	38		.65	8	15	20	33
.70	5	9	12	20		.70	4	8	11	18
.75	3	6	8	12		.75	3	5	7	11
.80	2	4	5	8		.80	2	4	5	7
.85	2	3	4	6		.85	2	3	3	5
.90	1	2	3	4		.90	1	2	3	4
.95	1	1	2	3		.95	1	1	2	3

		k = 6						k = 7		
			P^*						P^*	
π^*	.75	.90	.95	.99		π^*	.75	.90	.95	.99
.55	65	122	166	267		.55	61	112	150	238
.60	16	30	41	66		.60	15	28	37	59
.65	7	13	18	29		.65	7	12	16	26
.70	4	7	10	16		.70	4	7	9	14
.75	3	5	6	10		.75	3	4	6	9
.80	2	3	4	6		.80	2	3	4	6
.85	2	2	3	4		.85	2	2	3	4
.90	1	2	2	3		.90	1	2	2	3
.95	1	1	2	2		.95	1	1	2	2

Table L.1 (*Continued*)

		$k=8$					$k=9$		
			P^*					P^*	
π^*	.75	.90	.95	.99	π^*	.75	.90	.95	.99
.55	58	103	137	214	.55	54	95	126	195
.60	15	26	34	53	.60	14	24	31	48
.65	7	11	15	23	.65	6	11	14	21
.70	4	6	8	13	.70	4	6	8	12
.75	3	4	5	8	.75	2	4	5	7
.80	2	3	4	5	.80	2	3	3	5
.85	1	2	3	4	.85	1	2	2	3
.90	1	2	2	3	.90	1	1	2	2
.95	1	1	2	2	.95	1	1	1	2

		$k=10$					$k=12$		
			P^*					P^*	
π^*	.75	.90	.95	.99	π^*	.75	.90	.95	.99
.55	52	89	117	180	.55	47	79	102	155
.60	13	22	29	45	.60	12	20	26	39
.65	6	10	13	20	.65	5	9	11	17
.70	4	6	7	11	.70	3	5	6	9
.75	2	4	5	7	.75	2	3	4	6
.80	2	3	3	4	.80	2	2	3	4
.85	1	2	2	3	.85	1	2	2	3
.90	1	1	2	2	.90	1	1	2	2
.95	1	1	1	2	.95	1	1	1	2

		$k=14$					$k=16$		
			P^*					P^*	
π^*	.75	.90	.95	.99	π^*	.75	.90	.95	.99
.55	43	71	91	137	.55	39	64	82	123
.60	11	18	23	34	.60	10	16	21	31
.65	5	8	10	15	.65	5	7	9	14
.70	3	5	6	8	.70	3	4	5	8
.75	2	3	4	5	.75	2	3	3	5
.80	2	2	3	4	.80	1	2	2	3
.85	1	2	2	3	.85	1	2	2	2
.90	1	1	1	2	.90	1	1	1	2
.95	1	1	1	1	.95	1	1	1	1

Table L.1 (Continued)

	$k = 18$					$k = 20$			

	P*					P*			
π^*	.75	.90	.95	.99	π^*	.75	.90	.95	.99
.55	37	59	75	112	.55	34	55	69	102
.60	9	15	19	28	.60	9	14	17	26
.65	4	7	9	12	.65	4	6	8	11
.70	3	4	5	7	.70	3	4	5	6
.75	2	3	3	4	.75	2	2	3	4
.80	1	2	2	3	.80	1	2	2	3
.85	1	1	2	2	.85	1	1	2	2
.90	1	1	1	2	.90	1	1	1	2
.95	1	1	1	1	.95	1	1	1	1

Table L.2   Values of k to determine the conjectured least favorable (CLF) solution

	$k_0 = 3$					$k_0 = 4, 5, \ldots$			

	P*					P*			
π^*	.75	.90	.95	.99	π^*	.75	.90	.95	.99
.55	3	3	2	2	.55	4	3	2	2
.60	3	3	2	2	.60	4	3	2	2
.65	3	3	2	2	.65	4	3	2	2
.70	3	3	2	2	.70	4	3	2	2
.75	3	2	2	2	.75	3	2	2	2
.80	3	2	2	2	.80	3	2	2	2
.85	2	2	2	2	.85	4^a	2	2	2
.90	2	3^a	2	2	.90	2	3^a	2	2
.95	2	2	3^a	2	.95	2	2	3^a	2

[a]Entries not smooth because of rounding problems.

Tables for Selecting from k Normal Populations Those Better than a Control

Tables M.1–M.9 all relate to some aspect of the problem of selecting from k populations those better than a control or standard.

SOURCES FOR APPENDIX M

Tables M.1 and M.3 are adapted from Y. L. Tong (1969), On partitioning a set of normal populations by their locations with respect to a control, *The Annals of Mathematical Statistics*, **40**, 1300–1324, with permission.

Table M.4 is adapted from C. W. Dunnett (1964), New tables for multiple comparisons with a control, *Biometrics*, **20**, 482–491, with permission of the Biometric Society and the author.

Table M.6 is adapted from R. E. Schafer (1976), On selecting which of k populations exceed a standard, unpublished manuscript, with permission.

TABLE INTERPOLATION FOR APPENDIX M

Interpolation in Table M.3

It should be noted that the entries in Table M.3 for $\nu = \infty$ are in agreement with the corresponding entries in Table M.1 that apply to the case of a common known variance. These values for $\nu = \infty$ are very useful for interpolating to get values for any large values of ν that are not in the table. Here again we simply use linear interpolation with $1/\nu$ as was illustrated in the case of Table A.4.

Interpolation in Table M.4

To obtain the d value for values of ν not covered specifically in Table M.4, use linear interpolation with $1/\nu$. For the purpose of interpolating for values of ν larger than 120, it would be desirable to have more decimal places in the values of d for $\nu = \infty$, but the authors are presently not aware of any such exact table in the literature. Approximate entries for values of P^* other than .95 and .99 for $k = 1(2)19$ can be obtained by inverting the table for $\rho = \frac{1}{2}$ of Dunn, Kronmal, and Yee (1968), which was constructed by Monte Carlo methods, but the accuracy of the results would then be difficult to ascertain.

Table M.1 Case of common known variance and unknown mean μ_0 (values of b to determine the sample size)

k	P* .750	.900	.950	.975	.990
1	0.674	1,282	1.645	1.960	2.326
2	1.146	1.645	1.960	2.241	2.576
3	1.319	1.800	2.106	2.379	2.703
4	1.453	1.916	2.212	2.478	2.794
5	1.539	1.995	2.287	2.548	2.860
6	1.614	2.062	2.349	2.607	2.915
7	1.670	2.114	2.398	2.654	2.959
8	1.721	2.160	2.442	2.695	2.998
9	1.763	2.199	2.478	2.730	3.031
10	1.801	2.234	2.511	2.761	3.060
12	1.864	2.292	2.567	2.814	3.110
14	1.916	2.340	2.613	2.858	3.152
16	1.960	2.381	2.652	2.896	3.188
18	1.998	2.417	2.686	2.929	3.219
20	2.032	2.448	2.716	2.957	3.246

Table M.2 Values of B as the solution of $\Phi(B) = (P^*)^{1/k}$

k	P* .750	.900	.950	.975	.990
2	1.108	1.632	1.955	2.239	2.575
3	1.332	1.818	2.121	2.391	2.712
4	1.480	1.943	2.234	2.494	2.806
5	1.590	2.036	2.319	2.572	2.877
6	1.677	2.111	2.386	2.634	2.934
7	1.748	2.172	2.442	2.686	2.981
8	1.808	2.224	2.490	2.731	3.022
9	1.860	2.269	2.531	2.769	3.057
10	1.906	2.309	2.568	2.804	3.089
12	1.983	2.377	2.631	2.862	3.143
14	2.047	2.433	2.682	2.910	3.187
16	2.101	2.481	2.727	2.952	3.226
18	2.148	2.522	2.765	2.988	3.260
20	2.190	2.559	2.800	3.020	3.289
24	2.260	2.621	2.858	3.074	3.340

Remark. The entries for $k = 12, 16, 18, 20$, may be in error by one unit in the last digit shown. The remaining rows are correct to the last digit shown.

Table M.3 Case of common unknown variance
(values of b to determine the sample size)

	k = 2					k = 3				
			P*					P*		
ν	.750	.900	.950	.975	.990	.750	.900	.950	.975	.990
5	1.28547	2.00771	2.56513	3.15876	4.02790	1.49920	2.24189	2.82213	3.44437	4.36051
6	1.26035	1.93795	2.44344	2.96601	3.70523	1.46657	2.15646	2.67692	3.21856	3.98906
7	1.24291	1.89059	2.36222	2.83955	3.49824	1.44395	2.09861	2.58029	3.07086	3.75150
8	1.23009	1.85635	2.30424	2.75037	3.35462	1.42735	2.05686	2.51145	2.96695	3.58711
9	1.22027	1.83046	2.26079	2.68419	3.24933	1.41465	2.02534	2.45996	2.88996	3.46683
10	1.21251	1.81019	2.22705	2.63315	3.16892	1.40463	2.00070	2.42002	2.83069	3.37513
12	1.20102	1.78053	2.17806	2.55965	3.05435	1.38980	1.96468	2.36212	2.74547	3.24470
14	1.19293	1.75987	2.14423	2.50931	2.97673	1.37936	1.93963	2.32220	2.68721	3.15650
16	1.18693	1.74465	2.11947	2.47269	2.92071	1.37161	1.92120	2.29301	2.64487	3.09293
18	1.18229	1.73298	2.10056	2.44487	2.87839	1.36564	1.90707	2.27075	2.61273	3.04496
20	1.17860	1.72375	2.08566	2.42301	2.84531	1.36089	1.89590	2.25322	2.58751	3.00748
24	1.17311	1.71006	2.06367	2.39088	2.79692	1.35381	1.87936	2.22736	2.55046	2.95274
28	1.16921	1.70041	2.04822	2.36840	2.76325	1.34879	1.86770	2.20921	2.52457	2.91467
32	1.16631	1.69324	2.03678	2.35179	2.73848	1.34505	1.85904	2.19577	2.50545	2.88669
36	1.16405	1.68770	2.02796	2.33903	2.71948	1.34215	1.85236	2.18542	2.49076	2.86524
40	1.16226	1.68329	2.02095	2.32891	2.70446	1.33983	1.84704	2.17721	2.47912	2.84829
44	1.16079	1.67970	2.01526	2.32069	2.69228	1.33795	1.84271	2.17053	2.46967	2.83455
48	1.15957	1.67672	2.01053	2.31388	2.68221	1.33638	1.83911	2.16499	2.46184	2.82318
52	1.15854	1.67420	2.00655	2.30815	2.67374	1.33505	1.83608	2.16032	2.45525	2.81363
56	1.15766	1.67205	2.00315	2.30326	2.66651	1.33392	1.83349	2.15634	2.44963	2.80549
60	1.15689	1.67019	2.00021	2.29903	2.66029	1.33293	1.83125	2.15289	2.44478	2.79847
90	1.15334	1.66157	1.98661	2.27951	2.63157	1.32837	1.82086	2.13696	2.42236	2.76612
120	1.15158	1.65729	1.97987	2.26987	2.61743	1.32609	1.81571	2.12908	2.41129	2.75019
∞	1.14630	1.64457	1.95993	2.24140	2.57583	1.31930	1.80040	2.10574	2.37864	2.70340

	k = 4					k = 5				
			P*					P*		
ν	.750	.900	.950	.975	.990	.750	.900	.950	.975	.990
5	1.67149	2.42873	3.02640	3.67090	4.62385	1.78273	2.55442	3.16640	3.82816	4.80875
6	1.63178	2.32910	2.86038	3.41633	4.21075	1.73848	2.44550	2.98658	3.55432	4.36746
7	1.60427	2.26174	2.75012	3.25020	3.94718	1.70782	2.37189	2.86724	3.37579	4.08617
8	1.58409	2.21320	2.67169	3.13355	3.76519	1.68534	2.31886	2.78241	3.25053	3.89213
9	1.56866	2.17657	2.61310	3.04725	3.63226	1.66814	2.27887	2.71907	3.15792	3.75052
10	1.55647	2.14796	2.56770	2.98088	3.53105	1.65455	2.24764	2.67001	3.08673	3.64276
12	1.53845	2.10617	2.50195	2.88560	3.38732	1.63447	2.20203	2.59900	2.98458	3.48983
14	1.52577	2.07713	2.45668	2.82053	3.29029	1.62033	2.17035	2.55011	2.91488	3.38666
16	1.51636	2.05578	2.42361	2.77332	3.22044	1.60984	2.14707	2.51443	2.86432	3.31244
18	1.50910	2.03943	2.39840	2.73750	3.16779	1.60174	2.12924	2.48723	2.82598	3.25651
20	1.50333	2.02650	2.37856	2.70941	3.12669	1.59531	2.11514	2.46583	2.79591	3.21287
24	1.49473	2.00737	2.34932	2.66818	3.06669	1.58572	2.09429	2.43430	2.75181	3.14918
28	1.48864	1.99389	2.32881	2.63939	3.02501	1.57892	2.07961	2.41220	2.72102	3.10497
32	1.48409	1.98388	2.31363	2.61814	2.99439	1.57385	2.06870	2.39584	2.69831	3.07248
36	1.48056	1.97616	2.30195	2.60183	2.97093	1.56992	2.06029	2.38325	2.68087	3.04761
40	1.47775	1.97002	2.29267	2.58890	2.95240	1.56678	2.05360	2.37326	2.66705	3.02796
44	1.47546	1.96502	2.28514	2.57841	2.93738	1.56422	2.04815	2.36514	2.65584	3.01204
48	1.47355	1.96087	2.27889	2.56972	2.92497	1.56210	2.04364	2.35841	2.64656	2.99888
52	1.47194	1.95737	2.27362	2.56241	2.91453	1.56030	2.03982	2.35274	2.63875	2.98783
56	1.47056	1.95438	2.26913	2.55618	2.90564	1.55876	2.03657	2.34790	2.63209	2.97841
60	1.46937	1.95179	2.26525	2.55079	2.89798	1.55743	2.03375	2.34372	2.62634	2.97028
90	1.46382	1.93980	2.24729	2.52594	2.86267	1.55124	2.02070	2.32438	2.59979	2.93287
120	1.46106	1.93386	2.23840	2.51367	2.84529	1.54815	2.01423	2.31482	2.58669	2.91447
∞	1.45281	1.91622	2.21213	2.47751	2.79428	1.53895	1.99504	2.28654	2.54808	2.86045

	k = 6					k = 8				
			P*					P*		
ν	.750	.900	.950	.975	.990	.750	.900	.950	.975	.990
5	1.88216	2.66588	3.29014	3.96687	4.97154	2.02501	2.82937	3.47342	4.17376	5.21579
6	1.83356	2.54818	3.09743	3.67516	4.50437	1.97019	2.69895	3.26186	3.85575	4.71040
7	1.79988	2.46867	2.96961	3.48516	4.20682	1.93216	2.61082	3.12161	3.64884	4.38875
8	1.77517	2.41140	2.87879	3.35193	4.00177	1.90423	2.54735	3.02198	3.50382	4.16733
9	1.75626	2.36821	2.81101	3.25348	3.85222	1.88285	2.49948	2.94764	3.39672	4.00594
10	1.74132	2.33449	2.75851	3.17784	3.73848	1.86594	2.46210	2.89008	3.31446	3.88327
12	1.71923	2.28526	2.68257	3.06935	3.57717	1.84093	2.40754	2.80684	3.19655	3.70941
14	1.70367	2.25106	2.63031	2.99537	3.46843	1.82330	2.36964	2.74958	3.11618	3.59230
16	1.69212	2.22594	2.59218	2.94172	3.39023	1.81020	2.34180	2.70781	3.05793	3.50814
18	1.68321	2.20670	2.56313	2.90106	3.33134	1.80010	2.32048	2.67600	3.01379	3.44478
20	1.67612	2.19149	2.54027	2.86918	3.28540	1.79205	2.30364	2.65097	2.97921	3.39538
24	1.66556	2.16900	2.50660	2.82244	3.21839	1.78007	2.27872	2.61412	2.92850	3.32335
28	1.65807	2.15316	2.48300	2.78981	3.17188	1.77157	2.26118	2.58829	2.89312	3.27338
32	1.65248	2.14141	2.46554	2.76575	3.13773	1.76522	2.24816	2.56919	2.86704	3.23669
36	1.64815	2.13233	2.45211	2.74728	3.11158	1.76030	2.23811	2.55450	2.84702	3.20862
40	1.64470	2.12512	2.44145	2.73265	3.09093	1.75637	2.23013	2.54284	2.83117	3.18644
44	1.64188	2.11925	2.43278	2.72078	3.07420	1.75317	2.22363	2.53337	2.81831	3.16848
48	1.63954	2.11438	2.42560	2.71096	3.06037	1.75051	2.21823	2.52552	2.80766	3.15364
52	1.63755	2.11027	2.41956	2.70269	3.04875	1.74825	2.21369	2.51891	2.79871	3.14118
56	1.63586	2.10676	2.41439	2.69564	3.03886	1.74633	2.20980	2.51326	2.79107	3.13056
60	1.63439	2.10373	2.40993	2.68955	3.03032	1.74466	2.20644	2.50839	2.78448	3.12140
90	1.62757	2.08967	2.38931	2.66146	2.99103	1.73690	2.19088	2.48585	2.75406	3.07925
120	1.62417	2.08270	2.37912	2.64760	2.97171	1.73304	2.18316	2.47470	2.73905	3.05853
∞	1.61402	2.06202	2.34897	2.60676	2.91500	1.72150	2.16029	2.44177	2.69486	2.99774

Table M.3 (Continued)

<table>
<tr><td></td><td colspan="5">k = 10</td><td colspan="5">k = 12</td></tr>
<tr><td></td><td colspan="5">P*</td><td colspan="5">P*</td></tr>
<tr><td>ν</td><td>.750</td><td>.900</td><td>.950</td><td>.975</td><td>.990</td><td>.750</td><td>.900</td><td>.950</td><td>.975</td><td>.990</td></tr>
<tr><td>5</td><td>2.13209</td><td>2.95328</td><td>3.61302</td><td>4.33198</td><td>5.40324</td><td>2.21722</td><td>3.05255</td><td>3.72530</td><td>4.45962</td><td>5.55485</td></tr>
<tr><td>6</td><td>2.07255</td><td>2.81313</td><td>3.38705</td><td>3.99376</td><td>4.86846</td><td>2.15393</td><td>2.90461</td><td>3.48772</td><td>4.10505</td><td>4.99628</td></tr>
<tr><td>7</td><td>2.03122</td><td>2.71840</td><td>3.23724</td><td>3.77385</td><td>4.52823</td><td>2.10994</td><td>2.80455</td><td>3.33021</td><td>3.87465</td><td>4.64100</td></tr>
<tr><td>8</td><td>2.00083</td><td>2.65015</td><td>3.13081</td><td>3.61976</td><td>4.29419</td><td>2.07757</td><td>2.73244</td><td>3.21829</td><td>3.71320</td><td>4.39675</td></tr>
<tr><td>9</td><td>1.97754</td><td>2.59867</td><td>3.05142</td><td>3.50597</td><td>4.12366</td><td>2.05276</td><td>2.67803</td><td>3.13479</td><td>3.59398</td><td>4.21879</td></tr>
<tr><td>10</td><td>1.95913</td><td>2.55846</td><td>2.98995</td><td>3.41859</td><td>3.99408</td><td>2.03312</td><td>2.63554</td><td>3.07013</td><td>3.50245</td><td>4.08360</td></tr>
<tr><td>12</td><td>1.93185</td><td>2.49977</td><td>2.90105</td><td>3.29337</td><td>3.81051</td><td>2.00402</td><td>2.57348</td><td>2.97663</td><td>3.37128</td><td>3.89211</td></tr>
<tr><td>14</td><td>1.91262</td><td>2.45899</td><td>2.83991</td><td>3.20804</td><td>3.68690</td><td>1.98348</td><td>2.53036</td><td>2.91232</td><td>3.28191</td><td>3.76322</td></tr>
<tr><td>16</td><td>1.89832</td><td>2.42904</td><td>2.79531</td><td>3.14622</td><td>3.59811</td><td>1.96821</td><td>2.49868</td><td>2.86541</td><td>3.21717</td><td>3.67065</td></tr>
<tr><td>18</td><td>1.88728</td><td>2.40610</td><td>2.76134</td><td>3.09938</td><td>3.53128</td><td>1.95640</td><td>2.47442</td><td>2.82970</td><td>3.16813</td><td>3.60099</td></tr>
<tr><td>20</td><td>1.87850</td><td>2.38798</td><td>2.73463</td><td>3.06268</td><td>3.47919</td><td>1.94701</td><td>2.45525</td><td>2.80160</td><td>3.12971</td><td>3.54669</td></tr>
<tr><td>24</td><td>1.86540</td><td>2.36117</td><td>2.69529</td><td>3.00889</td><td>3.40325</td><td>1.93300</td><td>2.42689</td><td>2.76024</td><td>3.07340</td><td>3.46756</td></tr>
<tr><td>28</td><td>1.85610</td><td>2.34229</td><td>2.66774</td><td>2.97137</td><td>3.35058</td><td>1.92305</td><td>2.40693</td><td>2.73126</td><td>3.03413</td><td>3.41270</td></tr>
<tr><td>32</td><td>1.84915</td><td>2.32828</td><td>2.64736</td><td>2.94372</td><td>3.31193</td><td>1.91561</td><td>2.39211</td><td>2.70984</td><td>3.00519</td><td>3.37243</td></tr>
<tr><td>36</td><td>1.84377</td><td>2.31748</td><td>2.63168</td><td>2.92249</td><td>3.28235</td><td>1.90985</td><td>2.38068</td><td>2.69335</td><td>2.98297</td><td>3.34162</td></tr>
<tr><td>40</td><td>1.83948</td><td>2.30889</td><td>2.61924</td><td>2.90569</td><td>3.25899</td><td>1.90525</td><td>2.37160</td><td>2.68028</td><td>2.96539</td><td>3.31730</td></tr>
<tr><td>44</td><td>1.83597</td><td>2.30190</td><td>2.60914</td><td>2.89206</td><td>3.24008</td><td>1.90150</td><td>2.36420</td><td>2.66966</td><td>2.95112</td><td>3.29760</td></tr>
<tr><td>48</td><td>1.83305</td><td>2.29610</td><td>2.60077</td><td>2.88077</td><td>3.22445</td><td>1.89838</td><td>2.35807</td><td>2.66085</td><td>2.93932</td><td>3.28133</td></tr>
<tr><td>52</td><td>1.83059</td><td>2.29121</td><td>2.59371</td><td>2.87128</td><td>3.21132</td><td>1.89574</td><td>2.35289</td><td>2.65344</td><td>2.92939</td><td>3.26766</td></tr>
<tr><td>56</td><td>1.82848</td><td>2.28703</td><td>2.58769</td><td>2.86319</td><td>3.20014</td><td>1.89348</td><td>2.34847</td><td>2.64711</td><td>2.92092</td><td>3.25601</td></tr>
<tr><td>60</td><td>1.82666</td><td>2.28342</td><td>2.58250</td><td>2.85620</td><td>3.19049</td><td>1.89152</td><td>2.34465</td><td>2.64165</td><td>2.91361</td><td>3.24597</td></tr>
<tr><td>90</td><td>1.81816</td><td>2.26668</td><td>2.55846</td><td>2.82396</td><td>3.14612</td><td>1.88242</td><td>2.32695</td><td>2.61638</td><td>2.87988</td><td>3.19977</td></tr>
<tr><td>120</td><td>1.81393</td><td>2.25838</td><td>2.54658</td><td>2.80807</td><td>3.12430</td><td>1.87789</td><td>2.31818</td><td>2.60390</td><td>2.86326</td><td>3.17706</td></tr>
<tr><td>∞</td><td>1.80130</td><td>2.23378</td><td>2.51147</td><td>2.76125</td><td>3.06033</td><td>1.86435</td><td>2.29217</td><td>2.56700</td><td>2.81429</td><td>3.11048</td></tr>
</table>

Table M.3 (Continued)

	k = 16					k = 20				
			P*					P*		
ν	.750	.900	.950	.975	.990	.750	.900	.950	.975	.990
5	2.34734	3.20554	3.89898	4.65777	5.79095	2.44484	3.32111	4.03064	4.80857	5.97127
6	2.27832	3.04560	3.64352	4.27776	5.19528	2.37156	3.15216	3.76174	4.40917	5.34717
7	2.23027	2.93732	3.47406	4.03111	4.81657	2.32047	3.03766	3.58324	4.15022	4.95059
8	2.19487	2.85922	3.35361	3.85821	4.55641	2.28279	2.95502	3.45629	3.96860	4.67835
9	2.16769	2.80026	3.26370	3.73053	4.36687	2.25384	2.89261	3.36150	3.83445	4.47994
10	2.14617	2.75419	3.19408	3.63250	4.22289	2.23088	2.84381	3.28809	3.73145	4.32923
12	2.11422	2.68688	3.09338	3.49201	4.01901	2.19678	2.77248	3.18186	3.58382	4.11584
14	2.09165	2.64009	3.02410	3.39631	3.88182	2.17266	2.72288	3.10878	3.48325	3.97227
16	2.07484	2.60570	2.97357	3.32699	3.78331	2.15469	2.68640	3.05546	3.41040	3.86919
18	2.06185	2.57936	2.93509	3.27448	3.70920	2.14078	2.65846	3.01485	3.35522	3.79165
20	2.05150	2.55854	2.90483	3.23335	3.65144	2.12969	2.63637	2.98291	3.31200	3.73122
24	2.03605	2.52775	2.86027	3.17307	3.56729	2.11313	2.60369	2.93589	3.24866	3.64320
28	2.02507	2.50606	2.82906	3.13104	3.50896	2.10136	2.58068	2.90295	3.20450	3.58219
32	2.01687	2.48997	2.80598	3.10007	3.46616	2.09256	2.56360	2.87860	3.17196	3.53742
36	2.01050	2.47756	2.78823	3.07630	3.43341	2.08573	2.55042	2.85986	3.14699	3.50318
40	2.00543	2.46769	2.77415	3.05749	3.40756	2.08028	2.53995	2.84500	3.12722	3.47614
44	2.00128	2.45966	2.76271	3.04223	3.38663	2.07583	2.53142	2.83293	3.11119	3.45425
48	1.99783	2.45299	2.75323	3.02960	3.36933	2.07212	2.52435	2.82292	3.09792	3.43617
52	1.99491	2.44737	2.74525	3.01897	3.35481	2.06899	2.51838	2.81450	3.08676	3.42099
56	1.99241	2.44257	2.73843	3.00991	3.34244	2.06631	2.51329	2.80731	3.07724	3.40805
60	1.99025	2.43842	2.73255	3.00209	3.33177	2.06399	2.50888	2.80110	3.06903	3.39690
90	1.98019	2.41920	2.70534	2.96602	3.28269	2.05318	2.48847	2.77239	3.03114	3.34559
120	1.97518	2.40967	2.69190	2.94824	3.25857	2.04779	2.47836	2.75820	3.01246	3.32038
∞	1.96020	2.38142	2.65218	2.89587	3.18787	2.03170	2.44838	2.71629	2.95746	3.24648

Table M.4 Values of d for two-sided joint confidence statements
(two-sided multivariate t table with $\rho = \frac{1}{2}$)

$P^* = .95$

ν	1	2	3	4	5	6	7	8	9	10	11	12	15	20
5	2.57	3.03	3.29	3.48	3.62	3.73	3.82	3.90	3.97	4.03	4.09	4.14	4.26	4.42
6	2.45	2.86	3.10	3.26	3.39	3.49	3.57	3.64	3.71	3.76	3.81	3.86	3.97	4.11
7	2.36	2.75	2.97	3.12	3.24	3.33	3.41	3.47	3.53	3.58	3.63	3.67	3.78	3.91
8	2.31	2.67	2.88	3.02	3.13	3.22	3.29	3.35	3.41	3.46	3.50	3.54	3.64	3.76
9	2.26	2.61	2.81	2.95	3.05	3.14	3.20	3.26	3.32	3.36	3.40	3.44	3.53	3.65
10	2.23	2.57	2.76	2.89	2.99	3.07	3.14	3.19	3.24	3.29	3.33	3.36	3.45	3.57
11	2.20	2.53	2.72	2.84	2.94	3.02	3.08	3.14	3.19	3.23	3.27	3.30	3.39	3.50
12	2.18	2.50	2.68	2.81	2.90	2.98	3.04	3.09	3.14	3.18	3.22	3.25	3.34	3.45
13	2.16	2.48	2.65	2.78	2.87	2.94	3.00	3.06	3.10	3.14	3.18	3.21	3.29	3.40
14	2.14	2.46	2.63	2.75	2.84	2.91	2.97	3.02	3.07	3.11	3.14	3.18	3.26	3.36
15	2.13	2.44	2.61	2.73	2.82	2.89	2.95	3.00	3.04	3.08	3.12	3.15	3.23	3.33
16	2.12	2.42	2.59	2.71	2.80	2.87	2.92	2.97	3.02	3.06	3.09	3.12	3.20	3.30
17	2.11	2.41	2.58	2.69	2.78	2.85	2.90	2.95	3.00	3.03	3.07	3.10	3.18	3.27
18	2.10	2.40	2.56	2.68	2.76	2.83	2.89	2.94	2.98	3.01	3.05	3.08	3.16	3.25
19	2.09	2.39	2.55	2.66	2.75	2.81	2.87	2.92	2.96	3.00	3.03	3.06	3.14	3.23
20	2.09	2.38	2.54	2.65	2.73	2.80	2.86	2.90	2.95	2.98	3.02	3.05	3.12	3.22
24	2.06	2.35	2.51	2.61	2.70	2.76	2.81	2.86	2.90	2.94	2.97	3.00	3.07	3.16
30	2.04	2.32	2.47	2.58	2.66	2.72	2.77	2.82	2.86	2.89	2.92	2.95	3.02	3.11
40	2.02	2.30	2.44	2.54	2.62	2.68	2.73	2.77	2.81	2.85	2.87	2.90	2.97	3.06
60	2.00	2.27	2.41	2.51	2.58	2.64	2.69	2.73	2.77	2.80	2.83	2.86	2.92	3.00
120	1.98	2.24	2.38	2.47	2.55	2.60	2.65	2.69	2.73	2.76	2.79	2.81	2.87	2.95
∞	1.96	2.21	2.35	2.44	2.51	2.57	2.61	2.65	2.69	2.72	2.74	2.77	2.83	2.91

Table M.4 (*Continued*)

$P^* = .99$

ν	1	2	3	4	5	6	7	8	9	10	11	12	15	20
5	4.03	4.63	4.98	5.22	5.41	5.56	5.69	5.80	5.89	5.98	6.05	6.12	6.30	6.52
6	3.71	4.21	4.51	4.71	4.87	5.00	5.10	5.20	5.28	5.35	5.41	5.47	5.62	5.81
7	3.50	3.95	4.21	4.39	4.53	4.64	4.74	4.82	4.89	4.95	5.01	5.06	5.19	5.36
8	3.36	3.77	4.00	4.17	4.29	4.40	4.48	4.56	4.62	4.68	4.73	4.78	4.90	5.05
9	3.25	3.63	3.85	4.01	4.12	4.22	4.30	4.37	4.43	4.48	4.53	4.57	4.68	4.82
10	3.17	3.53	3.74	3.88	3.99	4.08	4.16	4.22	4.28	4.33	4.37	4.42	4.52	4.65
11	3.11	3.45	3.65	3.79	3.89	3.98	4.05	4.11	4.16	4.21	4.25	4.29	4.39	4.52
12	3.05	3.39	3.58	3.71	3.81	3.89	3.96	4.02	4.07	4.12	4.16	4.19	4.29	4.41
13	3.01	3.33	3.52	3.65	3.74	3.82	3.89	3.94	3.99	4.04	4.08	4.11	4.20	4.32
14	2.98	3.29	3.47	3.59	3.69	3.76	3.83	3.88	3.93	3.97	4.01	4.05	4.13	4.24
15	2.95	3.25	3.43	3.55	3.64	3.71	3.78	3.83	3.88	3.92	3.95	3.99	4.07	4.18
16	2.92	3.22	3.39	3.51	3.60	3.67	3.73	3.78	3.83	3.87	3.91	3.94	4.02	4.13
17	2.90	3.19	3.36	3.47	3.56	3.63	3.69	3.74	3.79	3.83	3.86	3.90	3.98	4.08
18	2.88	3.17	3.33	3.44	3.53	3.60	3.66	3.71	3.75	3.79	3.83	3.86	3.94	4.04
19	2.86	3.15	3.31	3.42	3.50	3.57	3.63	3.68	3.72	3.76	3.79	3.83	3.90	4.00
20	2.85	3.13	3.29	3.40	3.48	3.55	3.60	3.65	3.69	3.73	3.77	3.80	3.87	3.97
24	2.80	3.07	3.22	3.32	3.40	3.47	3.52	3.57	3.61	3.64	3.68	3.70	3.78	3.87
30	2.75	3.01	3.15	3.25	3.33	3.39	3.44	3.49	3.52	3.56	3.59	3.62	3.69	3.78
40	2.70	2.95	3.09	3.19	3.26	3.32	3.37	3.41	3.44	3.48	3.51	3.53	3.60	3.68
60	2.66	2.90	3.03	3.12	3.19	3.25	3.29	3.33	3.37	3.40	3.42	3.45	3.51	3.59
120	2.62	2.85	2.97	3.06	3.12	3.18	3.22	3.26	3.29	3.32	3.35	3.37	3.43	3.51
∞	2.58	2.79	2.92	3.00	3.06	3.11	3.15	3.19	3.22	3.25	3.27	3.29	3.35	3.42

Table M.5 Values of $(P^*)^{1/k}$

	P^*				
k	.750	.900	.950	.975	.990
2	.866	.949	.975	.987	.995
3	.909	.965	.983	.992	.997
4	.931	.974	.987	.994	.997
5	.944	.979	.990	.995	.998
6	.953	.983	.991	.996	.998
7	.960	.985	.993	.996	.999
8	.965	.987	.994	.997	.999
9	.969	.988	.994	.997	.999
10	.972	.990	.995	.997	.999

Table M.6 Values of Q and B for different values of c^* in selecting normal population variances better than a standard

c^*	1.1		1.2		1.3		1.4		1.5		1.6		1.7		1.8		1.9		2.0	
ν	Q	B	Q	B	Q	B	Q	B	Q	B	Q	B	Q	B	Q	B	Q	B	Q	B
1	0.520	0.454	0.539	0.453	0.556	0.451	0.572	0.448	0.586	0.445	0.599	0.441	0.612	0.438	0.623	0.434	0.634	0.430	0.644	0.426
2	0.533	0.692	0.563	0.690	0.590	0.686	0.615	0.681	0.637	0.676	0.658	0.670	0.677	0.664	0.694	0.658	0.710	0.651	0.724	0.644
3	0.542	0.787	0.580	0.784	0.615	0.780	0.646	0.775	0.674	0.768	0.699	0.762	0.722	0.754	0.742	0.747	0.761	0.739	0.778	0.732
4	0.550	0.838	0.595	0.835	0.635	0.830	0.671	0.824	0.703	0.817	0.731	0.810	0.756	0.802	0.779	0.794	0.799	0.786	0.817	0.778
5	0.557	0.869	0.607	0.866	0.652	0.861	0.692	0.854	0.727	0.847	0.758	0.840	0.785	0.832	0.808	0.823	0.829	0.815	0.847	0.806
6	0.563	0.890	0.618	0.886	0.667	0.881	0.710	0.875	0.748	0.868	0.780	0.860	0.808	0.852	0.832	0.843	0.853	0.834	0.871	0.825
7	0.568	0.905	0.628	0.902	0.681	0.896	0.727	0.890	0.766	0.883	0.799	0.875	0.828	0.866	0.852	0.857	0.873	0.848	0.891	0.839
8	0.573	0.917	0.637	0.913	0.693	0.908	0.741	0.901	0.782	0.894	0.817	0.885	0.845	0.877	0.870	0.868	0.890	0.859	0.907	0.849
9	0.578	0.925	0.646	0.922	0.705	0.916	0.755	0.910	0.797	0.902	0.832	0.894	0.861	0.885	0.885	0.876	0.904	0.867	0.920	0.858
10	0.582	0.933	0.654	0.929	0.715	0.924	0.767	0.917	0.810	0.909	0.845	0.901	0.874	0.892	0.897	0.883	0.916	0.874	0.932	0.864
11	0.586	0.939	0.661	0.935	0.725	0.929	0.778	0.923	0.822	0.915	0.857	0.907	0.886	0.898	0.909	0.889	0.927	0.879	0.941	0.870
12	0.590	0.944	0.669	0.940	0.735	0.934	0.789	0.928	0.833	0.920	0.868	0.911	0.896	0.902	0.918	0.893	0.936	0.884	0.949	0.874
13	0.594	0.948	0.675	0.944	0.743	0.938	0.799	0.932	0.843	0.924	0.878	0.915	0.906	0.906	0.927	0.897	0.943	0.888	0.956	0.878
14	0.598	0.951	0.682	0.947	0.751	0.942	0.808	0.935	0.852	0.927	0.887	0.919	0.914	0.910	0.935	0.900	0.950	0.891	0.962	0.881
15	0.601	0.954	0.688	0.951	0.759	0.945	0.816	0.938	0.861	0.930	0.895	0.922	0.922	0.913	0.941	0.903	0.956	0.894	0.967	0.884
16	0.604	0.957	0.694	0.953	0.767	0.948	0.824	0.941	0.869	0.933	0.903	0.924	0.928	0.915	0.947	0.906	0.961	0.896	0.971	0.886
17	0.608	0.960	0.699	0.956	0.774	0.950	0.832	0.943	0.876	0.935	0.910	0.927	0.935	0.918	0.953	0.908	0.966	0.898	0.975	0.889
18	0.611	0.962	0.705	0.958	0.780	0.952	0.839	0.945	0.883	0.937	0.916	0.929	0.940	0.920	0.957	0.910	0.970	0.900	0.978	0.891
19	0.614	0.964	0.710	0.960	0.787	0.954	0.846	0.947	0.890	0.939	0.922	0.931	0.945	0.921	0.962	0.912	0.973	0.902	0.981	0.892
20	0.617	0.965	0.715	0.961	0.793	0.956	0.852	0.949	0.896	0.941	0.927	0.932	0.950	0.923	0.965	0.914	0.976	0.904	0.984	0.894

22	0.944	0.907	0.952	0.864	0.959	0.804	0.964	0.725	0.968	0.622	0.897	0.987	0.907	0.981	0.916	0.972	0.926	0.958	0.935	0.937
24	0.946	0.917	0.954	0.875	0.961	0.815	0.967	0.734	0.971	0.628	0.899	0.990	0.909	0.985	0.919	0.977	0.928	0.964	0.938	0.945
26	0.948	0.925	0.956	0.884	0.963	0.825	0.969	0.742	0.973	0.633	0.901	0.993	0.911	0.988	0.921	0.981	0.930	0.970	0.940	0.952
28	0.950	0.933	0.958	0.893	0.965	0.834	0.971	0.750	0.975	0.638	0.902	0.994	0.913	0.991	0.922	0.985	0.932	0.974	0.941	0.958
30	0.952	0.939	0.960	0.901	0.967	0.843	0.972	0.758	0.976	0.643	0.904	0.996	0.914	0.993	0.924	0.987	0.934	0.978	0.943	0.963
32	0.953	0.945	0.961	0.908	0.968	0.850	0.974	0.765	0.978	0.647	0.905	0.997	0.915	0.994	0.925	0.990	0.935	0.982	0.944	0.968
34	0.954	0.951	0.962	0.915	0.969	0.858	0.975	0.772	0.979	0.652	0.906	0.997	0.916	0.995	0.926	0.991	0.936	0.984	0.945	0.972
36	0.955	0.955	0.963	0.921	0.970	0.865	0.976	0.778	0.980	0.656	0.907	0.998	0.917	0.996	0.927	0.993	0.937	0.987	0.946	0.975
38	0.956	0.960	0.964	0.927	0.971	0.871	0.977	0.785	0.981	0.660	0.908	0.998	0.918	0.997	0.928	0.994	0.938	0.989	0.947	0.978
40	0.957	0.963	0.965	0.932	0.972	0.878	0.978	0.791	0.982	0.664	0.909	0.999	0.919	0.998	0.929	0.995	0.939	0.990	0.948	0.981
44	0.958	0.970	0.966	0.941	0.974	0.889	0.979	0.802	0.983	0.671	0.910	0.999	0.920	0.998	0.930	0.997	0.940	0.993	0.950	0.985
48	0.960	0.975	0.968	0.949	0.975	0.899	0.981	0.812	0.985	0.679	0.911	1.000	0.922	0.999	0.932	0.998	0.941	0.995	0.951	0.989
52	0.961	0.980	0.969	0.956	0.976	0.908	0.982	0.822	0.986	0.685	0.912	1.000	0.923	0.998	0.933	0.998	0.942	0.996	0.952	0.991
56	0.962	0.983	0.970	0.961	0.977	0.916	0.983	0.831	0.987	0.692	0.913	1.000	0.923	0.999	0.933	0.999	0.943	0.997	0.953	0.993
60	0.962	0.986	0.970	0.966	0.978	0.923	0.983	0.840	0.987	0.698	0.914	1.000	0.924	0.999	0.934	0.999	0.944	0.998	0.953	0.995
64	0.963	0.988	0.971	0.971	0.978	0.930	0.984	0.847	0.988	0.704	0.915	1.000	0.925	0.999	0.935	0.999	0.945	0.998	0.954	0.996
68	0.964	0.990	0.972	0.974	0.979	0.936	0.985	0.855	0.989	0.710	0.915	1.000	0.925	1.000	0.935	1.000	0.945	0.999	0.955	0.997
72	0.964	0.992	0.972	0.977	0.979	0.941	0.985	0.862	0.989	0.715	0.916	1.000	0.926	1.000	0.936	1.000	0.946	0.999	0.955	0.997
76	0.965	0.993	0.973	0.980	0.980	0.946	0.986	0.868	0.990	0.721	0.916	1.000	0.926	1.000	0.936	1.000	0.946	0.999	0.956	0.998
80	0.965	0.995	0.973	0.983	0.980	0.951	0.986	0.874	0.990	0.726	0.917	1.000	0.927	1.000	0.937	1.000	0.947	1.000	0.956	0.998
84	0.965	0.995	0.974	0.985	0.981	0.955	0.987	0.880	0.990	0.731	0.917	1.000	0.927	1.000	0.937	1.000	0.947	1.000	0.956	0.999
88	0.966	0.996	0.974	0.987	0.981	0.958	0.987	0.886	0.991	0.736	0.918	1.000	0.927	1.000	0.938	1.000	0.947	1.000	0.957	0.999
92	0.966	0.997	0.974	0.988	0.981	0.962	0.987	0.891	0.991	0.740	0.918	1.000	0.928	1.000	0.938	1.000	0.948	1.000	0.957	0.999
96	0.966	0.997	0.974	0.990	0.982	0.965	0.987	0.896	0.991	0.745	0.919	1.000	0.928	1.000	0.938	1.000	0.948	1.000	0.957	0.999
100	0.967	0.998	0.975	0.991	0.982	0.968	0.988	0.900	0.992	0.749	0.919	1.000	0.928	1.000	0.938	1.000	0.948	1.000	0.958	1.000

Tables for Selecting the t Best Normal Populations

The entries in Table N.1 are the values of τ (or τ_t) required to attain the specified probability P^* of correctly selecting the t best populations with respect to means out of k normal populations with known variances.

SOURCE FOR APPENDIX N

Table N.1 is adapted from R. C. Milton (1966), Basic table for ranking and selection problems with normal populations, unpublished manuscript, with permission.

TABLE INTERPOLATION FOR APPENDIX N

In order to find the τ (or τ_t) value for a P^* value not covered specifically in Table N.1, it is again desirable to transform P^* into log $(1 - P^*)$ before doing any interpolation. In this case linear interpolation is on the border-line of being satisfactory and unsatisfactory. A better result is obtained if the τ values are also transformed into τ^2 before doing the linear interpolation. For example, suppose we wish to obtain the entry for $k = 6$, $t = 3$, $P^* = .950$ (assuming it is not given in the table). The two neighboring entries for $P^* = .900$ and $.975$ and the transformed values follow:

Table excerpt		Ready for interpolation	
P^*	τ	log $(1 - P^*)$	τ^2
.900	3.0627	-2.3026	9.3801
.950	$\boxed{\tau}$	-2.9957	$\boxed{\tau^2}$
.975	3.8429	-3.6889	14.7679

$$\frac{\tau^2 - 14.7679}{9.3801 - 14.7679} = \frac{-2.9957 - (-3.6889)}{-2.3026 - (-3.6889)} .$$

The required entry in the table on the right is $\tau^2 = 12.074$, and the desired result on the left is $\tau = \sqrt{12.074} = 3.475$. Since the correct entry is 3.477 the error is .002, or less than one-tenth of 1%.

Table N.1 Values of τ (or τ_t) for fixed P^*

$t = 2$

k	.750	.900	.950	.975	.990	.999
			P^*			
4	1.9037	2.6353	3.0808	3.4720	3.9323	4.9099
5	2.1474	2.8505	3.2805	3.6591	4.1058	5.0584
6	2.3086	2.9948	3.4154	3.7862	4.2244	5.1611
7	2.4277	3.1024	3.5164	3.8818	4.3140	5.2393
8	2.5215	3.1876	3.5968	3.9581	4.3858	5.3023
9	2.5984	3.2579	3.6633	4.0214	4.4454	5.3549
10	2.6634	3.3176	3.7198	4.0753	4.4964	5.4000

$t = 3$

k	.750	.900	.950	.975	.990	.999
			P^*			
6	2.3887	3.0627	3.4769	3.8429	4.2760	5.2042
7	2.5485	3.2051	3.6097	3.9679	4.3926	5.3052
8	2.6666	3.3113	3.7093	4.0620	4.4807	5.3820
9	2.7596	3.3955	3.7885	4.1371	4.5513	5.4440
10	2.8360	3.4649	3.8541	4.1995	4.6100	5.4957

$t = 4$

k	.750	.900	.950	.975	.990	.999
			P^*			
8	2.7074	3.3462	3.7412	4.0916	4.5078	5.4049
9	2.8249	3.4515	3.8398	4.1847	4.5949	5.4809
10	2.9174	3.5351	3.9183	4.2591	4.6642	5.5421

$t = 5$

k	.750	.900	.950	.975	.990	.999
			P^*			
10	2.9419	3.5562	3.9378	4.2771	4.6814	5.5562

Remark. By symmetry, for any given P^* the value of τ (or τ_t) for the pair (t, k) is the same as for the pair $(k - t, k)$.

APPENDIX O

Table of Critical Values of Fisher's F Distribution

In this table α is the probability that an F variable with ν_1 equal to numerator degrees of freedom and ν_2 equal to denominator degrees of freedom is greater than the corresponding table entry.

Appendix O (Continued)

α = .01

v_2 \ v_1	1	2	3	4	5	6	7	8	9	10	12	15	20	24	30	40	60	120	∞
1	4052	4999.5	5403	5625	5764	5859	5928	5982	6022	6056	6106	6157	6209	6235	6261	6287	6313	6339	6366
2	98.50	99.00	99.17	99.25	99.30	99.33	99.36	99.37	99.39	99.40	99.42	99.43	99.45	99.46	99.47	99.47	99.48	99.49	99.50
3	34.12	30.82	29.46	28.71	28.24	27.91	27.67	27.49	27.35	27.23	27.05	26.87	26.69	26.60	26.50	26.41	26.32	26.22	26.13
4	21.20	18.00	16.69	15.98	15.52	15.21	14.98	14.80	14.66	14.55	14.37	14.20	14.02	13.93	13.84	13.75	13.65	13.56	13.46
5	16.26	13.27	12.06	11.39	10.97	10.67	10.46	10.29	10.16	10.05	9.89	9.72	9.55	9.47	9.38	9.29	9.20	9.11	9.06
6	13.75	10.92	9.78	9.15	8.75	8.47	8.26	8.10	7.98	7.87	7.72	7.56	7.40	7.31	7.23	7.14	7.06	6.97	6.88
7	12.25	9.55	8.45	7.85	7.46	7.19	6.99	6.84	6.72	6.62	6.47	6.31	6.16	6.07	5.99	5.91	5.82	5.74	5.65
8	11.26	8.65	7.59	7.01	6.63	6.37	6.18	6.03	5.91	5.81	5.67	5.52	5.36	5.28	5.20	5.12	5.03	4.95	4.86
9	10.56	8.02	6.99	6.42	6.06	5.80	5.61	5.47	5.35	5.26	5.11	4.96	4.81	4.73	4.65	4.57	4.48	4.40	4.31
10	10.04	7.56	6.55	5.99	5.64	5.39	5.20	5.06	4.94	4.85	4.71	4.56	4.41	4.33	4.25	4.17	4.08	4.00	3.91
11	9.65	7.21	6.22	5.67	5.32	5.07	4.89	4.74	4.63	4.54	4.40	4.25	4.10	4.02	3.94	3.86	3.78	3.69	3.60
12	9.33	6.93	5.95	5.41	5.06	4.82	4.64	4.50	4.39	4.30	4.16	4.01	3.86	3.78	3.70	3.62	3.54	3.45	3.36
13	9.07	6.70	5.74	5.21	4.86	4.62	4.44	4.30	4.19	4.10	3.96	3.82	3.66	3.59	3.51	3.43	3.34	3.25	3.17
14	8.86	6.51	5.56	5.04	4.69	4.46	4.28	4.14	4.03	3.94	3.80	3.66	3.51	3.43	3.35	3.27	3.18	3.09	3.00
15	8.68	6.36	5.42	4.89	4.56	4.32	4.14	4.00	3.89	3.80	3.67	3.52	3.37	3.29	3.21	3.13	3.05	2.96	2.87
16	8.53	6.23	5.29	4.77	4.44	4.20	4.03	3.89	3.78	3.69	3.55	3.41	3.26	3.18	3.10	3.02	2.93	2.84	2.75
17	8.40	6.11	5.18	4.67	4.34	4.10	3.93	3.79	3.68	3.59	3.46	3.31	3.16	3.08	3.00	2.92	2.83	2.75	2.65
18	8.29	6.01	5.09	4.58	4.25	4.01	3.84	3.71	3.60	3.51	3.37	3.23	3.08	3.00	2.92	2.84	2.75	2.66	2.57
19	8.18	5.93	5.01	4.50	4.17	3.94	3.77	3.63	3.52	3.43	3.30	3.15	3.00	2.92	2.84	2.76	2.67	2.58	2.49
20	8.10	5.85	4.94	4.43	4.10	3.87	3.70	3.56	3.46	3.37	3.23	3.09	2.94	2.86	2.78	2.69	2.61	2.52	2.42
21	8.02	5.78	4.87	4.37	4.04	3.81	3.64	3.51	3.40	3.31	3.17	3.03	2.88	2.80	2.72	2.64	2.55	2.46	2.36
22	7.95	5.72	4.82	4.31	3.99	3.76	3.59	3.45	3.35	3.26	3.12	2.98	2.83	2.75	2.67	2.58	2.50	2.40	2.31
23	7.88	5.66	4.76	4.26	3.94	3.71	3.54	3.41	3.30	3.21	3.07	2.93	2.78	2.70	2.62	2.54	2.45	2.35	2.26
24	7.82	5.61	4.72	4.22	3.90	3.67	3.50	3.36	3.26	3.17	3.03	2.89	2.74	2.66	2.58	2.49	2.40	2.31	2.21

Appendix O (Continued)

α = .01

v_2 \ v_1	1	2	3	4	5	6	7	8	9	10	12	15	20	24	30	40	60	120	∞
25	7.77	5.57	4.68	4.18	3.85	3.63	3.46	3.32	3.22	3.13	2.99	2.85	2.70	2.62	2.54	2.45	2.36	2.27	2.17
26	7.72	5.53	4.64	4.14	3.82	3.59	3.42	3.29	3.18	3.09	2.96	2.81	2.66	2.58	2.50	2.42	2.33	2.23	2.13
27	7.68	5.49	4.60	4.11	3.78	3.56	3.39	3.26	3.15	3.06	2.93	2.78	2.63	2.55	2.47	2.38	2.29	2.20	2.10
28	7.64	5.45	4.57	4.07	3.75	3.53	3.36	3.23	3.12	3.03	2.90	2.75	2.60	2.52	2.44	2.35	2.26	2.17	2.06
29	7.60	5.42	4.54	4.04	3.73	3.50	3.33	3.20	3.09	3.00	2.87	2.73	2.57	2.49	2.41	2.33	2.23	2.14	2.03
30	7.56	5.39	4.51	4.02	3.70	3.47	3.30	3.17	3.07	2.98	2.84	2.70	2.55	2.47	2.39	2.30	2.21	2.11	2.01
40	7.31	5.18	4.31	3.83	3.51	3.29	3.12	2.99	2.89	2.80	2.66	2.52	2.37	2.29	2.20	2.11	2.02	1.92	1.80
60	7.08	4.98	4.13	3.65	3.34	3.12	2.95	2.82	2.72	2.63	2.50	2.35	2.20	2.12	2.03	1.94	1.84	1.73	1.60
120	6.85	4.79	3.95	3.48	3.17	2.96	2.79	2.66	2.56	2.47	2.34	2.19	2.03	1.95	1.86	1.76	1.66	1.53	1.38
∞	6.63	4.61	3.78	3.32	3.02	2.80	2.64	2.51	2.41	2.32	2.18	2.04	1.88	1.79	1.70	1.59	1.47	1.32	1.00

α = .05

v_2 \ v_1	1	2	3	4	5	6	7	8	9	10	12	15	20	24	30	40	60	120	∞
1	161.4	199.5	215.7	224.6	230.2	234.0	236.8	238.9	240.5	241.9	243.9	245.9	248.0	249.1	250.1	251.1	252.2	253.3	254.3
2	18.51	19.00	19.16	19.25	19.30	19.33	19.35	19.37	19.38	19.40	19.41	19.43	19.45	19.45	19.46	19.47	19.48	19.49	19.50
3	10.13	9.55	9.28	9.12	9.01	8.94	8.89	8.85	8.81	8.79	8.74	8.70	8.66	8.64	8.62	8.59	8.57	8.55	8.53
4	7.71	6.94	6.59	6.39	6.26	6.16	6.09	6.04	6.00	5.96	5.91	5.86	5.80	5.77	5.75	5.72	5.69	5.66	5.63
5	6.61	5.79	5.41	5.19	5.05	4.95	4.88	4.82	4.77	4.74	4.68	4.62	4.56	4.53	4.50	4.46	4.43	4.40	4.36
6	5.99	5.14	4.76	4.53	4.39	4.28	4.21	4.15	4.10	4.06	4.00	3.94	3.87	3.84	3.81	3.77	3.74	3.70	3.67
7	5.59	4.74	4.35	4.12	3.97	3.87	3.79	3.73	3.68	3.64	3.57	3.51	3.44	3.41	3.38	3.34	3.30	3.27	3.23
8	5.32	4.46	4.07	3.84	3.69	3.58	3.50	3.44	3.39	3.35	3.28	3.22	3.15	3.12	3.08	3.04	3.01	2.97	2.93
9	5.12	4.26	3.86	3.63	3.48	3.37	3.29	3.23	3.18	3.14	3.07	3.01	2.94	2.90	2.86	2.83	2.79	2.75	2.71

Appendix O (Continued)
$$\alpha = .05$$

ν_2 \ ν_1	1	2	3	4	5	6	7	8	9	10	12	15	20	24	30	40	60	120	∞
10	4.96	4.10	3.71	3.48	3.33	3.22	3.14	3.07	3.02	2.98	2.91	2.85	2.77	2.74	2.70	2.66	2.62	2.58	2.54
11	4.84	3.98	3.59	3.36	3.20	3.09	3.01	2.95	2.90	2.85	2.79	2.72	2.65	2.61	2.57	2.53	2.49	2.45	2.40
12	4.75	3.89	3.49	3.26	3.11	3.00	2.91	2.85	2.80	2.75	2.69	2.62	2.54	2.51	2.47	2.43	2.38	2.34	2.30
13	4.67	3.81	3.41	3.18	3.03	2.92	2.83	2.77	2.71	2.67	2.60	2.53	2.46	2.42	2.38	2.34	2.30	2.25	2.21
14	4.60	3.74	3.34	3.11	2.96	2.85	2.76	2.70	2.65	2.60	2.53	2.46	2.39	2.35	2.31	2.27	2.22	2.18	2.13
15	4.54	3.68	3.29	3.06	2.90	2.79	2.71	2.64	2.59	2.54	2.48	2.40	2.33	2.29	2.25	2.20	2.16	2.11	2.07
16	4.49	3.63	3.24	3.01	2.85	2.74	2.66	2.59	2.54	2.49	2.42	2.35	2.28	2.24	2.19	2.15	2.11	2.06	2.01
17	4.45	3.59	3.20	2.96	2.81	2.70	2.61	2.55	2.49	2.45	2.38	2.31	2.23	2.19	2.15	2.10	2.06	2.01	1.96
18	4.41	3.55	3.16	2.93	2.77	2.66	2.58	2.51	2.46	2.41	2.34	2.27	2.19	2.15	2.11	2.06	2.02	1.97	1.92
19	4.38	3.52	3.13	2.90	2.74	2.63	2.54	2.48	2.42	2.38	2.31	2.23	2.16	2.11	2.07	2.03	1.98	1.93	1.88
20	4.35	3.49	3.10	2.87	2.71	2.60	2.51	2.45	2.39	2.35	2.28	2.20	2.12	2.08	2.04	1.99	1.95	1.90	1.84
21	4.32	3.47	3.07	2.84	2.68	2.57	2.49	2.42	2.37	2.32	2.25	2.18	2.10	2.05	2.01	1.96	1.92	1.87	1.81
22	4.30	3.44	3.05	2.82	2.66	2.55	2.46	2.40	2.34	2.30	2.23	2.15	2.07	2.03	1.98	1.94	1.89	1.84	1.78
23	4.28	3.42	3.03	2.80	2.64	2.53	2.44	2.37	2.32	2.27	2.20	2.13	2.05	2.01	1.96	1.91	1.86	1.81	1.76
24	4.26	3.40	3.01	2.78	2.62	2.51	2.42	2.36	2.30	2.25	2.18	2.11	2.03	1.98	1.94	1.89	1.84	1.79	1.73
25	4.24	3.39	2.99	2.76	2.60	2.49	2.40	2.34	2.28	2.24	2.16	2.09	2.01	1.96	1.92	1.87	1.82	1.77	1.71
26	4.23	3.37	2.98	2.74	2.59	2.47	2.39	2.32	2.27	2.22	2.15	2.07	1.99	1.95	1.90	1.85	1.80	1.75	1.69
27	4.21	3.35	2.96	2.73	2.57	2.46	2.37	2.31	2.25	2.20	2.13	2.06	1.97	1.93	1.88	1.84	1.79	1.73	1.67
28	4.20	3.34	2.95	2.71	2.56	2.45	2.36	2.29	2.24	2.19	2.12	2.04	1.96	1.91	1.87	1.82	1.77	1.71	1.65
29	4.18	3.33	2.93	2.70	2.55	2.43	2.35	2.28	2.22	2.18	2.10	2.03	1.94	1.90	1.85	1.81	1.75	1.70	1.64
30	4.17	3.32	2.92	2.69	2.53	2.42	2.33	2.27	2.21	2.16	2.09	2.01	1.93	1.89	1.84	1.79	1.74	1.68	1.62
40	4.08	3.23	2.84	2.61	2.45	2.34	2.25	2.18	2.12	2.08	2.00	1.92	1.84	1.79	1.74	1.69	1.64	1.58	1.51
60	4.00	3.15	2.76	2.53	2.37	2.25	2.17	2.10	2.04	1.99	1.92	1.84	1.75	1.70	1.65	1.59	1.53	1.47	1.39
120	3.92	3.07	2.68	2.45	2.29	2.17	2.09	2.02	1.96	1.91	1.83	1.75	1.66	1.61	1.55	1.50	1.43	1.35	1.25
∞	3.84	3.00	2.60	2.37	2.21	2.10	2.01	1.94	1.88	1.83	1.75	1.67	1.57	1.52	1.46	1.39	1.32	1.22	1.00

Source Adapted from Table 18 of E. S. Pearson and H. O. Hartley, Eds. (1966), *Biometrika Tables for Statisticians*, Vol. 1, 3rd ed., Cambridge University Press, Cambridge, with permission of the Biometrika Trustees.

Tables for Complete Ordering Problems

Tables P.1 and P.2 apply to the problem of making a complete correct ordering of k normal populations with respect to means. Tables P.3, P.4, and P.5 apply to various problems of ordering k normal populations with respect to variances.

SOURCES FOR APPENDIX P

Table P.1 is adapted from R. J. Carroll and S. S. Gupta (1975), On the probabilities of rankings of k populations with applications, *Mimeo Series* No. 1024, Institute of Statistics, University of North Carolina, Chapel Hill, North Carolina, with permission.

Table P.2 is adapted from H. Freeman, A. Kuzmack, and R. Maurice (1967), Multivariate t and the ranking problem, *Biometrika*, **54**, 305–308, with permission of the Biometrika Trustees and the authors.

Table P.3 is adapted from R. E. Schafer and H. C. Rutemiller (1975), Some characteristics of a ranking procedure for population parameters based on chi-square statistics, *Technometrics*, **17**, 327–331, with permission.

Tables P.4 and P.5 are adapted from H. A. David (1956), The ranking of variances in normal populations, *Journal of the American Statistical Association*, **51**, 621–626, with permission.

Table P.1 The probability of a correct complete ordering of k normal populations with respect to means for given values of τ (or τ_t)

τ	2	3	4	5	6	7	8	9	10
0.00	.500	.167	.041	.008	.001	.000	.000	.000	.000
0.10	.528	.196	.056	.014	.003	.001	.000	.000	.000
0.20	.556	.228	.077	.023	.006	.002	.000	.000	.000
0.30	.584	.263	.101	.036	.012	.004	.001	.000	.000
0.40	.611	.299	.130	.052	.020	.008	.003	.001	.000
0.50	.638	.337	.162	.074	.033	.014	.006	.003	.001
0.60	.664	.376	.192	.100	.050	.025	.012	.006	.003
0.70	.690	.416	.237	.132	.073	.040	.022	.012	.006
0.80	.714	.456	.279	.168	.101	.060	.036	.021	.013
0.90	.738	.496	.324	.208	.134	.086	.055	.035	.022
1.00	.760	.536	.369	.252	.172	.117	.080	.054	.037
1.10	.782	.574	.415	.298	.214	.154	.110	.079	.057
1.20	.802	.612	.461	.346	.260	.195	.146	.110	.082
1.30	.821	.647	.506	.395	.308	.240	.187	.146	.114
1.40	.839	.681	.550	.444	.358	.288	.232	.187	.151
1.50	.855	.714	.593	.492	.408	.338	.281	.233	.193

τ	2	3	4	5	6	7	8	9	10
1.60	.871	.744	.633	.539	.458	.390	.332	.282	.240
1.70	.885	.772	.671	.584	.507	.441	.384	.334	.290
1.80	.898	.797	.707	.626	.555	.492	.436	.386	.342
1.90	.910	.821	.740	.667	.601	.541	.488	.439	.396
2.00	.921	.843	.770	.704	.644	.589	.538	.492	.450
2.10	.931	.862	.798	.739	.684	.633	.586	.543	.503
2.20	.940	.880	.824	.771	.722	.676	.632	.592	.554
2.30	.948	.856	.847	.800	.756	.715	.675	.638	.603
2.40	.955	.910	.867	.826	.788	.750	.715	.681	.649
2.50	.961	.923	.886	.850	.816	.783	.752	.721	.692
2.60	.967	.934	.902	.871	.841	.812	.785	.758	.732
2.70	.972	.944	.916	.890	.864	.839	.815	.791	.768
2.80	.976	.952	.929	.906	.884	.863	.842	.821	.801
2.90	.978	.960	.940	.921	.902	.883	.865	.847	.830
3.00	.983	.966	.950	.933	.917	.901	.886	.871	.856

τ	k								
	2	3	4	5	6	7	8	9	10
3.10	.986	.972	.958	.944	.931	.917	.904	.891	.878
3.20	.988	.977	.965	.953	.942	.931	.920	.909	.898
3.30	.990	.981	.971	.961	.952	.942	.933	.924	.915
3.40	.992	.984	.976	.968	.960	.953	.945	.937	.930
3.50	.993	.987	.980	.974	.968	.961	.955	.948	.942
3.60	.995	.989	.984	.979	.973	.968	.963	.958	.953
3.70	.996	.991	.987	.983	.978	.974	.970	.966	.962
3.80	.996	.993	.990	.986	.983	.979	.976	.972	.969
3.90	.997	.994	.991	.989	.986	.983	.981	.978	.975
4.00	.998	.996	.993	.991	.989	.987	.985	.982	.980
4.10	.998	.996	.995	.993	.991	.990	.988	.986	.984
4.20	.999	.997	.996	.994	.993	.992	.990	.989	.988
4.30	.999	.998	.997	.996	.995	.994	.992	.991	.990
4.40	.999	.998	.997	.997	.996	.995	.994	.993	.992

Table P.2 *Values of h to determine the common sample size in the second stage for attaining $P^* = .95$ for a correct complete ordering of k normal populations with respect to means when n is the common size of the first-stage sample*

n	k			
	3	4	5	6[†]
10	2.053	2.21	2.29	2.28
20	2.002	2.16	2.25	2.24
30	1.988	2.15	2.24	2.23
40	1.981	2.14	2.24	2.22
50	1.977	2.14	2.23	2.22
60	1.975	2.14	2.23	2.22
70	1.973	2.14	2.23	2.22
80	1.971	2.13	2.23	2.21
90	1.970	2.13	2.23	2.21
100	1.969	2.13	2.23	2.21
200	1.965	2.13	2.22	2.21
500	1.963	2.12	2.22	–

[†]The last digit of each entry for $k = 6$ is of questionable accuracy because we expect all the entries to increase as k increases for fixed n.

Table P.3 The probability of a correct complete ordering of k normal populations with respect to variances

$$k = 3$$

ν	1.1	1.2	1.3	1.4	1.5	1.6	1.7	1.8	1.9	2.0	2.5	3.0
1	.183	.206	.219	.233	.241	.253	.261	.278	.288	.298	.346	.388
2	.193	.221	.242	.265	.284	.311	.335	.352	.369	.386	.463	.523
3	.200	.234	.263	.294	.321	.352	.377	.403	.430	.454	.552	.621
4	.206	.241	.282	.318	.350	.386	.413	.452	.480	.504	.614	.692
5	.211	.252	.299	.335	.376	.419	.446	.490	.525	.553	.674	.754
6	.216	.263	.315	.355	.402	.445	.479	.520	.563	.589	.715	.792
7	.221	.272	.330	.373	.426	.474	.509	.554	.592	.625	.750	.828
8	.225	.279	.345	.394	.449	.497	.540	.588	.622	.660	.782	.859
9	.228	.288	.358	.409	.470	.521	.565	.613	.649	.687	.807	.880
10	.230	.296	.370	.424	.489	.542	.590	.635	.675	.713	.833	.900
12	.240	.316	.394	.454	.526	.578	.633	.676	.720	.760	.872	.928
14	.245	.329	.415	.483	.558	.616	.667	.716	.756	.796	.901	.952
16	.250	.342	.435	.508	.584	.648	.699	.748	.789	.828	.923	.967
18	.256	.352	.453	.529	.607	.673	.729	.776	.819	.852	.940	.976
20	.261	.365	.470	.554	.630	.703	.755	.800	.844	.872	.953	.983
24	.275	.387	.499	.592	.678	.748	.799	.848	.879	.905	.970	a
28	.286	.402	.527	.628	.715	.781	.837	.878	.903	.930	.983	a
32	.295	.427	.554	.659	.744	.812	.863	.900	.926	.944	.989	a
36	.303	.443	.580	.689	.772	.840	.887	.922	.941	.959	a	a
40	.309	.460	.604	.714	.799	.861	.905	.935	.957	.970	a	b
50	.330	.501	.654	.760	.849	.902	.941	.962	.977	.985	a	b
60	.351	.533	.700	.803	.886	.932	.959	.978	.986	a	a	b
70	.366	.563	.730	.840	.911	.947	.972	.987	a	a	b	b
80	.379	.589	.759	.863	.929	.964	.981	a	a	a	b	b
90	.393	.618	.786	.887	.945	.975	.988	a	a	a	b	b
100	.405	.639	.799	.905	.951	.978	a	a	a	a	b	b

[a] Exceeds .990.
[b] Exceeds .999.

Table P.3 (Continued)

k = 4

ν	1.1	1.2	1.3	1.4	1.5	1.6	1.7	1.8	1.9	2.0	2.5	3.0
1	.050	.058	.067	.074	.084	.092	.100	.110	.118	.123	.170	.207
2	.054	.073	.085	.100	.121	.138	.151	.170	.188	.207	.286	.353
3	.059	.080	.102	.120	.147	.173	.196	.221	.246	.272	.384	.470
4	.064	.085	.117	.141	.175	.206	.239	.270	.300	.330	.460	.559
5	.069	.094	.131	.161	.202	.240	.278	.313	.352	.384	.535	.640
6	.071	.103	.142	.180	.227	.272	.313	.354	.391	.432	.594	.711
7	.073	.109	.156	.198	.251	.301	.346	.391	.440	.478	.646	.752
8	.076	.113	.168	.213	.272	.329	.379	.430	.479	.518	.687	.795
9	.079	.121	.180	.232	.296	.353	.405	.467	.519	.561	.725	.830
10	.082	.127	.191	.248	.319	.376	.435	.493	.546	.595	.761	.857
12	.086	.143	.214	.282	.358	.429	.487	.549	.606	.654	.813	.898
14	.092	.154	.235	.315	.396	.468	.533	.598	.652	.704	.855	.928
16	.097	.165	.254	.342	.430	.508	.577	.642	.699	.748	.887	.950
18	.100	.175	.273	.368	.462	.545	.618	.677	.738	.783	.911	.964
20	.105	.188	.288	.394	.488	.576	.651	.711	.770	.810	.930	.974
24	.111	.210	.322	.439	.542	.642	.708	.775	.821	.860	.955	.986
28	.117	.228	.355	.485	.589	.685	.759	.814	.859	.893	.972	a
32	.122	.252	.387	.524	.633	.729	.791	.852	.889	.917	.981	a
36	.129	.267	.420	.556	.671	.767	.829	.880	.918	.939	.988	a
40	.138	.285	.445	.594	.708	.791	.860	.899	.936	.953	a	b
50	.156	.329	.511	.660	.778	.860	.908	.944	.962	.978	a	b
60	.172	.368	.566	.713	.829	.896	.939	.965	.979	.988	a	b
70	.186	.406	.617	.762	.867	.924	.959	.978	.988	a	b	b
80	.199	.436	.659	.802	.899	.945	.972	.986	a	a	b	b
90	.210	.472	.690	.836	.919	.962	.981	a	a	a	b	b
100	.220	.503	.710	.867	.925	.972	.983	a	a	a	b	b

[a]Exceeds .990.
[b]Exceeds .999.

$$k = 5$$

ν	1.1	1.2	1.3	1.4	1.5	1.6	1.7	1.8	1.9	2.0	2.5	3.0
1	.010	.014	.020	.022	.026	.029	.036	.041	.045	.048	.080	.108
2	.012	.021	.026	.035	.048	.055	.071	.080	.091	.106	.170	.229
3	.014	.025	.033	.050	.064	.083	.100	.123	.139	.162	.270	.358
4	.016	.027	.041	.060	.086	.109	.138	.165	.185	.217	.346	.452
5	.018	.034	.052	.078	.106	.134	.170	.204	.232	.271	.426	.547
6	.020	.036	.061	.090	.126	.162	.200	.237	.278	.321	.492	.616
7	.021	.040	.071	.104	.147	.190	.228	.276	.318	.368	.553	.678
8	.023	.042	.079	.111	.167	.210	.259	.314	.362	.409	.604	.731
9	.025	.045	.086	.128	.187	.235	.287	.347	.401	.453	.646	.774
10	.027	.049	.095	.142	.204	.261	.317	.382	.439	.492	.690	.810
12	.029	.061	.114	.169	.244	.307	.373	.441	.504	.559	.756	.866
14	.030	.068	.129	.197	.281	.350	.422	.490	.561	.618	.810	.902
16	.032	.074	.143	.228	.316	.394	.473	.547	.618	.670	.848	.931
18	.036	.082	.160	.251	.346	.431	.519	.591	.660	.717	.879	.951
20	.038	.090	.175	.273	.376	.469	.559	.630	.704	.752	.904	.965
24	.040	.110	.205	.319	.436	.542	.622	.704	.782	.812	.945	.982
28	.044	.124	.240	.366	.485	.560	.682	.758	.815	.860	.966	*a*
32	.048	.139	.269	.412	.533	.651	.734	.805	.857	.892	.978	*a*
36	.052	.156	.300	.448	.581	.694	.780	.840	.888	.919	.987	*a*
40	.056	.172	.326	.495	.628	.733	.817	.868	.912	.939	*a*	*a*
50	.064	.215	.396	.572	.713	.813	.879	.924	.952	.970	*a*	*b*
60	.077	.253	.454	.639	.776	.863	.918	.952	.973	.983	*a*	*b*
70	.087	.289	.514	.696	.825	.899	.945	.972	.985	*a*	*b*	*b*
80	.099	.319	.562	.745	.862	.928	.963	.982	*a*	*a*	*b*	*b*
90	.108	.360	.606	.786	.892	.950	.976	.990	*a*	*a*	*b*	*b*
100	.119	.388	.624	.820	.903	.964	.981	*a*	*a*	*a*	*b*	*b*

[a] Exceeds .990.
[b] Exceeds .999.

$$k = 6$$

ν	1.1	1.2	1.3	1.4	1.5	1.6	1.7	1.8	1.9	2.0	2.5	3.0
						c						
1	.002	.003	.005	.006	.008	.009	.011	.013	.017	.019	.035	.054
2	.003	.005	.010	.013	.019	.023	.031	.038	.046	.052	.102	.155
3	.004	.006	.013	.017	.028	.036	.050	.065	.078	.095	.182	.269
4	.004	.008	.018	.025	.041	.056	.072	.095	.119	.140	.262	.372
5	.005	.011	.023	.034	.052	.075	.098	.129	.156	.189	.342	.469
6	.005	.012	.028	.041	.067	.096	.122	.159	.199	.235	.408	.543
7	.006	.013	.034	.050	.081	.120	.150	.195	.239	.282	.474	.618
8	.006	.014	.039	.057	.098	.138	.179	.233	.278	.327	.527	.674
9	.007	.016	.043	.070	.114	.159	.206	.264	.319	.371	.578	.727
10	.008	.019	.048	.081	.131	.178	.233	.295	.353	.405	.628	.769
12	.009	.025	.059	.104	.164	.226	.285	.357	.423	.482	.706	.837
14	.009	.028	.070	.123	.199	.266	.345	.416	.485	.548	.768	.884
16	.010	.033	.081	.150	.229	.308	.399	.470	.541	.607	.816	.916
18	.011	.039	.095	.171	.259	.346	.441	.517	.600	.656	.854	.941
20	.012	.045	.109	.192	.290	.384	.483	.562	.644	.698	.884	.957
24	.013	.057	.138	.236	.347	.463	.560	.641	.716	.770	.926	.977
28	.015	.066	.162	.281	.404	.527	.624	.707	.777	.826	.955	.990
32	.017	.078	.191	.323	.458	.587	.683	.763	.824	.865	.970	[a]
36	.020	.090	.219	.366	.506	.635	.734	.805	.863	.901	.980	[a]
40	.024	.104	.243	.404	.556	.688	.775	.843	.895	.926	.987	[a]
50	.030	.140	.311	.491	.654	.772	.852	.906	.940	.962	[a]	[b]
60	.037	.171	.372	.568	.728	.834	.899	.944	.960	.981	[a]	[b]
70	.043	.205	.428	.632	.785	.877	.932	.965	.980	[b]	[b]	[b]
80	.050	.238	.486	.688	.829	.912	.955	.979	[a]	[b]	[b]	[b]
90	.057	.271	.534	.740	.865	.935	.964	.988	[a]	[b]	[b]	[b]
100	.064	.308	.567	.780	.876	.950	.972	.990	[a]	[b]	[b]	[b]

[a] Exceeds .990.

[b] Exceeds .999.

$$k = 7$$

ν	1.1	1.2	1.3	1.4	1.5	1.6	1.7	1.8	1.9	2.0	2.5	3.0
							c					
1	.000	.001	.002	.002	.002	.003	.004	.005	.006	.007	.016	.027
2	.001	.002	.003	.004	.006	.008	.013	.017	.022	.025	.061	.103
3	.001	.002	.004	.006	.012	.017	.026	.034	.044	.056	.127	.200
4	.001	.003	.006	.009	.019	.026	.040	.055	.073	.088	.192	.300
5	.001	.003	.008	.013	.027	.038	.059	.078	.104	.128	.268	.398
6	.001	.004	.010	.018	.036	.055	.079	.106	.139	.172	.336	.480
7	.002	.005	.013	.023	.045	.070	.101	.132	.176	.212	.398	.558
8	.002	.005	.016	.030	.058	.086	.122	.166	.209	.256	.465	.626
9	.002	.006	.019	.039	.070	.104	.152	.194	.248	.297	.520	.686
10	.002	.008	.022	.047	.083	.122	.172	.232	.283	.335	.571	.731
12	.002	.011	.030	.061	.109	.163	.219	.294	.355	.414	.656	.808
14	.002	.013	.036	.077	.136	.204	.272	.348	.423	.486	.729	.859
16	.003	.015	.047	.099	.164	.242	.323	.405	.482	.550	.781	.902
18	.003	.018	.055	.117	.194	.283	.372	.452	.539	.602	.825	.930
20	.004	.021	.064	.133	.225	.314	.415	.494	.589	.648	.862	.949
24	.004	.028	.084	.168	.283	.391	.490	.590	.671	.735	.917	.973
28	.006	.035	.111	.213	.340	.454	.567	.657	.735	.796	.947	.987
32	.009	.045	.135	.258	.394	.524	.628	.722	.793	.844	.964	a
36	.011	.054	.159	.302	.450	.586	.684	.775	.839	.882	.979	a
40	.013	.063	.181	.333	.495	.626	.733	.808	.876	.909	.986	a
50	.016	.089	.246	.429	.601	.729	.822	.889	.928	.955	a	b
60	.019	.117	.306	.506	.684	.801	.876	.932	.957	.975	a	b
70	.021	.145	.360	.577	.746	.849	.919	.957	.978	.986	b	b
80	.024	.170	.420	.643	.800	.895	.946	.975	.987	a	b	b
90	.027	.201	.469	.692	.841	.925	.961	.985	a	a	b	b
100	.030	.232	.487	.739	.854	.942	.969	.989	a	a	b	b

[a] Exceeds .990.
[b] Exceeds .999.

$$k = 8$$

ν	1.1	1.2	1.3	1.4	1.5	1.6	1.7	1.8	1.9	2.0	2.5	3.0
							c					
1	.000	.000	.000	.000	.001	.001	.001	.001	.002	.003	.008	.014
2	.000	.000	.001	.001	.003	.004	.006	.007	.010	.013	.035	.071
3	.000	.000	.001	.002	.006	.008	.013	.019	.025	.034	.088	.149
4	.000	.000	.002	.004	.009	.013	.023	.033	.046	.058	.147	.243
5	.000	.001	.003	.007	.014	.021	.035	.052	.069	.092	.211	.338
6	.000	.001	.004	.009	.019	.031	.050	.070	.096	.125	.277	.423
7	.000	.001	.005	.012	.026	.042	.066	.092	.129	.164	.342	.502
8	.000	.002	.006	.015	.034	.054	.085	.120	.160	.199	.405	.574
9	.001	.002	.008	.020	.042	.070	.108	.143	.196	.239	.468	.641
10	.001	.002	.010	.024	.054	.082	.126	.176	.228	.274	.518	.691
12	.001	.004	.017	.038	.076	.117	.170	.233	.298	.357	.614	.778
14	.001	.005	.022	.050	.097	.152	.218	.289	.364	.432	.690	.836
16	.001	.006	.027	.065	.120	.190	.268	.343	.425	.494	.750	.881
18	.001	.008	.033	.080	.146	.225	.317	.398	.481	.553	.801	.917
20	.001	.011	.040	.096	.176	.262	.359	.445	.537	.605	.844	.941
24	.002	.014	.054	.127	.222	.333	.438	.536	.630	.695	.906	.969
28	.002	.018	.071	.165	.278	.399	.520	.616	.705	.768	.938	.984
-32	.003	.023	.090	.207	.334	.472	.587	.689	.767	.825	.963	a
36	.003	.032	.110	.236	.381	.527	.648	.738	.816	.866	.974	a
40	.004	.037	.133	.276	.437	.582	.701	.796	.856	.897	.984	a
50	.005	.056	.192	.369	.550	.691	.800	.871	.919	.946	a	b
60	.007	.078	.249	.455	.640	.770	.861	.918	.952	.970	a	b
70	.010	.107	.301	.528	.712	.832	.906	.951	.974	.985	b	b
80	.012	.125	.360	.599	.768	.877	.936	.963	.986	.990	b	b
90	.015	.153	.412	.651	.821	.911	.959	.981	a	a	b	b
100	.018	.176	.447	.699	.848	.931	.963	.984	a	a	b	b

[a] Exceeds .990.
[b] Exceeds .999.

Table P.4   *Critical values for separation of normal population variances into groups (adjacent case) for* $\alpha = .01, .05$

ν	2	3	4	5	6	*k* 7	8	9	10	11	12
						$\alpha = .01$					
2	199.	153.	135.	126.	121.	118.	115.	114.	112.	111.	110.
4	23.2	18.1	15.9	14.8	14.0	13.6	13.2	13.0	12.8	12.6	12.5
5	14.9	11.9	10.4	9.68	9.21	8.87	8.64	8.47	8.33	8.21	8.11
6	11.1	8.86	7.83	7.27	6.93	6.69	6.52	6.38	6.26	6.15	6.06
7	8.89	7.18	6.35	5.90	5.62	5.44	5.32	5.21	5.12	5.03	4.96
8	7.50	6.11	5.42	5.05	4.82	4.66	4.55	4.47	4.40	4.34	4.28
9	6.54	5.40	4.82	4.49	4.29	4.15	4.06	3.99	3.93	3.88	3.83
10	5.85	4.85	4.34	4.06	3.88	3.75	3.67	3.60	3.55	3.50	3.46
12	4.91	4.14	3.73	3.49	3.34	3.24	3.17	3.11	3.07	3.03	3.00
15	4.07	3.49	3.17	2.99	2.87	2.78	2.72	2.68	2.64	2.61	2.59
20	3.32	2.91	2.67	2.53	2.43	2.36	2.32	2.28	2.26	2.24	2.22
30	2.63	2.36	2.20	2.10	2.03	1.98	1.94	1.92	1.90	1.89	1.88
60	1.96	1.82	1.72	1.67	1.63	1.60	1.58	1.56	1.55	1.54	1.54
∞	1.00	1.00	1.00	1.00	1.00	1.00	1.00	1.00	1.00	1.00	1.00
						$\alpha = .05$					
2	39.0	32.9	28.8	26.7	25.4	24.6	24.1	23.7	23.4	23.1	22.9
4	9.60	8.08	7.24	6.69	6.33	6.08	5.92	5.80	5.70	5.63	5.58
5	7.15	6.09	5.48	5.07	4.81	4.62	4.50	4.40	4.33	4.27	4.22
6	5.82	5.01	4.54	4.21	3.99	3.84	3.74	3.66	3.60	3.54	3.49
7	4.99	4.34	3.95	3.67	3.49	3.37	3.29	3.21	3.15	3.10	3.06
8	4.43	3.88	3.55	3.31	3.15	3.04	2.96	2.90	2.85	2.81	2.77
9	4.03	3.55	3.26	3.05	2.90	2.80	2.74	2.68	2.64	2.60	2.56
10	3.72	3.30	3.03	2.85	2.72	2.63	2.57	2.51	2.47	2.44	2.41
12	3.28	2.94	2.72	2.56	2.45	2.38	2.32	2.28	2.24	2.21	2.19
15	2.86	2.59	2.42	2.29	2.21	2.14	2.09	2.06	2.03	2.01	1.98
20	2.46	2.26	2.12	2.03	1.96	1.91	1.87	1.85	1.83	1.81	1.79
30	2.07	1.93	1.84	1.77	1.72	1.68	1.65	1.63	1.62	1.60	1.59
60	1.67	1.59	1.53	1.49	1.46	1.44	1.42	1.40	1.39	1.39	1.38
∞	1.00	1.00	1.00	1.00	1.00	1.00	1.00	1.00	1.00	1.00	1.00

Table P.5   *Critical values for separation of normal population variances into groups (maximum case) for* $\alpha = .01$

v	2	3	4	5	6	7	8	9	10	11	12
2	199.	224.	243.	258.	272.	283.	294.	303.	311.	319.	326.
3	47.5	52.9	57.0	60.2	62.9	65.4	67.5	69.1	70.7	72.2	73.7
4	23.2	25.5	27.3	28.7	29.8	30.9	31.8	32.5	33.2	33.9	34.5
5	14.9	16.4	17.4	18.2	18.9	19.4	19.9	20.4	20.8	21.1	21.5
6	11.1	12.1	12.8	13.3	13.8	14.1	14.5	14.7	15.0	15.3	15.5
7	8.89	9.64	10.2	10.6	10.9	11.2	11.4	11.6	11.8	12.0	12.2
8	7.50	8.10	8.52	8.84	9.08	9.29	9.49	9.65	9.81	9.94	10.1
9	6.54	7.04	7.38	7.64	7.84	8.03	8.18	8.31	8.44	8.55	8.65
10	5.85	6.27	6.55	6.78	6.95	7.11	7.24	7.35	7.45	7.55	7.64
12	4.91	5.22	5.45	5.62	5.75	5.87	5.97	6.06	6.13	6.21	6.27
15	4.07	4.31	4.47	4.60	4.69	4.78	4.86	4.92	4.98	6.03	5.10
20	3.32	3.48	3.60	3.68	3.75	3.82	3.87	3.91	3.95	3.99	4.02
30	2.63	2.73	2.81	2.86	2.91	2.95	2.98	3.01	3.03	3.06	3.08
60	1.96	2.02	2.05	2.09	2.11	2.13	2.14	2.16	2.17	2.18	2.19
∞	1.00	1.00	1.00	1.00	1.00	1.00	1.00	1.00	1.00	1.00	1.00

Tables for Subset Selection Problems

Tables Q.1–Q.6 apply to various aspects of subset selection problems. Tables Q.1 and Q.6 are for normal populations, and Tables Q.2–Q.5 are for binomial populations.

SOURCES FOR APPENDIX Q

Table Q.1 is adapted from S. S. Gupta (1965), On some multiple decision (selecting and ranking) rules, *Technometrics*, **7**, 225–245, with permission.

Table Q.2 is adapted from S. S. Gupta and M. Sobel (1960), Selecting a subset from binomial populations, in I. Olkin, S. G. Ghurye, W. Hoeffding, W. G. Madow, and H. B. Mann, Eds., *Contributions to Probability and Statistics, Essays in Honor of Harold Hotelling*, Stanford University Press, Stanford, California, pp. 224–248, with permission of the Board of Trustees of Leland Stanford Junior University.

Tables Q.3 and Q.4 are adapted from S. S. Gupta, M. J. Huyett, and M. Sobel (1957), Selecting and ranking problems with binomial populations, *Transactions of the American Society for Quality Control*, 635–644, with permission.

Table Q.6 is adapted from M. M. Desu and M. Sobel (1968), A fixed subset-size approach to a selection problem, *Biometrika*, **55**, 401–410, with permission of the Biometrika Trustees and the authors.

TABLE INTERPOLATION FOR APPENDIX Q

Interpolation in Table Q.2

Since the entries in Table Q.2 are integers that differ by small whole numbers, we have to interpolate between two successive entries only if they are unequal (we use the common entry when they are the same).

The type of "interpolation" we recommend here is essentially an approximate recomputation based on the normal approximation. Gupta and Sobel (1960) found this approximation to be conservative and to yield results that are never in error by more than one unit in d. Corresponding to any given k, P^*, and n, we first use k and P^* to find a τ value from Table A.1. For P^* not listed specifically in Table A.1, the interpolation method of Table A.1 can be used. Then the required d value is the smallest integer not less than $\frac{1}{2}(\tau\sqrt{n} - 1)$. For example, suppose that $k=3$, $P^*=.95$, and $n=40$. We first obtain $\tau=2.7101$ from Table A.1 and then compute $\frac{1}{2}(\tau\sqrt{n} - 1) = 8.07$, so that nine observations are required from each of the $k=3$ populations to meet the P^* requirement. (Note that we are selecting a subset and there is no δ^* in this requirement.) This entry does not appear in Table Q.2, which is an excerpt from the larger table in Gupta and Sobel (1960), but it does appear in the latter table and the results agree.

Interpolation in Table Q.3

The entries in Table Q.3 are again equivalent to the τ values in Table A.1, but because we are selecting a subset here it is desirable to include some values for a large number of populations k. Thus we have included entries for $k=20$, 40, and 50. If more accuracy is needed for $k \leqslant 25$, then the corresponding entry from Table A.1 can be used.

Interpolation in Table Q.4

The M value tabulated in Table Q.4 is the smallest integer not less than $np_0 - d_2\sqrt{np_0(1-p_0)}$. When n is large, the value of d_2 is found from Table M.2 as the entry corresponding to k, where k is the number of populations not including the standard or control.

Interpolation in Table Q.6

The notation λ used here is consistent with the notation in Desu and Sobel (1968); λ corresponds exactly to what we call τ. In fact, except for the fact that Table Q.6 does not cover $P^*=.750$, the special case $f=1$ corresponds exactly with the corresponding τ value in Table A.1 for $k=2$ through 9. Thus the entries for $P^*=.75$ and/or $k=10(5)25$ in Table A.1 can be used as Table Q.6 for the special case $f=1$.

In general, the same methods that are suggested for interpolation in Table A.1 should be used in Table Q.6. However, for different f values, better results can be obtained using a different transformation. We recommend using linear interpolation on log $(1-P^*)$ in all cases and an appropriate increasing function of λ depending on the value of f. For $f=1$ we find log λ (or log τ) gives good results. For $f=2$ it appears that λ^2 gives better results than log λ. For example, suppose we did not have the entry for $k = 5$, $f = 2$, $P^* = .975$. Using the transformations $\log(1 - P^*)$ and λ^2, we obtain the result as follows.

Table excerpt		Ready for interpolation	
P^*	λ	$\log(1 - P^*)$	λ^2
.950	2.2184	-2.9957	4.9213
.975	$\boxed{\lambda}$	-3.6889	$\boxed{\lambda^2}$
.990	3.0189	-4.6052	9.1138

$$\frac{\lambda^2 - 9.1138}{4.9213 - 9.1138} = \frac{-3.6889 - (-4.6052)}{-2.9957 - (-4.6052)},$$

$$\lambda^2 = 6.7270, \qquad \lambda = 2.5936.$$

Since the correct result is 2.5883, the error is 0.005 or about one-fifth of 1%. In this example the use of $\log \lambda$ gives the result 2.533, which is not as good as the preceding result.

If a large amount of interpolation is needed in these tables, we recommend trying some different transformations to see which gives better results. It is quite possible that there are better transformations than the ones suggested.

Table Q.1 Values of $E(S)$ for common spacing c/σ between normal population means

$k=2$

$c\sqrt{n}/\sigma$	P^* .90	.95	.99
0.5	1.77	1.88	1.97
1.0	1.69	1.82	1.95
1.5	1.58	1.72	1.90
2.0	1.44	1.59	1.82
3.0	1.20	1.32	1.58
4.0	1.06	1.12	1.31
5.0	1.01	1.03	1.11

$k=3$

$c\sqrt{n}/\sigma$	P^* .90	.95	.99
0.5	2.59	2.77	2.94
1.0	2.30	2.53	2.82
1.5	1.95	2.19	2.58
2.0	1.66	1.86	2.25
3.0	1.30	1.43	1.71
4.0	1.10	1.18	1.40
5.0	1.03	1.05	1.16

$k=4$

$c\sqrt{n}/\sigma$	P^* .90	.95	.99
0.5	3.33	3.60	3.88
1.0	2.70	3.03	3.53
1.5	2.12	2.39	2.92
2.0	1.75	1.96	2.37
3.0	1.36	1.49	1.77
4.0	1.14	1.22	1.44
5.0	1.04	1.07	1.20

$k=5$

$c\sqrt{n}/\sigma$	P^* .90	.95	.99
0.5	3.96	4.34	4.78
1.0	2.91	3.30	4.00
1.5	2.20	2.48	3.04
2.0	1.82	2.02	2.44
3.0	1.40	1.54	1.81
4.0	1.16	1.25	1.48
5.0	1.04	1.08	1.22

$k=10$

$c\sqrt{n}/\sigma$	P^* .90	.95	.99
0.5	5.50	6.30	7.72
1.0	3.29	3.71	4.52
1.5	2.44	2.72	3.26
2.0	1.99	2.19	2.60
3.0	1.51	1.65	1.92
4.0	1.24	1.34	1.57
5.0	1.08	1.13	1.30

$k=20$

$c\sqrt{n}/\sigma$	P^* .90	.95	.99
0.5	6.14	6.96	8.56
1.0	3.60	4.00	4.80
1.5	2.64	2.92	3.44
2.0	2.14	2.34	2.74
3.0	1.60	1.75	2.00
4.0	1.32	1.42	1.66
5.0	1.12	1.18	1.36

Remark. To find $E(S)$, multiply the common spacing by $\sqrt{n}$ and look for the result as a function of k and P^*.

Table Q.2 Values of d for selecting a subset using procedure SSB-I or SSB-Control-I

	P* = .90								
	k								
n	2	5	10	15	20	25	30	40	50
5	2	3	3	3	4	4	4	4	4
10	3	4	5	5	5	5	5	6	6
15	4	5	6	6	6	7	7	7	7
20	4	6	7	7	7	8	8	8	8
25	5	6	7	8	8	8	9	9	9
30	5	7	8	9	9	9	9	10	10
35	5	8	9	9	10	10	10	11	11
40	6	8	9	10	10	11	11	11	12
45	6	9	10	11	11	11	12	12	12
50	6	9	11	11	12	12	12	13	13

	P* = .95								
	k								
n	2	5	10	15	20	25	30	40	50
5	3	3	4	4	4	4	4	4	5
10	4	5	5	6	6	6	6	6	6
15	5	6	7	7	7	7	8	8	8
20	5	7	8	8	8	8	9	9	9
25	6	8	8	9	9	9	10	10	10
30	6	8	9	10	10	10	11	11	11
35	7	9	10	11	11	11	11	12	12
40	7	10	11	11	12	12	12	13	13
45	8	10	11	12	12	13	13	13	14
50	8	11	12	13	13	13	14	14	14

Table Q.3 Values of d_1 for selecting a subset using procedure SSB-II or SSB-Control-II

	P^*			
k	.75	.90	.95	.99
2	0.95	1.81	2.33	3.29
5	1.85	2.60	3.06	3.92
10	2.26	2.98	3.42	4.25
15	2.47	3.17	3.60	4.41
20	2.60	3.30	3.72	4.52
25	2.70	3.39	3.81	4.61
30	2.78	3.46	3.88	4.67
40	2.89	3.58	3.99	4.77
50	2.98	3.66	4.07	4.85

Table Q.4 Values of M for selecting a subset using procedure SSB-Standard

$k = 2$

$P^* = .90$

	p_0								
n	.10	.20	.30	.40	.50	.60	.70	.80	.90
5	0	0	0	1	1	2	2	3	4
10	0	0	1	2	3	4	5	6	8
15	0	1	2	4	5	7	8	10	12
20	0	2	3	5	7	9	11	14	16
25	1	3	5	7	9	12	15	17	21
50	2	6	11	16	20	26	31	36	42

$P^* = .95$

	p_0								
n	.10	.20	.30	.40	.50	.60	.70	.80	.90
5	0	0	0	0	1	1	2	2	3
10	0	0	1	2	2	3	5	6	7
15	0	1	2	3	4	6	7	9	11
20	0	1	3	4	6	8	11	13	16
25	0	2	4	6	8	11	14	17	20
50	2	6	10	14	19	24	30	35	41

$$k = 3$$
$$\dot{P}* = .90$$

n	.10	.20	.30	.40	p_0 .50	.60	.70	.80	.90
5	0	0	0	0	1	1	2	2	3
10	0	0	1	2	2	3	5	6	7
15	0	1	2	3	4	6	8	9	11
20	0	1	3	5	6	8	11	13	16
25	0	2	4	6	8	11	14	17	20
50	2	6	10	14	19	24	30	35	41

$$P* = .95$$

n	.10	.20	.30	.40	p_0 .50	.60	.70	.80	.90
5	0	0	0	0	0	1	1	2	3
10	0	0	0	1	2	3	4	5	7
15	0	0	1	2	4	5	7	9	11
20	0	1	2	4	6	8	10	12	15
25	0	1	3	5	8	10	13	16	19
50	1	5	9	13	18	23	29	34	41

$$k = 5$$
$$P* = .90$$

n	.10	.20	.30	.40	p_0 .50	.60	.70	.80	.90
5	0	0	0	0	0	1	1	2	3
10	0	0	0	1	2	3	4	5	7
15	0	0	1	2	4	5	7	9 ·	11
20	0	1	2	4	6	8	10	12	15
25	0	1	3	5	8	10	13	16	19
50	1	5	9	13	18	23	29	34	41

$$P* = .95$$

n	.10	.20	.30	.40	p_0 .50	.60	.70	.80	.90
5	0	0	0	0	0	1	1	2	3
10	0	0	0	1	2	3	4	5	6
15	0	0	1	2	3	5	6	8	10
20	0	1	2	3	5	7	9	12	15
25	0	1	3	5	7	9	12	15	19
50	1	4	8	12	17	22	28	33	40

Table Q.5 Values of 2 arcsin $\sqrt{p}$ in radians for $0<p<1$

p	.00	.01	.02	.03	.04	.05	.06	.07	.08	.09
.00	0.0000	0.2003	0.2838	0.3482	0.4027	0.4510	0.4949	0.5355	0.5735	0.6094
.10	0.6435	0.6761	0.7075	0.7377	0.7670	0.7954	0.8230	0.8500	0.8763	0.9021
.20	0.9273	0.9521	0.9764	1.0004	1.0239	1.0472	1.0701	1.0928	1.1152	1.1374
.30	1.1593	1.1810	1.2025	1.2239	1.2451	1.2661	1.2870	1.3079	1.3284	1.3490
.40	1.3694	1.3898	1.4101	1.4303	1.4505	1.4706	1.4907	1.5108	1.5308	1.5508
.50	1.5708	1.5908	1.6108	1.6308	1.6509	1.6710	1.6911	1.7113	1.7315	1.7518
.60	1.7722	1.7926	1.8132	1.8338	1.8546	1.8755	1.8965	1.9177	1.9391	1.9606
.70	1.9823	2.0042	2.0264	2.0488	2.0715	2.0944	2.1176	2.1412	2.1652	2.1895
.80	2.2143	2.2395	2.2653	2.2916	2.3186	2.3462	2.3746	2.4039	2.4341	2.4655
.90	2.4981	2.5322	2.5681	2.6061	2.6467	2.6906	2.7389	2.7934	2.8578	2.9413

Table Q.6 Values of λ for selecting a subset of fixed size f from k normal populations with a common known variance ($\lambda = \delta^* \sqrt{n} / \sigma$)

		P^*				
k	f	.900	.950	.975	.990	.999
2	1	1.8124	2.3262	2.7718	3.2900	4.3702
3	1	2.2302	2.7101	3.1284	3.6173	4.6450
	2	1.0919	1.5555	1.9565	2.4216	3.3876
4	1	2.4516	2.9162	3.3220	3.7970	4.7987
	2	1.5422	1.9797	2.3593	2.8010	3.7228
	3	0.7468	1.1912	1.5753	2.0204	2.9440
5	1	2.5997	3.0552	3.4532	3.9196	4.9048
	2	1.7926	2.2184	2.5883	3.0189	3.9193
	3	1.1964	1.6185	1.9844	2.4096	3.2954
	4	0.5280	0.9619	1.3368	1.7711	2.6720

P^*

k	f	.900	.950	.975	.990	.999
6	1	2.7100	3.1591	3.5517	4.0121	4.9855
	2	1.9639	2.3827	2.7466	3.1705	4.0574
	3	1.4536	1.8659	2.2235	2.6394	3.5071
	4	0.9709	1.3846	1.7431	2.1595	3.0266
	5	0.3708	0.7979	1.1669	1.5944	2.4808
7	1	2.7972	3.2417	3.6303	4.0860	5.0504
	2	2.0928	2.5068	2.8666	3.2859	4.1633
	3	1.6325	2.0388	2.3915	2.8017	3.6581
	4	1.2286	1.6335	1.9847	2.3929	3.2439
	5	0.8060	1.2143	1.5680	1.9789	2.8341
	6	0.2496	0.6719	1.0367	1.4593	2.3356
8	1	2.8691	3.3099	3.6953	4.1475	5.1046
	2	2.1955	2.6060	2.9627	3.3785	4.2489
	3	1.7686	2.1709	2.5201	2.9265	3.7750
	4	1.4099	1.8097	2.1564	2.5597	3.4007
	5	1.0619	1.4622	1.8093	2.2127	3.0532
	6	0.6772	1.0818	1.4322	1.8391	2.6860
	7	0.1516	0.5703	0.9319	1.3509	2.2197
9	1	2.9301	3.3679	3.7507	4.1999	5.1511
	2	2.2805	2.6882	3.0426	3.4556	4.3203
	3	1.8778	2.2771	2.6238	3.0272	3.8698
	4	1.5490	1.9452	2.2890	2.6888	3.5229
	5	1.2436	1.6392	1.9822	2.3811	3.2126
	6	0.9307	1.3277	1.6718	2.0718	2.9051
	7	0.5724	0.9740	1.3219	1.7259	2.5667
	8	0.0699	0.4857	0.8448	1.2610	2.1239

REMARK. Values of λ were rounded to the nearest fourth decimal place, that is, they were not rounded upward.

APPENDIX R

Tables for Gamma Selection Problems

The entries in Table R.1 are the values of M, the number of observations from each of k exponential distributions needed to satisfy the (δ^*, P^*) requirement for selecting the population with the largest θ value (scale parameter). The minimum number n of observations from each of k gamma populations, each with the same parameter r (degrees of freedom $2r$), is found from the relation $n = M/r$.

The entries in Table R.2 are the values of b needed to define the subset selection rule for a specified P^* in the problem of selecting a subset of k gamma populations, each with $2r$ degrees of freedom, that contains the population with the largest θ value (scale parameter) when samples of size n are taken from each gamma population. The value ν is computed from $\nu = 2rn$.

SOURCE FOR APPENDIX R

Table R.2 is adapted from S. S. Gupta (1963), On a selection and ranking procedure for gamma populations, *Annals of the Institute of Statistical Mathematics*, **14**, 199–216, with permission.

TABLE INTERPOLATION FOR APPENDIX R

Interpolation in Table R.1

The same methods that were used for interpolation in Table G.1 are applicable for Table R.1. In brief, we use linear interpolation on log δ^* and/or log $(1 - P^*)$ and/or $1/\sqrt{M}$. Thus if we did not have the entry for $k = 2$ and $P^* = .990$ corresponding to $\delta^* = 1.25$ we could use the entries 1.10 and 1.50 for δ^* to approximate it. The calculation would be as follows.

Table excerpt		Ready for interpolation	
$\delta*$	M	$\log \delta*$	$1/\sqrt{M}$
1.10	1192	0.0953	0.0289
1.25	$\boxed{M}$	0.2231	$\boxed{1/\sqrt{M}}$
1.50	67	0.4055	0.1222

$$\frac{1/\sqrt{M}-0.1222}{0.0289-0.1222}=\frac{0.2231-0.4055}{0.0953-0.4055}.$$

Thus $1/\sqrt{M}=0.0673$ and $M=221$. Since the correct answer is 218, the error is a little over 1%.

Interpolation in Table R.2

For the purpose of interpolation in Table R.2, it is desirable to transform the b values to $-(\log b)\sqrt{(\nu-1)/2}$ and then treat the resulting values as if they were τ values in Table A.1. In fact, if ν is large enough, the b value corresponding to any given $P*$ can be approximated by setting

$$\tau = -(\log b)\sqrt{\frac{(\nu-1)}{2}} \tag{1}$$

As the table now stands, the entries in each column are all approaching one so that an entry for $\nu=\infty$ is not very useful. If we had tabulated $-(\log b)\sqrt{(\nu-1)/2}$, then each column would be approaching the corresponding τ value in Table A.1. Thus the last entry for $k=2$, $P*=.75$ in the very first column, namely, .873, transforms into .9556 and the corresponding entry in Table A.1 is .9539. For example, suppose we want the entry in Table R.2 for $k=2$, $P*=.75$, $\nu=1000$. Then we can either interpolate in a transformed table using linear interpolation with $1/\nu$, or we can set the left-hand side of (1) equal to .9539 and solve (1) for b. In both cases we obtain the same result to four decimal places—namely, .9582—for the value of b. This shows that linear interpolation with $1/\nu$ and the τ values defined by (1) is adequate for most practical purposes.

The user can make use of the fact that each column in Table R.2 approaches 1, but this generally does not yield a satisfactory interpolation.

Table R.1 **Values of $M = nr$ needed to satisfy the (δ^*, P^*)**
requirement in selecting the gamma population with the
largest scale parameter

$k = 2$

δ^*	P^*						
	.750	.800	.850	.900	.950	.975	.990
1.10	101	157	238	363	597	847	1192
1.25	19	29	44	67	110	155	218
1.50	6	11	16	21	34	48	67
1.75	4	6	8	11	18	26	36
2.00	3	4	5	8	12	17	24
3.00	1	2	3	4	5	7	10
4.00	1	1	2	3	4	5	7
5.00	1	1	2	2	3	4	5

$k = 3$

δ^*	P^*						
	.750	.800	.850	.900	.950	.975	.990
1.10	227	302	402	549	810	1078	1441
1.25	42	56	74	101	149	198	264
1.50	13	18	23	31	45	61	80
1.75	7	10	13	16	24	32	43
2.00	5	6	8	11	16	21	28
3.00	2	3	4	5	7	9	12
4.00	2	2	3	3	5	6	8
5.00	1	2	2	3	4	5	6

$k = 4$

δ^*	P^*						
	.750	.800	.850	.900	.950	.975	.990
1.10	313	396	505	663	937	1216	1588
1.25	58	73	93	122	172	223	291
1.50	17	23	29	37	53	68	88
1.75	9	12	15	19	28	36	47
2.00	6	8	10	13	18	24	31
3.00	3	4	4	6	8	10	13
4.00	2	2	3	4	5	7	9
5.00	2	2	2	3	4	5	7

Table R.1 (Continued)

$k = 5$

δ^*	P^* .750	.800	.850	.900	.950	.975	.990
1.10	376	465	580	745	1029	1314	1692
1.25	69	86	107	137	188	240	310
1.50	20	27	33	42	58	74	94
1.75	11	14	18	22	31	39	50
2.00	7	9	11	14	20	25	33
3.00	3	4	5	6	8	11	14
4.00	2	3	3	4	6	7	9
5.00	2	2	3	3	4	5	7

$k = 6$

δ^*	P^* .750	.800	.850	.900	.950	.975	.990
1.10	427	520	640	809	1100	1390	1773
1.25	79	96	118	148	201	254	324
1.50	23	30	36	45	61	78	99
1.75	12	15	19	23	33	41	52
2.00	8	10	12	15	21	27	34
3.00	4	4	5	7	9	11	14
4.00	2	3	4	4	6	7	9
5.00	2	2	3	3	5	6	7

$k = 7$

δ^*	P^* .750	.800	.850	.900	.950	.975	.990
1.10	469	565	689	862	1158	1452	1839
1.25	86	104	126	158	212	266	336
1.50	25	32	39	48	65	81	102
1.75	13	17	21	25	35	43	54
2.00	9	11	13	16	22	28	35
3.00	4	5	6	7	9	12	15
4.00	3	3	4	5	6	8	10
5.00	2	2	3	4	5	6	7

$k = 8$

δ^*	P^*						
	.750	.800	.850	.900	.950	.975	.990
1.10	505	604	730	907	1207	1504	1895
1.25	93	111	134	166	221	275	346
1.50	29	34	41	51	67	84	105
1.75	14	18	23	27	36	45	56
2.00	9	11	14	17	23	29	36
3.00	4	5	6	7	10	12	15
4.00	3	3	4	5	6	8	10
5.00	2	3	3	4	5	6	8

$k = 9$

δ^*	P^*						
	.750	.800	.850	.900	.950	.975	.990
1.10	537	638	766	946	1250	1550	1943
1.25	99	117	141	173	229	284	355
1.50	30	36	43	53	70	87	108
1.75	15	19	24	28	37	46	57
2.00	10	12	14	18	24	30	37
3.00	4	5	6	7	10	12	15
4.00	3	3	4	5	6	8	10
5.00	2	3	3	4	5	6	8

$k = 10$

δ^*	P^*						
	.750	.800	.850	.900	.950	.975	.990
1.10	565	668	798	981	1287	1590	1985
1.25	104	123	146	180	236	291	363
1.50	32	38	45	55	72	89	110
1.75	16	20	24	29	38	47	59
2.00	10	12	15	18	24	30	38
3.00	4	5	6	8	10	13	16
4.00	3	3	4	5	7	8	10
5.00	2	3	3	4	5	6	8

Table R.2 *Values of b for selecting a subset of gamma (or chi-square) populations containing the one with the largest scale parameter*

		$k=2$				$k=3$		
		P^*				P^*		
v	.75	.90	.95	.99	.75	.90	.95	.99
2	.333	.111	.053	.010	.208	.072	.035	.007
4	.484	.244	.156	.063	.350	.183	.119	.048
6	.561	.327	.233	.118	.431	.260	.188	.097
8	.610	.386	.291	.166	.486	.317	.242	.140
10	.645	.430	.336	.206	.526	.360	.285	.178
12	.671	.466	.372	.241	.557	.396	.320	.210
14	.692	.494	.403	.270	.582	.426	.350	.239
16	.709	.519	.428	.297	.603	.451	.376	.264
18	.724	.539	.451	.320	.621	.472	.399	.287
20	.736	.558	.471	.340	.637	.492	.419	.307
30	.780	.622	.543	.419	.692	.561	.494	.386
40	.807	.664	.591	.473	.727	.607	.544	.440
50	.825	.694	.625	.513	.752	.640	.580	.481
60	.839	.717	.652	.545	.771	.666	.609	.514
70	.850	.735	.673	.570	.786	.686	.632	.540
80	.859	.750	.691	.592	.798	.703	.651	.562
90	.867	.762	.706	.610	.809	.718	.667	.581
100	.873	.773	.718	.626	.817	.730	.681	.598

		$k=4$				$k=5$		
		P^*				P^*		
v	.75	.90	.95	.99	.75	.90	.95	.99
2	.166	.059	.028	.006	.145	.052	.025	.005
4	.300	.159	.104	.043	.271	.145	.095	.039
6	.379	.232	.168	.087	.349	.215	.156	.082
8	.434	.286	.220	.128	.404	.268	.206	.121
10	.475	.329	.261	.164	.445	.310	.247	.156
12	.507	.364	.296	.196	.478	.345	.281	.187
14	.534	.394	.326	.224	.505	.374	.310	.214
16	.556	.419	.351	.249	.528	.400	.336	.238
18	.575	.441	.374	.271	.547	.422	.358	.260
20	.592	.460	.394	.291	.565	.441	.378	.280
30	.652	.532	.470	.369	.627	.514	.454	.358
40	.690	.579	.520	.424	.667	.562	.506	.412

		$k = 4$				$k = 5$		
		$P*$				$P*$		
v	.75	.90	.95	.99	.75	.90	.95	.99
50	.717	.614	.558	.465	.696	.597	.544	.454
60	.738	.641	.587	.497	.712	.620	.570	.486
70	.755	.662	.611	.524	.730	.642	.594	.513
80	.769	.680	.631	.547	.745	.661	.615	.536
90	.780	.695	.648	.567	.758	.677	.632	.556
100	.790	.708	.663	.583	.769	.691	.648	.573

		$k = 6$				$k = 7$		
		$P*$				$P*$		
v	.75	.90	.95	.99	.75	.90	.95	.99
2	.131	.047	.023	.004	.122	.044	.021	.004
4	.253	.135	.089	.037	.239	.128	.085	.035
6	.329	.203	.148	.078	.314	.195	.142	.074
8	.383	.255	.197	.116	.368	.246	.190	.112
10	.425	.297	.237	.150	.409	.287	.229	.146
12	.458	.332	.271	.180	.442	.321	.263	.175
14	.485	.361	.300	.207	.470	.350	.291	.202
16	.508	.386	.325	.231	.493	.376	.316	.226
18	.528	.408	.347	.253	.513	.398	.339	.247
20	.545	.428	.367	.273	.531	.417	.359	.267
30	.609	.500	.443	.350	.596	.490	.434	.344
40	.650	.549	.495	.404	.638	.539	.486	.398
50	.680	.585	.533	.446	.668	.576	.525	.440
60	.696	.607	.559	.478	.684	.597	.551	.471
70	.715	.630	.584	.505	.704	.621	.576	.499
80	.731	.650	.605	.528	.720	.641	.597	.522
90	.745	.666	.623	.548	.734	.657	.615	.542
100	.756	.680	.638	.565	.746	.672	.631	.559

		$k = 8$				$k = 9$		
		$P*$				$P*$		
v	.75	.90	.95	.99	.75	.90	.95	.99
2	.115	.041	.020	.004	.109	.039	.019	.004
4	.229	.123	.082	.034	.221	.119	.079	.032
6	.303	.188	.138	.072	.294	.183	.134	.070
8	.356	.239	.184	.109	.347	.232	.180	.106

		$k = 8$					$k = 9$		
			$P*$					$P*$	
ν	.75	.90	.95	.99	.75	.90	.95	.99
10	.397	.279	.223	.142	.388	.273	.218	.139
12	.430	.313	.256	.171	.421	.307	.251	.168
14	.458	.342	.285	.198	.448	.336	.279	.194
16	.482	.367	.310	.221	.472	.360	.304	.218
18	.502	.389	.332	.243	.492	.382	.326	.239
20	.520	.409	.352	.262	.510	.402	.346	.258
30	.585	.482	.428	.339	.576	.475	.422	.334
40	.628	.531	.480	.393	.619	.525	.474	.389
50	.659	.568	.518	.435	.651	.562	.513	.430
60	.674	.590	.544	.466	.666	.583	.538	.462
70	.695	.614	.569	.494	.687	.607	.564	.489
80	.711	.633	.591	.517	.704	.627	.585	.513
90	.725	.650	.609	.537	.718	.644	.604	.533
100	.738	.665	.625	.555	.731	.659	.620	.550

$k = 10$

		$P*$		
ν	.75	.90	.95	.99
2	.104	.038	.018	.004
4	.214	.116	.076	.032
6	.286	.178	.131	.069
8	.339	.228	.176	.104
10	.380	.268	.214	.136
12	.413	.301	.247	.165
14	.440	.330	.275	.191
16	.464	.355	.300	.214
18	.484	.376	.322	.236
20	.502	.396	.341	.255
30	.568	.469	.417	.331
40	.612	.519	.469	.385
50	.644	.556	.508	.427
60	.659	.577	.533	.457
70	.680	.602	.559	.485
80	.698	.622	.580	.509
90	.712	.639	.599	.529
100	.725	.654	.615	.547

APPENDIX S

Tables for Multivariate Selection Problems

Table S.1 is the ranking table for selection of the *largest* of k noncentral chi-square distributions with p degrees of freedom, noncentrality parameters $\theta_1, \theta_2, \ldots, \theta_k$, and a distance measure defined jointly by $\delta_1 = \theta_{[k]} - \theta_{[k-1]}$ and $\delta_2 = \theta_{[k]} / \theta_{[k-1]}$. The entry for a specified number k of multivariate normal populations, each with p variables (or components), δ_2^* and P^* is the value of $n\delta_1^*$ needed to satisfy the $(\delta_1^*, \delta_2^*, P^*)$ requirement when both $\delta_1 \geqslant \delta_1^*$ and $\delta_2 \geqslant \delta_2^*$; here n is the minimum common sample size needed per population for the problem of (1) selecting the population with the *largest* Mahalanobis 'distance' from the origin when the covariance matrices Σ_j are known or (2) selecting the population with the largest multiple correlation coefficient.

Table S.2 is the ranking table for selection of the *smallest* of k non-central chi-square distributions with p degrees of freedom, noncentrality parameters $\theta_1, \theta_2, \ldots, \theta_k$, and a distance measure defined jointly by $\delta_1 = \theta_{[2]} - \theta_{[1]}$ and $\delta_2 = \theta_{[2]} / \theta_{[1]}$. The entry for a specified number k of multivariate normal populations, each with p variables (or components), δ_2^* and P^* is the value of $n\delta_1^*$ needed to satisfy the $(\delta_1^*, \delta_2^*, P^*)$ requirement when both $\delta_1 \geqslant \delta_1^*$ and $\delta_2 \geqslant \delta_2^*$; here n is the minimum common sample size needed per population for the problem of selecting the population with the *smallest* Mahalanobis 'distance' from the origin when the covariance matrices Σ_j are known.

Corresponding entries in Tables S.1 and S.2 differ by at most 0.07% in all cases covered. For the special case $k = 2$, the results are exactly the same. As a result, for practical purposes Table S.1 can also be used for the problem of selecting the population with the smallest Mahalanobis 'distance' from the origin when the covariance matrices Σ_j are known.

516

Table S.3 is the ranking table for selecting the largest of k noncentral F distributions with p and $n-p$ degrees of freedom, noncentrality parameters $\theta_1, \theta_2, \ldots, \theta_k$, and a distance measure defined jointly by $\delta_1 = \theta_{[k]} - \theta_{[k-1]}$ and $\delta_2 = \theta_{[k]}/\theta_{[k-1]}$. The entry for a specified number k of multivariate normal populations, each with p variables (or components), δ_1^*, δ_2^*, and P^*, is the minimum common sample size n needed per population to satisfy the $(\delta_1^*, \delta_2^*, P^*)$ requirement when both $\delta_1 \geqslant \delta_1^*$, $\delta_2 \geqslant \delta_2^*$, for the problem of selecting the population with the *largest* Mahalanobis 'distance' from the origin when the covariance matrices Σ_j are unknown.

SOURCES FOR APPENDIX S

Tables S.1, S.2, and S.3 are adapted from M. H. Rizvi and R. C. Milton (1972), Multivariate tables related to Mahalanobis distance, unpublished manuscript, with permission.

Table S.1 *Values of $n\delta_1^*$ needed to satisfy the $(\delta_1^*, \delta_2^*, P^*)$ requirement for p variables in (1) selecting the population with the largest Mahalanobis 'distance' for Σ_j known, or (2) selecting the population with the largest multiple correlation coefficient.*

$$k = 2$$

δ_2^*	P^*	p 1	5	9	29
1.01	.90	1320	1320	1320	1320
	.95	2172	2172	2172	2172
	.99	4319	4319	4319	4319
1.05	.90	269.3	269.4	269.5	270.0
	.95	443.6	443.7	443.8	444.3
	.99	887.1	887.2	887.3	887.8
1.10	.90	137.9	138.1	138.3	139.2
	.95	227.1	227.3	227.5	228.5
	.99	454.3	454.5	454.7	455.6
1.15	.90	94.05	94.33	94.60	95.97
	.95	154.9	155.2	155.5	156.9
	.99	309.9	310.2	310.4	311.8
1.20	.90	72.11	72.48	72.84	74.59
	.95	118.8	119.2	119.5	121.3
	.99	237.6	238.0	238.3	240.1
1.25	.90	58.94	59.39	59.82	61.92
	.95	97.10	97.54	97.98	100.3
	.99	194.2	194.7	195.1	197.3
1.50	.90	32.52	33.32	34.08	37.48
	.95	53.56	54.37	55.15	58.77
	.99	107.1	108.0	108.7	112.6
1.75	.90	23.63	24.72	25.73	29.91
	.95	38.93	40.03	41.08	45.68
	.99	77.87	78.98	80.06	85.09
2.00	.90	19.14	20.48	21.66	26.33
	.95	31.54	32.89	34.15	39.42
	.99	63.08	64.46	65.78	71.72
2.50	.90	14.59	16.31	17.73	22.95
	.95	24.03	25.80	27.36	33.45
	.99	48.07	49.88	51.56	58.76
3.00	.90	12.26	14.28	15.85	21.35
	.95	20.19	22.28	24.05	30.61
	.99	40.39	42.5	44.50	52.51

$$k = 3$$

δ_2^*	P^*	p			
		1	5	9	29
1.01	.90	1998	1998	1998	1998
	.95	2948	2948	2948	2948
	.99	5220	5220	5220	5221
1.05	.90	407.8	407.9	408.0	408.5
	.95	602.1	602.2	602.3	602.8
	.99	1072	1072	1073	1073
1.10	.90	208.8	209.0	209.2	210.1
	.95	308.3	308.5	308.7	309.6
	.99	549.2	549.4	549.6	550.5
1.15	.90	142.4	142.7	143.0	144.3
	.95	210.3	210.6	210.8	212.2
	.99	374.6	374.9	375.2	376.6
1.20	.90	109.2	109.6	109.9	111.7
	.95	161.2	161.6	162.0	163.8
	.99	287.3	287.6	288.0	289.8
1.25	.90	89.25	89.69	90.13	92.27
	.95	131.8	132.2	132.7	134.8
	.99	234.8	235.2	235.7	237.9
1.50	.90	49.24	50.03	50.81	54.37
	.95	72.70	73.51	74.29	78.00
	.99	129.5	130.3	131.1	135.0
1.75	.90	35.78	36.87	37.90	42.41
	.95	52.84	53.94	55.00	59.78
	.99	94.14	95.25	96.33	101.44
2.00	.90	28.99	30.33	31.55	36.69
	.95	42.81	44.16	45.44	51.00
	.99	76.26	77.64	78.96	85.03
2.50	.90	22.09	23.82	25.34	31.24
	.95	32.62	34.39	35.99	42.56
	.99	58.12	59.92	61.62	69.04
3.00	.90	18.56	20.60	22.31	28.65
	.95	27.41	29.50	31.34	38.52
	.99	48.83	50.98	52.97	61.31

$$k = 4$$

δ_2^*	P^*	p			
		1	5	9	29
1.01	.90	2414	2414	2414	2414
	.95	3413	3413	3414	3414
	.99	5751	5751	5751	5751
1.05	.90	492.7	492.8	492.9	493.4
	.95	697.2	697.3	697.4	697.9
	.99	1182	1182	1182	1182
1.10	.90	252.3	252.5	252.7	253.6
	.95	357.0	357.2	357.4	358.3
	.99	605.1	605.3	605.5	606.4
1.15	.90	172.1	172.4	172.6	174.0
	.95	243.5	243.8	244.0	245.4
	.99	412.8	413.0	413.3	414.7
1.20	.90	131.9	132.3	132.7	134.4
	.95	186.7	187.1	187.4	189.2
	.99	316.5	316.9	317.2	319.0
1.25	.90	107.8	108.3	108.7	110.9
	.95	152.6	153.0	153.5	155.7
	.99	258.7	259.1	259.6	261.8
1.50	.90	59.49	60.29	61.07	64.69
	.95	84.18	84.99	85.77	89.51
	.99	142.7	143.5	144.3	148.2
1.75	.90	43.24	44.33	45.37	49.99
	.95	61.18	62.28	63.34	68.21
	.99	103.7	104.8	105.9	111.1
2.00	.90	35.03	36.36	37.61	42.93
	.95	49.57	50.92	52.20	57.89
	.99	84.03	85.40	86.73	92.85
2.50	.90	26.69	28.43	29.98	36.17
	.95	37.77	39.54	41.16	47.93
	.99	64.03	65.83	67.54	75.07
3.00	.90	22.43	24.47	26.23	32.94
	.95	31.74	33.83	35.70	43.15
	.99	53.80	55.95	57.95	66.45

$k = 5$

δ_2^*	P^*	1	5	9	29
				p	
1.01	.90	2715	2715	2715	2715
	.95	3746	3746	3746	3746
	.99	6128	6128	6128	6128
1.05	.90	554.1	554.2	554.3	554.8
	.95	765.2	765.3	765.4	765.9
	.99	1259	1259	1259	1260
1.10	.90	283.7	283.9	284.1	285.0
	.95	391.8	392.0	392.2	393.1
	.99	644.8	645.0	645.2	646.1
1.15	.90	193.5	193.8	194.1	195.4
	.95	267.2	267.5	267.8	269.2
	.99	439.8	440.1	440.4	441.8
1.20	.90	148.4	148.7	149.1	150.9
	.95	204.9	205.3	205.6	207.4
	.99	337.3	337.6	338.0	339.8
1.25	.90	121.3	121.7	122.2	124.3
	.95	167.5	167.9	168.4	170.5
	.99	275.7	276.1	276.6	278.8
1.50	.90	66.90	67.70	68.47	72.13
	.95	92.40	93.20	93.99	97.74
	.99	152.1	152.9	153.7	157.6
1.75	.90	48.62	49.71	50.75	55.45
	.95	67.15	68.25	69.31	74.22
	.99	110.5	111.6	112.7	117.9
2.00	.90	39.39	40.72	41.98	47.41
	.95	54.40	55.75	57.04	62.79
	.99	89.54	90.91	92.24	98.40
2.50	.90	30.02	31.75	33.32	36.69
	.95	41.46	43.22	44.86	51.75
	.99	68.23	70.04	71.75	79.35
3.00	.90	25.22	27.27	29.05	35.99
	.95	34.84	36.93	38.81	46.43
	.99	57.33	59.48	61.49	70.09

$$k = 6$$

δ_2^*	P^*	p			
		1	5	9	29
1.01	.90	2950	2950	2950	2950
	.95	4005	4005	4005	4006
	.99	6420	6420	6420	6420
1.05	.90	602.1	602.2	602.3	602.8
	.95	818.2	818.3	818.4	818.8
	.99	1319	1319	1319	1320
1.10	.90	308.3	308.5	308.6	309.6
	.95	418.9	419.1	419.3	420.2
	.99	675.6	675.8	676.0	676.9
1.15	.90	210.3	210.5	210.8	212.2
	.95	285.7	286.0	286.3	287.7
	.99	460.9	461.1	461.4	462.8
1.20	.90	161.2	161.6	162.0	163.7
	.95	219.1	219.5	219.8	221.6
	.99	353.4	353.7	354.1	355.9
1.25	.90	131.8	132.2	132.7	134.8
	.95	179.1	179.5	180.0	182.1
	.99	288.8	289.3	289.7	291.9
1.50	.90	72.70	73.49	74.27	77.94
	.95	98.79	99.59	100.38	104.15
	.99	159.3	160.1	160.9	164.8
1.75	.90	52.83	53.92	54.97	59.70
	.95	71.80	72.90	73.96	78.89
	.99	115.8	116.9	118.0	123.2
2.00	.90	42.80	44.14	45.39	50.89
	.95	58.17	59.52	60.81	66.61
	.99	93.82	95.19	96.52	102.70
2.50	.90	32.62	34.35	35.93	42.42
	.95	44.33	46.09	47.73	54.71
	.99	71.49	73.29	75.01	82.65
3.00	.90	27.41	29.45	31.26	38.35
	.95	37.25	39.34	41.23	48.96
	.99	60.07	62.22	64.23	72.90

$$k = 7$$

δ_2^*	P^*	p			
		1	5	9	29
1.01	.90	3143	3143	3143	3143
	.95	4217	4217	4217	4218
	.99	6659	6659	6659	6659
1.05	.90	641.5	641.6	641.7	642.2
	.95	861.5	861.6	861.7	862.2
	.99	1368	1368	1369	1369
1.10	.90	328.4	328.6	328.8	329.8
	.95	441.1	441.3	441.5	442.4
	.99	700.7	700.9	701.1	702.1
1.15	.90	224.0	224.3	224.6	226.0
	.95	300.9	301.1	301.4	302.8
	.99	478.0	478.3	478.6	479.9
1.20	.90	171.8	172.1	172.5	174.3
	.95	230.7	231.1	231.4	233.2
	.99	366.5	366.9	367.3	369.1
1.25	.90	140.4	140.8	141.3	143.4
	.95	188.6	189.0	189.4	191.6
	.99	299.6	300.0	300.5	302.7
1.50	.90	77.45	78.25	79.03	82.72
	.95	104.0	104.8	105.6	109.4
	.99	165.3	166.1	166.9	170.8
1.75	.90	56.29	57.38	58.42	63.20
	.95	75.60	76.70	77.76	82.72
	.99	120.1	121.2	122.3	127.5
2.00	.90	45.60	46.94	48.20	53.75
	.95	61.25	62.60	63.89	69.72
	.99	97.31	98.68	100.01	106.21
2.50	.90	34.75	36.48	38.07	44.65
	.95	46.67	48.43	50.08	57.12
	.99	74.15	75.95	77.67	85.35
3.00	.90	29.20	31.25	33.06	40.27
	.95	39.22	41.31	43.21	51.03
	.99	62.31	64.46	66.47	75.20

$$k = 8$$

δ_2^*	P^*	p			
		1	5	9	29
1.01	.90	3306	3306	3306	3307
	.95	4397	4397	4397	4397
	.99	6860	6860	6860	6860
1.05	.90	674.9	675.0	675.1	675.6
	.95	898.1	898.2	898.3	898.8
	.99	1410	1410	1410	1410
1.10	.90	345.5	345.7	345.9	346.9
	.95	459.9	460.0	460.2	461.2
	.99	722.0	722.2	722.4	723.3
1.15	.90	235.7	236.0	236.2	237.6
	.95	313.7	313.9	314.2	315.6
	.99	492.5	492.8	493.0	494.4
1.20	.90	180.7	181.1	181.4	183.2
	.95	240.5	240.9	241.2	243.0
	.99	377.6	378.0	378.4	380.2
1.25	.90	147.7	148.2	148.6	150.8
	.95	196.6	197.0	197.5	199.6
	.99	308.7	309.1	309.5	311.7
1.50	.90	81.49	82.28	83.06	86.76
	.95	108.4	109.2	110.0	113.8
	.99	170.3	171.1	171.9	175.8
1.75	.90	59.22	60.31	61.36	66.15
	.95	78.82	79.91	80.98	85.95
	.99	123.8	124.9	125.9	131.1
2.00	.90	47.98	49.31	50.58	56.17
	.95	63.85	65.20	66.50	72.36
	.99	100.3	101.6	103.0	109.2
2.50	.90	36.56	38.29	39.88	46.53
	.95	48.66	50.42	52.07	59.16
	.99	76.40	78.20	79.92	87.62
3.00	.90	30.72	32.77	34.59	41.89
	.95	40.89	42.98	44.89	52.77
	.99	64.20	66.34	68.36	77.13

$$k = 9$$

δ_2^*	P^*	p 1	5	9	29
1.01	.90	3448	3448	3448	3449
	.95	4552	4552	4552	4552
	.99	7034	7035	7035	7035
1.05	.90	703.9	704.0	704.1	704.6
	.95	929.9	930.0	930.1	930.6
	.99	1446	1446	1446	1446
1.10	.90	360.4	360.6	360.8	361.7
	.95	476.1	476.3	476.5	477.4
	.99	740.3	740.5	740.7	741.7
1.15	.90	245.8	246.1	246.4	247.7
	.95	324.8	325.0	325.3	326.7
	.99	505.0	505.3	505.6	507.0
1.20	.90	188.5	188.9	189.2	191.0
	.95	249.0	249.4	249.7	251.5
	.99	387.2	387.6	388.0	389.8
1.25	.90	154.1	154.5	154.9	157.1
	.95	203.5	204.0	204.4	206.6
	.99	316.5	317.0	317.4	319.6
1.50	.90	84.99	85.78	86.56	90.27
	.95	112.3	113.1	113.9	117.7
	.99	174.6	175.4	176.2	180.1
1.75	.90	61.77	62.85	63.90	68.72
	.95	81.60	82.70	83.76	88.75
	.99	126.9	128.0	129.1	134.3
2.00	.90	50.04	51.37	52.64	58.26
	.95	66.11	67.46	68.75	74.64
	.99	102.8	104.2	105.5	111.7
2.50	.90	38.13	39.86	41.46	48.16
	.95	50.38	52.14	53.79	60.92
	.99	78.34	80.14	81.86	89.59
3.00	.90	32.04	34.09	35.92	43.29
	.95	42.33	44.42	46.33	54.28
	.99	65.83	67.97	69.99	78.80

$$k = 10$$

δ_2^*	P^*	p 1	5	9	29
1.01	.90	3574	3574	3574	3574
	.95	4689	4689	4689	4689
	.99	7188	7188	7188	7188
1.05	.90	729.5	729.6	729.7	730.2
	.95	957.9	958.0	958.1	958.5
	.99	1477	1477	1477	1478
1.10	.90	373.5	373.7	373.9	374.8
	.95	490.4	490.6	490.8	491.8
	.99	756.5	756.7	756.9	757.9
1.15	.90	254.8	255.0	255.3	256.7
	.95	334.5	334.8	335.1	336.5
	.99	516.0	516.3	516.6	518.0
1.20	.90	195.3	195.7	196.1	197.8
	.95	256.5	256.9	257.2	259.0
	.99	395.7	396.1	396.4	398.2
1.25	.90	159.7	160.1	160.5	162.7
	.95	209.7	210.1	210.5	212.7
	.99	323.4	323.9	324.3	326.5
1.50	.90	88.08	88.87	89.65	93.37
	.95	115.7	116.5	117.2	121.0
	.99	178.4	179.2	180.0	183.9
1.75	.90	64.01	65.10	66.15	70.98
	.95	84.06	85.15	86.22	91.21
	.99	129.7	130.8	131.9	137.1
2.00	.90	51.86	53.19	54.46	60.11
	.95	68.10	69.45	70.74	76.64
	.99	105.1	106.4	107.8	114.0
2.50	.90	39.52	41.25	42.85	49.59
	.95	51.89	53.65	55.31	62.48
	.99	80.05	81.85	83.58	91.32
3.00	.90	33.21	35.25	37.09	44.52
	.95	43.61	45.69	47.61	55.60
	.99	67.27	69.41	71.43	80.27

Table S.2 *Values of $n\delta_1^*$ needed to satisfy the $(\delta_1^*, \delta_2^*, P^*)$ requirement for p variables in selecting the population with the smallest Mahalanobis 'distance' for Σ_j known*

$k = 3$

δ_2^*	P^*			p	
		1	5	9	29
1.01	.90	1999	1999	1999	2000
	.95	2952	2952	2952	2952
	.99	5259	5259	5259	5259
1.50	.90	49.24	50.05	50.83	54.44
	.95	72.70	73.52	74.31	78.05
	.99	129.5	130.3	131.1	135.0
3.00	.90	18.56	20.68	22.43	28.86
	.95	27.41	29.56	31.43	38.69
	.99	48.83	51.01	53.01	61.41

$k = 4$

δ_2^*	P^*			p	
		1	5	9	29
1.01	.90	2416	2416	2416	2416
	.95	3418	3419	3419	3419
	.99	5793	5793	5793	5793
1.50	.90	59.49	60.31	61.10	64.79
	.95	84.18	85.00	85.80	89.58
	.99	142.7	143.5	144.3	148.2
3.00	.90	22.43	24.59	26.41	33.28
	.95	31.74	33.91	35.83	43.42
	.99	53.80	55.99	58.02	66.61

Table S.2   (*Continued*)

k = 5

δ_2^*	P^*	p			
		1	5	9	29
1.01	.90	2717	2717	2717	2717
	.95	3752	3752	3752	3752
	.99	6173	6173	6173	6173
1.50	.90	66.90	67.72	68.52	73.25
	.95	92.40	93.21	94.01	97.83
	.99	152.1	152.9	153.7	157.6
3.00	.90	25.22	27.40	29.28	36.41
	.95	34.84	37.02	38.98	46.77
	.99	57.33	59.53	61.57	70.29

k = 10

δ_2^*	P^*	p			
		1	5	9	29
1.01	.90	3577	3577	3577	3577
	.95	4696	4696	4696	4696
	.99	7239	7239	7239	7239
1.50	.90	88.08	88.90	89.71	93.54
	.95	115.7	116.5	117.3	121.2
	.99	178.4	179.2	180.0	184.0
3.00	.90	33.21	35.44	37.41	45.19
	.95	43.61	45.83	47.85	56.14
	.99	67.27	69.48	71.56	80.60

Table S.3 *Minimum sample size n needed per population to satisfy the $(\delta_1^*, \delta_2^*, P^*)$ requirement in selecting the multivariate normal population with the largest Mahalanobis 'distance' for Σ_j unknown*

$$k = 2, \ p = 4$$

δ_2^* / $n\delta_1^*$	$P^* = .90$			$P^* = .95$			$P^* = .99$		
	1.50	2.00	3.00	1.50	2.00	3.00	1.50	2.00	3.00
20			29						
30		52	18			46			
40		35	15			29			
50		30	14		75	24			
60	98	27	13		58	21			74
70	83	25	13		50	20			56
80	75	24	12		45	19			48
90	70	24	12		42	18			43
100	66	23	12		40	18			40
110	64	23	12		39	17			37
120	61	22	12		37	17		98	34
130	60	22	12		36	17		98	34
140	58	22	12		36	17		92	33
150	57	21	11		35	16		87	32
160	56	21			34			84	
170	55	21		103	34			81	
180	55	21		101	34			78	
190	54	21		99	33			76	
200	54	21		97	33			74	
220	53			94					
240	52			91					
260	51			89					
280	51			88					
300	50			86					

Table S.3 (Continued)

$$k = 2, p = 10$$

$n\delta_1^*$	δ_2^* P*=.90 1.50	2.00	3.00	P*=.95 1.50	2.00	3.00	P*=.99 1.50	2.00	3.00
20			69						
30		81	31			91			
40		50	25			46			
50		41	22		100	36			
60		37	21		74	31			107
70	96	34	20		63	29			76
80	86	33	20		56	26			63
90	80	31	19		52	26			56
100	76	31	19		50	26			51
110	72	30	19		48	25			48
120	70	29	19		46	24			45
130	68	29	18		45	24			44
140	66	29	18		44	24		105	42
150	65	28	18		43	23		99	41
160	64	28			42			94	
170	63	28			42			91	
180	62	28		110	41			88	
190	62	28		107	41			85	
200	61	27		105	40			83	
220	60			102					
240	59			99					
260	58			97					
280	58			95					
300	57			94					

APPENDIX T

Excerpt of Table of Random Numbers

This table may be entered at any point and read in any direction and for any number of digits at a time.

10480	15011	01536	02011	81647	91646	69179	14194	62590	36207	20969	99570	91291	90700
22368	46573	25595	85393	30995	89198	27982	53402	93965	34095	52666	19174	39615	99505
24130	48360	22527	97265	76393	64809	15179	24830	49340	32081	30680	19655	63348	58629
42167	93093	06243	61680	07856	16376	39440	53537	71341	57004	00849	74917	97758	16379
37570	39975	81837	16656	06121	91782	60468	81305	49684	60672	14110	06927	01263	54613
77921	06907	11008	42751	27756	53498	18602	70659	90655	15053	21916	81825	44394	42880
99562	72905	56420	69994	98872	31016	71194	18738	44013	48840	63213	21069	10634	12952
96301	91977	05463	07972	18876	20922	94595	56869	69014	60045	18425	84903	42508	32307
89579	14342	63661	10281	17453	18103	57740	84378	25331	12566	58678	44947	05585	56941
85475	36857	53342	53988	53060	59533	38867	62300	08158	17983	16439	11458	18593	64952
28918	69578	88231	33276	70997	79936	56865	05859	90106	31595	01547	85590	91610	78188
63553	40961	48235	03427	49626	69445	18663	72695	52180	20847	12234	90511	33703	90322
09429	93969	52636	92737	88974	33488	36320	17617	30015	08272	84115	27156	30613	74952
10365	61129	87529	85689	48237	52267	67689	93394	01511	26358	85104	20285	29975	89868
07119	97336	71048	08178	77233	13916	47564	81056	97735	85977	29372	74461	28551	90707
51085	12765	51821	51259	77452	16308	60756	92144	49442	53900	70960	63990	75601	40719
02368	21382	52404	60268	89368	19885	55322	44819	01188	65255	64835	44919	05944	55157
01011	54092	33362	94904	31273	04146	18594	29852	71585	85030	51132	01915	92747	64951
52162	53916	46369	58586	23216	14513	83149	98736	23495	64350	94738	17752	35156	35749
07056	97628	33787	09998	42698	06691	76988	13602	51851	46104	88916	19509	25625	58104
48663	91245	85828	14346	09172	30168	90229	04734	59193	22178	30421	61666	99904	32812
54164	58492	22421	74103	47070	25306	76468	26384	58151	06646	21524	15227	96909	44592
32639	32363	05597	24200	13363	38005	94342	28728	35806	06912	17012	64161	18296	22851
29334	27001	87637	87308	58731	00256	45834	15398	46557	41135	10367	07684	36188	18510
02488	33062	28834	07351	19731	92420	60952	61280	50001	67658	32586	86679	50720	94953

81525 72295 04839 96423 24878 82651 66566 14778 76797 14780 13300 87074 79666 95725
29676 20591 68086 26432 46901 20849 89768 81536 86645 12659 92259 57102 80428 25280
00742 57392 39064 66432 84673 40027 32832 61362 98947 96067 64760 64584 96096 98253
05366 04213 25669 26422 44407 44048 37937 63904 45766 66134 75470 66520 34693 90449
91921 26418 64117 94305 26766 25940 39972 22209 71500 64568 91402 42416 07844 69618

00582 04711 87917 77341 42206 35126 74087 99547 81817 42607 43808 76655 62028 76630
00725 69884 62797 56170 86324 88072 76222 36086 84637 93161 76038 65855 77919 88006
69011 65795 95876 55293 18988 27354 26575 08625 40801 59920 29841 80150 12777 48501
25976 57948 29888 88604 67917 48708 18912 82271 65424 69774 33611 54262 85963 03547
09763 83473 73577 12908 30883 18317 28290 35797 05998 41688 34952 37888 38917 88050

91567 42595 27958 30134 04024 86385 29880 99730 55536 84855 29080 09250 79656 73211
17955 56349 90999 49127 20044 59931 06115 20542 18059 02008 73708 83517 36103 42791
46503 18584 18845 49618 02304 51038 20655 58727 28168 15475 56942 53389 20562 87338
92157 89634 94824 78171 84610 82834 09922 25417 44137 48413 25555 21246 35509 20468
14577 62765 35605 81263 39667 47358 56873 56307 61607 49518 89656 20103 77490 18062

98427 07523 33362 64270 01638 92477 66969 98420 04880 45585 46565 04102 46880 45709
34914 63976 88720 82765 34476 17032 87589 40836 32427 70002 70663 88863 77775 69348
70060 28277 39475 46473 23219 53416 94970 25832 69975 94884 19661 72828 00102 66794
53976 54914 06990 67245 68350 82948 11398 42878 80287 88267 47363 46634 06541 97809
76072 29515 40980 07391 58745 25774 22987 80059 39911 96189 41151 14222 60697 59583

90725 52210 83974 29992 65831 38857 50490 83765 55657 14361 31720 57375 56228 41546
64364 67412 33339 31926 14883 24413 59744 92351 97473 89286 35931 04110 23726 51900
08962 00358 31662 25388 61642 34072 81249 35648 56891 69352 48373 45578 78547 81788
95012 68379 93526 70765 10592 04542 76463 54328 02349 17247 28865 14777 62730 92277
15664 10493 20492 38391 91132 21999 59516 81652 27195 48223 46751 22923 32261 85653

16408 81899 04153 53381 79401 21438 83035 92350 36693 31238 59649 91754 72772 02338
18629 81953 05520 91962 04739 13092 97662 24822 94730 06496 35090 04822 86774 98289
73115 35101 47498 87637 99016 71060 88824 71013 18735 20286 23153 72924 35165 43040
57491 16703 23167 49323 45021 33132 12544 41035 80780 45393 44812 12515 98931 91202
30405 83946 23792 14422 15059 45799 22716 19792 09983 74353 68668 30429 70735 25499

16631 35006 85900 98275 32388 52390 16815 69298 82732 38480 73817 32523 41961 44437
96773 20206 42559 78985 05300 22164 24369 54224 35083 19687 11052 91491 00383 19746
38935 64202 14349 82674 66523 44133 00697 35552 35970 19124 63318 29686 03387 59846
31624 76384 17403 53363 44167 64486 64758 75366 76554 31601 12614 33072 60332 92325
78919 19474 23632 27889 47914 02584 37680 20801 72152 39339 34806 08930 85001 87820

APPENDIX U

Tables of Squares and Square Roots

For any number in the first column, the corresponding number in the second column is its square, the corresponding number in the third column is its square root, and the corresponding number in the fourth column is the square root of 10 times the number. In order to obtain the square or square root of any number with three significant digits, the decimal must be move appropriately. For example, since the square root of 1.23 is 1.10905, the square root of 123 is 11.0905 and the square root of 0.0123 is 0.110905. Similarly, since the square root of 12.3 is 3.50714, the square root of 1230 is 35.0714 and the square root of 0.123 is 0.350714.

N	N²	√N̄	√10N̄	N	N²	√N̄	√10N̄
1.00	1.0000	1.00000	3.16228	**1.50**	2.2500	1.22474	3.87298
1.01	1.0201	1.00499	3.17805	1.51	2.2801	1.22882	3.88587
1.02	1.0404	1.00995	3.19374	1.52	2.3104	1.23288	3.89872
1.03	1.0609	1.01489	3.20936	1.53	2.3409	1.23693	3.91152
1.04	1.0816	1.01980	3.22490	1.54	2.3716	1.24097	3.92428
1.05	1.1025	1.02470	3.24037	1.55	2.4025	1.24499	3.93700
1.06	1.1236	1.02956	3.25576	1.56	2.4336	1.24900	3.94968
1.07	1.1449	1.03441	3.27109	1.57	2.4649	1.25300	3.96232
1.08	1.1664	1.03923	3.28634	1.58	2.4964	1.25698	3.97492
1.09	1.1881	1.04403	3.30151	1.59	2.5281	1.26095	3.98748
1.10	1.2100	1.04881	3.31662	**1.60**	2.5600	1.26491	4.00000
1.11	1.2321	1.05357	3.33167	1.61	2.5921	1.26886	4.01248
1.12	1.2544	1.05830	3.34664	1.62	2.6244	1.27279	4.02492
1.13	1.2769	1.06301	3.36155	1.63	2.6569	1.27671	4.03733
1.14	1.2996	1.06771	3.37639	1.64	2.6896	1.28062	4.04969
1.15	1.3225	1.07238	3.39116	1.65	2.7225	1.28452	4.06202
1.16	1.3456	1.07703	3.40588	1.66	2.7556	1.28841	4.07431
1.17	1.3689	1.08167	3.42053	1.67	2.7889	1.29228	4.08656
1.18	1.3924	1.08628	3.43511	1.68	2.8224	1.29615	4.09878
1.19	1.4161	1.09087	3.44964	1.69	2.8561	1.30000	4.11096
1.20	1.4400	1.09545	3.46410	**1.70**	2.8900	1.30384	4.12311
1.21	1.4641	1.10000	3.47851	1.71	2.9241	1.30767	4.13521
1.22	1.4884	1.10454	3.49285	1.72	2.9584	1.31149	4.14729
1.23	1.5129	1.10905	3.50714	1.73	2.9929	1.31529	4.15933
1.24	1.5376	1.11355	3.52136	1.74	3.0276	1.31909	4.17133
1.25	1.5625	1.11803	3.53553	1.75	3.0625	1.32288	4.18330
1.26	1.5876	1.12250	3.54965	1.76	3.0976	1.32665	4.19524
1.27	1.6129	1.12694	3.56371	1.77	3.1329	1.33041	4.20714
1.28	1.6384	1.13137	3.57771	1.78	3.1684	1.33417	4.21900
1.29	1.6641	1.13578	3.59166	1.79	3.2041	1.33791	4.23084
1.30	1.6900	1.14018	3.60555	**1.80**	3.2400	1.34164	4.24264
1.31	1.7161	1.14455	3.61939	1.81	3.2761	1.34536	4.25441
1.32	1.7424	1.14891	3.63318	1.82	3.3124	1.34907	4.26615
1.33	1.7689	1.15326	3.64692	1.83	3.3489	1.35277	4.27785
1.34	1.7956	1.15758	3.66060	1.84	3.3856	1.35647	4.28952
1.35	1.8225	1.16190	3.67423	1.85	3.4225	1.36015	4.30116
1.36	1.8496	1.16619	3.68782	1.86	3.4596	1.36382	4.31277
1.37	1.8769	1.17047	3.70135	1.87	3.4969	1.36748	4.32435
1.38	1.9044	1.17473	3.71484	1.88	3.5344	1.37113	4.33590
1.39	1.9321	1.17898	3.72827	1.89	3.5721	1.37477	4.34741
1.40	1.9600	1.18322	3.74166	**1.90**	3.6100	1.37840	4.35890
1.41	1.9881	1.18743	3.75500	1.91	3.6481	1.38203	4.37035
1.42	2.0164	1.19164	3.76829	1.92	3.6864	1.38564	4.38178
1.43	2.0449	1.19583	3.78153	1.93	3.7249	1.38924	4.39318
1.44	2.0736	1.20000	3.79473	1.94	3.7636	1.39284	4.40454
1.45	2.1025	1.20416	3.80789	1.95	3.8025	1.39642	4.41588
1.46	2.1316	1.20830	3.82099	1.96	3.8416	1.40000	4.42719
1.47	2.1609	1.21244	3.83406	1.97	3.8809	1.40357	4.43847
1.48	2.1904	1.21655	3.84708	1.98	3.9204	1.40712	4.44972
1.49	2.2201	1.22066	3.86005	1.99	3.9601	1.41067	4.46094
1.50	2.2500	1.22474	3.87298	**2.00**	4.0000	1.41421	4.47214
N	N²	√N̄	√10N̄	N	N²	√N̄	√10N̄

N	N²	√N̄	√10N̄		N	N²	√N̄	√10N̄
2.00	4.0000	1.41421	4.47214		**2.50**	6.2500	1.58114	5.00000
2.01	4.0401	1.41774	4.48330		2.51	6.3001	1.58430	5.00999
2.02	4.0804	1.42127	4.49444		2.52	6.3504	1.58745	5.01996
2.03	4.1209	1.42478	4.50555		2.53	6.4009	1.59060	5.02991
2.04	4.1616	1.42829	4.51664		2.54	6.4516	1.59374	5.03984
2.05	4.2025	1.43178	4.52769		2.55	6.5025	1.59687	5.04975
2.06	4.2436	1.43527	4.53872		2.56	6.5536	1.60000	5.05964
2.07	4.2849	1.43875	4.54973		2.57	6.6049	1.60312	5.06952
2.08	4.3264	1.44222	4.56070		2.58	6.6564	1.60624	5.07937
2.09	4.3681	1.44568	4.57165		2.59	6.7081	1.60935	5.08920
2.10	4.4100	1.44914	4.58258		**2.60**	6.7600	1.61245	5.09902
2.11	4.4521	1.45258	4.59347		2.61	6.8121	1.61555	5.10882
2.12	4.4944	1.45602	4.60435		2.62	6.8644	1.61864	5.11859
2.13	4.5369	1.45945	4.61519		2.63	6.9169	1.62173	5.12835
2.14	4.5796	1.46287	4.62601		2.64	6.9696	1.62481	5.13809
2.15	4.6225	1.46629	4.63681		2.65	7.0225	1.62788	5.14782
2.16	4.6656	1.46969	4.64758		2.66	7.0756	1.63095	5.15752
2.17	4.7089	1.47309	4.65833		2.67	7.1289	1.63401	5.16720
2.18	4.7524	1.47648	4.66905		2.68	7.1824	1.63707	5.17687
2.19	4.7961	1.47986	4.67974		2.69	7.2361	1.64012	5.18652
2.20	4.8400	1.48324	4.69042		**2.70**	7.2900	1.64317	5.19615
2.21	4.8841	1.48661	4.70106		2.71	7.3441	1.64621	5.20577
2.22	5.9284	1.48997	4.71169		2.72	7.3984	1.64924	5.21536
2.23	4.9729	1.49332	4.72229		2.73	7.4529	1.65227	5.22494
2.24	5.0176	1.49666	4.73286		2.74	7.5076	1.65529	5.23450
2.25	5.0625	1.50000	4.74342		2.75	7.5625	1.65831	5.24404
2.26	5.1076	1.50333	4.75395		2.76	7.6176	1.66132	5.25357
2.27	5.1529	1.50665	4.76445		2.77	7.6729	1.66433	5.26308
2.28	5.1984	1.50997	4.77493		2.78	7.7284	1.66733	5.27257
2.29	5.2441	1.51327	4.78539		2.79	7.7841	1.67033	5.28205
2.30	5.2900	1.51658	4.79583		**2.80**	7.8400	1.67332	5.29150
2.31	5.3361	1.51987	4.80625		2.81	7.8961	1.67631	5.30094
2.32	5.3824	1.52315	4.81664		2.82	7.9524	1.67929	5.31037
2.33	5.4289	1.52643	4.82701		2.83	8.0089	1.68226	5.31977
2.34	5.4756	1.52971	4.83735		2.84	8.0656	1.68523	5.32917
2.35	5.5225	1.53297	4.84768		2.85	8.1225	1.68819	5.33854
2.36	5.5696	1.53623	4.85798		2.86	8.1796	1.69115	5.34790
2.37	5.6169	1.53948	4.86826		2.87	8.2369	1.69411	5.35724
2.38	5.6644	1.54272	4.87852		2.88	8.2944	1.69706	5.36656
2.39	5.7121	1.54596	4.88876		2.89	8.3521	1.70000	5.37587
2.40	5.7600	1.54919	4.89898		**2.90**	8.4100	1.70294	5.38516
2.41	5.8081	1.55252	4.90918		2.91	8.4681	1.70587	5.39444
2.42	5.8564	1.55563	4.91935		2.92	8.5264	1.70880	5.40370
2.43	5.9049	1.55885	4.92950		2.93	8.5849	1.71172	5.41295
2.44	5.9536	1.56205	4.93964		2.94	8.6436	1.71464	5.42218
2.45	6.0025	1.56525	4.94975		2.95	8.7025	1.71756	5.43139
2.46	6.0516	1.56844	4.95984		2.96	8.7616	1.72047	5.44059
2.47	6.1009	1.57162	4.96991		2.97	8.8209	1.72337	5.44977
2.48	6.1054	1.57480	4.97996		2.98	8.8804	1.72627	5.45894
2.49	6.2001	1.57797	4.98999		2.99	8.9401	1.72916	5.46809
2.50	6.2500	1.58114	5.00000		**3.00**	9.0000	1.73205	5.47723
N	N²	√N̄	√10N̄		N	N²	√N̄	√10N̄

N	N²	√N	√10N		N	N²	√N	√10N
3.00	9.0000	1.73205	5.47723		**3.50**	12.2500	1.87083	5.91608
3.01	9.0601	1.73494	5.48635		3.51	12.3201	1.87350	5.92453
3.02	9.1204	1.73781	5.49545		3.52	12.3904	1.87617	5.93296
3.03	9.1809	1.74069	5.50454		3.53	12.4609	1.87883	5.94138
3.04	9.2416	1.74356	5.51362		3.54	12.5316	1.88149	5.94979
3.05	9.3025	1.74642	5.52268		3.55	12.6025	1.88414	5.95819
3.06	9.3636	1.74929	5.53173		3.56	12.6736	1.88680	5.96657
3.07	9.4249	1.75214	5.54076		3.57	12.7449	1.88944	5.97495
3.08	9.4864	1.75499	5.54977		3.58	12.8164	1.89209	5.98331
3.09	9.5481	1.75784	5.55878		3.59	12.8881	1.89473	5.99166
3.10	9.6100	1.76068	5.56776		**3.60**	12.9600	1.89737	6.00000
3.11	9.6721	1.76352	5.57674		3.61	13.0321	1.90000	6.00833
3.12	9.7344	1.76635	5.58570		3.62	13.1044	1.90263	6.01664
3.13	9.7969	1.76918	5.59464		3.63	13.1769	1.90526	6.02495
3.14	9.8596	1.77200	5.60357		3.64	13.2496	1.90788	6.03324
3.15	9.9225	1.77482	5.61249		3.65	13.3225	1.91050	6.04152
3.16	9.9856	1.77764	5.62139		3.66	13.3956	1.91311	6.04949
3.17	10.0489	1.78045	5.63028		3.67	13.4689	1.91572	6.05805
3.18	10.1124	1.78326	5.63915		3.68	13.5424	1.91833	6.06630
3.19	10.1761	1.78606	5.64801		3.69	13.6161	1.92094	6.07454
3.20	10.2400	1.78885	5.65685		**3.70**	13.6900	1.92354	6.08276
3.21	10.3041	1.79165	5.66569		3.71	13.7641	1.92614	6.09098
3.22	10.3684	1.79444	5.67450		3.72	13.8384	1.92873	6.09918
3.23	10.4329	1.79722	5.68331		3.73	13.9129	1.93132	6.10737
3.24	10.4976	1.80000	5.69210		3.74	13.9876	1.93391	6.11555
3.25	10.5625	1.80278	5.70088		3.75	14.0625	1.93649	6.12372
3.26	10.6276	1.80555	5.70964		3.76	14.1376	1.93907	6.13188
3.27	10.6929	1.80831	5.71839		3.77	14.2129	1.94165	6.14003
3.28	10.7584	1.81108	5.72713		3.78	14.2884	1.94422	6.14817
3.29	10.8241	1.81384	5.73585		3.79	14.3641	1.94679	6.15630
3.30	10.8900	1.81659	5.74456		**3.80**	14.4400	1.94936	6.16441
3.31	10.9561	1.81934	5.75326		3.81	14.5161	1.95192	6.17252
3.32	11.0224	1.82209	5.76194		3.82	14.5924	1.95448	6.18061
3.33	11.0889	1.82483	5.77062		3.83	14.6689	1.95704	6.18870
3.34	11.1556	1.82757	5.77927		3.84	14.7456	1.95959	6.19677
3.35	11.2225	1.83030	5.78792		3.85	14.8225	1.96214	6.20484
3.36	11.2896	1.83303	5.79655		3.86	14.8996	1.96469	6.21289
3.37	11.3569	1.83576	5.80517		3.87	14.9769	1.96723	6.22093
3.38	11.4244	1.83848	5.81378		3.88	15.0544	1.96977	6.22896
3.39	11.4921	1.84120	5.82237		3.89	15.1321	1.97231	6.23699
3.40	11.5600	1.84391	5.83095		**3.90**	51.2100	1.97484	6.24500
3.41	11.6281	1.84662	5.83952		3.91	15.2881	1.97737	6.25300
3.42	11.6964	1.84932	5.84808		3.92	15.3664	1.97990	6.26099
3.43	11.7649	1.85203	5.85662		3.93	15.4449	1.98242	6.26897
3.44	11.8336	1.85472	5.86515		3.94	15.5236	1.98494	6.27694
3.45	11.9025	1.85742	5.87367		3.95	15.6025	1.98746	6.28490
3.46	11.9716	1.86011	5.88218		3.96	15.6816	1.98997	6.29285
3.47	12.0409	1.86279	5.89067		3.97	15.7609	1.99249	6.30079
3.48	12.1104	1.86548	5.89915		3.98	15.8404	1.99499	6.30872
3.49	12.1801	1.86815	5.90762		3.99	15.9201	1.99750	6.31644
3.50	12.2500	1.87083	5.91608		**4.00**	16.0000	2.00000	6.32456
N	N²	√N	√10N		N	N²	√N	√10N

N	N²	√N	√10N		N	N²	√N	√10N
4.00	16.0000	2.00000	6.32456		**4.50**	20.2500	2.12132	6.70820
4.01	16.0801	2.00250	6.33246		4.51	20.3401	2.12368	6.71565
4.02	16.1604	2.00499	6.34035		4.52	20.4304	2.12603	6.72309
4.03	16.2409	2.00749	6.34823		4.53	20.5209	2.12838	6.73053
4.04	16.3216	2.00998	6.35610		4.54	20.6116	2.13073	6.73795
4.05	16.4025	2.01246	6.36396		4.55	20.7025	2.13307	6.74537
4.06	16.4836	2.01494	6.37181		4.56	20.7936	2.13542	6.75278
4.07	16.5649	2.01742	6.37966		4.57	20.8849	2.13776	6.76018
4.08	16.6464	2.01990	6.38749		4.58	20.9764	2.14009	6.76757
4.09	16.7281	2.02237	6.39531		4.59	21.0681	2.14243	6.77495
4.10	16.8100	2.02485	6.40312		**4.60**	21.1600	2.14476	6.78233
4.11	16.8921	2.02731	6.41093		4.61	21.2521	2.14709	6.78970
4.12	16.9744	2.02978	6.41872		4.62	21.3444	2.14942	6.79706
4.13	17.0569	2.03224	6.42651		4.63	21.4369	2.15174	6.80441
4.14	17.1396	2.03470	6.43428		4.64	21.5296	2.15407	6.81175
4.15	17.2225	2.03715	6.44205		4.65	21.6225	2.15639	6.81909
4.16	17.3056	2.03961	6.44981		4.66	21.7156	2.15870	6.82642
4.17	17.3889	2.04206	6.45755		4.67	21.8089	2.16102	6.83374
4.18	17.4724	2.04450	6.46529		4.68	21.9024	2.16333	6.84105
4.19	17.5561	2.04695	6.47302		4.69	21.9961	2.16564	6.84836
4.20	17.6400	2.04939	6.48074		**4.70**	22.0900	2.16795	6.85565
4.21	17.7241	2.05183	6.48845		4.71	22.1841	2.17025	6.86294
4.22	17.8084	2.05426	6.49615		4.72	22.2784	2.17256	6.87023
4.23	17.8929	2.05670	6.50384		4.73	22.3729	2.17486	6.87750
4.24	17.9776	2.05913	6.51153		4.74	22.4676	2.17715	6.88477
4.25	18.0625	2.06155	6.51920		4.75	22.5625	2.17945	6.89202
4.26	18.1476	2.06398	6.52687		4.76	22.6576	2.18174	6.89928
4.27	18.2329	2.06640	6.53452		4.77	22.7529	2.18403	6.90652
4.28	18.3184	2.06882	6.54217		4.78	22.8484	2.18632	6.91375
4.29	18.4041	2.07123	6.54981		4.79	22.9441	2.18861	6.92098
4.30	18.4900	2.07364	6.55744		**4.80**	23.0400	2.19089	6.92820
4.31	18.5761	2.07605	6.66506		4.81	23.1361	2.19317	6.93542
4.32	18.6624	2.07846	6.57267		4.82	23.2324	2.19545	6.94262
4.33	18.7489	2.08087	6.58027		4.83	23.3289	2.19773	6.94982
4.34	18.8356	2.08327	6.58787		4.84	23.4256	2.20000	6.95701
4.35	18.9225	2.08567	6.59545		4.85	23.5225	2.20227	6.96419
4.36	19.0096	2.08806	6.60303		4.86	23.6196	2.20454	6.97137
4.37	19.0969	2.09045	6.61060		4.87	23.7169	2.20681	6.97854
4.38	19.1844	2.09284	6.61816		4.88	23.8144	2.20907	6.98570
4.39	19.2721	2.09523	6.62571		4.89	23.9121	2.21133	6.99285
4.40	19.3600	2.09762	6.63325		**4.90**	24.0100	2.21359	7.00000
4.41	19.4481	2.10000	6.64078		4.91	24.1081	2.21585	7.00714
4.42	19.5364	2.10238	6.64831		4.92	24.2064	2.21811	7.01427
4.43	19.6249	2.10476	6.65582		4.93	24.3049	2.22036	7.02140
4.44	19.7136	2.10713	6.66333		4.94	24.4036	2.22261	7.02851
4.45	19.8025	2.10950	6.67083		4.95	24.5025	2.22486	7.03562
4.46	19.8916	2.11187	6.67832		4.96	24.6016	2.22711	7.04273
4.47	19.9809	2.11424	6.68581		4.97	24.7009	2.22935	7.04982
4.48	20.0704	2.11660	6.69328		4.98	24.8004	2.23159	7.05691
4.49	20.1601	2.11896	6.70075		4.99	24.9001	2.23383	7.06399
4.50	20.2500	2.12132	6.70820		**5.00**	25.0000	2.23607	7.07107
N	N²	√N	√10N		N	N²	√N	√10N

N	N^2	$\sqrt{N}$	$\sqrt{10N}$	N	N^2	$\sqrt{10}$	$\sqrt{10N}$
5.00	25.0000	2.23607	7.07107	**5.50**	30.2500	2.34521	7.41620
5.01	25.1001	2.23830	7.07814	5.51	30.3601	2.34734	7.42294
5.02	25.2004	2.24054	7.08520	5.52	30.4704	2.34947	7.42967
5.03	25.3009	2.24277	7.09225	5.53	30.5809	2.35160	7.43640
5.04	25.4016	2.24499	7.09930	5.54	30.6916	2.35372	7.44312
5.05	25.5025	2.24722	7.10634	5.55	30.8025	2.35584	7.44983
5.06	25.6036	2.24944	7.11337	5.56	30.9136	2.35797	7.45654
5.07	25.7049	2.25167	7.12039	5.57	31.0249	2.36008	7.46324
5.08	25.8064	2.25389	7.12741	5.58	31.1364	2.36220	7.46994
5.09	25.9081	2.25610	7.13442	5.59	31.2481	2.36432	7.47663
5.10	26.0100	2.25832	7.14143	**5.60**	31.3600	2.36643	7.48331
5.11	26.1121	2.26053	7.14843	5.61	31.4721	2.36854	7.48999
5.12	26.2144	2.26274	7.15542	5.62	31.5844	2.37065	7.49667
5.13	26.3169	2.26495	7.16240	5.63	31.6969	2.37276	7.50333
5.14	26.4196	2.26716	7.16938	5.64	31.8096	2.37487	7.50999
5.15	26.5225	2.26936	7.17635	5.65	31.9225	2.37697	7.51665
5.16	26.6256	2.27156	7.18331	5.66	32.0356	2.37908	7.52330
5.17	26.7289	2.27376	7.19027	5.67	32.1489	2.38118	7.52994
5.18	26.8324	2.27596	7.19722	5.68	32.2624	2.38328	7.53658
5.19	26.9361	2.27816	7.20417	5.68	32.3761	2.38537	7.54321
5.20	27.0400	2.28035	7.21110	**5.70**	32.4900	2.38747	7.54983
5.21	27.1441	2.28254	7.21803	5.71	32.6041	2.38956	7.55645
5.22	27.2484	2.28473	7.22496	5.72	32.7184	2.39165	7.56307
5.23	27.3529	2.28692	7.23187	5.73	32.8329	2.39374	7.56968
5.24	27.4576	2.28910	7.23838	5.74	32.9476	2.39583	7.57628
5.25	27.5625	2.29129	7.24569	5.75	33.0625	2.39792	7.58288
5.26	27.6676	2.29347	7.25259	5.76	33.1776	2.40000	7.58947
5.27	27.7729	2.29565	7.25948	5.77	33.2929	2.40208	7.59605
5.28	27.8784	2.29783	7.26636	5.78	33.4084	2.40416	7.60263
5.29	27.9841	2.30000	7.27324	5.79	33.5241	2.40624	7.60920
5.30	28.0900	2.30217	7.28011	**5.80**	33.6400	2.40832	7.61577
5.31	28.1961	2.30434	7.28697	5.81	33.7561	2.41039	7.62234
5.32	28.3024	2.30651	7.29383	5.82	33.8724	2.41247	7.62889
5.33	28.4089	2.30868	7.30068	5.83	33.9889	2.41454	7.63544
5.34	28.5156	2.31084	7.30753	5.84	34.1056	2.41661	7.64199
5.35	28.6225	2.31301	7.31437	5.85	34.2225	2.41868	7.64853
5.36	28.7296	2.31517	7.32120	5.86	34.3396	2.42074	7.65506
5.37	28.8369	2.31733	7.32803	5.87	34.4569	2.42281	7.66159
5.38	28.9444	2.31948	7.33485	5.88	34.5744	2.42487	7.66812
5.39	29.0521	2.32164	7.34166	5.89	34.6921	2.42693	7.67463
5.40	29.1600	2.32379	7.34847	**5.90**	34.8100	2.42899	7.68115
5.41	29.2681	2.32594	7.35527	5.91	34.9281	2.43105	7.68765
5.42	29.3764	2.32809	7.36206	5.92	35.0464	2.43311	7.69415
5.43	29.4849	2.33024	7.36885	5.93	35.1649	2.43516	7.70065
5.44	29.5936	2.33238	7.37564	5.94	35.2836	2.43721	7.70714
5.45	29.7025	2.33452	7.38241	5.95	35.4025	2.43926	7.71362
5.46	29.8116	2.33666	7.38918	5.96	35.5216	2.44131	7.72010
5.47	29.9209	2.33880	7.39594	5.97	35.6409	2.44336	7.72658
5.48	30.0304	2.34094	7.40270	5.98	35.7604	2.44540	7.73305
5.49	30.1401	2.34307	7.40945	5.99	35.8801	2.44745	7.73951
5.50	30.2500	2.34521	7.41620	**6.00**	36.0000	2.44949	7.74597
N	N^2	$\sqrt{N}$	$\sqrt{10N}$	N	N^2	$\sqrt{N}$	$\sqrt{10N}$

N	N²	√N̄	√1̄0̄N̄	N	N²	√N̄	√1̄0̄N̄
6.00	36.0000	2.44949	7.74597	**6.50**	42.2500	2.54951	8.06226
6.01	36.1201	2.45153	7.75242	6.51	42.3801	2.55147	8.06846
6.02	36.2404	2.45357	7.75887	6.52	42.5104	2.55343	8.07465
6.03	36.3609	2.45561	7.76531	6.53	42.6409	2.55539	8.08084
6.04	36.4816	2.45764	7.77174	6.54	42.7716	2.55734	8.08703
6.05	36.6025	2.45967	7.77817	6.55	42.9025	2.55930	8.09321
6.06	36.7236	2.46171	7.78460	6.56	43.0336	2.56125	8.09938
6.07	36.8449	2.46374	7.79102	6.57	43.1649	2.56320	8.10555
6.08	36.9664	2.46577	7.79744	6.58	43.2964	2.56515	8.11172
6.09	37.0881	2.46779	7.80385	6.59	43.4281	2.56710	8.11788
6.10	37.2100	2.46982	7.81025	**6.60**	43.5600	2.56905	8.12404
6.11	37.3321	2.47184	7.81665	6.61	43.6921	2.57099	8.13019
6.12	37.4544	2.47386	7.82304	6.62	43.8244	2.57294	8.13634
6.13	37.5769	2.47588	7.82943	6.63	43.9569	2.57488	8.14248
6.14	37.6996	2.47790	7.83582	6.64	44.0896	2.57682	8.14862
6.15	37.8225	2.47992	7.84219	6.65	44.2225	2.57876	8.15475
6.16	37.9456	2.48193	7.84857	6.66	44.3556	2.58070	8.16088
6.17	38.0689	2.48395	7.85493	6.67	44.4889	2.58263	8.16701
6.18	38.1924	2.48596	7.86130	6.68	44.6224	2.58457	8.17313
6.19	38.3161	2.48797	7.86766	6.69	44.7561	2.58650	8.17924
6.20	38.4400	2.48998	7.87401	**6.70**	44.8900	2.58844	8.18535
6.21	38.5641	2.49199	7.88036	6.71	45.0241	2.59037	8.19146
6.22	38.6884	2.49399	7.88670	6.72	45.1584	2.59230	8.19756
6.23	38.8129	2.49600	7.89303	6.73	45.2929	2.59422	8.20366
6.24	38.9376	2.49800	7.89937	6.74	45.4276	2.59615	8.20975
6.25	39.0625	2.50000	7.90569	6.75	45.5625	2.59808	8.21584
6.26	39.1876	2.50200	7.91202	6.76	45.6976	2.60000	8.22192
6.27	39.3129	2.50400	7.91833	6.77	45.8329	2.60192	8.22800
6.28	39.4384	2.50599	7.92465	6.78	45.9684	2.60384	8.23408
6.29	39.5641	2.50799	7.93095	6.79	46.1041	2.60576	8.24015
6.30	39.6900	2.50998	7.93725	**6.80**	46.2400	2.60768	8.24621
6.31	39.8161	2.51197	7.94355	6.81	46.3761	2.60960	8.25227
6.32	39.9424	2.51396	7.94984	6.82	46.5124	2.61151	8.25833
6.33	40.0689	2.51595	7.95613	6.83	46.6489	2.61343	8.26438
6.34	40.1956	2.51794	7.96241	6.84	46.7856	2.61534	8.27043
6.35	40.3225	2.51992	7.96869	6.85	46.9225	2.61725	8.27647
6.36	40.4496	2.52190	7.97496	6.86	47.0596	2.61916	8.28251
6.37	40.5769	2.52389	7.98123	6.87	47.1969	2.62107	8.28855
6.38	40.7044	2.52587	7.98749	6.88	47.3344	2.62298	8.29458
6.39	40.8321	2.52784	7.99375	6.89	47.4721	2.62488	8.30060
6.40	40.9600	2.52982	8.00000	**6.90**	47.6100	2.62679	8.30662
6.41	41.0881	2.53180	8.00625	6.91	47.7481	2.62869	8.31264
6.42	41.2164	2.53377	8.01249	6.92	47.8864	2.63059	8.31865
6.43	41.3449	2.53574	8.01873	6.93	48.0249	2.63249	8.32466
6.44	41.4736	2.53772	8.02496	6.94	48.1636	2.63439	8.33067
6.45	41.6025	2.53969	8.03119	6.95	48.3025	2.63629	8.33667
6.46	41.7316	2.54165	8.03741	6.96	48.4416	2.63818	8.34266
6.47	41.8609	2.54362	8.04363	6.97	48.5809	2.64008	8.34865
6.48	41.9904	2.54558	8.04984	6.98	48.7204	2.64197	8.35464
6.49	42.1201	2.54755	8.05605	6.99	48.8601	2.64386	8.36062
6.50	42.2500	2.54951	8.06226	**7.00**	49,0000	2.64575	8.36660
N	N²	√N̄	√1̄0̄N̄	N	N²	√N̄	√1̄0̄N̄

N	N^2	$\sqrt{N}$	$\sqrt{10N}$	N	N^2	$\sqrt{N}$	$\sqrt{10N}$
7.00	49.0000	2.64575	8.36660	**7.50**	56.2500	2.73861	8.66025
7.01	49.1401	2.64764	8.37257	7.51	56.4001	2.74044	8.66603
7.02	49.2804	2.64953	8.37854	7.52	56.5504	2.74226	8.67179
7.03	49.4209	2.65141	8.38451	7.53	56.7009	2.74408	8.67756
7.04	49.5616	2.65330	8.39047	7.54	56.8516	2.74591	8.68332
7.05	49.7025	2.65518	8.39643	7.55	57.0025	2.74773	8.68907
7.06	49.8436	2.65707	8.40238	7.56	57.1536	2.74955	8.69483
7.07	49.9849	2.65895	8.40833	7.57	57.3049	2.75136	8.70057
7.08	50.1264	2.66083	8.41427	7.58	57.4564	2.75318	8.70632
7.09	50.2681	2.66271	8.42021	7.59	57.6081	2.75500	8.71206
7.10	50.4100	2.66458	8.42615	**7.60**	57.7600	2.75681	8.71780
7.11	50.5521	2.66646	8.43208	7.61	57.9121	2.75862	8.72353
7.12	50.6944	2.66833	8.43801	7.62	58.0644	2.76043	8.72926
7.13	50.8369	2.67021	8.44393	7.63	58.2169	2.76225	8.73499
7.14	50.9796	2.67208	8.44985	7.64	58.3696	2.76405	8.74071
7.15	51.1225	2.67395	8.45577	7.65	58.5225	2.76586	8.74643
7.16	51.2656	2.67582	8.46168	7.66	58.6756	2.76767	8.75214
7.17	51.4089	2.67769	8.46759	7.67	58.8289	2.76948	8.75785
7.18	51.5524	2.67955	8.47349	7.68	58.9824	2.77128	8.76356
7.19	51.6961	2.68142	8.47939	7.69	59.1361	2.77308	8.76926
7.20	51.8400	2.68328	8.48528	**7.70**	59.2900	2.77489	8.77496
7.21	51.9841	2.68514	8.49117	7.71	59.4441	2.77669	8.78066
7.22	52.1284	2.68701	8.49706	7.72	59.5984	2.77849	8.78635
7.23	52.2729	2.68887	8.50294	7.73	59.7529	2.78029	8.79204
7.24	52.4176	2.69072	8.50882	7.74	59.9076	2.78209	8.79773
7.25	52.5625	2.69258	8.51469	7.75	60.0625	2.78388	8.80341
7.26	52.7076	2.69444	8.52056	7.76	60.2176	2.78568	8.80909
7.27	52.8529	2.69629	8.52643	7.77	60.3729	2.78747	8.81476
7.28	52.9984	2.69815	8.53229	7.78	60.5284	2.78927	8.82043
7.29	53.1441	2.70000	8.53815	7.79	60.6841	2.79106	8.82610
7.30	53.2900	2.70185	8.54400	**7.80**	60.8400	2.79285	8.83176
7.31	53.4361	2.70370	8.54985	7.81	60.9961	2.79464	8.83742
7.32	53.5824	2.70555	8.55570	7.82	61.1524	2.79643	8.84308
7.33	53.7289	2.70740	8.56154	7.83	61.3089	2.79821	8.84873
7.34	53.8756	2.70924	8.56738	7.84	61.4656	2.80000	8.85438
7.35	54.0225	2.71109	8.57321	7.85	61.6225	2.80179	8.86002
7.36	54.1696	2.71293	8.57904	7.86	61.7796	2.80357	8.86566
7.37	54.3169	2.71477	8.58487	7.87	61.9369	2.80535	8.87130
7.38	54.4644	2.71662	8.59069	7.88	62.0944	2.80713	8.87694
7.39	54.6121	2.71846	8.59651	7.89	62.2521	2.80891	8.88257
7.40	54.7600	2.72029	8.60233	**7.90**	62.4100	2.81069	8.88819
7.41	54.9081	2.72213	8.60814	7.91	62.5681	2.81247	8.89382
7.42	55.0564	2.72397	8.61394	7.92	62.7264	2.81425	8.89944
7.43	55.2049	2.72580	8.61974	7.93	62.8849	2.81603	8.90505
7.44	55.3536	2.72764	8.62554	7.94	63.0436	2.81780	8.91067
7.45	55.5025	2.72947	8.63134	7.95	63.2025	2.81957	8.91628
7.46	55.6516	2.73130	8.63713	7.96	63.3616	2.82135	8.92188
7.47	55.8009	2.73313	8.64292	7.97	63.5209	2.82312	8.92749
7.48	55.9504	2.73496	8.64870	7.98	63.6804	2.82489	8.93308
7.49	56.1001	2.73679	8.65448	7.99	63.8401	2.82666	8.93868
7.50	56.2500	2.73861	8.66025	**8.00**	64.0000	2.82843	8.94427
N	N^2	$\sqrt{N}$	$\sqrt{10N}$	N	N^2	$\sqrt{N}$	$\sqrt{10N}$

N	N²	√N	√10N		N	N²	√N	√10N
8.00	64.0000	2.82843	8.94427		**8.50**	72.2500	2.91548	9.21954
8.01	64.1601	2.83019	8.94986		8.51	72.4201	2.91719	9.22497
8.02	64.3204	2.83196	8.95545		8.52	72.5904	2.91890	9.23038
8.03	64.4809	2.83373	8.96103		8.53	72.7609	2.92062	9.23580
8.04	64.6416	2.83549	8.96660		8.54	72.9316	2.92233	9.24121
8.05	64.8025	2.83725	8.97218		8.55	73.1025	2.92404	9.24662
8.06	64.9636	2.83901	8.97775		8.56	73.2736	2.92575	9.25203
8.07	65.1249	2.84077	8.98332		8.57	73.4449	2.92746	9.25743
8.08	65.2864	2.84253	8.98888		8.58	73.6164	2.92916	9.26283
8.09	65.4481	2.84429	8.99444		8.59	73.7881	2.93087	9.26823
8.10	65.6100	2.84605	9.00000		**8.60**	73.9600	2.93258	9.27362
8.11	65.7721	2.84781	9.00555		8.61	74.1321	2.93428	9.27901
8.12	65.9344	2.84956	9.01110		8.62	74.3044	2.93598	9.28440
8.13	66.0969	2.85132	9.01665		8.63	74.4769	2.93769	9.28978
8.14	66.2596	2.85307	9.02219		8.64	74.6496	2.93939	9.29516
8.15	66.4225	2.85482	9.02774		8.65	74.8225	2.94109	9.30054
8.16	66.5856	2.85657	9.03327		8.66	74.9956	2.94279	9.30591
8.17	66.7489	2.85832	9.03881		8.67	75.1689	2.94449	9.31128
8.18	66.9124	2.86007	9.04434		8.68	75.3424	2.94618	9.31665
8.19	67.0761	2.86182	9.04986		8.69	75.5161	2.94788	9.32202
8.20	67.2400	2.86356	9.05539		**8.70**	75.6900	2.94958	9.32738
8.21	67.4041	2.86531	9.06091		8.71	75.8641	2.95127	9.33274
8.22	67.5684	2.86705	9.06642		8.72	76.0384	2.95296	9.33809
8.23	67.7329	2.86880	9.07193		8.73	76.2129	2.95466	9.34345
8.24	67.8976	2.87054	9.07744		8.74	76.3876	2.95635	9.34880
8.25	68.0625	2.87228	9.08295		8.75	76.5625	2.95804	9.35414
8.26	68.2276	2.87402	9.08845		8.76	76.7376	2.95973	9.35949
8.27	68.3929	2.87576	9.09395		8.77	76.9129	2.96142	9.36483
8.28	68.5584	2.87750	9.09945		8.78	77.0884	2.96311	9.37017
8.29	68.7241	2.87924	9.10494		8.79	77.2641	2.96479	9.37550
8.30	68.8900	2.88097	9.11045		**8.80**	77.4400	2.96648	9.38083
8.31	69.0561	2.88271	9.11592		8.81	77.6161	2.96816	9.38616
8.32	69.2224	2.88444	9.12140		8.82	77.7924	2.96985	9.39149
8.33	69.3889	2.88617	9.12688		8.83	77.9689	2.97153	9.39681
8.34	69.5556	2.88791	9.13236		8.84	78.1456	2.97321	9.40213
8.35	69.7225	2.88964	9.13783		8.85	78.3225	2.97489	9.40744
8.36	69.8896	2.89137	9.14330		8.86	78.4996	2.97658	9.41276
8.37	70.0569	2.89310	9.14877		8.87	78.6769	2.97825	9.41807
8.38	70.2244	2.89482	9.15423		8.88	78.8544	2.97993	9.42338
8.39	70.3921	2.89655	9.15969		8.89	79.0321	2.98161	9.42868
8.40	70.5600	2.89828	9.16515		**8.90**	79.2100	2.98329	9.43398
8.41	70.7281	2.90000	9.17061		8.91	79.3881	2.98496	9.43928
8.42	70.8964	2.90172	9.17606		8.92	79.5664	2.98664	9.44458
8.43	71.0649	2.90345	9.18150		8.93	79.7449	2.98831	9.44987
8.44	71.2336	2.90517	9.18695		8.94	79.9236	2.98998	9.45516
8.45	71.4025	2.90689	9.19239		8.95	80.1025	2.99166	9.46044
8.46	71.5716	2.90861	9.19783		8.96	80.2816	2.99333	9.46573
8.47	71.7409	2.91033	9.20326		8.97	80.4609	2.99500	9.47101
8.48	71.9104	2.91204	9.20869		8.98	80.6404	2.99666	9.47629
8.49	72.0801	2.91376	9.21412		8.99	80.8201	2.99833	9.48156
8.50	72.2500	2.91548	9.21954		**9.00**	81.0000	3.00000	9.48683
N	N²	√N	√10N		N	N²	√N	√10N

N	N²	√N	√10N	N	N²	√N	√10N
9.00	81.0000	3.00000	9.48683	**9.50**	90.2500	3.08221	9.74679
9.01	81.1801	3.00167	9.49210	9.51	90.4401	3.08383	9.75192
9.02	81.3604	3.00333	9.49737	9.52	90.6304	3.08545	9.75705
9.03	81.5409	3.00500	9.50263	9.53	90.8209	3.08707	9.76217
9.04	81.7216	3.00666	9.50789	9.54	91.0116	3.08869	9.76729
9.05	81.9025	3.00832	9.51315	9.55	91.2025	3.09031	9.77241
9.06	82.0836	3.00998	9.51840	9.56	91.3936	3.09192	9.77753
9.07	82.2649	3.01164	9.52365	9.57	91.5849	3.09354	9.78264
9.08	82.4464	3.01330	9.52890	9.58	91.7764	3.09516	9.78775
9.09	82.6281	3.01496	9.53415	9.59	91.9681	3.09677	9.79285
9.10	82.8100	3.01662	9.53939	**9.60**	92.1600	3.09839	9.79796
9.11	82.9921	3.01828	9.54463	9.61	92.3521	3.10000	9.80306
9.12	83.1744	3.01993	9.54987	9.62	92.5444	3.10161	9.80816
9.13	83.3569	3.02159	9.55510	9.63	92.7369	3.10322	9.81326
9.14	83.5396	3.02324	9.56033	9.64	92.9296	3.10483	9.81835
9.15	83.7225	3.02490	9.56556	9.65	93.1225	3.10644	9.82344
9.16	83.9056	3.02655	9.57079	9.66	93.3156	3.10805	9.82853
9.17	84.0889	3.02820	9.57601	9.67	93.5089	3.10966	9.83362
9.18	84.2724	3.02985	9.58123	9.68	93.7024	3.11127	9.83870
9.19	84.4561	3.03150	9.58645	9.69	93.8961	3.11288	9.84378
9.20	84.6400	3.03315	9.59166	**9.70**	94.0900	3.11448	9.84886
9.21	84.8241	3.03480	9.59687	9.71	94.2841	3.11609	9.85393
9.22	85.0084	3.03645	9.60208	9.72	94.4784	3.11769	9.85901
9.23	85.1929	3.03809	9.60729	9.73	94.6729	3.11929	9.86408
9.24	85.3776	3.03974	9.61249	9.74	94.8676	3.12090	9.86914
9.25	85.5625	3.04138	9.61769	9.75	95.0625	3.12250	9.87421
9.26	85.7476	3.04302	9.62289	9.76	95.2576	3.12410	9.87927
9.27	85.9329	3.04467	9.62808	9.77	95.4529	3.12570	9.88433
9.28	86.1184	3.04631	9.63328	9.78	95.6484	3.12730	9.88939
9.29	86.3041	3.04795	9.63846	9.79	95.8441	3.12890	9.89444
9.30	86.4900	3.04959	9.64365	**9.80**	96.0400	3.13050	9.89949
9.31	86.6761	3.05123	9.64883	9.81	96.2361	3.13209	9.90454
9.32	86.8624	3.05287	9.65401	9.82	96.4324	3.13369	9.90959
9.33	87.0489	3.05450	9.65919	9.83	96.6289	3.13528	9.91464
9.34	87.2356	3.05614	9.66437	9.84	96.8256	3.13688	9.91968
9.35	87.4225	3.05778	9.66954	9.85	97.0225	3.13847	9.92472
9.36	87.6096	3.05941	9.67471	9.86	97.2196	3.14006	9.92975
9.37	87.7969	3.06105	9.67988	9.87	97.4169	3.14166	9.93479
9.38	87.9844	3.06268	9.68504	9.88	97.6144	3.14325	9.93982
9.39	88.1721	3.06431	9.69020	9.89	97.8121	3.14484	9.94485
9.40	88.3600	3.06594	9.69536	**9.90**	98.0100	3.14643	9.94987
9.41	88.5481	3.06757	9.70052	9.91	98.2081	3.14802	9.95490
9.42	88.7364	3.06920	9.70567	9.92	98.4064	3.14960	9.95992
9.43	88.9249	3.07083	9.71082	9.93	98.6049	3.15119	9.96494
9.44	89.1136	3.07246	9.71597	9.94	98.8036	3.15278	9.96995
9.45	89.3025	3.07409	9.72111	9.95	99.0025	3.15436	9.97497
9.46	89.4916	3.07571	9.72625	9.96	99.2016	3.15595	9.97998
9.47	89.6809	3.07734	9.73139	9.97	99.4009	3.15753	9.98499
9.48	89.8704	3.07896	9.73653	9.98	99.6004	3.15911	9.98999
9.49	90.0601	3.08058	9.74166	9.99	99.8001	3.16070	9.99500
9.50	90.2500	3.08221	9.74679	**10.0**	100.000	3.16228	10.0000
N	N²	√N	√10N	N	N²	√N	√10N

Bibliography

Alam, K. and M. H. Rizvi (1966). Selection from multivariate normal populations. *Annals of the Institute of Statistical Mathematics*, **18**, 307–318.

Alam, K., M. H. Rizvi, and H. Solomon (1976). Selection of largest multiple correlation coefficients: Exact sample size case. *The Annals of Statistics*, **4**, 614–620.

Alam, K. and J. R. Thompson (1972). On selecting the least probable multinomial event. *The Annals of Mathematical Statistics*, **43**, 1981–1990.

Alam, K. and J. R. Thompson (1973). A problem of ranking and estimation with Poisson processes. *Technometrics*, **15**, 801–808.

Bahadur, R. R. (1950). On a problem in the theory of k populations. *The Annals of Mathematical Statistics*, **21**, 362–375.

Bahadur, R. R. and H. Robbins (1950). The problem of the greater mean. *The Annals of Mathematical Statistics*, **21**, 469–487.

Barr, D. R. and M. H. Rizvi (1966). Ranking and selection problems of uniform distributions. *Trabajos de Estadística*, **17** (Cuadernos 2 and 3), 15–31.

Bechhofer, R. E. (1954). A single-sample multiple decision procedure for ranking means of normal populations with known variances. *The Annals of Mathematical Statistics*, **25**, 16–39.

Bechhofer, R. E. (1958). A sequential multiple-decision procedure for selecting the best of several normal populations with a common unknown variance, and its use with various experimental designs. *Biometrics*, **14**, 408–429.

Bechhofer, R. E. (1970). On ranking the players in a 3-player tournament, in M. L. Puri, Ed., *Nonparametric Techniques in Statistical Inference*, Cambridge University Press, Cambridge, England, pp. 545–549.

Bechhofer, R. E., C. W. Dunnett, and M. Sobel (1954). A two-sample multiple decision procedure for ranking means of normal populations with a common unknown variance. *Biometrika*, **41**, 170–176.

Bechhofer, R. E., S. A. Elmaghraby, and N. Morse (1959). A single-sample multiple-decision procedure for selecting the multinomial event which has the largest probability. *The Annals of Mathematical Statistics*, **30**, 102–119.

Bechhofer, R. E., J. Kiefer, and M. Sobel (1968). *Sequential Identification and Ranking Procedures*. University of Chicago Press, Chicago, Illinois.

Bechhofer, R. E. and M. Sobel (1954). A single-sample multiple-decision procedure for ranking variances of normal populations. *The Annals of Mathematical Statistics*, **25**, 273–289.

Becker, W. A. (1961). Comparing entries in random sample tests. *Poultry Science*, **40**, 1507–1514.

Becker, W. A. (1962). Ranking all-or-none traits in random sample tests. *Poultry Science*, **41**, 1437–1438.

Becker, W. A. (1964). Changes in performance of entries in random sample tests. *Poultry Science*, **43**, 716–722.

Bishop, T. A. and E. Dudewicz (1976). Complete ranking of reliability related distributions. *Technical Report* 125, Department of Statistics, The Ohio State University, Columbus, Ohio.

Bland, R. P. and T. L. Bratcher (1968). Bayesian ranking binomial populations. *SIAM Journal of Applied Mathematics*, **16**, 843–850.

Blumenthal, S., T. Christie, and M. Sobel (1971). Best sequential tests for an identification problem with partially overlapping distributions. *Sankhyā, A*, **33**, 185–192.

Bratcher, T. L. and P. Bhalla (1974). On the properties of an optimal selection procedure. *Communications in Statistics*, **3**, 191–196.

Bross, I. D. J. (1949). Maximum utility statistics. *Ph.D. Dissertation*, North Carolina State College, Raleigh, North Carolina.

Bühlmann, H. and P. J. Huber (1963). Pairwise comparison and ranking in tournaments. *The Annals of Mathematical Statistics*, **34**, 501–510.

Cacoullos, T. and M. Sobel (1966). An inverse sampling procedure for selecting the most probable event in a multinomial distribution, in P. R. Krishnaiah, Ed., *Multivariate Analysis: Proceedings of an International Symposium*. Academic Press, New York, pp. 423–455.

Carroll, R. J. (1976). On the uniformity of sequential ranking procedures. *Institute of Statistics Mimeo Series* No. 1067, University of North Carolina, Chapel Hill, North Carolina.

Carroll, R. J. and S. S. Gupta (1975). On the probabilities of rankings of k populations with applications. *Institute of Statistics Mimeo Series* No. 1024, University of North Carolina, Chapel Hill; North Carolina. *Journal of Statistical Computation and Simulation* 5(1977), 145-157.

Carroll, R. J., S. S. Gupta, and D. Y. Huang (1975). On selection procedures for the t best populations and some related problems. *Communications in Statistics*, **4**, 987–1008.

Chen, H. J. (1974). Percentage points of multivariate t distribution and their application. *Technical Report* 74-14, Department of Mathematical Sciences, Memphis State University, Memphis, Tennessee.

Chen, H. J. and E. J. Dudewicz (1973). Interval estimation of a ranked mean of *k* normal populations with unequal sample size, *Technical Report* 73-21, Department of Mathematical Sciences, Memphis State University, Memphis, Tennessee.

Chen, H. J. and E. J. Dudewicz (1976). Procedures for fixed-width interval estimation of the largest normal mean. *Journal of the American Statistical Association*, **71**, 752–756.

Chen, H. J. and A. W. Montgomery (1975). A table for interval estimation of the largest mean of *k* normal populations. *Biometrische Zeitschrift*, **17**, 411–414.

Chernoff, H. and J. Yahav (1976). A subset selection problem employing a new criterion. *Technical Report* No. 4, Department of Mathematics, Massachusetts Institute of Technology, Cambridge, Massachusetts.

Chiu, W. K. (1974). Selecting the *n* populations with largest means from *k* normal populations with unknown variances. *Australian Journal of Statistics*, **16**, 144–147.

Chow, Y. S., S. Moriguti, H. Robbins, and E. Samuels (1964). The optimal selection based on relative rank (the "secretary" problem). *Israel Journal of Mathematics*, **2**, 81–90.

Chow, Y. S., H. Robbins, and D. Siegmund (1971). *Great Expectations: The Theory of Optimal Stopping*. Houghton Mifflin, Boston, Massachusetts.

Dalal, S. R. and V. Srinavasan (1976). Determining sample size for pre-testing comparative effectiveness of advertising copies. Unpublished manuscript. To appear in *Management Science*.

David, H. A. (1956). The ranking of variances in normal populations. *Journal of the American Statistical Association*, **51**, 621–626.

David, H. A. (1963). *The Method of Paired Comparisons*. Charles Griffin, London, England.

Deely, J. J. (1965). Multiple decision procedures from an empirical Bayes approach. *Ph.D. Thesis (Mimeo Series* No. 45), Department of Statistics, Purdue University, West Lafayette, Indiana.

Deely, J. J. and S. S. Gupta (1968). On the properties of subset selection procedures. *Sankhyā, A*, **30**, 37–50.

Desu, M. M. (1970). A selection problem. *The Annals of Mathematical Statistics*, **41**, 1596–1603.

Desu, M. M. and M. Sobel (1968). A fixed subset-size approach to a selection problem. *Biometrika*, **55**, 401–410.

Desu, M. M. and M. Sobel (1971). Nonparametric procedures for selecting fixed-size subsets, in S. S. Gupta and J. Yackel, Eds., *Statistical Decision Theory and Related Topics*. Academic Press, New York, pp. 255–273.

Deverman, J. N. (1969). A general selection procedure relative to *t* best populations. *Ph.D. Dissertation*, Department of Statistics, Purdue University, West Lafayette, Indiana.

Deverman, J. N. and S. S. Gupta (1969). On a selection procedure concerning the t best populations. *Technical Report*, Sandia Laboratories, Livermore, California.

Dudewicz, E. J. (1968). A categorized bibliography on multiple-decision (ranking and selection) procedures. Department of Statistics, University of Rochester, Rochester, New York.

Dudewicz, E. J. (1970). Confidence intervals for ranked means. *Naval Research Logistics Quarterly*, **17**, 69–78.

Dudewicz, E. J. (1972a). Confidence intervals for power, with special reference to medical trials. *The Australian Journal of Statistics*, **14**, 211–216.

Dudewicz, E. J. (1972b). Two-sided confidence intervals for ranked means. *Journal of the American Statistical Association*, **67**, 462–464.

Dudewicz, E. J. (1976). *Introduction to Statistics and Probability*. Holt, Rinehart and Winston, New York.

Dudewicz, E. J. and S. R. Dalal (1975). Allocation of observations in ranking and selection with unequal variances. *Sankhyā*, B, **37**, 28–78.

Dudewicz, E. J., J. S. Ramberg, and H. J. Chen (1975). New tables for multiple comparisons with a control (unknown variances). *Biometrische Zeitschrift*, **17**, 13–26.

Dudewicz, E. J., and Y. L. Tong (1971). Optimum confidence intervals for the largest location parameter, in S. S. Gupta and J. Yackel, Eds., *Statistical Decision Theory and Related Topics*. Academic Press, New York, pp. 363–376.

Dunn, O. J., R. A. Kronmal, and W. J. Yee (1968). Tables of the multivariate t distribution. *Technical Report*, School of Public Health, University of California at Los Angeles, California.

Dunnett, C. W. (1955). A multiple comparison procedure for comparing several treatments with a control. *Journal of the American Statistical Association*, **50**, 1096–1121.

Dunnett, C. W. (1960). On selecting the largest of k normal population means (with discussion), *Journal of the Royal Statistical Society*, B, **22**, 1–40.

Dunnett, C. W. (1964). New tables for multiple comparisons with a control. *Biometrics*, **20**, 482–491.

Dunnett, C. W. and M. Sobel (1954). A bivariate generalization of Student's t distribution, with tables for certain special cases. *Biometrika*, **41**, 153–169.

Eaton, M. L. (1967). The generalized variance, testing and ranking problem. *The Annals of Mathematical Statistics*, 941–943.

de Finetti, B. (1972). *Probability, Induction and Statistics*. John Wiley, New York.

Freeman, H., A. Kuzmack, and R. Maurice (1967). Multivariate t and the ranking problem. *Biometrika*, **54**, 305–308.

Frischtak, R. M. (1973). Statistical multiple-decision procedures for some multivariate selection problems. *Technical Report* No. 187, Department of Operations Research, Cornell University, Ithaca, New York.

Girshick, M. A. (1946). Contributions to sequential analysis: I. *The Annals of Mathematical Statistics*, **17**, 123–143.

Gnanadesikan, M. and S. S. Gupta (1970). A selection procedure for multivariate normal distributions in terms of the generalized variance. *Technometrics*, **12**, 103–117.

Goel, P. K. (1972). A note on the non-existence of subset selection procedure for Poisson populations. *Mimeo Series* No. 303, Department of Statistics, Purdue University, West Lafayette, Indiana.

Goel, P. K. and H. Rubin (1975). On selecting a subset containing the best population—a Bayesian approach. *Mimeo Series* No. 432, Department of Statistics, Purdue University, West Lafayette, Indiana.

Govindarajulu, Z. and A. P. Gore (1971). Selection procedures with respect to measures of association, in S. S. Gupta and J. Yackel, Eds., *Statistical Decision Theory and Related Topics*, Academic Press, New York, pp. 313–345.

Govindarajulu, Z. and C. Harvey (1974). Bayesian procedures for ranking and selection problems. *Annals of the Institute of Statistical Mathematics*, **26**, 35–53.

Grundy, P. M., D. H. Rees, and M. J. R. Healy (1954). Decision between two alternatives—how many experiments? *Biometrics*, **10**, 317–323.

Gupta, S. S. (1956). On a decision rule for a problem in ranking means. *Ph.D. Dissertation* (*Institute of Statistics Mimeo Series* No. 150), University of North Carolina, Chapel Hill, North Carolina.

Gupta, S. S. (1963). On a selection and ranking procedure for gamma populations. *Annals of the Institute of Statistical Mathematics*, **14**, 199–216.

Gupta, S. S. (1965). On some multiple decision (selection and ranking) rules. *Technometrics*, **7**, 225–245.

Gupta, S. S. (1977). Personal communication.

Gupta, S. S. and D. Y. Huang (1975). Subset selection procedures for the entropy function associated with the binomial populations. *Sankhyā, A*, **37**.

Gupta, S. S. and D. Y. Huang (1976). On ranking and selection procedures and tests of homogeneity for binomial populations, in S. Ikeda, Ed., *Essays in Probability and Statistics*, Tokyo, Japan, pp. 501–533.

Gupta, S. S., M. J. Huyett, and M. Sobel (1957). Selecting and ranking problems with binomial populations. *Transactions of the American Society for Quality Control*, 635–644.

Gupta, S. S. and K. Nagel (1971). On some contributions to multiple decision theory, in S. S. Gupta and J. Yackel, Eds., *Statistical Decision Theory and Related Topics*, Academic Press, New York, pp. 79–102.

Gupta, S. S., K. Nagel, and S. Panchapakesan (1973). On the order statistics from equally correlated normal random variables. *Biometrika*, **60**, 403–413.

Gupta, S. S. and S. Panchapakesan (1972). On multiple procedures. *Journal of Mathematical and Physical Sciences*, **6**, 1–72.

Gupta, S. S. and M. Sobel (1957). On a statistic which arises in selection and ranking problems. *The Annals of Mathematical Statistics*, **28**, 957–967.

Gupta, S. S. and M. Sobel (1958). On selecting a subset which contains all populations better than a standard. *The Annals of Mathematical Statistics*, **29**, 235–244.

Gupta, S. S. and M. Sobel (1960). Selecting a subset containing the best of several binomial populations, in I. Olkin, S. G. Ghurye, W. Hoeffding, W. G. Madow, and H. B. Mann, Eds., *Contributions to Probability and Statistics, Essays in Honor of Harold Hotelling*, Stanford University Press, Stanford, California, pp. 224–248.

Gupta, S. S. and M. Sobel (1962a). On selecting a subset containing the population with the smallest variance. *Biometrika*, **49**, 495–507.

Gupta, S. S. and M. Sobel (1926b). On the smallest of several correlated F statistics. *Biometrika*, **49**, 509–523.

Gupta, S. S. and W. Y. Wong (1975). Subset selection procedures for finite schemes in information theory. *Mimeo Series* No. 422, Department of Statistics, Purdue University, West Lafayette, Indiana.

Guttman, I. (1961). Best populations and tolerance regions. *Annals of the Institute of Statistical Mathematics*, **13**, 9–26.

Guttman, I. and R. C. Milton (1969). Procedures for a best population problem when the criterion of bestness involves a fixed tolerance region. *Annals of the Institute of Statistical Mathematics*, **21**, 149–161.

Guttman, I. and G. C. Tiao (1964). A Bayesian approach to some best population problems. *The Annals of Mathematical Statistics*, **35**, 825–835.

Hall, W. J. (1954). An optimum property of Bechhofer's single-sample multiple decision procedure for ranking means and some extensions. *Mimeo Series* No. 118, Institute of Statistics, University of North Carolina, Chapel Hill, North Carolina.

Hall, W. J. (1959). The most economical character of Bechhofer and Sobel decision rules. *The Annals of Mathematical Statistics*, **30**, 964–969.

Hoel, P., M. Sobel, and G. H. Weiss (1972). A two-stage procedure for choosing the better of two binomial populations. *Biometrika*, **59**, 317–322.

Huber, P. J. (1963a). A remark on a paper of Trawinski and David entitled "Selection of the best treatment in a paired comparison experiment." *The Annals of Mathematical Statistics*, **34**, 92–94.

Huber, P. J. (1963b). Pairwise comparison and ranking: Optimum properties of the sum procedure. *The Annals of Mathematical Statistics*, **34**, 511–520.

Kleijnen, J. P. C. (1975). *Statistical Techniques in Simulation*, Part II. Marcel Dekker, New York.

Kleijnen, J. P. C. and T. H. Naylor (1969). The use of multiple ranking procedures to analyze simulations of business and economic systems. *Proceedings of the Business and Economics Section of the American Statistical Association* (New York meeting), Washington, D.C., pp. 605–615.

Kleijnen, J. P. C., T. H. Naylor, and T. G. Seaks (1972). The use of multiple ranking procedures to analyze simulations of management systems: A tutorial. *Management Science*, Application Series, **18**, B 245–257.

Krishnaiah, P. R. and J. V. Armitage (1966). Tables for the multivariate t-distribution. *Sankhyā, B*, **28**, Parts 1 and 2 (part of *Aerospace Research Laboratories Report* 65-199, Wright-Patterson Air Force Base, Dayton, Ohio).

Krishnaiah, P. R. and M. H. Rizvi (1966). Some procedures for selection of multivariate normal populations better than a control, in P. R. Krishnaiah, Ed., *Multivariate Analysis*, Academic Press, New York, pp. 477–490.

Lehmann, E. L. (1963). A class of selection procedures based on ranks. *Mathematische Annalen*, **150**, 268–275.

Leone, F. C., L. S. Nelson, and R. B. Nottingham (1961). The folded normal distribution. *Technometrics*, **3**, 543–550.

Lin, W. T. (1976). An application of multiple-ranking procedures to analyze simulations of accounting systems. Unpublished manuscript.

Mahamunulu (=Desu), D. M. (1967). Some fixed-sample ranking and selection problems. *The Annals of Mathematical Statistics*, **38**, 1079–1091.

McCabe, G. P. and J. N. Arvesen (1974). Subset selection procedure for regression variables. *Journal of Statistical Computation and Simulation*, **3**, 137–146.

Miller, R. G., Jr. (1966). *Simultaneous Statistical Inference*. McGraw-Hill, New York.

Milton, R. C. (1963). Tables of the equally correlated multivariate normal probability integral. *Technical Report* No. 27, Department of Statistics, University of Minnesota, Minneapolis, Minnesota.

Milton, R. C. (1966). Basic table for ranking and selection problems with normal populations. Unpublished manuscript.

Mosteller, F. (1948). A k-sample slippage test for an extreme population. *The Annals of Mathematical Statistics*, **19**, 58–65.

Mosteller, F. and J. Tukey (1950). Significance levels for a k-sample slippage test. *The Annals of Mathematical Statistics*, **21**, 120–123.

Ofosu, J. B. (1972). A minimax procedure for selecting the population with the largest (smallest) scale parameter. *Calcutta Statistical Association Bulletin*, **16**, 143–154.

Ofosu, J. B. (1974). Optimum sample size for a problem in choosing the population with the largest mean: Some comments on Somerville's paper. *Journal of the American Statistical Association*, **69**, 270. Rejoinder by P. N. Somerville, ibid., 270–271.

Olkin, I., M. Sobel, and Y. L. Tong (1976). Estimating the true probability of a correct selection for location and scale families. *Technical Report* 110, Department of Statistics, Stanford University, Stanford, California.

Paulson, E. (1952a). An optimum solution to the k-sample slippage problem for the normal distribution. *The Annals of Mathematical Statistics*, **23**, 610–616.

Paulson, E. (1952b). On the comparison of several experimental categories with a control. *The Annals of Mathematical Statistics*, **23**, 239–246.

Paulson, E. (1964). A sequential procedure for selecting the population with the largest mean from k normal populations. *The Annals of Mathematical Statistics*, **35**, 174–180.

Puri, M. L. and P. S. Puri (1969). Multiple decision procedures based on ranks for certain problems in analysis of variance. *The Annals of Mathematical Statistics*, **40**, 619–632.

Puri, P. S. and M. L. Puri (1968). Selection procedures based on ranks: Scale parameter case. *Sankhyā, A,* **30**, 291–302.

Raiffa, H. and R. Schlaifer (1961). *Applied Statistical Decision Theory.* Harvard University Press, Boston, Massachusetts.

Randles, R. H., J. S. Ramberg, and R. V. Hogg (1973). An adaptive procedure for selecting the population with the largest location parameter. *Technometrics,* **15**, 769–789.

Rizvi, M. H. (1971). Some selection problems involving folded normal distribution. *Technometrics,* **13**, 355–369.

Rizvi, M. H. and K. M. L. Saxena (1972). Distribution-free interval estimation of the largest α-quantile. *Journal of the American Statistical Association,* **67**, 196–198.

Rizvi, M. H. and K. M. L. Saxena (1974). On interval estimation and simultaneous selection of ordered location or scale parameters. *The Annals of Statistics,* **2**, 1340–1345.

Rizvi, M. H. and M. Sobel (1967). Nonparametric procedures for selecting a subset containing the population with the largest α-quantile. *The Annals of Mathematical Statistics,* **38**, 1788–1803.

Rizvi, M. H., M. Sobel, and G. G. Woodworth (1968). Nonparametric ranking procedures for comparison with a control. *The Annals of Mathematical Statistics,* **39**, 2075–2093.

Rizvi, M. H. and H. Solomon (1973). Selection of largest multiple correlation coefficients: Asymptotic case. *Journal of the American Statistical Association,* **68**, 184–188. Corrigenda (1974), **69**, 288.

Rizvi, M. H. and G. G. Woodworth (1970). On selection procedures based on rank: Counterexamples concerning the least favorable configurations. *The Annals of Mathematical Statistics,* **41**, 1942–1951.

Robbins, H. (1964). The empirical Bayes approach to statistical decision problems. *The Annals of Mathematical Statistics,* **35**, 1–20.

Robbins, H., M. Sobel, and N. Starr (1968). A sequential procedure for selecting the best of k populations. *The Annals of Mathematical Statistics,* **39**, 88–92.

Schafer, R. E. (1974). A single-sample complete ordering procedure for certain populations, in F. Proschan and R. J. Serfling, Eds., *Reliability and Biometry: Statistical Analysis of Lifelength,* Society for Industrial and Applied Mathematics, Philadelphia, Pennsylvania, pp. 597–617.

Schafer, R. E. (1976). On selecting which of k populations exceed a standard. Unpublished manuscript to appear in C. P. Tsokos, Ed., Papers Presented at the Conference on the Theory and Applications of Reliability (December 1975) at University of South Florida, Academic Press, New York.

Schafer, R. E. and H. C. Rutemiller (1975). Some characteristics of a ranking procedure for population parameters based on chi-square statistics. *Technometrics,* **17**, 327–331.

Sobel, M. (1967). Nonparametric procedures for selecting the t populations with the largest α-quantiles. *The Annals of Mathematical Statistics*, **38**, 1804–1816.

Sobel, M. (1968). Selecting a subset containing at least one of the t best populations. *Technical Report* No. 105, Department of Statistics, University of Minnesota, Minneapolis, Minnesota.

Sobel, M. (1969). Selecting a subset containing at least one of the t best populations, in P. R. Krishnaiah, Ed., *Multivariate Analysis-II*. Academic Press, New York, pp. 515–540.

Sobel, M. and M. Huyett (1957). Selecting the best one of several binomial populations. *Bell System Technical Journal*, **36**, 537–576.

Sobel, M. and Y. L. Tong (1971). Optimal allocation of observations for partitioning a set of normal populations in comparison with a control. *Biometrika*, **58**, 177–181.

Sobel, M. and G. H. Weiss (1970). Inverse sampling and other selection procedures for tournaments with 2 or 3 players, in M. L. Puri, Ed., *Nonparametric Techniques in Statistical Inference*, Cambridge University Press, Cambridge, England, pp. 515–544.

Sobel, M. and G. H. Weiss (1972a). Play-the-winner rule and inverse sampling for selecting the best of $k \geqslant 3$ binomial populations. *The Annals of Mathematical Statistics*, **43**, 1808–1826.

Sobel, M. and G. H. Weiss (1972b). Recent results on using the play the winner sampling rule with binomial selection problems, in L. LeCam, J. Neyman, and E. L. Scott, Eds., *Proceedings of the Sixth Berkeley Symposium on Mathematical Statistics and Probability*, Vol. 1, University of California Press, Berkeley, California, pp. 717–736.

Soller, M. and J. Putter (1964). On the probability that the best chicken stock will come out best in a single random sample test. *Poultry Science*, **43**, 1425–1427.

Soller, M. and J. Putter (1965). Probability of correct selection of sires having the highest transmitting ability. *Journal of Dairy Science*, **48**, 747–748.

Stein, C. (1945). A two-sample test for a linear hypothesis whose power is independent of the variance. *The Annals of Mathematical Statistics*, **16**, 243–258.

Tamhane, A. C. (1976). A three-stage elimination type procedure for selecting the largest normal mean (common unknown variance case). Unpublished manuscript.

Tong, Y. L. (1967). On partitioning a set of normal populations by their locations with respect to a control using a single-stage, a two-stage, and a sequential procedure. *Ph.D. Dissertation (Technical Report* No. 90), Department of Statistics, University of Minnesota, Minneapolis, Minnesota.

Tong, Y. L. (1969). On partitioning a set of normal populations by their locations with respect to a control. *The Annals of Mathematical Statistics*, **40**, 1300–1324.

Tong, Y. L. (1971). Interval estimation of the ordered means of two normal populations based on hybrid estimators. *Biometrika*, **58**, 605–613.

Trawinski, B. J. and H. A. David (1963). Selection of the best treatment in a paired comparison experiment. *The Annals of Mathematical Statistics*, **34**, 75–91.

Truax, D. R. (1953). An optimum slippage test for the variance of k normal distributions. *The Annals of Mathematical Statistics*, **24**, 669–674.

Wald, A. (1939). Contributions to the theory of statistical estimation and testing hypotheses. *The Annals of Mathematical Statistics*, **10**, 299–326.

Wald, A. (1945a). Sequential tests of statistical hypotheses. *The Annals of Mathematical Statistics*, **16**, 117–186.

Wald, A. (1945b). Statistical decision functions which minimize the maximum risk. *The Annals of Mathematical Statistics*, **16**, 265–280.

Wald, A. (1947a). *Sequential Analysis*. John Wiley, New York.

Wald, A. (1947b). Foundations of a general theory of sequential decision functions. *Econometrica*, **15**, 279–313.

Wald, A. (1949). Statistical decision functions. *The Annals of Mathematical Statistics*, **20**, 165–205.

Wald, A. (1950). *Statistical Decision Functions*. John Wiley, New York.

Wetherill, G. B. and J. B. Ofosu (1974). Selection of the best k normal populations. *Applied Statistics*, **23**, 253–277.

References for Applications

Anderson, R. C. (1962). Failure imagery in the fantasy of eighth graders as a function of three conditions of induced arousal. *Journal of Educational Psychology*, **53**, 292–298.

Barnett, R. N. and W. J. Youden (1970). A revised scheme for the comparison of quantitative methods. *American Journal of Clinical Pathology*, **54**, 454–462.

Becker, W. A. (1961). Comparing entries in random sample tests. *Poultry Science*, **40**, 1507–1514.

Becker, W. A. (1962). Ranking all-or-none traits in random sample tests. *Poultry Science*, **41**, 1437–1438.

Becker, W. A. (1964). Changes in performance of entries in random sample tests. *Poultry Science*, **43**, 716–722.

Berger, P. K. and J. E. Sullivan (1970). Instructional set, interview context and the incidence of "don't-know" responses. *Journal of Applied Psychology*, **54**, 414–416.

The Birmingham News (April 7, 1976), Birmingham, Alabama, p. 2.

Birren, F. (1945). *Selling with Color*. McGraw-Hill, New York.

Black, J. M. and W. Z. Olson (1947). Durability of temperature-setting and intermediate temperature setting resin glues cured to different degrees in yellow birch plywood. U.S. Department of Agriculture Forest Products *Laboratory Report* 1537, Washington, D. C.

Bradley, R. A. (1954). Incomplete block rank analysis: On the appropriateness of the model for a method of paired comparisons. *Biometrics*, **10**, 375–390.

Brown, W. R. J., W. G. Howe, J. E. Jackson, and R. H. Morris (1956). Multivariate normality of the colormatching process. *Journal of the Optical Society of America*, **46**, 46–49.

Carrier, N. A., K. D. Orton, and L. F. Malpass (1962). Responses of Bright, Normal, and EMH children to an orally-administered Children's Manifest Anxiety Scale. *Journal of Educational Psychology*, **53**, 271–274.

The Chronicle of Higher Education, **12** No. 8 (April 19, 1976), Washington, D.C., 14.

Cochran, W. G. and G. M. Cox (1957). *Experimental Designs*, John Wiley, New York.

Cooley, W. W. and P. R. Lohnes (1962). *Multivariate Procedures for the Behavioral Sciences*. John Wiley, New York.

Crutcher, H. L. and L. W. Falls (1976). Multivariate normality. National Aeronautics and Space Administration *Technical Note* D-8226, Washington, D.C.

Dalal, S. R. and V. Srinivasan (1976). Determining sample size for pre-testing comparative effectiveness of advertising copies. Unpublished manuscript. To appear in *Management Science*.

David, H. A. (1963). *The Method of Paired Comparisons*. Charles Griffin, London, England.

DiPaola, P. P. (1945). Use of correlation in quality control. *Industrial Quality Control*, **1**, No. 7 (July), 10–12.

Dunlap, J. W. (1950). The effect of color in direct mail advertising. *Journal of Applied Psychology*, **34**, 280–281.

Dunnett, C. W. (1955). A multiple comparison procedure for comparing several treatments with a control. *Journal of the American Statistical Association*, **50**, 1096–1121.

Dunnett, C. W. (1964). New tables for multiple comparisons with a control. *Biometrics*, **20**, 482–491.

Dunnett, C. W. (1972). Drug screening: The never-ending search for new and better drugs, in J. Tanur, Ed., *Statistics: A Guide to the Unknown*. Holden-Day, San Francisco, California, pp. 23–33.

Ehrlich, P. R. and A. H. Ehrlich (1970). *Population Resources Environment: Issues in Human Ecology*. Freeman, San Francisco, California.

Fichter, L. S. (1976). The theory and practice of using confidence regions to test the equality of more than two means. *Ph.D. Dissertation*, The University of Alabama, University, Alabama.

Fisher, R. A. (1936). The use of multiple measurements in taxonomic problems. *Annals of Eugenics*, **7**, 179–188.

Fowler, M. J., M. J. Sullivan and B. R. Ekstrand (1973). Sleep and memory. *Science*, **179** (January 19, 1973), 302–304.

Gatewood, R. D. and R. Perloff (1973). An experimental investigation of three methods of providing weight and price information to consumers. *Journal of Applied Psychology*, **57**, 81–85.

Grant, E. L. and R. S. Leavenworth (1972). *Statistical Quality Control*. McGraw-Hill, New York.

Griffith, B. A., A. E. R. Westman, and B. H. Lloyd (1948). Analysis of variance. *Industrial Quality Control*, **4**, No. 6 (May), 13–22.

Gulliksen, H. (1956). A least squares solution for paired comparisons with incomplete data. *Psychometrika*, **21**, 125–144.

Harris, M. L. and C. W. Harris (1973). *A Structure of Concept Attainment Abilities*. Wisconsin Research and Development Center for Cognitive Learning, Madison, Wisconsin.

Houston, Michael J. (1972). The effect of unit-price on choices of brand and size in economic shopping. *Journal of Marketing*, **36** (July), 51–54.

International Conference on Family Planning Programs, Geneva, 1965 (1966). *Family Planning and Population Programs*. University of Chicago Press, Chicago, Illinois.

Jones, L. V. and R. D. Bock (1957). Methodology of preference measurement. *Final Report*, Quartermaster Food and Container Institute for the Armed Forces, pp. 1–202.

Jones, R. R. and W. J. Burns (1970). In-country training for the Peace Corps. *Journal of Applied Psychology*, **54**, 533–537.

Kaneko, I. (1975). A reduction theorem for the linear complementarity problem with a x-matrix. *Technical Report* 7575-24, Department of Industrial Engineering, University of Wisconsin, Madison, Wisconsin. To appear in *Journal of Linear Algebra and Applications*.

Kleijnen, J. P. C., T. H. Naylor, and T. G. Seaks (1972). The use of multiple ranking procedures to analyze simulations of management systems: A tutorial. *Management Science*, Application Series, **18**, B, 245–257.

Leone, F. C., L. S. Nelson, and R. B. Nottingham (1961). The folded normal distribution. *Technometrics*, **3**, 543–550.

Locke, E. A. and R. J. Whiting (1974). Sources of satisfaction and dissatisfaction among solid waste management employees. *Journal of Applied Psychology*, **59**, 145–156.

McGill, W. J. (1963). Stochastic latency mechanism, in R. D. Luce, R. R. Bush, and E. Galanter, Eds., *Handbook of Mathematical Psychology*, John Wiley, New York.

Morgan, C. J. (1965). *Physiological Psychology*. McGraw-Hill, New York.

Mosteller, F. (1951). Remarks on the method of paired comparisons: III. A test of significance for paired comparisons when equal standard deviations and equal correlations are assumed. *Psychometrika*, **16**, 207–218.

Mueller, C. G. (1950). Theoretical relationships among some measures of conditioning. *Proceedings of the National Academy of Sciences*, **56**, 123–134.

New York Post (January 23, 1976), p. 4.

O'Connell, D. C. (1970). Facilitation of recall by linguistic structure in nonsense strings. *Psychological Bulletin*, **74**, 441–452.

Osgood, C. E. (1957). A behavioristic analysis of perception and language as cognitive phenomena, in J. S. Bruner, Ed., *Contemporary Approaches to Cognition*, Harvard University Press, Cambridge, Massachusetts.

Preston, F. W. (1956). A quality control chart for the weather. *Industrial Quality Control*, **10**, No. 12 (April), 4–6.

Proschan, F. (1963). Theoretical explanation of observed decrease failure rate. *Technometrics*, **5**, 375–384.

Rao, C. R. (1948). The utilization of multiple measurements in problems of biological classification. *Journal of the Royal Statistical Society*, B, **10**, 159–193.

Rizvi, M. H. and H. Solomon (1973). Selection of the largest multiple correlation coefficients: Asymptotic case. *Journal of the American Statistical Association*, **68**, 184–188. Corrigenda (1974), **69**, 288.

Rulon, P. J., D. V. Tiedeman, M. M. Tatsuoka, and C. R. Langmuir (1967). *Multivariate Statistics for Personnel Classification*. John Wiley, New York.

Science News (1974), **105**, No. 1, 29.

Soller, M. and J. Putter (1964). On the probability that the best chicken stock will come out best in a single random sample test. *Poultry Science*, **43**, 1425–1427.

Soller, M. and J. Putter (1965). Probability of correct selection of sires having highest transmitting ability. *Journal of Dairy Science*, **48**, 747–748.

Snedecor, G. W. and W. G. Cochran (1967). *Statistical Methods*. Iowa State University Press, Ames, Iowa.

Sudman, S. and R. Ferber (1971). Experiments in obtaining consumer expenditures by diary methods. *Journal of the American Statistical Association*, **66**, 725–735.

Tatsuoka, M. M. (1971). *Multivariate Analysis: Techniques for Educational and Psychological Research*. John Wiley, New York.

Tintner, G. (1946). Some applications of multivariate analysis to economic data. *Journal of the American Statistical Association*, **41**, 472–500.

U.S. Bureau of the Census (1960). U.S. Census of population: 1960, subject reports, age at first marriage. *Final Report* PC(2)–4D.

U.S. Bureau of the Census (1960). U.S. Census of population: 1960, subject reports, marital status, *Final Report* PC(2)

Villars, D. S. (1951). *Statistical Design and Analysis of Experiments for Development Research*. Wm. C. Brown, Dubuque, Iowa.

Wernimont, G. (1947). Quality control in the chemical laboratory. *Industrial Quality Control*, **3**, No. 6 (May), 5–11.

Wilk, M. B., R. Gnanadesikan, and M. J. Huyett (1962). Estimation of parameters of the gamma distribution using order statistics. *Biometrika*, **49**, 525–546.

Williams, C. B. (1940). A note on the statistical analysis of sentence length as a criterion of literary style. *Biometrika*, **31**, 356–361.

Yule, G. U. (1939). On sentence length as a statistical characteristic of style in prose, with applications to two cases of disputed authorship. *Biometrika*, **30**, 363–390.

Index for Data and Examples

557

Name Index

561

Subject Index